Modular Forms on Schiermonnikoog

Modular forms are functions with an enormous amount of symmetry which play a
central role in number theory connecting it with analysis and geometry. They have
played a prominent role in mathematics since the 19th century and their study
continues to flourish today. They pop up in string theory and played a decisive role in
the proof of Fermat's Last Theorem. Modular forms formed the inspiration to
Langlands' conjectures and are expected to play an important role in the description of
the cohomology of varieties defined over number fields.

This collection of up-to-date articles originated from the conference
"Modular Forms" held on the Island of Schiermonnikoog in the Netherlands
in the autumn of 2006.

Bas Edixhoven is Professor of Mathematics in the Mathematical
Institute at Leiden University.

Gerard van der Geer is Professor of Algebra in the Korteweg-de Vries
Institute for Mathematics at the University of Amsterdam.

Ben Moonen is Associate Professor in the Korteweg-de Vries
Institute for Mathematics at the University of Amsterdam.

Modular Forms on Schiermonnikoog

BAS EDIXHOVEN
GERARD VAN DER GEER
BEN MOONEN

CAMBRIDGE
UNIVERSITY PRESS

CAMBRIDGE UNIVERSITY PRESS

Cambridge, New York, Melbourne, Madrid, Cape Town, Singapore, São Paulo, Delhi

Cambridge University Press
The Edinburgh Building, Cambridge CB2 8RU, UK

Published in the United States of America by Cambridge University Press, New York

www.cambridge.org
Information on this title: www.cambridge.org/9780521493543

First published 2008

Printed in the United Kingdom at the University Press, Cambridge

A catalogue record for this publication is available from the British Library

ISBN-13 978-0-521-49354-3 hardback

Contents

Preface

This volume grew out of a very succesful conference on Modular Forms that was held in October 2006 on the Dutch island of Schiermonnikoog and that was organised with financial support from the *Foundation Compositio Mathematica*. For some of the participants the journey to the island was a long one, but once on the island this was soon forgotten, and we look back at a very pleasant conference in beautiful surroundings. We thank the *Foundation Compositio Mathematica* for making this conference possible.

The present volume contains, in addition to an introduction by the editors, sixteen refereed papers, not necessarily related to lectures at the conference. We thank all authors and all referees for their contributions.

Bas Edixhoven
Gerard van der Geer
Ben Moonen

Contributors

Bas Edixhoven
Mathematisch Instituut, Universiteit Leiden,
P.O. Box 9512, 2300 RA Leiden, The Netherlands
edix@math.leidenuniv.nl

Gerard van der Geer
Korteweg-de Vries Instituut, Universiteit van Amsterdam, Plantage
Muidergracht 24, 1018 TV Amsterdam, The Netherlands
geer@science.uva.nl

Ben Moonen
Korteweg-de Vries Instituut, Universiteit van Amsterdam, Plantage
Muidergracht 24, 1018 TV Amsterdam, The Netherlands
bmoonen@science.uva.nl

Siegfried Böcherer
Kunzenhof 4B, 79117 Freiburg, Germany
boecherer@t-online.de

Kathrin Bringmann
School of Mathematics, University of Minnesota, Minneapolis,
MN 55455, U.S.A.
bringman@math.umn.edu

Jan Hendrik Bruinier
Technische Universität Darmstadt, Fachbereich Mathematik,
Schlossgartenstrasse 7, D–64289 Darmstadt, Germany
bruinier@mathematik.tu-darmstadt.de

Luis Dieulefait
Departament d'Algebra i Geometria, Universitat de Barcelona,
Gran Via de les Corts Catalanes 585, 08007 – Barcelona, Spain
ldieulefait@ub.edu

Bernhard Heim
Max-Planck Institut für Mathematik, Vivatsgasse 7, 53111 Bonn,
Germany
heim@mpim-bonn.mpg.de

Tomoyoshi Ibukiyama
Department of Mathematics, Graduate School of Science, Osaka
University, Machikaneyama 1-16, Toyonaka, Osaka, 560-0043 Japan
ibukiyam@math.wani.osaka-u.ac.jp

Robin de Jong
Mathematisch Instituut, Universiteit Leiden, P.O. Box 9512,
2300 RA Leiden, The Netherlands
rdejong@math.leidenuniv.nl

Hidenori Katsurada
Muroran Institute of Technology, 27-1 Mizumoto, Muroran,
050-8585, Japan
hidenori@mmm.muroran-it.ac.jp

Andrew Knightly
Department of Mathematics and Statistics,
University of Maine, Neville Hall, Orono, ME 04469-5752, USA
knightly@math.umaine.edu

Winfried Kohnen
Mathematisches Institut, Universität Heidelberg,
Im Neuenheimer Feld 288, D–69120 Heidelberg, Germany
winfried@mathi.uni-heidelberg.de

Charles Li
Department of Mathematics, Chinese University of
Hong Kong, Shatin NT, Hong Kong, Peoples Republic of China
charles@charlesli.org

Yuri I. Manin
Max-Planck Institut für Mathematik, Vivatsgasse 7,
53111 Bonn, Germany
manin@mpim-bonn.mpg.de

Department of Mathematics, Northwestern University,
2033 Sheridan Road, Evanston, IL 60208-2730, USA
manin@math.northwestern.edu

Matilde Marcolli
Max-Planck Institut für Mathematik, Vivatsgasse 7,
53111 Bonn, Germany
marcolli@mpim-bonn.mpg.de

Rainer Schulze-Pillot
Fachrichtung Mathematik, Universität des Saarlandes,
Postfach 151150, 66041 Saarbrücken, Germany
schulzep@math.uni-sb.de

Nils-Peter Skoruppa
Fachbereich Mathematik, Universität Siegen, Walter-Flex-Straße 3,
57068 Siegen, Germany
nils.skoruppa@uni-siegen.de

Rainer Weissauer
Mathematisches Institut, Universität Heidelberg, Im Neuenheimer
Feld 288, 69120 Heidelberg, Germany
weissauer@mathi.uni-heidelberg.de

Martin H. Weissman
Department of Mathematics, University of California,
Santa Cruz, U.S.A.
weissman@ucsc.edu

Gabor Wiese
Institut für Experimentelle Mathematik, Ellernstraße 29, 45326
Essen, Germany
gabor@pratum.net

Modular Forms

Bas Edixhoven, Gerard van der Geer and Ben Moonen

> *There are five fundamental operations in mathematics: addition,*
> *subtraction, multiplication, division and modular forms*

—M. Eichler[1]

Modular functions played a prominent role in the mathematics of the 19th century, where they appear in the theory of elliptic functions, i.e., elements of the function field of an elliptic curve, but also in the theory of binary quadratic forms. The term seems to stem from Dirichlet, but the functions are clearly present in the works of Gauss, Abel and Jacobi. They play an important role in the work of Kronecker, Eisenstein and Weierstrass, and later in that century they appear as central themes in the work of Poincaré and Klein. The theory of Riemann surfaces developed by Riemann became an important tool, and Klein and Fricke studied and popularized the Riemann surfaces defined by congruence subgroups of the modular group $SL(2, \mathbb{Z})$.

Modular forms appear as theta functions in the work of Jacobi in the 1820's, and, up to a factor $q^{1/24}$, already in Euler's identity

$$\prod_{n \geq 1}(1 - q^n) = \sum_{k \in \mathbb{Z}}(-1)^k q^{k(3k-1)/2} .$$

They show up in a natural way in the expansions of elliptic functions and as such they were studied by Eisenstein, but the concept of modular forms was formalized only later. Apparently, it was Klein who introduced the term "Modulform", cf. page 144 of Klein-Fricke [12].

One had to wait till Hecke for the first systematic study of modular forms on $SL(2, \mathbb{Z})$ and its congruence subgroups. The first appearance of the word "Modulform" in Hecke's work seems to be in [11].

---

[1] Apocryphal statement ascribed to Martin Eichler, March 29, 1912–October 7, 1992.

A crucial point in our story came when Hecke introduced the "averaging" operators that bear his name and that give essential arithmetic information on modular forms. Given (in modern terminology) a Hecke eigenform f on $\Gamma_1(N)$ with Fourier series $\sum a(n)q^n$, normalised by the condition $a(1) = 1$, Hecke could interpret the Fourier coefficient $a(n)$ as the eigenvalue of his operator $T(n)$. This also enabled him to express the Dirichlet series $L(f, s) = \sum_{n \geq 1} a(n)n^{-s}$ as an Euler product $\prod_p (1 - a(p)p^{-s} + \varepsilon(p)p^{k-1-2s})^{-1}$, where k is the weight of f and $\varepsilon \colon (\mathbb{Z}/N\mathbb{Z})^\times \to \mathbb{C}^\times$ its character. Thus he generalized a result of Mordell, who had proved in 1917 the multiplicativity of the Ramanujan τ-function that gives the Fourier coefficients of the weight 12 cusp form Δ. (This property of the τ-function had been observed by Ramanujan in 1916.) Though the eigenvalues of eigenforms showed a definite arithmetic flavour, it remained at that time a mystery why there should be arithmetic information in the Fourier coefficients of eigenforms. Hecke did not know, at that time, that the space of cusp forms of a given weight and level possesses a basis of eigenforms for the Hecke operators $T(n)$ with n prime to the level. But a little later Petersson defined an inner product with respect to which these $T(n)$ are normal, and with this it followed that such a basis exists. Hecke also proved, using the Mellin transform, that the Dirichlet series $L(f, s)$ associated to a cusp form f of weight k on $\Gamma_1(N)$ has an analytic continuation to a holomorphic function on the whole complex plane and satisfies a functional equation relating $L(f, s)$ to $L(g, k - s)$, where $g(\tau) = \tau^k f(-1/N\tau)$.

The second important step that Hecke made was to characterise the Dirichlet series $\sum_{n>0} a(n)n^{-s}$ of the form $L(f, s)$ with f a cusp form of weight k on $\mathrm{SL}(2, \mathbb{Z})$ by regularity conditions and a functional equation relating $L(f, s)$ to $L(f, k - s)$. Indeed, a Fourier series $f = \sum_{n \geq 1} a(n)q^n$ that is holomorphic on the upper half plane is a cusp form of weight k on $\mathrm{SL}(2, \mathbb{Z})$ precisely when $f(-1/\tau) = \tau^k f(\tau)$. This so-called converse theorem generalized a theorem of Hamburger, saying that a sufficiently regular Dirichlet series that satisfies the functional equation of the Riemann zeta function is in fact a multiple of the Riemann zeta function.

The L-function that Hecke associates to a cusp form has its roots in earlier work of Gauss, Dirichlet and Riemann. But although Hecke was working at the same mathematics department (in Hamburg) as Artin, who was then working on his Artin L-series for representations of the Galois group of a number field, it seems that neither of them appreciated the link between the two types of L-functions. This may seem odd to us, but it is good to realize that the moment that the link was recognized in its full conjectural setting represents a second turning point in our history. Indeed, looking from a large distance one may distinguish two turning points for the history of modular forms in the 20th century: Hecke's introduction of the Hecke operators and his converse theorem, and Langlands's letter of January 1967 to Weil, in which he laid out a grand program in which modular forms are an incarnation of non-abelian class field theory. Langlands's letter pointed out the common source for the L-series of Hecke and Artin, and brought the two types of L-functions together in a larger framework. We will come to that later.

But at the time that Hecke revolutionized the topic, it also lost its prominence, as novel developments in topology and algebra started to attract more attention. This was a time when many new concepts appeared, like the notions of algebraic topology and homology theory, and when new algebraic structures like rings and algebras were studied. These notions completely changed the face of mathematics at the time. Klein writes in this connection: "Es hat sich hier ein merkwürdiger Umschwung vollzogen. Als ich studierte, galten die Abelschen Funktionen—in Nachwirkung der Jacobischen Tradition—als der unbestrittene Gipfel der Mathematik, und jeder von uns hatte den selbstverständlichen Ehrgeiz, hier selbst weiter zu kommen. Und jetzt? Die junge Generation kennt die Abelschen Funktionen kaum mehr."(Vorlesungen über die Entwicklung der Mathematik, VII.)

In retrospect these developments, like the construction of homology and cohomology, the emergence of new algebraic structures and the development of an algebraic foundation for algebraic geometry, were the necessary ingredients for the later growth of the theory of modular forms.

The fact that there was a shift of focus to new topics in mathematics does not mean that the theory of modular forms came to a standstill. Throughout the 20th century there have been new ideas and generalizations, broadening but also deepening the subject. Some of these generalizations dealt with the extension of the notion of modular forms to other groups. An example of this is the step from $SL(2, \mathbb{Z})$ to the group $SL(2, O_K)$ with O_K the ring of integers of a totally real field, the Hilbert modular group. Hilbert was inspired by Kronecker's "Jugendtraum" about generating abelian extensions of imaginary quadratic fields. The Kronecker-Weber theorem says that all abelian extensions of $\mathbb{Q}$ are contained in the field generated, over $\mathbb{Q}$, by all roots of unity, i.e., by the torsion points of the circle group. It was also found that for an imaginary quadratic field K, the values of a suitable elliptic function at the torsion points of an elliptic curve with complex multiplication by O_K could be used to generate abelian extensions of K. Hilbert envisioned an analogue of the Kronecker-Weber theorem and the theory of complex multiplication for abelian extensions of CM-fields (totally imaginary quadratic extensions of totally real number fields). He devoted to this the 12th of his famous

Mathematische Probleme, presented at the ICM 1900 in Paris.[2] As part of his investigations, Hilbert had worked out a theory of modular functions for totally real fields, more precisely for modular functions for the action of SL(2, O_K) on the product of $n = [K : \mathbb{Q}]$ upper half planes. He wrote an unpublished manuscript about it, and under his guidance his student Blumenthal wrote his *Habilitationsschrift* about the basics of the theory. Hecke, also a student of Hilbert, wrote his thesis about it, this time with the purpose of setting up a theory of abelian extensions of quartic CM-fields. After these beginnings this development seemed to dry up, and though impressive progress has been made, Hilbert's 12th problem is to date unsolved. But recently two new ideas have been launched: Manin's "Alterstraum", and Darmon's "Stark-Heegner points".

In the years after Hecke the number of mathematicians involved in modular forms shrank to a small group, including Eichler, Maass, Petersson and Rankin, but they continued to contribute. In 1946 Maass, working under difficult circumstances in postwar Germany, showed that one could sacrifice holomorphicity by considering eigenfunctions of the Laplacian $y^2(\partial^2/\partial x^2 + \partial^2/\partial y^2)$ that are invariant under the modular group.

In another direction, Siegel generalized the notion of the modular group inspired by his quantitative theory of representations of quadratic forms by quadratic forms, and also by the theory of period matrices of Riemann surfaces; see [19]. He made a detailed study of the symplectic group and its geometry, thus picking up a thread left by Riemann and neglected by many, Scorza being one of the exceptions. In his groundbreaking paper of 1857, Riemann had introduced the period matrix of a Riemann surface of genus g, and had shown that it can be normalized in the form of a complex symmetric g by g matrix with positive definite imaginary part. Siegel considered the so-called Siegel upper half space $\mathbb{H}_g$ of all such period matrices, on which the symplectic group acts by fractional linear transformations. He determined a fundamental domain and its natural volume, studied the function field of the quotient space Sp$(2g, \mathbb{Z})\backslash\mathbb{H}_g$, and he introduced the notion of a (Siegel) modular form. Siegel's main motivation was his desire to describe in a quantitive way the representations of integral quadratic forms by other quadratic forms. His central result can be expressed as an equality of a theta series with an Eisenstein series for the Siegel modular group.

In the 1950's and 1960's another vast generalization of the theory of modular forms was conceived by the introduction of the general notion of automorphic form, and of the subsequent adèlisation of this concept. According to Borel

---

[2] See [15] for further historical information on Hilbert's 12th problem and Kronecker's "Jugendtraum".

and Jacquet in [3] and [4], it had first been observed by Gelfand and Fomin that modular forms and other automorphic forms on the upper half plane and other bounded symmetric domains can be viewed as smooth vectors in representations of the ambient Lie group G on suitable spaces of functions on G that are invariant under the discrete subgroup Γ. A general definition was given by Harish-Chandra in [9] for a semisimple connected Lie group G, a discrete subgroup Γ and a maximal compact subgroup K. An automorphic form on G with respect to K and Γ is then a left-Γ-invariant and right-K-finite smooth function $f: G \to \mathbb{C}$, finite under the center of the enveloping algebra of the Lie algebra of G, and satisfying a certain growth condition. The consideration of the system of all congruence subgroups of a connected reductive group G over $\mathbb{Q}$ then led to the notion of automorphic forms on the group $G(\mathbb{A})$ of adèlic points of G, and this notion was then further generalised to connected reductive groups over global fields F. An important consequence of this point of view is that the space of automorphic forms on $G(\mathbb{A})$ can be studied as a representation of the group $G(\mathbb{A}_f)$, as well as of K and the Lie algebra of $G(\mathbb{R})$. Irreducible representations thus obtained are called automorphic representations; they can be decomposed as restricted tensor products of irreducible representations of the local groups $G(F_v)$. In a precise way, these local representations generalise the systems of eigenvalues of a Hecke eigenform. Especially the Russian school contributed to the early development in this direction (Gelfand, Graev, Piatetskii-Shapiro,...). The necessary theory of algebraic groups, arithmetic subgroups and adèle groups had been developed in the meantime, see for example [5].

On another stage but also during the 1950's and 1960's, weight two modular forms for congruence subgroups of $SL(2, \mathbb{Z})$ were related to differential forms on modular curves, and hence to the Jacobian varieties of modular curves, also in positive characteristic. Advances in algebraic geometry made it possible to study the reduction of curves and Jacobians at almost all primes. This led to the identification of the (partial) Hasse-Weil zeta functions of modular curves with a product of L-functions of such modular forms, at least at almost all primes (Eichler, Shimura; see [17]), thus proving the meromorphic continuation and the existence of a functional equation for these zeta functions. Kuga and Shimura were even able to do the same for forms of higher weight on the unit group of a quaternion algebra (see [13]).

In particular, the Hasse-Weil L-functions of elliptic curves over $\mathbb{Q}$ occurring as isogeny factor of the Jacobian of a modular curve were identified (up to finitely many Euler factors) with L-functions of modular forms. Deuring proved in 1955 that the L-function of an elliptic curve with complex multiplication is a product of two Hecke L-functions associated to

"Grössencharaktere". In the same year, Taniyama [20] raised the question whether the meromorphicity and functional equation of the Hasse-Weil zeta functions of elliptic curves over number fields could be proved by finding suitable automorphic forms (see [18], where Shimura evokes a vivid portrait of their interaction in that time). Taniyama's idea was that the expected functional equation should imply modularity for the associated Fourier series along the lines of Hecke who characterized modular forms on SL(2, $\mathbb{Z}$) by the functional equation of their associated Dirichlet series.

In [24] Weil extended Hecke's argument by showing that if for sufficiently many Dirichlet characters χ the Dirichlet series $\sum \chi(n)a(n)n^{-s}$ associated to a function f on the upper half plane given by a Fourier series $\sum a(n)q^n$ have a suitable continuation to $\mathbb{C}$ and satisfy an explicitly given functional equation then f is a modular form on a congruence subgroup $\Gamma_0(N)$ (with N determined by the functional equations). At the end Weil states the modularity question for an elliptic curve E over $\mathbb{Q}$ in a precise form: the complete Hasse-Weil L-function is defined, as well as the conductor of E, and the expected functional equations. It was this paper of Weil that drew renewed attention to the modularity question for elliptic curves over $\mathbb{Q}$.

In January 1967 Langlands wrote a letter [14] to Weil that marked the start of the "Langlands program". The main idea of this program is that the L-function associated to a Galois representation should coincide with the L-function that can be associated to some "algebraic" automorphic representation (generalising algebraic Hecke characters on idèle groups), and therefore has an analytic continuation and satisfies a functional equation. For example, the Artin L-function for an irreducible continuous n-dimensional complex representation of the Galois group of a number field F should be the L-function associated to an automorphic cuspidal representation of GL(n, $\mathbb{A}_F$). This leads to conjectural correspondences, both global and local, between Galois representations and automorphic representations, characterised by being compatible with suitable L-factors and ε-factors. Also compatible systems of l-adic representations can be taken into account, and general reductive groups G over number fields F are considered. The Langlands dual group ^LG is introduced in order to formulate the natural (conjectural) relations between automorphic representations on different reductive groups: the functoriality principle. In collaboration with Jacquet, Langlands gave support for the functoriality principle by working it out and establishing the Jacquet-Langlands correspondence for the group GL(2) and its inner twists (unit groups of quaternion algebras). Here, trace formulas (Selberg) play the main role. The Langlands program constitutes a grand framework for number theory, representation theory and algebraic geometry, and has become one of the focal points in pure mathematics.

By that time the new methods of algebraic geometry, after the revolution in that field led by Grothendieck, came to play their role in the theory of modular forms. Eichler and Shimura had shown that the space of modular forms of weight $k \geq 2$ and level N can be interpreted as the $(k-1, 0)$-part of the Hodge decomposition of the cohomology of a suitable local system on the modular curve $X_1(N)$: the $k-2$ symmetric power of the rank two local system given by the fiberwise cohomology of the universal family of elliptic curves. In 1968 Deligne showed that the l-adic étale cohomology of a non-singular projective model of the $k-2$ power of the universal elliptic curve over the j-line provides the Galois representations then conjecturally associated to modular forms. Here Deligne had to deal with the technical difficulties caused by the presence of cusps. As a consequence of his results, the Ramanujan conjecture on the absolute value of the Fourier coefficients of these modular forms would follow from the Weil conjectures on the cohomology of non-singular projective varieties over finite fields. Six years later Deligne himself proved the last open part of these conjectures, and Ramanujan's conjecture followed. A clear link between modular forms and Galois representations was established.

The new methods of algebraic geometry were also needed strongly to overcome the hurdles in extending results for GL(2) to other groups, like the symplectic group. The main reason for this is that the associated modular varieties are of higher dimension. Moreover, the fact that the spaces that are considered are usually not complete presents serious obstacles. Satake showed how the quotient space $\mathrm{Sp}(2g, \mathbb{Z})\backslash \mathbb{H}_g$ can be compactified by adding the orbits of the rational boundary components $\mathbb{H}_i$ with $0 \leq i \leq g$ in $\overline{\mathbb{H}}_g$, thus obtaining a normal analytic space, which however for $g > 1$ is very singular. Baily and Borel generalized his construction to the so-called Baily-Borel compactification where the quotient of a bounded symmetric domain under an arithmetic subgroup is compactified to a projective variety that contains the original as a quasi-projective open subvariety. The embedding in projective space is given by modular forms of an appropriate weight. In other words, the homogeneous coordinate rings of these compactifications are the graded algebras of modular forms. These Baily-Borel compactifications are in general very singular. Igusa constructed a smooth compactification of $\mathrm{Sp}(2g, \mathbb{Z})\backslash \mathbb{H}_g$ for $g \leq 3$ by blowing up the Satake compactification along the ideal of the boundary. Mumford launched a big program to construct smooth compactifications by toroidal methods. A drawback of these compactifications is that they are not canonical, but depend on combinatorial data (cone decompositions).

Around the same time Hirzebruch discovered how to resolve the singularities of Hilbert modular surfaces. The singularities were resolved by cycles of rational curves. This provided a lot of information about Hilbert modular forms for real quadratic fields. In particular, it led to a geometric interpretation of the inverse of the Doi-Naganuma lifting from elliptic modular forms to Hilbert modular forms and the Fourier coefficients of the elliptic modular forms were interpreted by Hirzebruch and Zagier as the intersection numbers of modular curves on these modular surfaces.

It is interesting to note the parallel to the situation of a century earlier, when the modern theory of Riemann surfaces was brought into play in order to understand the spaces on which the modular functions live. But for a mature arithmetic theory of modular forms the full force of the newly developed algebraic geometry was needed. Here we think of Grothendieck's theory of moduli functors and their representability and Mumford's results in geometric invariant theory for the moduli spaces of abelian varieties, the compactification theory of Mumford c.s. and a version over the integers that was provided by Chai and Faltings.

Another instance where the power of algebraic geometry was brought to bear is the beautiful theorem of Gross and Zagier relating derivatives of L-functions of modular forms at the center of the critical strip to the heights of Heegner points on modular curves.

Around 1985, Frey came up with the idea and some arguments that the modularity conjecture should contradict the marvelous properties of the "ABC-elliptic curve" over $\mathbb{Q}$ associated by Hellegouarch to a hypothetical solution of the Fermat equation. Hence, Fermat's last theorem should be a consequence of the modularity conjecture, which therefore attracted much attention. Soon thereafter, in [16], Serre formulated a conjecture on irreducible odd continuous representations $\rho \colon \mathrm{Gal}(\overline{\mathbb{Q}}/\mathbb{Q}) \to \mathrm{GL}(2, \overline{\mathbb{F}}_p)$, where odd means that the determinant of complex conjugation equals -1. The precise form of this conjecture was to make clear the "epsilon" that was needed apart from the modularity conjecture to prove Fermat's last theorem. Serre conjectured that every such ρ can be obtained from a normalised eigenform $f = \sum a(n)q^n$ of weight $k(\rho)$ on $\Gamma_1(N(\rho))$, with $k(\rho)$ and $N(\rho)$ given in terms of the ramification of ρ. The pair $(N(\rho), k(\rho))$ was intended to be the minimal possible. After a first step by Mazur, Ribet was able to establish the "epsilon", and this motivated Wiles to set out to prove a form of the modularity conjecture that would suffice to prove Fermat's last theorem. The realization of this by Wiles in 1994, with the help of Taylor, is certainly one of the triumphs of 20th century mathematics and of the theory of modular forms.

Wiles's breakthrough, now about fifteen years ago, was based on the study of deformations of Galois representations, a theory initiated by Mazur. The most striking of his results is that completions of Hecke algebras can often be interpreted as universal deformation rings, where one considers deformations whose ramification is suitably restricted at all primes.

Since then, these deformation theoretic methods have been generalised and have led to spectacular progress. The full modularity conjecture for elliptic curves over $\mathbb{Q}$ was proved in [6]. Here, the formulation of the restrictions on the ramification uses the local Langlands correspondence for GL(2), as well as Fontaine's theory of p-adic Galois representations. Fontaine and Mazur have conjectured that all irreducible continuous p-adic representations of the Galois group of $\mathbb{Q}$ that are unramified at almost all primes and are everywhere potentially semi-stable should come from geometry, and, according to the Langlands program, from automorphic representations. Breuil started investigating the possibility of a p-adic local Langlands correspondence for p-adic Galois representations of p-adic fields; here the question is what one should put on the automorphic side. Taylor obtained potential modularity results for two-dimensional p-adic Galois representations over totally real fields, thereby proving meromorphic continuation and functional equation for the associated L-functions; see [22] and [23], and [21] (the long version). It came as a surprise to many that these methods could be applied in the GL(n)-case, when in March of 2006 Taylor, Clozel, Harris and Shepherd-Barron announced their proof of the Sato-Tate conjecture for elliptic curves over $\mathbb{Q}$ with multiplicative reduction at at least one prime (see the preprints on Taylor's home page). Dieulefait and Wintenberger noticed that Taylor's potential modularity results made it possible to construct compatible systems of l-adic representations even in cases where modularity was not known. This led to the proof, in 2007, by Khare and Wintenberger, using important results of Kisin, of Serre's conjecture that is mentioned above (see the preprints on Khare's home page). All this is more than many had been inclined to hope.

Of course there has been progress on the subject of modular forms and the Langlands program that is independent of Wiles's breakthrough. In this respect we should mention the work of Drinfeld, who proved the global Langlands correspondence for GL(2) in the function field case in the 1970's, and Lafforgue, who generalised that to GL(n) in 2002. In the 1970's, Mazur did groundbreaking work concerning rational points on modular curves and their Jacobians. Harris and Taylor proved the local Langlands correspondence for GL(n) over p-adic fields around 2000, using the geometry of certain Shimura varieties. And there must be much more, that we, the editors of this volume, are not aware of because of our own limited background. For example, it is clear that in the results described above, base change results are used, trace formulas, properties of L-functions of pairs, fundamental lemmas and what not. There are important developments that we have not even mentioned, like the work of Borcherds. We hope that readers will enjoy this introduction nevertheless, and will excuse us for any omissions.

For an algebraic geometer the main lure of modular forms may come from from the fact that algebraic varieties defined over a number field are a natural source for modular forms. Indeed, according to Langlands the corresponding Galois representations should all come from automorphic representations. The developments of the recent years have thus tied modular forms very closely to arithmetic algebraic geometry and this has been fruitful to both algebraic geometry and the theory of modular forms. But further progress certainly requires a better understanding of modular forms on other groups than GL(2). The groups that correspond to modular varieties parametrizing algebro-geometric objects offer maybe the best hopes, as algebraic geometry may bring further clues. But even well-studied moduli spaces still seem far beyond our grasp. For example, what automorphic forms occur in the cohomology of the moduli space M_g of curves?

The goals as formulated by the Langlands conjectures may seem very distant, but recent developments as in the work of Laumon-Ngô and Ngô on the "fundamental lemma" yield the prospect of rapid advances in the near future. Apart from that modular forms appear again and again at unexpected places, for example in new developments in mathematical physics like string theory, showing that the topic is still full of life.

Bibliography

[1] A. Ash, D. Mumford, M. Rapoport, Y. Tai: Smooth compactification of locally symmetric varieties. *Lie Groups: History, Frontiers and Applications*, Vol. IV. Math. Sci. Press, Brookline, Mass., 1975.

[2] W. Baily, A. Borel: Compactification of arithmetic quotients of bounded symmetric domains. *Ann. of Math.* **84** (1966), pp. 442–528.

[3] A. Borel: Introduction to automorphic forms. *Algebraic Groups and Discontinuous Subgroups (Proc. Sympos. Pure Math., Boulder, Colo., 1965)*, pp. 199–210, Amer. Math. Soc., Providence, R.I., 1966.

[4] A. Borel, H. Jacquet: Automorphic forms and automorphic representations. With a supplement "On the notion of an automorphic representation" by R. P. Langlands. *Proc. Sympos. Pure Math., XXXIII, Automorphic forms, representations and L-functions (Proc. Sympos. Pure Math., Oregon State Univ., Corvallis, Ore., 1977)*, Part 1, pp. 189–207, Amer. Math. Soc., Providence, R.I., 1979.

[5] Proceedings of Symposia in Pure Mathematics. Vol. IX: Algebraic groups and discontinuous subgroups. Proceedings of the Symposium in Pure Mathematics of the American Mathematical Society held at the University of Colorado, Boulder, Colorado (July 5-August 6, 1965). Edited by Armand Borel and George D. Mostow. American Mathematical Society, Providence, R.I., 1966.

[6] C. Breuil, B. Conrad, F. Diamond, R. Taylor: On the modularity of elliptic curves over $\mathbb{Q}$: wild 3-adic exercises. *J. Amer. Math. Soc.* **14** (2001), no. 4, 843–939.

[7] P. Deligne: Formes modulaires et représentations ℓ-adiques. Sém. Bourbaki 1968/9, no. 355. *Lecture Notes in Math.* **179** (1971), p. 139–172.

[8] G. Faltings, C-L. Chai: Degeneration of abelian varieties. *Ergebnisse der Math.* **22**. Springer Verlag 1990.

[9] Harish-Chandra: Automorphic forms on a semisimple Lie group. *Proc. Nat. Acad. Sci. U.S.A.* **45** (1959) 570–573.

[10] H. Hasse: Zetafunktion und L-Funktionen zu einem arithmetischen Funktionenkörper vom Fermatschen Typus. (German) *Abh. Deutsch. Akad. Wiss. Berlin. Kl. Math. Nat. 1954* (1954), no. 4, 70 pp. (1955).

[11] E. Hecke: Analytische Funktionen und algebraische Zahlen, II. *Hamb. Math. Abh.* **3**, 213-236 (1924).

[12] Klein and Fricke: Vorlesungen über die Theorie der elliptischen Functionen. Teubner, Leipzig, 1890.

[13] M. Kuga, G. Shimura: On the zeta function of a fibre variety whose fibres are abelian varieties. *Ann. of Math. (2)* **82** (1965), 478–539.

[14] R. P. Langlands: Letter to André Weil, January 1967. Available at www.sunsite.ubc.ca/DigitalMathArchive/Langlands/intro.html

[15] N. Schappacher: On the history of Hilbert's twelfth problem: a comedy of errors. *Matériaux pour l'histoire des mathématiques au XXe siècle (Nice, 1996)*, 243–273, Sémin. Congr., 3, Soc. Math. France, Paris, 1998.

[16] J-P. Serre: Sur les représentations modulaires de degré 2 de $\mathrm{Gal}(\overline{\mathbb{Q}}/\mathbb{Q})$. *Duke Math. J.* **54** (1987), no. 1, 179–230.

[17] G. Shimura: On the zeta-functions of the algebraic curves uniformized by certain automorphic functions. *J. Math. Soc. Japan* **13** (1961), 275–331.

[18] G. Shimura: Yutaka Taniyama and his time. Very personal recollections. *Bull. London Math. Soc.* **21** (1989), no. 2, 186–196.

[19] C.L. Siegel: Über die analytische Theorie der quadratischen Formen. *Annals of Math.* **36** (1935), 527–606.

[20] Y. Taniyama: Jacobian varieties and number fields. *Proceedings of the international symposium on algebraic number theory, Tokyo & Nikko, 1955*, pp. 31–45. Science Council of Japan, Tokyo, 1956.

[21] R. Taylor: Galois representations. *Ann. Fac. Sci. Toulouse Math. (6)* **13** (2004), no. 1, 73–119.

[22] R. Taylor: Remarks on a conjecture of Fontaine and Mazur. *J. Inst. Math. Jussieu* **1** (2002), no. 1, 125–143.

[23] R. Taylor: On the meromorphic continuation of degree two L-functions. *Doc. Math. 2006, Extra Vol.*, 729–779.

[24] A. Weil: Über die Bestimmung Dirichletscher Reihen durch Funktionalgleichungen. *Math. Ann.* **168** (1967), 149–156.

On the basis problem for Siegel modular forms with level

Siegfried Böcherer, Hidenori Katsurada
and Rainer Schulze-Pillot

Introduction

The simplest and most familiar examples of holomorphic Siegel modular forms are theta series attached to lattices in a positive definite quadratic space. The basis problem for Siegel (cuspidal) modular forms asks whether one can get all cusp forms (with quadratic or trivial character) for a given congruence subgroup of type $\Gamma_0^n(N)$ as linear combinations of such theta series attached to lattices of level dividing N. For level $N = 1$ this was solved affirmatively provided that the weight is large enough [2]. For higher levels and degree 1 there are results due to the deep work of Waldspurger [21]. Following the lines of thought of these works in the case of arbitrary degree n and level N, one has to consider, for a given cusp form f of degree n and weight k, an integral of type

$$\Lambda_E^{2n,k}(f)(z) := \int_{\Gamma_0^n(N)\backslash \mathbb{H}_n} f(w)\, \overline{E^{2n}\left(\left(\begin{smallmatrix} w & 0 \\ 0 & -\bar{z} \end{smallmatrix}\right)\right)}\, \det(w)^k\, d^*w\,,$$

where E^{2n} is an appropriate Eisenstein series of degree $2n$ and weight k, and where d^*w denotes the invariant volume element on $\mathbb{H}_n$. If E^{2n} can be expressed as a linear combination of theta series this leads to a similar expression for $\Lambda_E^{2n,k}(f)$ and (after evaluation of the integral) finally to an expression of the desired type for a multiple of f itself.

In the case of level one there is only one choice of E^{2n}, for level $N \neq 1$ there are in general many possibilities for E^{2n}. The most ambitious choice is to take E^{2n} to be the genus theta series of a fixed genus of quadratic forms; if successful this gives an expression for f as linear combination of theta series of quadratic forms in this genus. There will in general be delicate problems related with the bad primes and there seems at the moment not much hope to do this in general (see however [4] where some cases are done from this point of view with considerable pain; the same kind of question is adressed in [19], where Shimura ultimately sticks to the case of principal congruence subgroups).

A variant of this approach has been taken by Kuang in [15], where more generally Hilbert Siegel modular forms are treated: In the case of weight k divisible by 4 with $k > 2n+1$, squarefree level N and f of primitive nebentype modulo N, Kuang chooses an Eisenstein series which is a linear combination of inhomogeneous theta series attached to even unimodular quadratic forms (i.e., theta series of non trivial cosets of lattices on the unique positive definite quadratic space of dimension $2k$ which carries even unimodular lattices). This Eisenstein series is obtained from the Siegel-Weil theorem and can be viewed as a modified genus theta series, the individual inhomogeneous theta series occurring in it may (and will) not have the same transformation type as the given f. He then obtains an expression for f as a linear combination of such inhomogeneous theta series which is somewhat analogous to Eichler's Brandt matrices with character (Kuang asserts his result for $k > n + 1$ but uses $k > 2n + 1$ in the proof since E^{2n} is assumed to be absolutely convergent).

In this work we have a different aim: We stick to theta series of full lattices (having the same transformation type as the given f) but choose an Eisenstein series which involves several genera of quadratic forms. The reason for this is that there is one type of Eisenstein series for which we can compute $\Lambda_E(f)$ in a particularly simple fashion, namely

$$F(Z) := \sum_{C,D} \chi(\det(C)) \det(CZ + D)^{-k},$$

where χ is any Dirichlet character modulo $N > 1$.

The integral was computed explicitly in [6, 8]. It can be zero, if the level is not squarefree; we will be interested only in the squarefree case. As essential new tools (not available when [6, 8] were written) we use

- the injectivity of the Hecke operators $U(p)$ acting on spaces of modular forms for $\Gamma_0^n(N)$ with N squarefree [3];
- the fact that the Eisenstein series F is a linear combination of genus theta series for genera of positive quadratic forms of levels dividing N (if the weight k is large enough and the character is quadratic or trivial) [12, 5].

The first of these statements allows us (when combined with the information from [6, 8] about the contribution of the bad primes, where $U(p)$ appears in the "numerator") to deduce the *injectivity* of the map $\Lambda_F^{2n,k}$ on the space of cusp forms in question, provided that the weight is large enough. It was somewhat surprising for us that we do not need any kind of newform argument here. The second statement implies that the *image* of $\Lambda_F^{2n,k}$ consists of linear combinations of theta series.

Our results (see section 3 for precise statements) say that Siegel cusp forms of squarefree level and quadratic nebentypus are linear combinations of appropriate theta series, provided that the weight is large enough. Here "appropriate" means that the theta series in question have level (dividing) N and they also have the correct nebentypus. We have to allow all (or at least sufficiently many) such theta series; in particular, we cannot fix the genus of the quadratic forms (the quadratic space however can and will be fixed). In this way we can avoid more delicate questions about the bad primes (these questions can be indeed quite delicate, as can be seen from Waldspurger's work on the case ellipic modular forms of primitive nebentypus mod N and quadratic forms of discriminant N and of discriminant N^{2k-1}, see [21, Théorème 3]). A second reason for dealing with several genera is that this allows to solve the basis problem for the full space of modular forms (not only cusp forms!), see section 4.

It should be clear (at least to experts) that our methods immediately (but with more burden concerning terminology) carry over to theta series with harmonic coefficients and also to the case of vector valued modular forms. We will briefly describe the main modifications in section 5.

1 Preliminaries

For basic facts about Siegel modular forms we refer to [1, 10, 13]. The group $\mathrm{GSp}(n, \mathbb{R})$ acts on the upper half space $\mathbb{H}_n$ in the usual way. Let $\rho \colon \mathrm{GL}(n, \mathbb{C}) \to V_\rho$ be a finite dimensional (irreducible) polynomial representation. Then we define the slash-operator for functions $f \colon \mathbb{H}_n \to V_\rho$ and $M = \left(\begin{smallmatrix} A & B \\ C & D \end{smallmatrix} \right)$ by

$$(f \mid_\rho M)(Z) := \left(\sqrt{\mu(M)} \right)^{\sum \lambda_i} \cdot \rho(CZ + D)^{-1} \cdot f\left((AZ + B)(CZ + D)^{-1}\right).$$

Here $\mu(M)$ denotes the similitude factor of M and $(\lambda_1, \ldots, \lambda_n)$ is the weight of the representation ρ. In the case of a one-dimensional representation $\rho = \det^k$ we write just $f \mid_k$ instead of $f \mid_{\det^k}$. For a natural number $N > 1$ we put

$$\Gamma_0^n(N) := \left\{ \left(\begin{smallmatrix} A & B \\ C & D \end{smallmatrix} \right) \in \mathrm{Sp}(n, \mathbb{Z}) \ \middle| \ C \equiv 0 \bmod N \right\}.$$

We view a Dirichlet character modulo N as a character of $\Gamma_0^n(N)$ by $\chi(M) := \chi\left(\det(D)\right)$. Then the space of Siegel modular forms of degree n, weight ρ and character χ for the congruence subgroup $\Gamma_0^n(N)$ is the space of all holomorphic functions $f : \mathbb{H} \to V_\rho$ that satisfy

$$f \mid_\rho M = \chi(M) \cdot f$$

for all $M \in \Gamma_0^n(N)$ (with the ususal vanishing conditions in the cusps for $n = 1$). We denote the space of these modular forms by $[\Gamma_0^n(N), \rho, \chi]$ and the subspace of cusp forms by $[\Gamma_0^n(N), \rho, \chi]_0$.

Several times we have to use embeddings of small symplectic groups into bigger ones. For

$$g = \begin{pmatrix} a & b \\ c & d \end{pmatrix} \in \mathrm{Sp}(n) \quad \text{and} \quad h = \begin{pmatrix} A & B \\ C & D \end{pmatrix} \in \mathrm{Sp}(m)$$

we denote by $g \times h$ the element of $\mathrm{Sp}(n + m)$ defined by

$$g \times h = \begin{pmatrix} a & 0 & b & 0 \\ 0 & A & 0 & B \\ c & 0 & d & 0 \\ 0 & C & 0 & D \end{pmatrix}.$$

2 The pullback formula for F

Here we just recall the result of a computation done in [6, 8]. We can use a more general framework for the moment: Let $N > 1$ be arbitrary, χ an arbitrary Dirichlet character mod N and k a positive number with $\chi(-1) = (-1)^k$. Then for a complex number s with $k + 2\Re(s) > n + 1$ the degree n Eisenstein series

$$F_k^n(Z, \chi, s) := \sum_{C,D} \chi\big(\det(C)\big) \det(CZ + D)^{-k} \frac{\det(Y)^s}{|\det(CZ + D)|^{2s}}$$

converges absolutely and uniformly in domains of type $\Im(Z) \geq \lambda \cdot 1_n$ with $\lambda > 0$. For a cusp form $f \in [\Gamma_0^n(N), k, \chi]_0$ we consider

$$\Lambda_F^{2n,k}(f)(z, s) := \int_{\Gamma_0^n(N) \backslash \mathbb{H}_n} f(w) \, \overline{F_k^{2n}\left(\begin{pmatrix} w & 0 \\ 0 & -\bar{z} \end{pmatrix}, \chi, \bar{s} \right)} \det(w)^k \, d^*w .$$

The properties of this map can be formulated completely linearly, but we prefer to assume that f is an eigenform for all Hecke operators coming from Hecke pairs $\big(\mathrm{Sp}(n, \mathbb{Q}_p), \mathrm{Sp}(n, \mathbb{Z}_p)\big)$ for all "good primes" p (i.e., coprime to N). Then we can associate to such an eigenform f the standard L-function

$$L^N(f, s) := \prod_{(p, N) = 1} \frac{1}{\big(1 - \chi(p)p^{-s}\big)}$$

$$\cdot \prod_{i=1}^{n} \frac{1}{\big(1 - \chi(p)\alpha_i(p)p^{-s}\big)\big(1 - \chi(p)\alpha_i(p)^{-1}p^{-s}\big)} \cdot$$

Here the α_i denote the Satake parameters attached to the eigenform f. It is well known that this (partial) Euler product converges absolutely for $\Re(s) \gg 0$ and has a meromorphic continuation to the whole complex plane. We also need Hecke operators for the bad primes; we describe them in greater detail: Let

$$D = \begin{pmatrix} d_1 & & \\ & \ddots & \\ & & d_n \end{pmatrix} \qquad \text{with} \quad d_i \mid d_{i+1}$$

be an (integral) elementary divisor matrix with $\det(D) \mid N^\infty$. Then

$$\mathrm{GL}(n, \mathbb{Z}) \cdot D \cdot \mathrm{GL}(n, \mathbb{Z}) \longmapsto \Gamma_0^n(N) \cdot \begin{pmatrix} D^{-1} & 0 \\ 0 & D \end{pmatrix} \cdot \Gamma_0^n(N)$$

induces an embedding of a $\mathrm{GL}(n)$-Hecke algebra into the Hecke algebra of the pair $\big(\Gamma_0^n(N), \mathrm{Sp}(n, \mathbb{Z}[\frac{1}{N}])\big)$. For D as above, we define the Hecke operator $T(D)$ on $[\Gamma_0^n(N), k, \chi]_0$ by

$$f \mid T(D) := \sum_i \chi\big(\det(D)\det(\alpha_i)\big) \, f \mid_k \begin{pmatrix} \alpha_i & \beta_i \\ \gamma_i & \delta_i \end{pmatrix}$$

where

$$\Gamma_0^n(N) \cdot \begin{pmatrix} D^{-1} & 0 \\ 0 & D \end{pmatrix} \cdot \Gamma_0^n(N) = \bigcup_i \Gamma_0^n(N) \cdot \begin{pmatrix} \alpha_i & \beta_i \\ \gamma_i & \delta_i \end{pmatrix} \cdot$$

(One actually chooses representatives with $\gamma_i = 0$.)

Furthermore we need the Hecke operators $U(p)$, which are best explained by their action on Fourier expansions of modular forms: for $f(z) = \sum_T a(T)e^{2\pi i \, \mathrm{tr}(T \cdot z)}$ we have

$$\big(f \mid U(N)\big)(z) = \sum_T a(N \cdot T)e^{2\pi i \, \mathrm{tr}(T \cdot z)} \cdot$$

Then the formulas (2.37) and (3.23) from [8] give, for an eigenfunction f of all good Hecke operators, the formula ($\Re(s) \gg 0$)

$$\Lambda_F^{2n,k}(f)(z, s) = \Omega(s) \times \frac{N^{\frac{n(n+1)-nk}{2}}}{\mathcal{L}^N(k+2s, \chi)}$$

$$\times L^N\left(f \mid_k \left(\begin{smallmatrix} 0 & -1 \\ N & 0 \end{smallmatrix}\right), k+2s-n\right)$$

$$\times \sum_D f \mid_k \left(\begin{smallmatrix} 0 & -1 \\ N & 0 \end{smallmatrix}\right) \mid U(N) \mid T(D)(z) \, \det(D)^{-k-2s}.$$

Here D runs over all elementary divisor matrices with $\det(D) \mid N^\infty$, the factor $\Omega(s)$ is essentially a Γ-factor, namely

$$\Omega(s) = (-1)^{\frac{nk}{2}} 2^{\frac{n(n+1)}{2}+1-2ns} \pi^{\frac{n(n+1)}{2}} \cdot \frac{\Gamma_n\left(1+s-\frac{n}{2}\right)\Gamma_n\left(1+s-\frac{n(n+1)}{2}\right)}{\Gamma_n\left(k+s\right)\Gamma_n\left(k+s-\frac{n}{2}\right)},$$

with

$$\Gamma_n(s) = \pi^{\frac{n(n-1)}{4}} \prod_{i=1}^{n} \Gamma\left(s - \frac{i-1}{2}\right),$$

and $\mathcal{L}^N(s, \chi)$ comes from a normalizing factor of the Eisenstein series:

$$\mathcal{L}^N(s, \chi) = L^N(s, \chi) \prod_{i=1}^{n} L^N\left(2s - 2i, \chi^2\right).$$

To analyse this formula, we should mention the following well known facts

- The $U(N)$ commute with all $T(D)$.
- The $T(D)$ are *weakly* multiplicative, i.e., $T(D_1 \cdot D_2) = T(D_1) \circ T(D_2)$ if $\det(D_1)$ and $\det(D_2)$ are coprime.
- There is Tamagawa's rationality theorem (e.g. [18, Theorem 3.212]): For $p \mid N$ write $T_p(i_1, \ldots, i_n)$ instead of

$$T\left(\begin{pmatrix} p^{i_1} & & \\ & \ddots & \\ & & p^{i_n} \end{pmatrix}\right),$$

and define

$$\pi(p)_{n,i} := T_p(\underbrace{1, \ldots 1}_{i}; \underbrace{0, \ldots, 0}_{n-i}).$$

Then

$$\sum_{0 \leq i_1 \leq \cdots \leq i_n} T_p(i_1, \ldots, i_n) X^{i_1 + \cdots + i_n} = \frac{1}{\sum_{i=0}^{n} (-1)^i p^{\frac{i(i-1)}{2}} \pi(p)_{n,i} X^i} \cdot$$

- The operator $U(N)$ is injective on $[\Gamma_0^n(N), k, \chi]_0$, see [3]. We may therefore consider its inverse $U(N)^{-1}$ on this space.

Finally let us denote by W_N the isomorphism

$$[\Gamma_0^n(N), k, \chi]_0 \cong [\Gamma_0^n(N), k, \overline{\chi}]_0$$

defined by the "Fricke involution"

$$f \mapsto f \mid W_N := f \mid_k \begin{pmatrix} 0 & -1 \\ N & 0 \end{pmatrix},$$

and by $T_N(s)$ the endomorphism of $[\Gamma_0^n(N), k, \chi]_0$ defined by the following finite sum of Hecke operators:

$$f \mapsto f \mid T_N(s) := f \mid \left(\prod_{p \mid N} \sum_{i=0}^{n} (-1)^i p^{\frac{i(i-1)}{2}} \pi(p)_{n,i} p^{-is} \right).$$

By the facts mentioned above we see that this endomorphism $T_N(s)$ is—at least in the range of convergence—the inverse of the map

$$f \mapsto \sum_{\det(D) \mid N^\infty} f \mid T(D) \det(D)^{-s}.$$

Then the integral formula from above can be rewritten as

$$\Lambda_F^{2n,k} \left(f \mid T_N(k + 2s) \mid U(N)^{-1} \mid W_N^{-1}, s \right)$$

$$= \Omega(s) \cdot \frac{N^{\frac{n(n+1)-nk}{2}}}{\mathcal{L}^N(k + 2s, \chi)} \cdot L^N(f, k + 2s - n) \cdot f(z).$$

Only the following simple very special consequence of the formula above will be needed later on.

Proposition 2.1. *Let χ be a quadratic character. Assume that k is large enough such that*

- *The Γ-factor $\Omega(s)$ has neither a pole nor a zero in $s = 0$*
- *For all eigenforms $f \in [\Gamma_0^n(N), k, \chi]_0$ of the Hecke operators at the good places, the Euler product $L^N(f, s)$ converges absolutely in $s = k - n$.*

Then

$$f \mapsto \Lambda_F^{2n,k}\left(f \mid T_N(k)^{-1} \mid U(N)^{-1} \mid W_N^{-1}, 0\right)$$

defines an automorphism Λ of the space $[\Gamma_0^n(N), k, \chi]_0$

Remark 2.2. The conditions on k are certainly satisfied for $k > 2n + 1$ (standard elementary estimate [1]); more sophisticated estimates show that $k > \frac{5n}{3} + 1$ is sufficient [9]. Any progress towards Ramanujan-Petersson will improve this bound.

3 Theta series

The theta series in which we are interested come from lattices L in positive definite quadratic spaces (V, q) over $\mathbb{Q}$ of dimension $m = 2k$; we always assume that $q(L) \subset 2\mathbb{Z}$. After choosing a $\mathbb{Z}$-basis of L we can associate to L a symmetric positive definite even matrix S. We will freely switch between the language of lattices and the language of matrices. Our theta series of degree n will be the usual ones:

$$\vartheta^n(L, z) = \vartheta^n(S, z) = \sum_{X \in \mathbb{Z}^{(m,n)}} e^{\pi i \ \mathrm{tr}(X^t S X z)}.$$

We start with the following elementary observation.

Observation 3.1. *Let N be squarefree, let χ be a quadratic character modulo N, and assume that $F_k^{2n}(Z, \chi, 0)$ is a linear combination*

$$F_k^{2n}(Z, \chi, 0) = \sum_i a_i \vartheta^{2n}(L_i, Z)$$

of theta series attached to appropriate lattices L_i in positive definite quadratic spaces. Here "appropriate" means that the rank of L_i is $2k$, the levels of the L_i divide N and the nebentypus characters fit, i.e. $\chi = \left(\frac{(-1)^k \det(L_i)}{\cdot}\right)$ for all i. Under these assumptions, using $\vartheta^{2n}\left(L, \left(\begin{smallmatrix} w & 0 \\ 0 & z \end{smallmatrix}\right)\right) = \vartheta^n(L, w) \cdot \vartheta^n(L, z)$ for $w, z \in \mathbb{H}_n$, we obtain for all $f \in [\Gamma_0^n(N), k, \chi]_0$

$$\Lambda(f) = \sum_i \overline{a_i} \langle f, \vartheta^n(L_i) \rangle \cdot \vartheta^n(L_i),$$

where $\langle \ , \ \rangle$ denotes the Petersson inner product.

At this point we do not care about the nature of the coefficients a_i; in [12] and more generally in [5] it is shown that the assumptions above are true for $k \geq 2n + 1$ (using Hecke summation in the definition of F_k^{2n} for $k = 2n + 1$); in fact, a stronger result is shown there: the linear combination of theta series is a linear combination of genus theta series (the coefficients are not specified but can be computed explicitly).

Combining the proposition 2.1 and the above observation, we obtain the following result.

Theorem 3.2. *Assume that N is squarefree and $k \geq 2n + 1$; then all cusp forms in $[\Gamma_0^n(N), k, \chi]$ are linear combinations of theta series.*

Remark 3.3. Analysing the proofs in [12] and [5], we can get a slight refinement of the Theorem: Assume that L is a lattice of level dividing N in a positive definite quadratic space V of dimension $m = 2k$ such that $\chi = \left(\frac{(-1)^k \det(L)}{\cdot} \right)$. Furthermore assume that for all $p \mid N$ the local space V_p has Witt index $\geq 2n$, equivalently the local orthogonal group $O(V_p)$ has rank at least $2n$; this implies $k \geq 2n$. For the local lattice L_p this means that it has in its Jordan splitting at least $2n$ hyperbolic planes (possibly scaled with powers of p); globally we see that there are at least $(2n + 1)^\nu$ genera of lattices of level dividing N in this space V (with $\nu :=$ number of primes dividing N); these genera can be obtained locally from L_p by changing the scaling of the hyperbolic planes. For any choice of $(2n + 1)^\nu$ such genera, the Eisenstein series $F_k^{2n}(Z, \chi, 0)$ is a linear combination of the genus theta series of these genera. The space $[\Gamma_0^n(N), k, \chi]_0$ therefore consists of linear combinations of theta series from these genera. The existence of a lattice L as described above is assured if

$$k \geq \begin{cases} 2n & 4 \mid k \\ 2n + 1 & \text{otherwise} \end{cases}$$

4 The case of noncuspidal modular forms

We have to recall the machinery of Siegel's ϕ-operator and (at least implicitly) the decomposition of $[\Gamma_0^n(N), k, \chi]$ into components coming from cusp forms of lower degree. For level one such a decomposition was first given by Maaß [16] and Klingen [13] using Poincaré series and Klingen-Eisenstein series respectively. The case of congruence subgroups is (at least technically) more complicated, see e.g. [14] or [17], where some aspects are covered. We do not know an appropriate reference for our purposes (especially for Klingen-Eisenstein series for congruence subgroups) so we describe the setup for $\Gamma_0^n(N)$ in an ad-hoc manner. As before, N should be squarefree with ν prime factors.

4.1 Siegel's ϕ - operator

For $M \in \mathrm{Sp}(n, \mathbb{Z})$ and $f \in [\Gamma_0^n(N), k, \chi]$ we define

$$\phi_M^j(f)(z) := \lim_{\lambda \to \infty} (f \mid_k M) \left(\begin{pmatrix} z & 0 \\ 0 & i\lambda \cdot 1_j \end{pmatrix} \right).$$

This operator is compatible in an obvious way with the action of the maximal parabolic subgroup $C_{n,n-j}$ defined by

$$C_{n,n-j} := \left\{ \begin{pmatrix} A & B \\ C & D \end{pmatrix} \in \mathrm{Sp}(n) \,\middle|\, (C, D) = \begin{pmatrix} C_1 & C_2 & D_1 & D_2 \\ 0 & 0 & 0 & D_4 \end{pmatrix} \right\}$$

where C_1 and D_1 are block matrices of size $n - j$. We also need the group homomorphism $* : C_{n,n-j} \to \mathrm{Sp}(n - j)$ defined by

$$* : M = \begin{pmatrix} A & B \\ C & D \end{pmatrix} \mapsto M^* := \begin{pmatrix} A_1 & B_1 \\ C_1 & D_1 \end{pmatrix}.$$

The compatibility alluded to above then means, with $M \in \mathrm{Sp}(n, \mathbb{Z})$ and $g \in C_{n,n-j}(\mathbb{Z})$, that

$$\phi_M^j(f \mid_k g) = \phi_{Mg}^j(f) = \phi_M^j(f) \mid_k g^*.$$

Therefore the operator ϕ_M^j depends (essentially) only on the double coset

$$\Gamma_0^n(N) \backslash M / C_{n,n-j}(\mathbb{Z}) \in \Gamma_0^n(N) \backslash \mathrm{Sp}(n, \mathbb{Z}) / C_{n,n-j}(\mathbb{Z}).$$

The number of these double cosets is $(j + 1)^\nu$. We may (and will) choose representatives in the form $M = 1_{2n-2j} \times M_j$ with $M_j \in \mathrm{Sp}(j, \mathbb{Z})$. It is then sufficient to choose the M_j as representatives of

$$\Gamma_0^j(N) \backslash \mathrm{Sp}(j, \mathbb{Z}) / C_{j,0}(\mathbb{Z}).$$

We get a linear map

$$\phi_M^j : [\Gamma_0^n(N), k, \chi] \longrightarrow [(M^{-1}\Gamma_0(N)M \cap C_{n,n-j}(\mathbb{Z}))^*, k, \chi_M],$$

where $\chi_M(g) := \chi(MgM^{-1})$ for $g \in M^{-1}\Gamma_0^n(N)M$. If M is of the special form $M = 1_{2n-2j} \times M_j$ then the target space of ϕ_M^j is simply $[\Gamma_0^{n-j}(N), k, \chi]$.

We recall that f is cuspidal if $\phi_M(f) = 0$ for all M. We get a filtration

$$[\Gamma_0^n(N), k, \chi]_0 \subset [\Gamma_0^n(N), k, \chi]_1 \subset \cdots \subset [\Gamma_0^n(N), k, \chi]_n = [\Gamma_0^n(N), k, \chi]$$

with

$$[\Gamma_0^n(N), k, \chi]_i = \left\{ f \in [\Gamma_0(N), k, \chi] \,\middle|\, \phi_M^{i+1} = 0 \quad \text{for all } M \right\}.$$

4.2 Sections for Siegel's ϕ-operators

To construct all modular forms of degree n from cusp forms it is then sufficient to construct for all decompositions $n = r + j$ with $0 \leq r \leq n$ and a set $\mathcal{M}_j$ of representatives of $\Gamma_0^j(N) \backslash \mathrm{Sp}(j, \mathbb{Z}) / C_{j,0}(\mathbb{Z})$ linear maps

$$\mathcal{E}_M^{r,n} : [\Gamma_0^r(N), k, \chi]_0 \longmapsto [\Gamma_0^n(N), k, \chi] \qquad (M \in \mathcal{M})$$

such that for any two elements $M, M' \in \mathcal{M}_j$ we have

$$(\phi_{1_{2r} \times M'}^{n-r} \circ \mathcal{E}_M^{r,n})(f) = \begin{cases} \mathcal{E}^{r,r}(f) & \text{if} \quad M = M' \\ 0 & \text{if} \quad M \neq M'. \end{cases}$$

Here $\mathcal{E}^{r,r}$ can be any automorphism of $[\Gamma_0^r(N), k, \chi]_0$. It is then easy to see that the image of $\mathcal{E}^{r,n}$ lies in the subspace $[\Gamma_0^n(N), k, \chi]_{n-r}$ and

$$[\Gamma_0^n(N), k, \chi]_r = \sum_{i \geq n-r} \sum_M \mathrm{Image}(\mathcal{E}_M^{i,n}).$$

A straightforward choice of such maps is given by Klingen-Eisenstein series (this construction is described in [13, 10] for level one), defined for $f \in [\Gamma_0^n(N), k, \chi]_0$ and $M \in \mathcal{M}_j$ by

$$f \mapsto \sum_{g \in \sim \backslash (1_{2r} \times M^{-1}) \cdot \Gamma_0^n(N)} \chi\big((1_{2r} \times M) \cdot g\big) \cdot \tilde{f} \mid_k g$$

with

$$\tilde{f}(Z) := f(z)$$

for $Z = \begin{pmatrix} z & * \\ * & * \end{pmatrix} \in \mathbb{H}_n$ and $z \in \mathbb{H}_r$. The equivalence $\sim$ in the summation above means "modulo the action of the group"

$$(1_{2r} \times M^{-1}) \Gamma_0^n(N)(1_{2r} \times M) \cap C_{n,r}.$$

The Klingen-Eisenstein series are easily seen to satisfy the requirement above, with the automorphism $\mathcal{E}^{r,r}$ being the identity.

4.3 Generalizing the map $\Lambda_F^{2r,k}$

The disadvantage of the construction using Klingen-Eisenstein series is that it does *not* provide a link to theta series. Therefore we define such maps $\mathcal{E}_M^{r,n}$ by using a generalization of the Eisenstein series of type F: Starting from $0 \leq r \leq n$ and an element $M \in \mathcal{M}_j$ with $r + j = n$ we put

$$\tilde{M} := \begin{pmatrix} 0_{2r} & -1_{2r} \\ 1_{2r} & 0_{2r} \end{pmatrix} \times M^{-1} \in \mathrm{Sp}(2r + j, \mathbb{Z}).$$

Then we define for $k > n + r + 1$ the "Siegel Eisenstein series of degree $2r + j = n + r$ attached to the cusp $\tilde{M}$" by

$$E_k^{n+r}(Z, \chi, \tilde{M}) := \sum_{\gamma \in \sim \backslash \tilde{M} \cdot \Gamma_0^{n+r}(N)} \chi(\tilde{M}^{-1}\gamma) j(\gamma, Z)^{-k},$$

where the equivalence is with respect to the group

$$C_{n+r,0} \cap \left(\tilde{M} \Gamma_0^{n+r}(N) \tilde{M}^{-1} \right).$$

We point out here that our favourite Eisenstein series $F^{2r}(Z, \chi)$ is the special case $j = 0$ and hence

$$\tilde{M} := \begin{pmatrix} 0_{2r} & -1_{2r} \\ 1_{2r} & 0_{2r} \end{pmatrix}$$

in the construction above.

We may then apply the ϕ-operator termwise to the series defining the Eisenstein series and we obtain for

$$\tilde{M} = 1_{4r} \times M \quad \text{and} \quad \tilde{M}' = \begin{pmatrix} 0_{2r} & -1_{2r} \\ 1_{2r} & 0_{2r} \end{pmatrix} \times M'$$

with $M, M' \in \mathcal{M}_j$:

$$\phi_{\tilde{M}}^j \circ E_k^{n+r}(*, \chi, \tilde{M}') = \begin{cases} F_k^{2r}(*, \chi) & \text{if} \quad M = M' \\ 0 & \text{if} \quad M \neq M'. \end{cases}$$

Then the maps $[\Gamma_0^r(N), k, \chi]_0 \longmapsto [\Gamma_0^n(N); k, \chi]$ given by $f \mapsto g$ with

$$g(z) = \int f(w) \overline{E_k^{n+r}\left(\begin{pmatrix} w & 0 \\ 0 & -\bar{z} \end{pmatrix}, \chi, \tilde{M} \right)} \det(w)^k d^*w$$

(with $z \in \mathbb{H}_n$) have all the properties requested above from the maps $\mathcal{E}_M^{r,n}$. Of course we need here that for $r = n$ this map coincides with the automorphism $\Lambda_F^{2r,k}$ considered in section 2.

Implicitly we used above the absolute convergence of the Siegel Eisenstein series $E_k^{n+r}(Z, \chi, \tilde{M})$ for all $0 \leq r < n$; therefore we get in the theorem below the additional condition $k > 2n$. We have now to remember from [12, 5] that for k large enough *all* the Siegel Eisenstein series are linear combinations of theta series, therefore the image of the map above lies in the space of theta series (for all M). We obtain

Theorem 4.1. *Suppose that* $k \geq 2n + 1$. *Then all modular forms in* $[\Gamma_0^n(N), k, \chi]$ *are linear combinations of theta series.*

Remark 4.2. It should be possible to get these results in a more group-theoretic way by a careful analysis of the double cosets

$$\left(\tilde{M} \Gamma_0^{2r+j}(N) \tilde{M}^{-1} \right)_\infty \backslash \tilde{M} \cdot \Gamma_0^{2r+j}(N) / \Gamma_0^r(N) \times \Gamma_0^{r+j}(N)$$

5 Holomorphic differential operators

The calculus of holomorphic differential operators as described in [11] and already used in [7] allows us to extend our results to the case of vector-valued modular forms and theta series with harmonic coefficients.

For a polynomial representation $\rho = \rho_0 \otimes \det^k : \mathrm{GL}(n, \mathbb{C}) \to V_\rho$ we consider polynomial functions $P : \mathbb{C}^{(m,n)} \to V_\rho$ satisfying

$$P(XA) = \rho(A^t) P(X)$$

for all $X \in \mathbb{C}^{(m,n)}$ and $A \in \mathrm{GL}(n, \mathbb{C})$, and

$$\sum_{\substack{1 \le i \le m \\ 1 \le j \le n}} \frac{\partial^2 P}{\partial^2 x_{ij}} = 0 .$$

The theta series we consider here are then of type

$$\theta^n(S, P)(z) := \sum_{X \in \mathbb{Z}^{(m,n)}} P(\sqrt{S} \cdot X) e^{\pi i \, \mathrm{tr}(X^t S X z)} .$$

We denote by $\theta^n(m, \rho, N, \chi)$ the vector space generated by (V_ρ- valued) theta series for positive definite quadratic forms of rank $m = 2k$ and level dividing N with character χ; this is a subspace of $[\Gamma_0^n(N), \rho, \chi]$.

We need the differential operators here only for the "convergent case" (i.e., we apply them to an Eisenstein series of degree $2n$, weight k with $s = 0$ and $k > 2n + 1$). We give a very short summary of the main facts needed here. For details we refer to [11, 7]. For $\rho = \det^k \otimes \rho_0$ as above there is a holomorphic differential operator $\mathcal{D}_{k,\rho}$ acting on functions on $\mathbb{H}_{2n}$, which is a polynomial in the partial derivatives, evaluated for $z_2 = 0$. This operator can be viewed as a generalization of the restriction map used in section 3. It maps C^∞-functions F on $\mathbb{H}_{2n}$ to $V_\rho \otimes V_\rho$ valued functions on $\mathbb{H}_n \times \mathbb{H}_n$ and satisfies for all $M \in \mathrm{Sp}(n, \mathbb{R})$

$$\mathcal{D}_{k,\rho}(F \mid_k M \times 1_{2n}) = (\mathcal{D}_{k,\rho} F) \mid_\rho^{(w)} M$$

$$\mathcal{D}_{k,\rho}(F \mid_k 1_{2n} \times M) = (\mathcal{D}_{k,\rho} F) \mid_\rho^{(z)} M .$$

In these formulas, we use the upper indices w and z to indicate the variable to which $M \in \mathrm{Sp}(n, \mathbb{R})$ is applied.

Now we can proceed as in the previous section: Suppose that $k > 2n + 1$ and let ρ_0 be an irreducible polynomial representation of $GL(n, \mathbb{C})$; we put $\rho = \det^k \otimes \rho_0$. As usual we equip V_ρ with a scalar product $\{,\}$ with respect to which $U(n, \mathbb{C})$ acts by unitary operators on V_ρ. Then for $f \in [\Gamma_0^n(N), \rho, \chi]_0$ we define (with $w = u + iv \in \mathbb{H}_n$)

$$\Lambda_F^{2n,k,\rho}(f)(z)$$

$$:= \int_{\Gamma_0^n(N) \backslash \mathbb{H}_n} \left\{ \rho(\sqrt{v}) f(w), \rho(\sqrt{v})(DF_k^{2n})\left(\begin{pmatrix} w & 0 \\ 0 & -\bar{z} \end{pmatrix}, \chi \right) \right\} d^*w .$$

Then there is a nonzero constant $\Omega = \Omega(k, \rho, N)$ such that for all $f \in [\Gamma_0^n, \rho, \chi]_0$ that are eigenfunctions for all "good" Hecke operators,

$$\Lambda_F^{2n,k,\rho}(f)(z) = \Omega(k, \rho, N) \frac{1}{\mathcal{L}^N(k, \chi)}$$

$$\times L^N(f \mid_\rho \begin{pmatrix} 0 & -1 \\ N & 0 \end{pmatrix}, k - n)$$

$$\times \sum_D f \mid_\rho \begin{pmatrix} 0 & -1 \\ N & 0 \end{pmatrix} \mid U(N) \cdot T(D)(z) \det(D)^{-k} .$$

Then as in section 3

$$f \mapsto \Lambda_F^{2n,k,\rho}(f \mid T_N(k) \mid U(N)^{-1} \mid W_N^{-1})$$

defines an automorphism Λ of $[\Gamma_0^n(N), \rho, k]_0$ for $k > 2n + 1$.

On the other hand, the modular form defined by

$$z \mapsto \int_{\Gamma_0^n(N) \backslash \mathbb{H}_n} \left\{ \rho(\sqrt{v}) f(w), \rho(\sqrt{v})(D\vartheta^{2n}(L, \left(\begin{pmatrix} w & 0 \\ 0 & -\bar{z} \end{pmatrix} \right)) \right\} d^*w$$

is a theta series with harmonic coefficients, which we denote by $\theta^n(L, f)$.

For all eigenforms f of the good Hecke operators we have the identity

$$\Lambda(f) = \sum_i \overline{a_i} \theta^n(L_i, f) ,$$

and in particular we get the following result.

Theorem 5.1. *If* $\frac{m}{2} > 2n + 1$ *then we have*

$$[\Gamma_0^n(N), \rho, \chi]_0 \subset \theta^n(m, \rho, N, \chi) .$$

Remark 5.2. Concerning the bound $k > 2n + 1$ and the quadratic forms needed in the theorem above, the remark following the theorem of section 3 also applies here.

Remark 5.3.   We can also give a version of the theorem above including noncusp forms. For this we need more complicated differential operators and we need the more general ϕ-operators as described in Weissauer's paper [20]. We omit the details.

Bibliography

[1] Andrianov, A.: Quadratic Forms and Hecke Operators. Grundlehren der Mathematik 286, Springer 1987.

[2] Böcherer, S.: Siegel modular forms and theta series. In: Proc. Symp. Pure Math. 49, vol. 2, Amer. Math. Soc., Providence, RI, 1989, pp. 3–17.

[3] Böcherer, S.: On the Hecke operator $U(p)$. J. Math. Kyoto Univ. 50 (2005), 807–829.

[4] Böcherer, S.: On Eisenstein series of degree 2 for squarefree levels and the genus version of the basis problem. In: Automorphic Forms and Zeta Functions, World Scientific Publishing Co. Pte. Ltd., Hackensack, NJ, 2006, pp. 43–70.

[5] Böcherer, S., Hironaka, Y., Sato, F.: Linear independence of local densities and its application to Siegel modular forms. Preprint 2007.

[6] Böcherer, S., Schulze-Pillot, R.: Siegel modular forms and theta series attached to definite quaternion algebras. Nagoya Math. J. 121 (1991), 35–96.

[7] Böcherer, S., Schulze-Pillot, R.: Siegel modular forms and theta series attached to definite quaternion algebras II. Nagoya Math. J. 147 (1997), 71–106.

[8] Böcherer, S., Schmidt, C.G.: p-adic measures attached to Siegel modular forms. Ann. Inst. Fourier 50 (2000), 1375–1443.

[9] Duke, W., Howe, R., Li, J.-S.: Estimating Hecke eigenvalues of Siegel modular forms. Duke Math. J. 67 (1992), 219–240.

[10] Freitag, E.: Siegelsche Modulfunktionen. Grundlehren der Mathematik 254, Springer 1983.

[11] Ibukiyama, T.: On differential operators on automorphic forms and invariant pluriharmonic polynomials. Comm. Math. Univ. St. Pauli 48 (1999), 103–118.

[12] Katsurada, H., Schulze-Pillot, R.: Genus Theta Series, Hecke operators and the basis problem for Eisenstein series. In: Automorphic Forms and Zeta Functions, World Scientific Publishing Co. Pte. Ltd., Hackensack, NJ, 2006, pp. 234–261.

[13] Klingen, H.: Introductory Lectures on Siegel modular forms. Cambridge University Press 1990.

[14] Koecher, M.: Zur Theorie der Modulfunktionen n-ten Grades II. Math. Ann. 61 (1955), 455–466.

[15] Kuang, J.: On the linear representability of Hilbert-Siegel modular forms by theta series. Am. J. Math. 116 (1994), 921–994.

[16] Maaß, H.: Über die Darstellung der Modulformen n-ten Grades durch Poincarésche Reihen. Math. Ann. 123 (1951), 125–151.

[17] Satake, I.: Surjectivité de l'operateur Φ. Séminaire Cartan 10, no. 2 (1957–1958), Exp. 16 (online version: URL stable: http://www.numdam.org/item?id= SHC_1957-1958__10_2_A7_0)

[18] Shimura, G.: Introduction to the Arithmetic Theory of Automorphic Functions. Princeton University Press 1971.

[19] Shimura, G.: Euler Products and Eisenstein Series. Regional Conference Series in Math. 93. Amer. Math. Soc., Providence, RI, 1997.

[20] Weissauer, R.: Vektorwertige Modulformen kleinen Gewichts. Journ. reine angew. Math. 343 (1983), 184–202.

[21] Waldspurger, J.-L.: Engendrement par des séries thêta de certains espaces de formes modulaires. Invent. Math. 50 (1978), 135–168.

Mock theta functions, weak Maass forms, and applications

Kathrin Bringmann

1 Introduction

The goal of this article is to provide an overview on mock theta functions and their connection to weak Maass forms.

The theory of modular forms has important applications to many areas of mathematics, e.g. quadratic forms, elliptic curves, partitions as well as other areas throughout mathematics. Let me explain this with the example of partitions. If $p(n)$ denotes the number of partitions of an integer n, then by Euler, we have

$$P(q) := \sum_{n=0}^{\infty} p(n) \, q^{24n-1} = \frac{1}{\eta(24z)},$$

where $\eta(z)$ is Dedekind's η-functions, a weight $\frac{1}{2}$ cusp form ($q = e^{2\pi i z}$ throughout). The theory of modular forms can be employed to show many important properties of $p(n)$. For example Rademacher used the circle method to prove that if n is a positive integer, then

$$p(n) = \frac{2\pi}{(24n-1)^{3/4}} \sum_{k=1}^{\infty} \frac{A_k(n)}{k} \cdot I_{\frac{3}{2}} \left(\frac{\pi \sqrt{24n-1}}{6k} \right). \tag{1.1}$$

Here $I_s(x)$ is the usual I-Bessel function of order s. Furthermore, if $k \geq 1$ and n are integers and $e(x) := e^{2\pi i x}$, then define

$$A_k(n) := \sum_{h \pmod{k}^*} \omega_{h,k} \, e^{-\frac{2\pi i h n}{k}},$$

where h runs through all primitive elements modulo k, and where

$$\omega_{h,k} := \exp\left(\pi i s(h,k)\right).$$

Here

$$s(h, k) := \sum_{\mu \pmod k} \left(\left(\frac{\mu}{k}\right)\right) \left(\left(\frac{h\mu}{k}\right)\right)$$

with

$$((x)) := \begin{cases} x - \lfloor x \rfloor - \frac{1}{2} & \text{if } x \in \mathbb{R} \setminus \mathbb{Z}, \\ 0 & \text{if } x \in \mathbb{Z}. \end{cases}$$

Moreover $p(n)$ satisfies some nice congruence properties. The most famous ones are the Ramanujan congruences:

$$p(5n + 4) \equiv 0 \pmod 5, \tag{1.2}$$

$$p(7n + 5) \equiv 0 \pmod 7, \tag{1.3}$$

$$p(11n + 6) \equiv 0 \pmod{11}. \tag{1.4}$$

In a celebrated paper Ono [32] treated these kinds of congruences systematically. Combining Shimura's theory of modular forms of half-integral weight with results of Serre on modular forms modulo ℓ he showed that for any prime $\ell \geq 5$ there exist infinitely many non-nested arithmetic progressions of the form $An + B$ such that

$$p(An + B) \equiv 0 \pmod \ell.$$

Moreover partitions are related to Eulerian series. For example we have

$$\sum_{n=0}^{\infty} p(n) q^n = 1 + \sum_{n=1}^{\infty} \frac{q^{n^2}}{(1 - q)^2 (1 - q^2)^2 \cdots (1 - q^n)^2}. \tag{1.5}$$

Other examples that relate Eulerian series to modular forms are the Rogers-Ramanujan identities

$$1 + \sum_{n=1}^{\infty} \frac{q^{n^2}}{(1 - q)(1 - q^2) \cdots (1 - q^n)} = \frac{1}{\prod_{n=1}^{\infty} (1 - q^{5n-1})(1 - q^{5n-4})},$$

$$1 + \sum_{n=1}^{\infty} \frac{q^{n^2+n}}{(1 - q)(1 - q^2) \cdots (1 - q^n)} = \frac{1}{\prod_{n=1}^{\infty} (1 - q^{5n-2})(1 - q^{5n-3})}.$$

$$\tag{1.6}$$

Mock theta functions, which can also be defined as Eulerian series, stand out of this context. For example the mock theta function $f(q)$, defined by Ramanujan [34] in his last letter to Hardy, is given by

$$f(q) := 1 + \sum_{n=1}^{\infty} \frac{q^{n^2}}{(1 + q)^2 (1 + q^2)^2 \cdots (1 + q^n)^2}. \tag{1.7}$$

Even if (1.5) and (1.7) have a similar shape $P(q)$ is modular whereas $f(q)$ is not. The mock theta functions were mysterious objects for a long time, for example there was even a discussion how to rigorously define them. Despite those problems they have applications in vast areas of mathematics (e.g. [2, 3, 4, 18, 20, 28, 39] just to mention a few). In this context one has to understand Dyson's quote given 1987 at the Ramanujan's Centary Conference:

"The mock theta-functions give us tantalizing hints of a grand synthesis still to be discovered. Somehow it should be possible to build them into a coherent group-theoretical structure, analogous to the structure of modular forms which Hecke built around the old theta-functions of Jacobi. This remains a challenge for the future."

Zwegers [39, 40] made a first step towards solving this challenge. He observed that Ramanujan's mock theta functions can be interpreted as part of a real analytic vector valued modular form. The author and Ono build on those results to relate functions like $f(q)$ to weak Maass forms (see Section 2 for the definition of a weak Maass form). It turns out that $f(q)$ is the "holomorphic part" of a weak Maass form, the "non-holomorphic part" is a Mordell type integral involving weight $\frac{3}{2}$ theta functions. This is the special case of an infinite family of weak Maass forms that arise from Dyson's rank generating functions (see Section 3). This new theory has a wide range of applications. For example we obtain exact formulas for Ramanujan's mock theta function $f(q)$ (see Section 4), congruences for Dyson's ranks (see Section 5), asymptotics and inequalities for ranks (see Section 6) and identities for rank differences that involve modular forms (see Section 8). In Section 7 we show furthermore a correspondence between weight $\frac{3}{2}$ weak Maass forms and weight $\frac{1}{2}$ theta functions.

This paper is an extended version of a talk given at the conference "Modular forms" held in October 2006 in Schiermonnikoog. The author thanks the organizers B. Edixhoven, G. van der Geer, and B. Moonen for a stimulating atmosphere.

Acknowledgement. The author thanks the referee for many helpful comments.

2 General facts on weak Maass forms

Here we recall basic facts on weak Maass forms, first studied by Bruinier and Funke [17]. For $k \in \frac{1}{2}\mathbb{Z} \setminus \mathbb{Z}$ and $z = x + iy$ with $x, y \in \mathbb{R}$, the weight k hyperbolic Laplacian is given by

$$\Delta_k := -y^2 \left(\frac{\partial^2}{\partial x^2} + \frac{\partial^2}{\partial y^2} \right) + iky \left(\frac{\partial}{\partial x} + i \frac{\partial}{\partial y} \right).$$

If v is odd, then define ϵ_v by

$$\epsilon_v := \begin{cases} 1 & \text{if } v \equiv 1 \pmod 4, \\ i & \text{if } v \equiv 3 \pmod 4. \end{cases}$$

A *(harmonic) weak Maass form of weight k on a subgroup* $\Gamma \subset \Gamma_0(4)$ is any smooth function $f : \mathbb{H} \to \mathbb{C}$ satisfying the following:

(i) For all $A = \begin{pmatrix} a & b \\ c & d \end{pmatrix} \in \Gamma$ and all $z \in \mathbb{H}$, we have

$$f(Az) = \left(\frac{c}{d}\right)^{2k} \epsilon_d^{-2k} (cz+d)^k \, f(z).$$

(ii) We have that $\Delta_k f = 0$.

(iii) The function $f(z)$ has at most linear exponential growth at all the cusps of Γ.

In a similar manner one defines weak Maass forms on $\Gamma_0(4N)$ (N a positive integer) with Nebentypus χ (a Dirichlet character) by requiring

$$f(Az) = \chi(d) \left(\frac{c}{d}\right)^{2k} \epsilon_d^{-2k} (cz+d)^k \, f(z)$$

instead of (i). Harmonic weak Maass forms have Fourier expansions of the form

$$f(z) = \sum_{n=n_0}^{\infty} \gamma_y(n) q^{-n} + \sum_{n=n_1}^{\infty} a(n) q^n,$$

with $n_0, n_1 \in \mathbb{Z}$. The $\gamma_y(n)$ are functions in y, the imaginary part of z. We refer to $\sum_{n=n_0}^{\infty} \gamma_y(n) q^{-n}$ as the *non-holomorphic part* of $f(z)$, and we refer to $\sum_{n=n_1}^{\infty} a(n) q^n$ as its *holomorphic part*.

Moreover we need the anti-linear differential operator ξ_k defined by

$$\xi_k(g)(z) := 2iy^k \overline{\frac{\partial}{\partial \bar{z}} g(z)}.$$

If g is a harmonic weak Maass form of weight k for the group Γ, then $\xi_k(g)$ is a weakly holomorphic modular form (i.e, a modular form with poles at most at the cusps of Γ) of weight $2 - k$ on Γ. Furthermore, ξ_k has the property that its kernel consists of those weight k weak Maass forms which are weakly holomorphic modular forms.

3 Dyson's ranks and weak Maass forms

In order to explain the Ramanujan congruences Dyson introduced the so-called rank of a partition [23]. The *rank* of a partition is defined to be its largest part minus the number of its parts. In his famous paper Dyson conjectured that ranks could be used to "explain" the congruences (1.2) and (1.3) with modulus 5 and 7. More precisely, he conjectured that the partitions of $5n + 4$ (resp. $7n + 5$) form 5 (resp. 7) groups of equal size when sorted by their ranks modulo 5 (resp. 7). In 1954, Atkin and Swinnerton-Dyer proved Dyson's rank conjecture [7].

To study ranks, it is natural to investigate a generating function. If $N(m, n)$ denotes the number of partitions of n with rank m, then it is well known that

$$R(w; q) := 1 + \sum_{n=1}^{\infty} \sum_{m=-\infty}^{\infty} N(m, n) w^m q^n = 1 + \sum_{n=1}^{\infty} \frac{q^{n^2}}{(wq; q)_n (w^{-1}q; q)_n},$$

where

$$(a; q)_n = (a)_n := (1 - a)(1 - aq) \cdots (1 - aq^{n-1}).$$

If we let $w = 1$ we recover $P(q)$ in its Eulerian form (1.5), i.e., (up to a q-power) a weight $-\frac{1}{2}$ modular form. Moreover $R(-1; q)$ is the generating function for the number of partitions with even rank minus the number of partitions with odd rank and equals

$$R(-1; q) = 1 + \sum_{n=1}^{\infty} \frac{q^{n^2}}{(1 + q)^2 (1 + q^2)^2 \cdots (1 + q^n)^2} = f(q)$$

with $f(q)$ as in (1.7). The author and Ono [12] showed that the functions $R(w; q)$ for $w \neq 1$ a root of unity are the holomorphic parts of weak Maass forms. To make this statement more precise suppose that $0 < a < c$ are integers, and let $\zeta_c := e^{\frac{2\pi i}{c}}$. If $f_c := \frac{2c}{\gcd(c,6)}$, then define for $\tau \in \mathbb{H}$ the weight $\frac{3}{2}$ cuspidal theta function $\Theta\left(\frac{a}{c}; \tau\right)$ by

$$\Theta\left(\tfrac{a}{c}; \tau\right) := \sum_{m \,(\mathrm{mod}\, f_c)} (-1)^m \sin\left(\frac{a\pi(6m + 1)}{c}\right) \cdot \theta\left(6m + 1, 6f_c; \tfrac{\tau}{24}\right),$$

where

$$\theta(\alpha, \beta; \tau) := \sum_{n \equiv \alpha \,(\mathrm{mod}\, \beta)} n e^{2\pi i \tau n^2}.$$

Moreover let $\ell_c := \mathrm{lcm}(2c^2, 24)$ and define

$$D\left(\tfrac{a}{c}; q\right) = D\left(\tfrac{a}{c}; z\right) := -S_1\left(\tfrac{a}{c}; z\right) + q^{-\frac{\ell_c}{24}} R(\zeta_c^a; q^{\ell_c}), \tag{3.1}$$

where the period integral $S_1\left(\frac{a}{c}; z\right)$ is given by

$$S_1\left(\tfrac{a}{c}; z\right) := \frac{-i \sin\left(\frac{\pi a}{c}\right) \ell_c^{\frac{1}{2}}}{\sqrt{3}} \int_{-\bar{z}}^{i\infty} \frac{\Theta\left(\frac{a}{c}; \ell_c \tau\right)}{\sqrt{-i(\tau + z)}} \, d\tau. \tag{3.2}$$

Theorem 3.1. (i) *If* $0 < a < c$, *then* $D\left(\frac{a}{c}; z\right)$ *is a weak Maass form of weight* $\frac{1}{2}$ *on* $\Gamma_c := \left\langle \left(\begin{smallmatrix} 1 & 1 \\ 0 & 1 \end{smallmatrix}\right), \left(\begin{smallmatrix} 1 & 0 \\ \ell_c^2 & 1 \end{smallmatrix}\right) \right\rangle$.

(ii) *If* c *is odd, then* $D\left(\frac{a}{c}; z\right)$ *is a weak Maass form of weight* $\frac{1}{2}$ *on* $\Gamma_1\left(6 f_c^2 \ell_c\right)$.

If $\frac{a}{c} = \frac{1}{2}$, it turns out, using results of Zwegers [39], that $D\left(\frac{1}{2}; z\right)$ is a weak Maass form on $\Gamma_0(144)$ with Nebentypus $\chi_{12}(\cdot) := \left(\frac{12}{\cdot}\right)$.

A similar phenomenon as in Theorem 3.1 occurs in the case of overpartitions. Recall that an *overpartition* is a partition, where the first occurance of a summand may be overlined. For a non-negative integer n we denote by $\overline{p}(n)$ the number of overpartitions of n. We have the generating function [21]

$$\overline{P}(q) := \sum_{n \geq 0} \overline{p}(n) q^n = \frac{\eta(2z)}{\eta(z)^2}, \tag{3.3}$$

which is a weight $-\frac{1}{2}$ modular form. Moreover the generating function for $\overline{N}(m, n)$, the number of overpartitions of n with rank m, is given by

$$\mathcal{O}(w; q) := 1 + \sum_{n=1}^{\infty} \overline{N}(m, n) w^m q^n = \sum_{n=0}^{\infty} \frac{(-1)_n q^{\frac{1}{2}n(n+1)}}{(wq; q)_n \left(\frac{q}{w}; q\right)_n}.$$

In particular the case $w = 1$ gives by (3.3) a modular form. Moreover it turns out [10] that for $w \notin \{-1, 1\}$ a root of unity $\mathcal{O}(w; q)$ is the holomorphic part of a weight $\frac{1}{2}$ weak Maass form. In contrast to the case of usual partitions one obtains in the case $w = -1$ the holomorphic part of a weight $\frac{3}{2}$ weak Maass form.

Sketch of proof of Theorem 3.1. We first determine the transformation law of $R\left(\zeta_c^a; q\right)$. For this we define certain related functions. For $q := e^{2\pi i z}$ let

$$M\left(\frac{a}{c}; q\right) := \frac{1}{(q; q)_\infty} \sum_{n=-\infty}^{\infty} \frac{(-1)^n q^{n+\frac{a}{c}}}{1 - q^{n+\frac{a}{c}}} \cdot q^{\frac{3}{2}n(n+1)},$$

$$M_1\left(\frac{a}{c}; q\right) := \frac{1}{(q; q)_\infty} \sum_{n=-\infty}^{\infty} \frac{(-1)^{n+1} q^{n+\frac{a}{c}}}{1 + q^{n+\frac{a}{c}}} \cdot q^{\frac{3}{2}n(n+1)},$$

$$N\left(\frac{a}{c}; q\right) := \frac{(1 - \zeta_c^a)}{(q; q)_\infty} \sum_{n \in \mathbb{Z}} \frac{(-1)^n q^{\frac{n}{2}(3n+1)}}{1 - \zeta_c^a q^n},$$

$$N_1\left(\frac{a}{c}; q\right) := \sum_{n \in \mathbb{Z}} \frac{(-1)^n q^{\frac{3n}{2}(n+1)}}{1 - \zeta_c^a q^{n+\frac{1}{2}}}.$$

As an abuse of notation we also write $M\left(\frac{a}{c};z\right)$ instead of $M\left(\frac{a}{c};q\right)$ and in the same way we treat the other functions. One can show [27] that

$$R\left(\zeta_c^a;q\right) = N\left(\frac{a}{c};q\right).$$

Moreover for $0 \le b < c$, define $M(a,b,c;z)$ by

$$M(a,b,c;q) := \frac{1}{(q;q)_\infty} \sum_{n=-\infty}^{\infty} \frac{(-1)^n q^{n+\frac{a}{c}}}{1-\zeta_c^b q^{n+\frac{a}{c}}} \cdot q^{\frac{3}{2}n(n+1)}.$$

In addition, if $\frac{b}{c} \notin \left\{0, \frac{1}{2}, \frac{1}{6}, \frac{5}{6}\right\}$, then define the integer $k(b,c)$ by

$$k(b,c) := \begin{cases} 0 & \text{if } 0 < \frac{b}{c} < \frac{1}{6}, \\ 1 & \text{if } \frac{1}{6} < \frac{b}{c} < \frac{1}{2}, \\ 2 & \text{if } \frac{1}{2} < \frac{b}{c} < \frac{5}{6}, \\ 3 & \text{if } \frac{5}{6} < \frac{b}{c} < 1, \end{cases} \tag{3.4}$$

and let

$$N(a,b,c;q) := -\frac{i\zeta_{2c}^a \, q^{-\frac{b}{2c}}}{2\,(q;q)_\infty} \sum_{n=-\infty}^{\infty} \frac{(-1)^n q^{\frac{n}{2}(3n+1)-k(b,c)n}}{1-\zeta_c^a q^{n-\frac{b}{c}}}.$$

Remark. The above defined functions can also be rewritten in terms of the functions

$$T_k(x;q) := \frac{1}{(q;q)_\infty} \sum_{n \in \mathbb{Z}} \frac{(-1)^n \, q^{\frac{n}{2}(3n+2k+1)}}{1-xq^n}$$

with $k \in \mathbb{Z}$. These functions can all be expressed in terms of T_0:

$$T_m(x;q) - xT_{m+1}(x;q) = (-1)^m \chi_3(1-m)q^{-\frac{m}{6}(m+1)}$$

with

$$\chi_3(m) := \begin{cases} 1 & \text{if } m \equiv 1 \pmod{3}, \\ -1 & \text{if } m \equiv -1 \pmod{3}, \\ 0 & \text{if } m \equiv 0 \pmod{3}. \end{cases}$$

Also

$$T_m\left(x^{-1};q\right) = -xT_{-m}(x;q).$$

Moreover we need the following Mordell type integrals.

$$J\left(\tfrac{a}{c};\alpha\right) := \int_0^\infty e^{-\frac{3}{2}\alpha x^2} \frac{\cosh\left(\left(\frac{3a}{c}-2\right)\alpha x\right) + \cosh\left(\left(\frac{3a}{c}-1\right)\alpha x\right)}{\cosh(3\alpha x/2)}\, dx,$$

$$J_1\left(\tfrac{a}{c};\alpha\right) := \int_0^\infty e^{-\frac{3}{2}\alpha x^2} \frac{\sinh\left(\left(\frac{3a}{c}-2\right)\alpha x\right) - \sinh\left(\left(\frac{3a}{c}-1\right)\alpha x\right)}{\sinh(3\alpha x/2)}\, dx,$$

$$J(a,b,c;\alpha) := \int_{-\infty}^\infty e^{-\frac{3}{2}\alpha x^2 + 3\alpha x \frac{a}{c}} \frac{\left(\zeta_c^b e^{-\alpha x} + \zeta_c^{2b} e^{-2\alpha x}\right)}{\cosh\left(3\alpha x/2 - 3\pi i \frac{b}{c}\right)}\, dx.$$

Modifying an argument of Watson [38] one can show using contour integration.

Lemma 3.2. *Suppose that $0 < a < c$ are coprime integers, and that α and β have the property that $\alpha\beta = \pi^2$. If $q := e^{-\alpha}$ and $q_1 := e^{-\beta}$, then we have*

$$q^{\frac{3a}{2c}\left(1-\frac{a}{c}\right)-\frac{1}{24}} \cdot M\left(\tfrac{a}{c};q\right) = \sqrt{\tfrac{\pi}{2\alpha}}\, \csc\left(\tfrac{a\pi}{c}\right) q_1^{-\frac{1}{6}} \cdot N\left(\tfrac{a}{c};q_1^4\right) - \sqrt{\tfrac{3\alpha}{2\pi}} \cdot J\left(\tfrac{a}{c};\alpha\right),$$

and

$$q^{\frac{3a}{2c}\left(1-\frac{a}{c}\right)-\frac{1}{24}} \cdot M_1\left(\tfrac{a}{c};q\right) = -\sqrt{\tfrac{2\pi}{\alpha}}\, q_1^{\frac{4}{3}} \cdot N_1\left(\tfrac{a}{c};q_1^2\right) - \sqrt{\tfrac{3\alpha}{2\pi}} \cdot J_1\left(\tfrac{a}{c};\alpha\right).$$

If moreover $\frac{b}{c} \notin \left\{\frac{1}{2}, \frac{1}{6}, \frac{5}{6}\right\}$, then

$$q^{\frac{3a}{2c}\left(1-\frac{a}{c}\right)-\frac{1}{24}} \cdot M(a,b,c;q) =$$

$$\sqrt{\tfrac{8\pi}{\alpha}} \cdot e^{-2\pi i \frac{a}{c} k(b,c) + 3\pi i \frac{b}{c}\left(\frac{2a}{c}-1\right)} \zeta_c^{-b} q_1^{\frac{4b}{c} k(b,c) - \frac{6b^2}{c^2} - \frac{1}{6}} \cdot N(a,b,c;q_1^4)$$

$$- \sqrt{\tfrac{3\alpha}{8\pi}} \cdot \zeta_{2c}^{-5b} \cdot J(a,b,c;\alpha).$$

Remarks. (1) The case $b = 0$ is contained in [27].

(2) It is nowadays more common to write modular transformation laws in terms of τ and $-\frac{1}{\tau}$ than in q and q_1.

The above transformation laws allow us to construct an infinite family of a vector valued weight $\frac{1}{2}$ weak Maass forms (see [12] for the definition of vector valued weak Maass form). For simplicity we assume for the remainder of this section that c is odd. Using the functions

$$\mathcal{N}\left(\tfrac{a}{c};q\right) = \mathcal{N}\left(\tfrac{a}{c};z\right) := \csc\left(\tfrac{a\pi}{c}\right) \cdot q^{-\frac{1}{24}} \cdot N\left(\tfrac{a}{c};q\right),$$

$$\mathcal{M}\left(\tfrac{a}{c};q\right) = \mathcal{M}\left(\tfrac{a}{c};z\right) := 2q^{\frac{3a}{2c}\cdot\left(1-\frac{a}{c}\right)-\frac{1}{24}} \cdot M\left(\tfrac{a}{c};q\right),$$

we define the vector valued (holomorphic) function $F\left(\frac{a}{c};z\right)$ by

$$F\left(\tfrac{a}{c};z\right) := \left(F_1\left(\tfrac{a}{c};z\right), F_2\left(\tfrac{a}{c};z\right)\right)^T$$
$$= \left(\sin\left(\tfrac{\pi a}{c}\right)\mathcal{N}\left(\tfrac{a}{c};\ell_c z\right), \sin\left(\tfrac{\pi a}{c}\right)\mathcal{M}\left(\tfrac{a}{c};\ell_c z\right)\right)^T.$$

Similarly, define the vector valued (non-holomorphic) function $G\left(\frac{a}{c};z\right)$ by

$$G\left(\tfrac{a}{c};z\right) = \left(G_1\left(\tfrac{a}{c};z\right), G_2\left(\tfrac{a}{c};z\right)\right)^T$$
$$:= \left(2\sqrt{3}\sin\left(\tfrac{\pi a}{c}\right)\sqrt{-i\ell_c z}\,J\left(\tfrac{a}{c};-2\pi i\ell_c z\right), \frac{2\sqrt{3}\sin\left(\tfrac{\pi a}{c}\right)}{i\ell_c z}\,J\left(\tfrac{a}{c};\tfrac{2\pi i}{\ell_c z}\right)\right)^T.$$

Following a method of Zwegers [39], which uses the Mittag-Leffler partial fraction decomposition, we can realize the function $G\left(\frac{a}{c};z\right)$ as a vector valued theta integral.

Lemma 3.3. *For* $z \in \mathbb{H}$, *we have*

$$G\left(\tfrac{a}{c};z\right) = \frac{i\ell_c^{\frac{1}{2}}\sin\left(\tfrac{\pi a}{c}\right)}{\sqrt{3}}\int_0^{i\infty}\frac{\left((-i\ell_c\tau)^{-\frac{3}{2}}\Theta\left(\tfrac{a}{c};\tfrac{-1}{\ell_c\tau}\right), \Theta\left(\tfrac{a}{c};\ell_c\tau\right)\right)^T}{\sqrt{-i(\tau+z)}}\,d\tau.$$

We next determine the necessary modular transformation properties of the vector

$$S\left(\tfrac{a}{c};z\right) = \left(S_1\left(\tfrac{a}{c};z\right), S_2\left(\tfrac{a}{c};z\right)\right)$$
$$:= \frac{-i\sin\left(\tfrac{\pi a}{c}\right)\ell_c^{\frac{1}{2}}}{\sqrt{3}}\cdot\int_{-\bar{z}}^{i\infty}\frac{\left(\Theta\left(\tfrac{a}{c};\ell_c\tau\right), (-i\ell_c\tau)^{-\frac{3}{2}}\Theta\left(\tfrac{a}{c};-\tfrac{1}{\ell_c\tau}\right)\right)^T}{\sqrt{-i(\tau+z)}}\,d\tau.$$

Lemma 3.4. *We have*

$$S\left(\tfrac{a}{c};z+1\right) = S\left(\tfrac{a}{c};z\right),$$
$$\frac{1}{\sqrt{-i\ell_c z}}\cdot S\left(\tfrac{a}{c};-\tfrac{1}{\ell_c^2 z}\right) = \begin{pmatrix}0 & 1\\ 1 & 0\end{pmatrix}\cdot S\left(\tfrac{a}{c};z\right) + G\left(\tfrac{a}{c};z\right).$$

Combining the above and using the transformation law for $\Theta\left(\frac{a}{c};\tau\right)$ [35], one can now conclude that $D\left(\frac{a}{c};z\right)$ satisfies the correct transformation law under the stated group. To see that $D\left(\frac{a}{c};z\right)$ is annihilated by $\Delta_{\frac{1}{2}}$, we write

$$\Delta_{\frac{1}{2}} = -4y^{\frac{3}{2}}\frac{\partial}{\partial z}\sqrt{y}\frac{\partial}{\partial\bar{z}}. \tag{3.5}$$

Since $q^{-\frac{\ell_c}{24}} R(\zeta_b^a; q^{\ell_c})$ is a holomorphic function in z, it is thus clearly annihilated by $\Delta_{\frac{1}{2}}$. Moreover

$$\frac{\partial}{\partial \bar{z}} \left(S_1 \left(\tfrac{a}{c}; z\right)\right) = -\frac{\sin\left(\frac{\pi a}{c}\right) \ell_c^{\frac{1}{2}}}{\sqrt{6y}} \cdot \Theta\left(\tfrac{a}{c}; -\ell_c \bar{z}\right).$$

Hence, we find that $\sqrt{y} \frac{\partial}{\partial \bar{z}} \left(D\left(\frac{a}{c}; z\right)\right)$ is anti-holomorphic, and therefore by (3.5) annihilated by $\Delta_{\frac{1}{2}}$. Using that $\Theta\left(\frac{a}{c}; \tau\right)$ is a weight $\frac{3}{2}$ cusp form it is not hard to conclude that $D\left(\frac{a}{c}; z\right)$ has at most linear exponential growth at the cusps.   □

4 The Andrews-Dragonette-Conjecture

One can use the theory of weak Maass form to obtain exact formulas for the coefficients of the mock theta function $f(q)$ which we denote by $\alpha(n)$ [11]. Recall that

$$f(q) = R(-1; q) = 1 + \sum_{n=1}^{\infty} (N_e(n) - N_o(n)) q^n$$

$$= 1 + \sum_{n=1}^{\infty} \frac{q^{n^2}}{(1+q)^2(1+q^2)^2 \cdots (1+q^n)^2},$$

where $N_e(n)$ (resp. $N_o(n)$) denotes the number of partitions of n with even (resp. odd) rank. It is a classical problem to find exact formulas for $N_e(n)$ and $N_o(n)$. Since by (1.1) we have an exact formula for the partition function $p(n)$ this is equivalent to the problem of determining exact formulas for $\alpha(n)$. Ramanujan's last letter to Hardy includes the claim that

$$\alpha(n) = (-1)^{n-1} \frac{\exp\left(\pi\sqrt{\frac{n}{6} - \frac{1}{144}}\right)}{2\sqrt{n - \frac{1}{24}}} + O\left(\frac{\exp\left(\frac{1}{2}\pi\sqrt{\frac{n}{6} - \frac{1}{144}}\right)}{\sqrt{n - \frac{1}{24}}}\right). \quad (4.1)$$

Typical of his writings, Ramanujan did not give a proof. Dragonette finally showed (4.1) in her 1951 Ph.D. thesis [22] written under the direction of Rademacher. In his 1964 Ph.D. thesis, also written under Rademacher, Andrews [2] improved upon Dragonette's work, and showed that

$\alpha(n) =$

$$\pi(24n-1)^{-\frac{1}{4}} \sum_{k=1}^{\lceil\sqrt{n}\rceil} \frac{(-1)^{\lfloor\frac{k+1}{2}\rfloor} A_{2k}\left(n - \frac{k(1+(-1)^k)}{4}\right)}{k} I_{\frac{1}{2}}\left(\frac{\pi\sqrt{24n-1}}{12k}\right) + O(n^\epsilon).$$

Moreover they made the

Conjecture (Andrews-Dragonette). *If n is a positive integer, then*

$\alpha(n) =$

$$\pi(24n-1)^{-\frac{1}{4}} \sum_{k=1}^{\infty} \frac{(-1)^{\lfloor\frac{k+1}{2}\rfloor} A_{2k}\left(n - \frac{k(1+(-1)^k)}{4}\right)}{k} \cdot I_{\frac{1}{2}}\left(\frac{\pi\sqrt{24n-1}}{12k}\right). \quad (4.2)$$

The author and Ono [11] used the theory of weak Maass forms to prove this conjecture.

Theorem 4.1. *The Andrews-Dragonette Conjecture is true.*

Sketch of Proof. From Section 3 we know that $D\left(\frac{1}{2}; z\right)$ is a weight $\frac{1}{2}$ weak Maass form on $\Gamma_0(144)$ with Nebentypus character χ_{12}. We will construct a Maass-Poincaré series which we will show equals $D\left(\frac{1}{2}; z\right)$. The Andrews-Dragonette Conjecture can be concluded by computing the coefficients of the Poincaré series. For $s \in \mathbb{C}$, $k \in \frac{1}{2} + \mathbb{Z}$, and $y \in \mathbb{R} \setminus \{0\}$, let

$$\mathcal{M}_s(y) := |y|^{-\frac{k}{2}} M_{\frac{k}{2}\,\mathrm{sgn}(y),\, s-\frac{1}{2}}(|y|),$$

$$\mathcal{W}_s(y) := |y|^{-\frac{1}{4}} W_{\frac{1}{4}\,\mathrm{sgn}(y),\, s-\frac{1}{2}}(|y|),$$

where $M_{\nu,\mu}(z)$ and $W_{\nu,\mu}(z)$ are the standard Whittaker functions. Furthermore, let

$$\varphi_{s,k}(z) := \mathcal{M}_s\left(-\frac{\pi y}{6}\right) e\left(-\frac{x}{24}\right).$$

It is straightforward to confirm that $\varphi_{s,k}(z)$ is an eigenfunction of Δ_k. Moreover for matrices $\left(\begin{smallmatrix} a & b \\ c & d \end{smallmatrix}\right) \in \Gamma_0(2)$, with $c \geq 0$, define

$$\chi\left(\left(\begin{smallmatrix} a & b \\ c & d \end{smallmatrix}\right)\right) :=$$

$$\begin{cases} e\left(-\frac{b}{24}\right) & \text{if } c = 0, \\ i^{-1/2}(-1)^{\frac{1}{2}(c+ad+1)} e\left(-\frac{a+d}{24c} - \frac{a}{4} + \frac{3dc}{8}\right) \cdot \omega_{-d,c}^{-1} & \text{if } c > 0. \end{cases}$$

Define the Poincaré series $P_k(s; z)$ by

$$P_k(s; z) := \frac{2}{\sqrt{\pi}} \sum_{M \in \Gamma_\infty \backslash \Gamma_0(2)} \chi(M)^{-1} (cz + d)^{-k} \varphi_{s,k}(Mz), \qquad (4.3)$$

were $\Gamma_\infty := \left\{ \pm \left(\begin{smallmatrix} 1 & n \\ 0 & 1 \end{smallmatrix} \right) : n \in \mathbb{Z} \right\}$. The series $P_{\frac{1}{2}} \left(1 - \frac{k}{2}; z \right)$ is absolute convergent for $k < \frac{1}{2}$ and annihilated by $\Delta_{\frac{k}{2}}$. The function $P_{\frac{1}{2}} \left(\frac{3}{4}; z \right)$ can be analytically continued by its Fourier expansion, which bases on a modification of an argument of Hooley involving the interplay between solutions of quadratic congruences and the representation of integers by quadratic forms. This calculation is lengthy, and is carried out in detail in [11]. We compute the Fourier expansion of $P_{\frac{1}{2}} \left(\frac{3}{4}; z \right)$ as

$$P_{\frac{1}{2}} \left(\frac{3}{4}; z \right) =$$
$$\left(1 - \pi^{-\frac{1}{2}} \cdot \Gamma \left(\frac{1}{2}, \frac{\pi y}{6} \right) \right) \cdot q^{-\frac{1}{24}} + \sum_{n=-\infty}^{0} \gamma_y(n) q^{n-\frac{1}{24}} + \sum_{n=1}^{\infty} \beta(n) q^{n-\frac{1}{24}},$$

where for positive integers n we have

$$\beta(n) = \pi (24n - 1)^{-\frac{1}{4}}$$
$$\cdot \sum_{k=1}^{\infty} \frac{(-1)^{\lfloor \frac{k+1}{2} \rfloor} A_{2k} \left(n - \frac{k(1+(-1)^k)}{4} \right)}{k} I_{\frac{1}{2}} \left(\frac{\pi \sqrt{24n-1}}{12k} \right)$$

and for non-positive integers n we have

$$\gamma_y(n) = \pi^{\frac{1}{2}} |24n - 1|^{-\frac{1}{4}} \cdot \Gamma \left(\frac{1}{2}, \frac{\pi |24n-1| \cdot y}{6} \right)$$
$$\cdot \sum_{k=1}^{\infty} \frac{(-1)^{\lfloor \frac{k+1}{2} \rfloor} A_{2k} \left(n - \frac{k(1+(-1)^k)}{4} \right)}{k} \cdot J_{\frac{1}{2}} \left(\frac{\pi \sqrt{|24n-1|}}{12k} \right).$$

Here the incomplete gamma function $\Gamma(a; x)$ is defined by

$$\Gamma(a; x) := \int_x^\infty e^{-t} t^{a-1} \, dt. \qquad (4.4)$$

To finish the proof, we have to show that $\alpha(n) = \beta(n)$. For this we let

$$P(z) := P_{\frac{1}{2}} \left(\frac{3}{4}; 24z \right) = P_{nh}(z) + P_h(z),$$
$$M(z) := D \left(\frac{1}{2}; z \right) = M_{nh}(z) + M_h(z)$$

canonically decomposed into a non-holomorphic and a holomorphic part. In particular

$$P_h(z) = q^{-1} + \sum_{n=1}^{\infty} \beta(n) \, q^{24n-1},$$

$$M_h(z) = q^{-1} f(q^{24}) = q^{-1} + \sum_{n=1}^{\infty} \alpha(n) \, q^{24n-1}.$$

The function $P(z)$ and $M(z)$ are weak Maass forms of weight $\frac{1}{2}$ for $\Gamma_0(144)$ with Nebentypus χ_{12}. We first prove that $P_{nh}(z) = M_{nh}(z)$. For this compute that $\xi_{\frac{1}{2}}(P(z))$ and $\xi_{\frac{1}{2}}(M(z))$ are holomorphic modular forms of weight $\frac{3}{2}$ with Nebentypus χ_{12} with the property that their non-zero Fourier coefficients are supported on arithmetic progression congruent to 1 (mod 24). Choose a constant c such that the coefficients up to q^{24} of $\xi_{\frac{1}{2}}(P(z))$ and $c\xi_{\frac{1}{2}}(M(z))$ agree and since

$$\dim_{\mathbb{C}}\left(M_{\frac{3}{2}}\left(\Gamma_0(144), \chi_{12}\right)\right) = 24,$$

$\xi_{\frac{1}{2}}(P(z)) = c\xi_{\frac{1}{2}}(M(z))$, which implies that $P_{nh}(z) = cM_{nh}(z)$. Thus the function $H(z) := P(z) - cM(z)$ is a weakly holomorphic modular form. We have to show that $c = 1$. For this we apply the inversion $z \mapsto -\frac{1}{z}$. By work of Zwegers [39] this produces a nonholomorphic part unless $c = 1$. Since $H(z)$ is weakly holomorphic we conclude that $c = 1$. To be more precise, the Poincaré series considered here is a component of a vector valued weak Maass form whose transformation law is known by work of Zwegers (see also [26], where such a vector valued Poincaré series is constructed). Estimating the coefficients of $P(z)$ and $M(z)$ against $n^{\frac{3}{4}+\epsilon}$, one obtains that $H(z)$ is a holomorphic modular form of weight $\frac{1}{2}$ on $\Gamma_0(144)$ with Nebentypus χ_{12}. Since this space is trivial we obtain $H(z) = 0$ which employs the claim. $\qquad\square$

5 Congruences for Dyson's rank generating functions

In this section we prove an infinite family of congruences for Dyson's ranks which generalizes partitions congruences [12].

Theorem 5.1. *Let t be a positive odd integer, and let $Q \nmid 6t$ be prime. If j is a positive integer, then there are infinitely many non-nested arithmetic progressions $An + B$ such that for every $0 \leq r < t$ we have*

$$N(r, t; An + B) \equiv 0 \pmod{Q^j}.$$

Remarks. (1) The congruences in Theorem 5.1 may be viewed as a combinatorial decomposition of the partition function congruence

$$p(An + B) \equiv 0 \pmod{\mathcal{Q}^j}.$$

(2) Congruences for $t = \mathcal{Q}^j$ were shown in [9].

Sketch of proof. First observe that

$$\sum_{n=0}^{\infty} N(r, t; n)q^n = \frac{1}{t}\sum_{n=0}^{\infty} p(n)q^n + \frac{1}{t}\sum_{j=1}^{t-1} \zeta_t^{-rj} \cdot R(\zeta_t^j; q). \tag{5.1}$$

Using the results from Section 3 we can conclude that

$$\sum_{n=0}^{\infty}\left(N(r, t; n) - \frac{p(n)}{t}\right)q^{\ell_t n - \frac{\ell_t}{24}}$$

is the holomorphic part of a weak Maass form of weight $\frac{1}{2}$ on $\Gamma_1\left(6f_t^2\ell_t\right)$.

We wish to apply certain quadratic twists which 'kill' the non–holomorphic part of $D\left(\frac{a}{c}; q\right)$. For this we compute on which arithmetic progressions it is supported. This will enable us to use results on congruences for half integer weight modular forms. We obtain

$$D\left(\tfrac{a}{c}; z\right) = q^{-\frac{\ell_c}{24}} + \sum_{n=1}^{\infty}\sum_{m=-\infty}^{\infty} N(m, n)\zeta_c^{am}q^{\ell_c n - \frac{\ell_c}{24}}$$

$$- \frac{2\sin\left(\frac{\pi a}{c}\right)}{\sqrt{\pi}} \cdot \left\{ \sum_{m \pmod{f_c}} (-1)^m \sin\left(\frac{a\pi(6m+1)}{c}\right) \right.$$

$$\left. \times \sum_{n \equiv 6m+1 \pmod{6f_c}} \Gamma\left(\tfrac{1}{2}; \tfrac{\ell_c n^2 y}{6}\right) q^{-\frac{\ell_c n^2}{24}} \right\}.$$

In particular the non-holomorphic part of $D\left(\frac{a}{c}; z\right)$ is supported on certain fixed arithmetic progression. Generalizing the theory of twists of modular forms to twists of weak Maass forms, one can show.

Proposition 5.2. *If $0 \le r < t$ are integers, where t is odd, and $\mathcal{P} \nmid 6t$ is prime, then*

$$\sum_{\substack{n \ge 1 \\ \left(\frac{24\ell_t n - \ell_t}{\mathcal{P}}\right) = -\left(\frac{-\ell_t}{\mathcal{P}}\right)}} \left(N(r, t; n) - \frac{p(n)}{t}\right) q^{\ell_t n - \frac{\ell_t}{24}}$$

is a weight $\frac{1}{2}$ weakly holomorphic modular form on $\Gamma_1\left(6f_t^2\ell_t\mathcal{P}^4\right)$.

To prove Theorem 5.1, we shall employ a recent general result of Treneer [36]. We use the following fact, which generalized Serre's results on p-adic modular forms.

Proposition 5.3. *Suppose that $f_1(z), f_2(z), \ldots, f_s(z)$ are half-integral weight cusp forms where*

$$f_i(z) \in S_{\lambda_i + \frac{1}{2}}(\Gamma_1(4N_i)) \cap \mathcal{O}_K[[q]],$$

and where $\mathcal{O}_K$ is the ring of integers of a fixed number field K. If $\mathcal{Q}$ is prime and $j \geq 1$ is an integer, then the set of primes L for which

$$f_i(z) \mid T_{\lambda_i}(L^2) \equiv 0 \pmod{\mathcal{Q}^j},$$

for each $1 \leq i \leq s$, has positive Frobenius density. Here $T_{\lambda_i}(L^2)$ denotes the usual L^2 index Hecke operator of weight $\lambda_i + \frac{1}{2}$.

Now suppose that $\mathcal{P} \nmid 6t\mathcal{Q}$ is prime. By Proposition 5.2, for every $0 \leq r < t$

$$F(r, t, \mathcal{P}; z) = \sum_{n=1}^{\infty} a(r, t, \mathcal{P}; n)q^n$$

$$:= \sum_{\left(\frac{24\ell_t n - \ell_t}{\mathcal{P}}\right) = -\left(\frac{-\ell_t}{\mathcal{P}}\right)} \left(N(r, t; n) - \frac{p(n)}{t}\right) q^{\ell_t n - \frac{\ell_t}{24}} \qquad (5.2)$$

is a weakly holomorphic modular form of weight $\frac{1}{2}$ on $\Gamma_1\left(6f_t^2\ell_t\mathcal{P}^4\right)$. Furthermore, by the work of Ahlgren and Ono [1], it is known that

$$P(t, \mathcal{P}; z) = \sum_{n=1}^{\infty} p(t, \mathcal{P}; n)q^n := \sum_{\left(\frac{24\ell_t n - \ell_t}{\mathcal{P}}\right) = -\left(\frac{-\ell_t}{\mathcal{P}}\right)} p(n)q^{\ell_t n - \frac{\ell_t}{24}} \qquad (5.3)$$

is a weakly holomorphic modular form of weight $-\frac{1}{2}$ on $\Gamma_1\left(24\ell_t\mathcal{P}^4\right)$.

Now since $\mathcal{Q} \nmid 24f_t^2\ell_t\mathcal{P}^4$, a generalization of a result of Treneer (see Theorem 3.1 of [36]), implies that there is a sufficiently large integer m for which

$$\sum_{\mathcal{Q}\nmid n} a(r, t, \mathcal{P}; \mathcal{Q}^m n)q^n,$$

for all $0 \leq r < t$, and

$$\sum_{\mathcal{Q}\nmid n} p(t, \mathcal{P}; \mathcal{Q}^m n)q^n$$

are all congruent modulo $\mathcal{Q}^j$ to forms in the graded ring of half-integral weight cusp forms with algebraic integer coefficients on $\Gamma_1\left(24f_t^2\ell_t\right)$. Applying Proposition 5.3 to these $t+1$ forms gives that a positive proportion of primes L

have the property that these $t + 1$ half-integral weight cusp forms modulo Q^j are annihilated by the index L^2 half-integral weight Hecke operators. Theorem 5.1 now follows *mutatis mutandis* as in the proof of Theorem 1 of [32]. □

6 Asymptotics for Dyson's rank partition functions

We obtain asymptotic formulas for Dyson's rank generating functions [8]. As an application, we solve a conjecture of Andrews and Lewis on inequalities between ranks. We write

$$R(\zeta_c^a; q) =: 1 + \sum_{n=1}^{\infty} A\left(\tfrac{a}{c}; n\right) q^n.$$

Let k and h be coprime integers, h' defined by $hh' \equiv -1 \pmod{k}$ if k is odd and by $hh' \equiv -1 \pmod{2k}$ if k is even, $k_1 := \frac{k}{\gcd(k,c)}$, $c_1 := \frac{c}{\gcd(k,c)}$, and $0 < l < c_1$ is defined by the congruence $l \equiv ak_1 \pmod{c_1}$. Furthermore we define, for $n, m \in \mathbb{Z}$, the following sums of Kloosterman type

$$D_{a,c,k}(n, m) := (-1)^{ak+l} \sum_{h \pmod{k}^*} \omega_{h,k} \cdot e^{\frac{2\pi i}{k}(nh+mh')},$$

$$B_{a,c,k}(n, m) := (-1)^{ak+1} \sin\left(\tfrac{\pi a}{c}\right)$$

$$\times \sum_{h \pmod{k}^*} \frac{\omega_{h,k}}{\sin\left(\tfrac{\pi a h'}{c}\right)} \cdot e^{-\frac{3\pi i a^2 k_1 h'}{c}} \cdot e^{\frac{2\pi i}{k}(nh+mh')},$$

where for $B_{a,c,k}(n, m)$ we require that $c | k$. Moreover, for $c \nmid k$, let

$$\delta_{c,k,r} := \begin{cases} -\left(\tfrac{1}{2}+r\right)\tfrac{l}{c_1} + \tfrac{3}{2}\left(\tfrac{l}{c_1}\right)^2 + \tfrac{1}{24} & \text{if } 0 < \tfrac{l}{c_1} < \tfrac{1}{6}, \\ -\tfrac{5l}{2c_1} + \tfrac{3}{2}\left(\tfrac{l}{c_1}\right)^2 + \tfrac{25}{24} - r\left(1 - \tfrac{l}{c_1}\right) & \text{if } \tfrac{5}{6} < \tfrac{l}{c_1} < 1, \\ 0 & \text{otherwise,} \end{cases} \tag{6.1}$$

and for $0 < \tfrac{l}{c_1} < \tfrac{1}{6}$ or $\tfrac{5}{6} < \tfrac{l}{c_1} < 1$ define $m_{a,c,k,r}$ to be

$$\frac{1}{2c_1^2}\left(-3a^2k_1^2 + 6lak_1 - ak_1c_1 - 3l^2 + lc_1 - 2ark_1c_1 + 2lc_1r\right)$$

if $0 < \tfrac{l}{c_1} < \tfrac{1}{6}$, and

$$\frac{1}{2c_1^2}\left(-6ak_1c_1 - 3a^2k_1^2 + 6lak_1 + ak_1c_1 + 6lc_1\right.$$

$$\left. - 3l^2 - 2c_1^2 - lc_1 + 2ark_1c_1 + 2c_1(c_1 - l)r\right)$$

if $\tfrac{5}{6} < \tfrac{l}{c_1} < 1$.

Using the Circle Method, we obtain [8] the following asymptotic formulas for the coefficients $A\left(\tfrac{a}{c}; n\right)$.

Theorem 6.1. *If* $0 < a < c$ *are coprime integers and c is odd, then for positive integers n we have that*

$$
A\left(\tfrac{a}{c}; n\right) = \frac{4\sqrt{3}i}{\sqrt{24n-1}} \sum_{\substack{1 \le k \le \sqrt{n} \\ c|k}} \frac{B_{a,c,k}(-n,0)}{\sqrt{k}} \cdot \sinh\left(\frac{\pi\sqrt{24n-1}}{6k}\right)
$$

$$
+ \frac{8\sqrt{3} \cdot \sin\left(\frac{\pi a}{c}\right)}{\sqrt{24n-1}} \sum_{\substack{1 \le k \le \sqrt{n} \\ c \nmid k \\ r \ge 0 \\ \delta_{c,k,r} > 0}} \frac{D_{a,c,k}(-n, m_{a,c,k,r})}{\sqrt{k}} \sinh\left(\frac{\pi\sqrt{2\delta_{c,k,r}(24n-1)}}{\sqrt{3}k}\right)
$$

$$
+ O_c\left(n^\epsilon\right).
$$

Using (5.1) one can conclude asymptotics for $N(a, c; n)$ from Thm. 6.1.

Corollary 6.2. *For integers* $0 \le a < c$, *where c is an odd integer, we have*

$$
N(a, c; n) = \frac{2\pi}{c \cdot \sqrt{24n-1}} \sum_{k=1}^{\infty} \frac{A_k(n)}{k} \cdot I_{\frac{3}{2}}\left(\frac{\pi\sqrt{24n-1}}{6k}\right)
$$

$$
+ \frac{1}{c} \sum_{j=1}^{c-1} \zeta_c^{-aj} \left(\frac{4\sqrt{3}i}{\sqrt{24n-1}} \sum_{c|k} \frac{B_{j,c,k}(-n,0)}{\sqrt{k}} \sinh\left(\frac{\pi}{6k}\sqrt{24n-1}\right) \right.
$$

$$
+ \frac{8\sqrt{3}\sin\left(\frac{\pi j}{c}\right)}{\sqrt{24n-1}} \sum_{\substack{k,r \\ c \nmid k \\ \delta_{c,k,r} > 0}} \frac{D_{j,c,k}(-n, m_{j,c,k,r})}{\sqrt{k}} \sinh\left(\sqrt{\frac{2\delta_{c,k,r}(24n-1)}{3}}\frac{\pi}{k}\right) \right)
$$

$$
+ O_c\left(n^\epsilon\right).
$$

This corollary implies a conjecture of Andrews and Lewis. In [6, 31] they showed

$$
\begin{array}{ll}
N(0, 2; 2n) < N(1, 2; 2n) & \text{if } n \ge 1, \\
N(0, 4; n) > N(2, 4; n) & \text{if } 26 < n \equiv 0, 1 \pmod 4, \\
N(0, 4; n) < N(2, 4; n) & \text{if } 26 < n \equiv 2, 3 \pmod 4.
\end{array}
$$

Moreover, they conjectured (see Conjecture 1 of [6]).

Conjecture (Andrews and Lewis). *For all* $n > 0$, *we have*

$$
\begin{array}{ll}
N(0, 3; n) < N(1, 3; n) & \text{if } n \equiv 0 \text{ or } 2 \pmod 3, \\
N(0, 3; n) > N(1, 3; n) & \text{if } n \equiv 1 \pmod 3.
\end{array} \tag{6.2}
$$

A careful analysis of the asymptotics in Corollary 6.2 gives the following theorem.

Theorem 6.3. *The Andrews-Lewis Conjecture is true for all positive integers* $n \notin \{3, 9, 21\}$ *in which case we have equality in* (6.2).

Sketch of proof of Theorem 6.1. We use the Hardy Littlewood method. By Cauchy's Theorem we have for $n > 0$

$$A\left(\tfrac{a}{c}; n\right) = \frac{1}{2\pi i} \int_C \frac{N\left(\tfrac{a}{c}; q\right)}{q^{n+1}} \, dq,$$

where C is an arbitrary path inside the unit circle surrounding 0 counterclockwise. Now let

$$\vartheta'_{h,k} := \frac{1}{k(k_1 + k)}, \qquad \vartheta''_{h,k} := \frac{1}{k(k_2 + k)},$$

where $\frac{h_1}{k_1} < \frac{h}{k} < \frac{h_2}{k_2}$ are adjacent Farey fractions in the Farey sequence of order $N := \lfloor n^{1/2} \rfloor$. We make the substitution $q = e^{-\frac{2\pi}{n} + 2\pi i t}$ $(0 \le t \le 1)$ and then decompose the path of integration into paths along the Farey arcs $-\vartheta'_{h,k} \le \Phi \le \vartheta''_{h,k}$, where $\Phi = t - \frac{h}{k}$ and $0 \le h \le k \le N$ with $(h, k) = 1$. One obtains

$$A\left(\tfrac{a}{c}; n\right) = \sum_{h,k} e^{-\frac{2\pi i h n}{k}} \int_{-\vartheta'_{h,k}}^{\vartheta''_{h,k}} N\left(\tfrac{a}{c}; e^{\frac{2\pi i}{k}(h+iz)}\right) \cdot e^{\frac{2\pi n z}{k}} \, d\Phi,$$

where $z = \frac{k}{n} - k\Phi i$. One can conclude from the transformation law of $N\left(\tfrac{a}{c}; q\right)$ (in a modified version of Lemma 3.2) that

$$A\left(\tfrac{a}{c}; n\right) = \sum_1 + \sum_2 + \sum_3,$$

where $\sum_1$ is defined by

$$i \sin\left(\tfrac{\pi a}{c}\right) \sum_{\substack{h,k \\ c|k}} \left\{ \omega_{h,k} \frac{(-1)^{ak+1}}{\sin\left(\tfrac{\pi a h'}{c}\right)} \cdot e^{-\frac{3\pi i a^2 k_1 h'}{c} - \frac{2\pi i h n}{k}} \right.$$

$$\left. \times \int_{-\vartheta'_{h,k}}^{\vartheta''_{h,k}} z^{-\frac{1}{2}} \cdot e^{\frac{2\pi z}{k}\left(n - \frac{1}{24}\right) + \frac{\pi}{12kz}} N\left(\tfrac{ah'}{c}; q_1\right) d\Phi \right\},$$

$\sum_2$ is given by

$$-4i \sin\left(\tfrac{\pi a}{c}\right) \sum_{\substack{h,k \\ c \nmid k}} \left\{ \omega_{h,k} (-1)^{ak+l} e^{-\frac{2\pi i h' s a}{c} - \frac{3\pi i h' a^2 k_1}{c c_1} + \frac{6\pi i h' l a}{c c_1} - \frac{2\pi i h n}{k}} \right.$$

$$\left. \int_{-\vartheta'_{h,k}}^{\vartheta''_{h,k}} z^{-\frac{1}{2}} \cdot e^{\frac{2\pi z}{k}\left(n - \frac{1}{24}\right) + \frac{\pi}{12kz}} \cdot q_1^{\frac{sl}{c_1} - \frac{3l^2}{2c_1^2}} \cdot N\left(ah', \tfrac{lc}{c_1}, c; q_1\right) d\Phi \right\},$$

and

$$\sum_3 := 2\sin^2\left(\tfrac{\pi a}{c}\right) \sum_{h,k} \left\{ \frac{\omega_{h,k}}{k} \cdot e^{-\frac{2\pi i h n}{k}} \sum_{v \;(\mathrm{mod}\; k)} (-1)^v \, e^{-\frac{3\pi i h' v^2}{k} + \frac{\pi i h' v}{k}} \right.$$

$$\left. \int_{-\vartheta'_{h,k}}^{\vartheta''_{h,k}} e^{\frac{2\pi z}{k}\left(n-\frac{1}{24}\right)} \cdot z^{\frac{1}{2}} \cdot I_{a,c,k,v}(z) d\Phi \right\}.$$

Here $q_1 := e^{\frac{2\pi i}{k}\left(h' + \frac{i}{\tau}\right)}$ and

$$I_{a,c,k,v}(z) := \int_{\mathbb{R}} e^{-\frac{3\pi z x^2}{k}} \cdot H_{a,c}\left(\tfrac{\pi i v}{k} - \tfrac{\pi i}{6k} - \tfrac{\pi z x}{k}\right) dx$$

with

$$H_{a,c}(x) := \frac{\cosh(x)}{\sinh\left(x + \frac{\pi i a}{c}\right) \cdot \sinh\left(x - \frac{\pi i a}{c}\right)}.$$

Two important steps are the estimation of $I_{a,c,k,v}(z)$ and certain Kloosterman sums.

Lemma 6.4. *We have*

$$z^{\frac{1}{2}} \cdot I_{a,c,k,v}(z) \ll k \cdot n^{\frac{1}{4}} \cdot g_{a,c,k,v},$$

where $g_{a,c,k,v} := \left(\min\left(6kc\left\{\tfrac{v}{k} - \tfrac{1}{6k} + \tfrac{a}{c}\right\}, 6kc\left\{\tfrac{v}{k} - \tfrac{1}{6k} - \tfrac{a}{c}\right\}\right)\right)^{-1}$, *with* $\{x\} := x - \lfloor x \rfloor$ *for* $x \in \mathbb{R}$.

Lemma 6.5. *For* $n, m \in \mathbb{Z}$, $0 \le \sigma_1 < \sigma_2 \le k$, $D \in \mathbb{Z}$ *with* $(D, k) = 1$, *we have*

$$\sum_{\substack{h \;(\mathrm{mod}\; k)^* \\ \sigma_1 \le Dh' \le \sigma_2}} \omega_{h,k} \cdot e^{\frac{2\pi i}{k}(hn + h'm)} \ll \gcd(24n+1, k)^{\frac{1}{2}} \cdot k^{\frac{1}{2}+\epsilon}.$$

If $c|k$, *then we have*

$$(-1)^{ak+1} \sin\left(\tfrac{\pi a}{c}\right) \sum_{\substack{h \;(\mathrm{mod}\; k)^* \\ \sigma_1 \le Dh' \le \sigma_2}} \frac{\omega_{h,k}}{\sin\left(\frac{\pi a h'}{c}\right)} \cdot e^{-\frac{3\pi i a^2 k_1 h'}{c}} \cdot e^{\frac{2\pi i}{k}(hn + h'm)}$$

$$\ll \gcd(24n+1, k)^{\frac{1}{2}} \cdot k^{\frac{1}{2}+\epsilon}.$$

To estimate $\sum_1$, we write

$$N\left(\tfrac{ah'}{c}; q_1\right) =: 1 + \sum_{r \in \mathbb{N}} a(r) \cdot e^{\frac{2\pi i m_r h'}{k}} \cdot e^{-\frac{2\pi r}{kz}},$$

where m_r is a sequence in $\mathbb{Z}$ and the coefficients $a(r)$ are independent of $a, c, k,$ and h. We treat the constant term and the term coming from from

$r \geq 1$ seperately since they contribute to the main term and to the error term, respectively. We denote the associated sums by S_1 and S_2, respectively and first estimate S_2. Throughout we need the easily verified fact that $\text{Re}(z) = \frac{k}{n}$, $\text{Re}\left(\frac{1}{z}\right) > \frac{k}{2}$, $|z|^{-\frac{1}{2}} \leq n^{\frac{1}{2}} \cdot k^{-\frac{1}{2}}$, and $\vartheta'_{h,k} + \vartheta''_{h,k} \leq \frac{2}{k(N+1)}$. Decompose

$$\int_{-\vartheta'_{h,k}}^{\vartheta''_{h,k}} = \int_{-\frac{1}{k(N+k)}}^{\frac{1}{k(N+k)}} + \int_{-\frac{1}{k(k_1+k)}}^{-\frac{1}{k(N+k)}} + \int_{\frac{1}{k(N+k)}}^{\frac{1}{k(k_2+k)}} \tag{6.3}$$

and denote the associated sums by S_{21}, S_{22}, and S_{23}, respectively. Furthermore in S_{22} (and similarly S_{23}) we write

$$\int_{-\frac{1}{k(k+k_1)}}^{-\frac{1}{k(N+k)}} = \sum_{l=k_1+k}^{N+k-1} \int_{-\frac{1}{kl}}^{-\frac{1}{k(l+1)}}.$$

We have that

$$k_1 \equiv -h' \pmod{k}, \quad k_2 \equiv h' \pmod{k},$$
$$N - k < k_1 \leq N, \quad N - k < k_2 \leq N.$$

Using Lemma 6.5, we can show that $\sum_1$ equals

$$i \sin\left(\frac{\pi a}{c}\right) \sum_{\substack{h,k \\ c|k}} \omega_{h,k} \cdot \frac{(-1)^{ak+1}}{\sin\left(\frac{\pi a h'}{c}\right)} \cdot e^{-\frac{3\pi i a^2 k_1 h'}{c} - \frac{2\pi i h n}{k}}$$

$$\times \int_{-\vartheta'_{h,k}}^{\vartheta''_{h,k}} z^{-\frac{1}{2}} \cdot e^{\frac{2\pi z}{k}\left(n-\frac{1}{24}\right) + \frac{\pi}{12 k z}} \, d\Phi + O\left(n^{\epsilon}\right).$$

In a similar (but more complicated) manner we prove that $\sum_2$ equals

$$2 \sin\left(\frac{\pi a}{c}\right) \sum_{\substack{k,r \\ c \nmid k \\ \delta_{c,k,r} > 0}} (-1)^{ak+l} \sum_h \omega_{h,k} e^{\frac{2\pi i}{k}(-nh+m_{a,c,k,r}h')}$$

$$\times \int_{-\vartheta'_{h,k}}^{\vartheta''_{h,k}} z^{-\frac{1}{2}} e^{\frac{2\pi z}{k}\left(n-\frac{1}{24}\right) + \frac{2\pi}{kz}\delta_{c,k,r}} \, d\Phi + O\left(n^{\epsilon}\right).$$

Using Lemma 6.4 we estimate $\sum_3$ in a similar way. In $\sum_1$ and $\sum_2$ we next write

$$\int_{-\vartheta'_{h,k}}^{\vartheta''_{h,k}} = \int_{-\frac{1}{kN}}^{\frac{1}{kN}} - \int_{-\frac{1}{kN}}^{-\frac{1}{k(k+k_1)}} - \int_{\frac{1}{k(k+k_2)}}^{\frac{1}{kN}}$$

and denote the associated sums by S_{11}, S_{12}, and S_{13}, respectively. The sums S_{12} and S_{13} contribute to the error terms which can be bounded as before. To finish the proof, we have to estimate integrals of the shape

$$I_{k,r} := \int_{-\frac{1}{kN}}^{\frac{1}{kN}} z^{-\frac{1}{2}} \cdot e^{\frac{2\pi}{k}\left(z\left(n-\frac{1}{24}\right)+\frac{r}{z}\right)} d\Phi.$$

One can show that

$$I_{k,r} = \frac{1}{ki} \int_{\Gamma} z^{-\frac{1}{2}} \cdot e^{\frac{2\pi}{k}\left(z\left(n-\frac{1}{24}\right)+\frac{r}{z}\right)} dz + O\left(n^{-\frac{3}{4}}\right),$$

where Γ denotes the circle through $\frac{k}{n} \pm \frac{i}{N}$ and tangent to the imaginary axis at 0. Making the substitution $t = \frac{2\pi r}{kz}$ and using the Hankel formula, we obtain

$$I_{k,r} = \frac{4\sqrt{3}}{\sqrt{k\,(24n-1)}} \sinh\left(\sqrt{\frac{2r\,(24n-1)}{3}}\,\frac{\pi}{k}\right) + O\left(n^{-\frac{3}{4}}\right)$$

from which we can easily conclude the theorem.   □

7 Correspondences for weight $\frac{3}{2}$ weak Maass forms

In [13] we classify those weak Maass forms whose holomorphic part arise from basic hypergeometric series, and we obtain a one-to-one correspondence

$$\{\Theta(\chi; z)\} \quad \longleftrightarrow \quad \{M_\chi(z)\}$$

between holomorphic weight $\frac{1}{2}$ theta functions $\Theta(\chi; z)$ and certain weak Maass forms $M_\chi(z)$ of weight $\frac{3}{2}$. To state our result, define the basic hypergeometric-type series

$$F(\alpha, \beta, \gamma, \delta, \epsilon, \zeta) := \sum_{n=0}^{\infty} \frac{(\alpha; \zeta)_n \delta^{n^2} \epsilon^n}{(\beta; \zeta)_n (\gamma; \zeta)_n}$$

and by differentiation of a special case of the Rogers and Fine identity [25]

$$F_{RF}(\alpha, \beta) := \frac{\alpha}{2} \cdot \frac{\partial}{\partial \alpha} \left(\frac{1}{1+\alpha} F(\alpha, -\alpha\beta, 0, 1, \alpha, \beta) \right) = \sum_{n=1}^{\infty} (-1)^n n \alpha^{2n} \beta^{n^2}.$$

Moreover let $\Theta_0(z) := \sum_{n \in \mathbb{Z}} q^{n^2}$ be the classical Jacobi theta function. For a non-trivial even Dirichlet character χ with conductor b define

$$\Theta(\chi; z) := \sum_{n \in \mathbb{Z}} \chi(n) q^{n^2},$$

which is a weight $\frac{1}{2}$ modular form on $\Gamma_0(4b^2)$. Moreover let the non-holomorphic theta integral $N_\chi(z)$ be given by

$$N_\chi(z) := -\frac{ib^2}{\pi} \int_{-\bar{z}}^{i\infty} \frac{\Theta(\chi; \tau)}{(-i(\tau+z))^{\frac{3}{2}}} \, d\tau.$$

Lastly, define the function Q_χ by

$$Q_\chi(z) := \frac{4b\sqrt{2}}{\Theta_0(b^2 z)} \sum_{a \pmod b} \chi(a) \sum_{j \pmod b} \zeta_b^{ja} L\left(q^{2a} \zeta_b^j, 2b; q\right), \qquad (7.1)$$

where $L(w, d; q)$ is defined by

$$\sum_{n\in\mathbb{Z}} n\frac{q^{n^2} w^n}{1 - q^{dn}} = \sum_{k\geq 0} \left(F_{RF}\left(w^{\frac{1}{2}} q^{\frac{kd}{2}}, -q\right) + F_{RF}\left(w^{-\frac{1}{2}} q^{\frac{(k+1)d}{2}}, -q\right)\right).$$

$$(7.2)$$

In [13], we show:

Theorem 7.1. *If χ is a non-trivial even Dirichlet character with conductor b, then*

$$M_\chi(z) := Q_\chi(z) - N_\chi(z)$$

is a weight $\frac{3}{2}$ weak Maass form on $\Gamma_0(4b^2)$ with Nebentypus χ.

Corollary 7.2. *We have that*

$$y^{3/2} \cdot \overline{\frac{\partial}{\partial \bar{z}} M_\chi(z)} = -\frac{ib^2}{2\sqrt{2\pi}} \cdot \Theta(\chi; z).$$

Since the Serre-Stark Basis Theorem asserts that the spaces of holomorphic weight $\frac{1}{2}$ modular forms have explicit bases of theta functions, Corollary 7.2 gives a bijection between weight $\frac{3}{2}$ weak Maass forms and weight $\frac{1}{2}$ modular forms.

Sketch of proof of Theorem 7.1. We only give a sketch of proof here; details can be found in [13]. For $0 < a < b$, we let

$$\psi_{a,b}(z) := 2 \sum_{m\geq 1} \sum_{\substack{-bm\leq d<0 \\ d\equiv \pm a \pmod b}} (d + bm) \, e^{2\pi i (b^2 m^2 - d^2) z}.$$

Note for fixed b, the functions $\psi_{a,b}$ are determined by representatives $a \in \mathbb{Z}/b\mathbb{Z}$. These q-series will be associated to the period integral

$$\mathcal{N}(a, b; z) := -\frac{ib^2}{\pi} \int_{-\bar{z}}^{i\infty} \frac{\Theta_{a,b}(\tau)}{(-i(z+\tau))^{\frac{3}{2}}} \, d\tau,$$

where

$$\Theta_{a,b}(z) := \sum_{\substack{n\in\mathbb{Z} \\ n\equiv a \pmod{b}}} e^{2\pi i n^2 z}.$$

More precisely, we define the functions $\mathcal{M}(a, b; z)$ by

$$\mathcal{M}(a, b; z) := 2^{\frac{3}{2}} b^2 \frac{\psi_{a,b}(z)}{\Theta_0(b^2 z)} - \mathcal{N}(a, b; z).$$

We first show the following theorem.

Theorem 7.3. *If $0 < a < b$ are coprime positive integers, where $b \geq 4$ is even, then $\mathcal{M}(a, b; z)$ is a weight $\frac{3}{2}$ weak Maass form on $\Gamma(4b^2)$.*

In order to prove Theorem 7.3, we define functions that are "dual" to $\psi_{a,b}$ under the involution $z \mapsto -\frac{1}{z}$:

$$\phi_{a,b}(z) := 2 \sum_{n=1}^{\infty} n \cdot e^{\pi i n^2 z} \frac{\left(1 - e^{4\pi i n z}\right)}{\left(1 - 2\cos\left(\frac{2\pi a}{b}\right) e^{2\pi i n z} + e^{4\pi i n z}\right)}.$$

As with $\psi_{a,b}$, the function $\phi_{a,b}$ depends only on the residue class of a (mod b). We show.

Lemma 7.4. *Assuming the hypotheses above, we have*

$$\frac{\phi_{a,b}(z)}{\Theta_0\left(\frac{z}{2}\right)} = -\frac{1}{b} \cdot (-iz)^{-3/2} \cdot \frac{\psi_{a,b}\left(-\frac{1}{2b^2 z}\right)}{\Theta_0\left(-\frac{1}{2z}\right)} + 2\sqrt{-iz} \int_{\mathbb{R}} \frac{u \, e^{\pi i u^2 z}}{1 - \zeta_c^a e^{\pi i u^2 z}} \, du.$$

Next define the Mordell-type integral

$$\mathcal{I}_{a,b}(z) := 2\sqrt{-iz} \int_{\mathbb{R}} \frac{u \, e^{\pi i u^2 z}}{1 - \zeta_c^a e^{\pi i u^2 z}} \, du.$$

This integral can be rewritten as a theta integral.

Lemma 7.5. *We have*

$$\mathcal{I}_{a,b}(z) = \frac{1}{2\pi} \int_0^{\infty} \frac{\Theta_{a,b}\left(-\frac{1}{2iub^2}\right) \cdot u^{-\frac{1}{2}}}{(-i(iu + z))^{\frac{3}{2}}} \, du.$$

To prove Theorem 7.3, we first show that $\mathcal{M}(a, b; z)$ is annihilated by the weight $\frac{3}{2}$ Laplace operator. Now we show that $\mathcal{M}(a, b; z)$ obeys the weight $\frac{3}{2}$ transformation laws with respect to $\Gamma(4b^2)$. To see this, we use the fact that by work of Shimura $\Theta_{a,b}(\tau)$ is a weight $\frac{1}{2}$ modular form on $\Gamma(4b^2)$. By definition, the period integral $\mathcal{N}(a, b; z)$ inherits the transformation properties of $\Theta_{a,b}(\tau)$ with a shift in weight. To finish the proof of Theorem 7.1, observe that

$$\psi_{a,b}(z) = \frac{2}{b} \sum_{j \pmod b} \zeta_b^{ja} L\left(\zeta_b^j q^{2a}, 2b; q\right)$$

$$= \frac{2}{b} \sum_{j \pmod b} \zeta_b^{ja} \sum_{k \geq 0} \left(F_{RF}\left(\zeta_b^{\frac{j}{2}} q^{kb+a}, -q\right) + F_{RF}\left(\zeta_b^{-\frac{j}{2}} q^{b-a+kb}, -q\right) \right).$$

This yields

$$Q_\chi(z) = \frac{2^{\frac{3}{2}} b^2}{\Theta_0(b^2 z)} \sum_{a \pmod b} \chi(a) \psi_{a,b}(z).$$

Now the claim follows using that $\mathcal{N}_\chi(z) = \sum_{a=1}^{b-1} \chi(a) \mathcal{N}(a, b; z)$.   □

8 Identities for rank differences

There are a lot of identities that relate modular forms and Eulerian series, e.g. the Roger Ramanujan identities (1.6). Let $f_0(q)$, $f_1(q)$, $\Phi(q)$, and $\Psi(q)$ denote the mock theta functions

$$f_0(q) := \sum_{n=0}^{\infty} \frac{q^{n^2}}{(-q)_n}, \qquad \Phi(q) := -1 + \sum_{n=0}^{\infty} \frac{q^{5n^2}}{(q; q^5)_{n+1}(q^4; q^5)_n},$$

$$f_1(q) := \sum_{n=0}^{\infty} \frac{q^{n^2+n}}{(-q)_n}, \qquad \Psi(q) := -1 + \sum_{n=0}^{\infty} \frac{q^{5n^2}}{(q^2; q^5)_{n+1}(q^3; q^5)_n}.$$

The mock theta conjectures of Ramanujan are a list of ten identities involving these functions. Andrews and Garvan [5] proved that these conjectures are equivalent to the following pair of identities that essentially express two weight $\frac{1}{2}$ modular forms as linear combinations of Eulerian series:

$$\frac{(q^5; q^5)_\infty (q^5; q^{10})_\infty}{(q; q^5)_\infty (q^4; q^5)_\infty} = f_0(q) + 2\Phi(q^2), \tag{8.1}$$

$$\frac{(q^5; q^5)_\infty (q^5; q^{10})_\infty}{(q^2; q^5)_\infty (q^3; q^5)_\infty} = f_1(q) + 2q^{-1}\Psi(q^2). \tag{8.2}$$

These were shown by Hickerson in 1988 [28]. He proved later on several more identities of this type [29], and, more recently, Choi and Yesilyurt have obtained even further such identities (for example, see [18, 19, 37]) using methods similar to those of Hickerson. The difficulty in proving the mock theta identities lies in the fact that mock theta functions are not modular forms. Moreover identities for ranks are known, e.g.

$$N(r, t; n) = N(r - t, t; n)$$

or more complicated identities [7]

$$N(1, 7; 7n + 1) = N(2, 7; 7n + 1) = N(3, 7; 7n + 1).$$

Furthermore Atkin and Swinnerton-Dyer [7] proved some very surprising identities such as

$$-\frac{(q; q^7)_\infty^2 (q^6; q^7)_\infty^2 (q^7; q^7)_\infty^2}{(q; q)_\infty}$$

$$= \sum_{n=0}^{\infty} (N(0, 7; 7n + 6) - N(1, 7; 7n + 6)) q^n.$$

This identity expresses a weight $\frac{1}{2}$ modular form as a linear combination of Eulerian series. From the new perspective described in Section 3 that the rank generating functions are the holomorphic parts of weak Maass forms, the mock theta conjectures arise naturally in the theory of Maass forms and arise from linear relations between the non-holomorphic parts of independent Maass forms [16].

Theorem 8.1. *Suppose that $t \geq 5$ is prime, $0 \leq r_1, r_2 < t$ and $0 \leq d < t$. Then the following are true:*

(i) *If $\left(\frac{1-24d}{t}\right) = -1$, then*

$$\sum_{n=0}^{\infty} (N(r_1, t; tn + d) - N(r_2, t; tn + d)) q^{24(tn+d)-1}$$

is a weight $\frac{1}{2}$ weakly holomorphic modular form on $\Gamma_1(576t^{10})$.

(ii) *Suppose that $\left(\frac{1-24d}{t}\right) = 1$. If $r_1, r_2 \not\equiv \pm\frac{1}{2}(1 + \alpha) \pmod{t}$, for any $0 \leq \alpha < 2t$ satisfying $1 - 24d \equiv \alpha^2 \pmod{2t}$, then*

$$\sum_{n=0}^{\infty} (N(r_1, t; tn + d) - N(r_2, t; tn + d)) q^{24(tn+d)-1}$$

is a weight $\frac{1}{2}$ weakly holomorphic modular form on $\Gamma_1\left(576t^6\right)$.

Theorem 8.1 is optimal in a way that for all other pairs r_1 and r_2 (apart from trivial cases) that

$$\sum_{n=0}^{\infty} (N(r_1, t; tn + d) - N(r_2, t, tn + d)) q^{24(tn+d)-1}$$

is the holomorphic part of a weak Maass form which has a non-vanishing non-holomorphic part.

Theorem 8.2. *Suppose that* $t > 1$ *is an odd integer. If* $0 \leq r_1, r_2 < t$ *are integers, and* $\mathcal{P} \nmid 6t$ *is prime, then*

$$\sum_{\substack{n \geq 1 \\ (\frac{24l_t n - l_t}{\mathcal{P}}) = -(\frac{-l_t}{\mathcal{P}})}} (N(r_1, t; n) - N(r_2, t; n)) q^{l_t n - \frac{l_t}{24}}$$

is a weight $\frac{1}{2}$ *weakly holomorphic modular form on* $\Gamma_1(6 f_t^2 l_t \mathcal{P}^4)$.

Since Theorem 8.2 follows easily from Section 3, we only consider Theorem 8.1

Proof of Theorem 8.1. Using the results from Sections 3 and 5, one can reduce the claim to the identity

$$\sum_{j=1}^{t-1} \left(\zeta_t^{-r_1 j} - \zeta_t^{-r_2 j} \right) \sin\left(\frac{\pi j}{t} \right) \sin\left(\frac{\pi j \alpha}{t} \right) = 0$$

which can be easily verified using that $\sin(x) = \frac{1}{2i} \left(e^{ix} - e^{-ix} \right)$.   □

Bibliography

[1] S. Ahlgren and K. Ono, *Congruence properties for the partition function*, Proc. Natl. Acad. Sci., USA **98**, No. 23 (2001), pages 12882–12884.

[2] G. E. Andrews, *On the theorems of Watson and Dragonette for Ramanujan's mock theta functions*, Amer. J. Math. **88** No. 2 (1966), pages 454–490.

[3] G. E. Andrews, *Partitions with short sequences and mock theta functions*, Proc. Natl. Acad. Sci. USA, **102** No. 13 (2005), pages 4666–4671.

[4] G. E. Andrews, F. Dyson, and D. Hickerson, *Partitions and indefinite quadratic forms*, Invent. Math. **91** No. 3 (1988), pages 391–407.

[5] G. E. Andrews and F. Garvan, *Ramanujan's "lost notebook". VI: The mock theta conjectures*, Adv. in Math. **73** (1989), pages 242–255.

[6] G. E. Andrews and R. P. Lewis, *The ranks and cranks of partitions moduli 2, 3 and 4*, J. Number Th. **85** (2000), pages 74–84.

[7] A. O. L. Atkin and H. P. F. Swinnerton-Dyer, *Some properties of partitions*, Proc. London Math. Soc. **66** No. 4 (1954), pages 84-106.

[8] K. Bringmann, *Asymptotics for rank partition functions*, Trans. Amer. Math. Soc., accepted for publication.

[9] K. Bringmann, *On certain congruences for Dyson's ranks*, submitted for publication.

[10] K. Bringmann and J. Lovejoy, *Dyson's rank, overpartitions, and weak Maass forms*, Int. Math. Res. Not. IMRN 2007, no. 19, Art. ID rnm063.

[11] K. Bringmann and K. Ono, *The f(q) mock theta function conjecture and partition ranks*, Invent. Math. **165** (2006), pages 243–266.

[12] K. Bringmann and K. Ono, *Dysons ranks and Maass forms*, Ann. of Math., accepted for publication.

[13] K. Bringmann, A. Folsom, and K. Ono, *q-series and weight 3/2 Maass forms*, in preparation.

[14] K. Bringmann and K. Ono, *Arithmetic properties of coefficients of half-integral weight Maass-Poincaré series*, Math. Ann. **337** (2007), pages 591–612.

[15] K. Bringmann and K. Ono, *Lifting elliptic cusp forms to Maass forms with an application to partitions*, Proc. Nat. Acad. Sc. **104** (2007), pages 3725–3731.

[16] K. Bringmann, K. Ono, and R. Rhoades, *Eulerian series as modular forms*, J. Amer. Math. Soc., accepted for publication.

[17] J. H. Bruinier and J. Funke, *On two geometric theta lifts*, Duke Math. J. **125** (2004), pages 45–90.

[18] Y.-S. Choi, *Tenth order mock theta functions in Ramanujan's lost notebook*, Invent. Math. **136** (1999), pages 497–569.

[19] Y.-S. Choi, *Tenth order mock theta functions in Ramanujan's lost notebook, II*, Adv. in Math. **156** (2000), pages 180–285.

[20] H. Cohen, *q-identities for Maass waveforms*, Invent. Math. **91** No. 3 (1988), pages 409–422.

[21] S. Corteel and J. Lovejoy, *Overpartitions*, Trans. Amer. Math. Soc. **356** (2004), pages 1623–1635.

[22] L. Dragonette, *Some asymptotic formulae for the mock theta series of Ramanujan*, Trans. Amer. Math. Soc. **72** No. 3 (1952), pages 474–500.

[23] F. Dyson, *Some guesses in the theory of partitions*, Eureka (Cambridge) **8** (1944), pages 10–15.

[24] M. Eichler, *On the class number of imaginary quadratic fields and the sums of divisors of natural numbers*, J. Indian Math. Soc. (N.S.) **19** (1955), pages 153–180.

[25] N. J. Fine, *Basic hypergeometric series and applications*, Math. Surveys and Monographs, Vol. 27, Amer. Math. Soc., Providence, 1988.

[26] S. Garthwaite, *The coefficients of the $\omega(q)$ mock theta function*, Int. J. Number Theory, accepted for publication.

[27] B. Gordon and R. McIntosh, *Modular transformations of Ramanujan's fifth and seventh order mock theta functions*, Ramanujan J. **7** (2003), pages 193–222.

[28] D. Hickerson, *A proof of the mock theta conjectures*, Invent. Math. **94** (1988), pages 639–660.

[29] D. Hickerson, *On the seventh order mock theta functions*, Invent. Math. **94** (1988), pages 661–677.

[30] F. Hirzebruch and D. Zagier, *Intersection numbers of curves on Hilbert modular surfaces and modular forms with Nebentypus*, Invent. Math. **36** (1976), pages 57–113.

[31] R. P. Lewis, *The ranks of partitions modulo 2*, Discuss. Math. **167/168** (1997), pages 445–449.

[32] K. Ono, *Distribution of the partition function modulo m*, Ann. of Math. **151** (2000), pages 293–307.

[33] K. Ono, *The web of modularity: Arithmetic of the coefficients of modular forms and q-series*, CBMS Regional Conference, **102**, Amer. Math. Soc., Providence, R. I., 2004.

[34] S. Ramanujan, *The lost notebook and other unpublished papers*, Narosa, New Delhi, 1988.

[35] G. Shimura, *On modular forms of half integral weight*, Ann. of Math. **97** (1973), pages 440–481.

[36] S. Treneer, *Congruences for the coefficients of weakly holomorphic modular forms*, Proc. London Math. Soc. **93** (2006), pages 304–324.

[37] H. Yesilyurt, *Four identities related to third order mock theta functions in Ramanujan's lost notebook*, Adv. in Math. **190** (2005), pages 278–299.

[38] G. N. Watson, *The final problem: An account of the mock theta functions*, J. London Math. Soc. **2** (2) (1936), pages 55–80.

[39] S. P. Zwegers, *Mock ϑ-functions and real analytic modular forms*, q-series with applications to combinatorics, number theory, and physics (Ed. B. C. Berndt and K. Ono), Contemp. Math. **291**, Amer. Math. Soc., (2001), pages 269–277.

[40] S. P. Zwegers, *Mock theta functions*, Ph.D. Thesis, Universiteit Utrecht, 2002.

Sign changes of coefficients of half integral weight modular forms

Jan Hendrik Bruinier and Winfried Kohnen

Abstract

For a half integral weight modular form f we study the signs of the Fourier coefficients $a(n)$. If f is a Hecke eigenform of level N with quadratic character, and t is a fixed square-free positive integer with $a(t) \neq 0$, we show that for all but finitely many primes p the sequence $(a(tp^{2m}))_m$ has infinitely many signs changes. Moreover, we prove similar (partly conditional) results for arbitrary cusp forms f which are not necessarily Hecke eigenforms.

Mathematics Subject Classification: 11F30, 11F37

1 Introduction

Fourier coefficients of cusp forms, in particular their signs, are quite mysterious. Let f be a non-zero elliptic cusp form of positive real weight κ, with multiplier v and with real Fourier coefficients $a(n)$ for $n \in \mathbb{N}$. Using the theory of L-functions, it was shown under quite general conditions in [KKP] that the sequence $(a(n))_{n \in \mathbb{N}}$ has infinitely many sign changes, i.e., there are infinitely many n such that $a(n) > 0$ and there are infinitely many n such that $a(n) < 0$.

This is especially interesting when κ is an integer and f is a Hecke eigenform of level N, and so the $a(n)$ are proportional to the Hecke eigenvalues. For recent work in this direction we refer to e.g. [KoSe], [IKS], [KSL].

In the present note we shall consider the case of half-integral weight $\kappa = k + 1/2, k \in \mathbb{N}$, and level N divisible by 4. Note that this case is distinguished through the celebrated works of Shimura [Sh] and Waldspurger [Wa] in the following way. First, for each square-free positive integer t, there exists a linear lifting from weight $k + 1/2$ to even integral weight $2k$ determined by the coefficients $a(tn^2)$ (where $n \in \mathbb{N}$), see [Ni], [Sh]. In particular, through these liftings, the theory of Hecke eigenvalues is the same as that in the integral weight case. Secondly, if f is a Hecke eigenform, then the *squares* $a(t)^2$ are essentially proportional to the central critical values of the Hecke L-function of F twisted with the quadratic character $\chi_{t,N} = (\frac{(-1)^k N^2 t}{\cdot})$, see [Wa]. Here F is a Hecke eigenform of weight $2k$ corresponding to f under the Shimura correspondence.

These facts motivate the following questions. First, it is natural to ask for sign changes of the sequence $(a(tn^2))_{n \in \mathbb{N}}$ where t is a fixed positive square-free integer. We start with a conditional result here, namely if the Dirichlet L-function associated to $\chi_{t,N}$ has no zeros in the interval $(0, 1)$ (Chowla's conjecture, see [Ch]), then the sequence $(a(tn^2))_{n \in \mathbb{N}}$ –if not identically zero– changes sign infinitely often (Theorem 2.1). Note that by work of Conrey and Soundararajan [CS], Chowla's conjecture is true for a positive proportion of positive square-free integers t. If f is a Hecke eigenform, we can in fact prove an unconditional and much better result on the sign changes of the sequence $(a(tp^{2m}))_{m \in \mathbb{N}}$ where p is a prime not dividing N (see Theorem2.2).

Secondly, one may ask for sign changes of the sequence $(a(t))_t$, where t runs through the square-free integers only. This question is more difficult to treat and was already raised more than two decades ago in [KZ]. Numerical calculations seem to suggest not only that there are infinitely many sign changes, but also that "half" of these coefficients are positive and "half" of them are negative.

It seems quite difficult to prove any general theorem here, and we can only prove a result that seems to point into the right direction (see Theorem 2.4): Under the (clearly necessary) assumption for $k = 1$ that f is contained in the orthogonal complement of the space of unary theta functions, there exist infinitely many positive square-free integers t and for each such t a natural number n_t, such that the sequence $(a(tn_t^2))_t$ has infinitely many sign changes. Note that we do not require f to be an eigenform. We in fact prove a slightly stronger result (Theorem 2.5).

As an immediate application, in the integral weight as well as in the half-integral weight case, we may consider representation numbers of quadratic forms. Let Q be a positive definite integral quadratic form, and for a positive integer n let $r_Q(n)$ be the number of integral representations of n by Q. Then the associated theta series is the sum of a modular form lying in the space of Eisenstein series and a corresponding cusp form. An infinity of sign changes of the coefficients of the latter (the "error term" for $r_Q(n)$) means that $r_Q(n)$ for infinitely many n is larger (respectively less) than the corresponding Eisenstein coefficient (the "main term" for $r_Q(n)$).

Exact statements of our results are given in Section 2, while Section 3 contains their proofs. These are based on the existence of the Shimura lifts, the theory of L-functions, and on results on quadratic twists proved in [Br]. In Section 4 some numerical examples are given.

2 Notation and statement of results

We denote by $\mathbb{N}$ the set of positive integers. The set of square-free positive integers is denoted by $\mathbb{D}$. Throughout we write $q = e^{2\pi i z}$ for z in the upper complex half plane $\mathbb{H}$.

Let k be a positive integer. Let N be a positive integer divisible by 4, and let χ be a Dirichlet character modulo N. We write χ^* for the Dirichlet character modulo N given by $\chi^*(a) = \left(\frac{-4}{a}\right)^k \chi(a)$. Moreover, we write $S_{k+1/2}(N, \chi)$ for the space of cusp forms of weight $k + 1/2$ for the group $\Gamma_0(N)$ with character χ in the sense of Shimura [Sh].

If m and r are positive integers and ψ is an odd primitive Dirichlet character modulo r, then the unary theta function

$$\theta_{\psi,m}(z) = \sum_{n \in \mathbb{Z}} \psi(n) n q^{mn^2}$$

belongs to $S_{3/2}\left(N, \left(\frac{-4m}{\cdot}\right)\psi\right)$ for all N divisible by $4r^2m$, cf. [Sh]. Let $S_{3/2}^*(N, \chi)$ be the orthogonal complement with respect to the Petersson scalar product of the subspace of $S_{3/2}(N, \chi)$ spanned by such theta series. For $k \geq 2$ we simply put $S_{k+1/2}^*(N, \chi) = S_{k+1/2}(N, \chi)$. It is well known that the Shimura lift maps $S_{k+1/2}^*(N, \chi)$ to the space $S_{2k}(N/2, \chi^2)$ of cusp forms of integral weight $2k$ for $\Gamma_0(N/2)$ with character χ^2.

Throughout this section, let $f = \sum_{n=1}^{\infty} a(n) q^n \in S_{k+1/2}^*(N, \chi)$ be a nonzero cup form with Fourier coefficients $a(n) \in \mathbb{R}$. In our first result we consider the coefficients $a(tn^2)$ for fixed $t \in \mathbb{D}$ and varying n.

Theorem 2.1. *Let $t \in \mathbb{D}$ such that $a(t) \neq 0$, and write $\chi_{t,N}$ for the quadratic character $\chi_{t,N} = \left(\frac{(-1)^k N^2 t}{\cdot}\right)$. Assume that the Dirichlet L-function $L(s, \chi_{t,N})$ has no zeros in the interval $(0, 1)$. Then the sequence $(a(tn^2))_{n \in \mathbb{N}}$ has infinitely many sign changes.*

For Hecke eigenforms, we prove the following unconditional result.

Theorem 2.2. *Suppose that the character χ of f is real, and suppose that f is an eigenform of all Hecke operators $T(p^2)$ with corresponding eigenvalues λ_p for p coprime to N. Let $t \in \mathbb{D}$ such that $a(t) \neq 0$. Then for all but finitely* *many primes p coprime to N the sequence $(a(tp^{2m}))_{m \in \mathbb{N}}$ has infinitely many sign changes.*

Remark 2.3. Let K_f be the number field generated by the Hecke eigenvalues λ_p of f. The number of exceptional primes in Theorem 2.2 is bounded by r, where 2^r is the highest power of 2 dividing the degree of K_f over $\mathbb{Q}$.

Next, we consider the coefficients $a(tn^2)$ for varying $t \in \mathbb{D}$.

Theorem 2.4. *For every $t \in \mathbb{D}$ there is an $n_t \in \mathbb{N}$ such that the sequence $(a(tn_t^2))_{t \in \mathbb{D}}$ has infinitely many sign changes.*

In the special case when f is a Hecke eigenform, Theorem 2.4 is an easy consequence of Theorem 2.2. However, we do not assume this in Theorem 2.4. Notice that the statement is obviously wrong for the theta functions $\theta_{\psi,m}$. We shall also prove the following slightly stronger statement.

Theorem 2.5. *Let $p_1, \ldots, p_r$ be distinct primes not dividing N and let $\varepsilon_1, \ldots, \varepsilon_r \in \{\pm 1\}$. Write $\mathbb{D}'$ for the set of $t \in \mathbb{D}$ satisfying $(\frac{t}{p_j}) = \varepsilon_j$ for $j = 1, \ldots, r$. Then for every $t \in \mathbb{D}'$ there is an $n_t \in \mathbb{N}$ such that the sequence $(a(tn_t^2))_{t \in \mathbb{D}'}$ has infinitely many sign changes.*

3 Proofs

Here we prove the results of the previous section.

Proof of Theorem 2.1. Put

$$A(n) := \sum_{d \mid n} \chi_{t,N}(d) d^{k-1} a(\frac{n^2}{d^2} t). \tag{3.1}$$

According to [Sh, Ni], the series

$$F(z) := \sum_{n \geq 1} A(n) q^n$$

is in $S_{2k}(N/2, \chi^2)$ and is non-zero due to our assumption $a(t) \neq 0$. Note that (3.1) is equivalent to the Dirichlet series identity

$$\sum_{n \geq 1} a(tn^2) n^{-s} = \frac{1}{L(s - k + 1, \chi_{t,N})} \cdot L(F, s) \tag{3.2}$$

in the range of absolute convergence, where $L(F, s)$ is the Hecke L-function attached to F.

Now suppose that $a(tn^2) \geq 0$ for all but finitely many n. Then by a classical theorem of Landau, either the Dirichlet series on the left hand side of (3.2) has a singularity at the real point of its line of convergence or must converge everywhere (see e.g. [KKP]).

By our hypothesis, $L(s, \chi_{t,N})$ has no real zeros for $\Re(s) > 0$. Hence the series on the left hand side of (3.2) converges for $\Re(s) > k - 1$. In particular, we have

$$a(tn^2) \ll_\epsilon n^{k-1+\epsilon} \qquad (\epsilon > 0).$$

From (3.1) we therefore deduce that

$$A(n) \ll_\epsilon \sum_{d \mid n} d^{k-1} \left(\frac{n}{d}\right)^{k-1+\epsilon} \ll_\epsilon n^{k-1+2\epsilon} \qquad (\epsilon > 0).$$

Consequently, the Rankin-Selberg Dirichlet series

$$R_F(s) = \sum_{n \geq 1} A(n)^2 n^{-s}$$

must be convergent for $\Re(s) > 2k - 1$. However, it is well-known that the latter has a pole at $s = 2k$ with residue $c_k \|F\|^2$, where $c_k > 0$ is a constant depending only on k, and $\|F\|^2$ is the square of the Petersson norm of F. Since $F \neq 0$, we obtain a contradiction. This proves the claim. □

Proof of Theorem 2.2 and Remark 2.3. We use the same notation as in the proof of Theorem 2.1. Since f is an eigenfunction of $T(p^2)$, the function F is an eigenfunction under the usual Hecke operator $T(p)$ with eigenvalue λ_p. Since $\chi^2 = 1$, the eigenvalue λ_p is real. One has

$$\sum_{m \geq 1} a(tp^{2m}) p^{-ms} = a(t) \frac{1 - \chi_{t,N}(p) p^{k-1-s}}{1 - \lambda_p p^{-s} + p^{2k-1-2s}} \qquad (3.3)$$

for $\Re(s)$ sufficiently large, which is the local variant of (3.2).

The denominator of the right-hand side of (3.3) factors as

$$1 - \lambda_p p^{-s} + p^{2k-1-2s} = (1 - \alpha_p p^{-s})(1 - \beta_p p^{-s})$$

where $\alpha_p + \beta_p = \lambda_p$ and $\alpha_p \beta_p = p^{2k-1}$. Explicitly one has

$$\alpha_p, \beta_p = \frac{\lambda_p \pm \sqrt{\lambda_p^2 - 4p^{2k-1}}}{2}. \qquad (3.4)$$

Now assume that $a(tp^{2m}) \geq 0$ for almost all m. Then by Landau's theorem the Dirichlet series on the left hand side of (3.3) either converges everywhere or has a singularity at the real point of its abscissa of convergence. The first alternative clearly is impossible, since the right-hand side of (3.3) has a pole for $p^s = \alpha_p$ or $p^s = \beta_p$.

Thus the second alternative must hold, and in particular α_p or β_p must be real. By Deligne's theorem (previously the Ramanujan-Petersson conjecture), we have

$$\lambda_p^2 \leq 4p^{2k-1},$$

hence in combination with (3.4) we find that

$$\lambda_p = \pm 2p^{k-1/2}.$$

In particular we conclude that $\sqrt{p}$ is contained in the number field K_f generated by the Hecke eigenvalues of f.

Since for different primes $p_1, \ldots, p_r$ the degree of the field extension

$$\mathbb{Q}(\sqrt{p_1}, \ldots, \sqrt{p_r})/\mathbb{Q}$$

is 2^r, we deduce our assertion.   □

Throughout this section, let $f = \sum_{n=1}^{\infty} a(n)q^n \in S_{k+1/2}(N, \chi)$ be an arbitrary non-zero cup form with Fourier coefficients $a(n) \in \mathbb{R}$. For the proof of Theorem 2.4 we need the following three propositions.

Proposition 3.1. *There exist infinitely many $n \in \mathbb{N}$ such that $a(n)$ is negative.*

Proof. This result is proved in [KKP] in much greater generality. For the convenience of the reader we sketch the argument in the present special case.

Assume that there exist only finitely many $n \in \mathbb{N}$ such that $a(n) < 0$. Then the Hecke L-function of f (which is entire, cf. [Sh]) converges for all $s \in \mathbb{C}$ by Landau's theorem. Consequently,

$$a(n) \ll_C n^C$$

for all $C \in \mathbb{R}$.

This implies that the Rankin L-function of f also converges for all $s \in \mathbb{C}$. Arguing as at the end of the proof of Theorem 2.1 we find that f vanishes identically, contradicting the assumption $f \neq 0$.   □

Proposition 3.2. *Let p be a prime not dividing N, and let $\varepsilon \in \{\pm 1\}$. Assume that $a(n) \geq 0$ for all positive integers n with $(\frac{n}{p}) = \varepsilon$. Then f is an eigenform of the Hecke operator $T(p^2)$ with eigenvalue*

$$-\varepsilon \chi^*(p)(p^k + p^{k-1}).$$

Proof. We consider the cusp form

$$\tilde{f} := \sum_{\substack{n \geq 1 \\ (\frac{n}{p}) = \varepsilon}} a(n)q^n.$$

According to [Br], Section 2 (iii) and (iv), this form belongs to the space $S_{k+1/2}(Np^2, \chi)$. By assumption, $\tilde{f}$ has non-negative real coefficients. Using Proposition 3.1 we find that $\tilde{f} = 0$. Consequently, $a(n) = 0$ for all positive integers n with $(\frac{n}{p}) = \varepsilon$. Now the assertion follows from [Br], Lemma 1.   □

Proposition 3.3. *Let p be a prime not dividing N, and let $\varepsilon \in \{\pm 1\}$. Assume that f is contained in $S_{k+1/2}^*(N, \chi)$. Then there exist positive integers n, n' such that*

$$\left(\frac{n}{p}\right) = \varepsilon \quad and \ a(n) < 0,$$

$$\left(\frac{n'}{p}\right) = \varepsilon \quad and \ a(n') > 0.$$

Proof. Suppose that $a(n) \geq 0$ for all $n \in \mathbb{N}$ with $(\frac{n}{p}) = \varepsilon$. Then, by Proposition 3.2, f is an eigenform of $T(p^2)$ with eigenvalue $\lambda_p = -\varepsilon \chi^*(p)(p^k + p^{k-1})$. Using the Shimura lift, we see that λ_p is also an eigenvalue of the integral weight Hecke operator $T(p)$ on $S_{2k}(N/2, \chi^2)$. But it is easy to see that any eigenvalue λ of this Hecke operator satisfies the bound $|\lambda| < p^k + p^{k-1}$, see e.g. [Ko2] (or use the stronger Deligne bound). We obtain a contradiction.

Finally, replacing f by $-f$ we deduce the existence of n' with the claimed properties.   □

Proof of Theorem 2.4. Suppose that there exist finitely many square-free $t_1, \ldots, t_h \in \mathbb{N}$ such that $a(tn^2) \leq 0$ for all square-free integers t different from t_ν, $\nu = 1, \ldots, h$, and all $n \in \mathbb{N}$. Choose a prime p coprime to N such that

$$\left(\frac{t_\nu}{p}\right) = 1, \quad \text{for all } \nu = 1, \ldots, h.$$

Then $a(n) \leq 0$ for all $n \in \mathbb{N}$ with $(\frac{n}{p}) = -1$. But this contradicts Proposition 3.3. Hence there exist infinitely many square-free $t \in \mathbb{N}$ for which there is an $n_t \in \mathbb{N}$ such that $a(tn_t^2) > 0$.

Finally, replacing f by $-f$ we deduce the existence of infinitely many square-free $t \in \mathbb{N}$ for which there is an $n_t \in \mathbb{N}$ such that $a(tn_t^2) < 0$.   □

Proof of Theorem 2.5. The assertion follows by combining Theorem 2.4 and Proposition 3.3.   □

4 Examples

Let $f \in S_{k+1/2}^*(N, \chi_0)$ be a cusp form with trivial character χ_0, square-free level, and real coefficients $a(n)$. We suppose that f is contained in the plus

Table 1. *The proportion of positive coefficients of δ*

X	10	10^2	10^3	10^4	10^5	10^6
$R_{\text{tot}}^+(\delta, X)$	0.600	0.520	0.518	0.504600	0.499600	0.499822
$R_{\text{fund}}^+(\delta, X)$	0.667	0.548	0.515	0.501643	0.500016	0.499836

space, that is, $a(n) = 0$ when $(-1)^k n \equiv 2, 3 \pmod 4$, see [KZ], [Ko1]. For a positive number X, we define the quantity

$$R_{\text{tot}}^+(f, X) = \frac{\#\{n \le X; \quad a(n) > 0\}}{\#\{n \le X; \quad a(n) \ne 0\}}.$$

Moreover, in view of Waldspurger's theorem, it is natural to consider the coefficients $a(d)$ especially for *fundamental* discriminants d. Therefore we put

$$R_{\text{fund}}^+(f, X) = \frac{\#\{d \le X; \quad d \text{ fundamental discriminant and } a(d) > 0\}}{\#\{d \le X; \quad d \text{ fundamental discriminant and } a(d) \ne 0\}}.$$

The numerical experiments below suggest that

$$\lim_{X \to \infty} R_{\text{tot}}^+(f, X) = 1/2, \qquad \lim_{X \to \infty} R_{\text{fund}}^+(f, X) = 1/2.$$

First we consider the Delta function $\Delta(z) = q \cdot \prod_{n \ge 1}(1 - q^n)^{24}$. A cusp form of weight $13/2$ of level 4 in the plus space corresponding to Δ under the Shimura lift is given by

$$\delta(z) = \frac{1}{8\pi i}\left(2E_4(4z)\theta'(z) - E_4'(4z)\theta(z)\right) \in S_{13/2}^+(4, \chi_0),$$

see [KZ]. Here $E_4(z) = 1 + 240\sum_{n \ge 1}\sigma_3(n)q^n$ is the classical Eisenstein series of weight 4 and $\theta(z) = \sum_{n \in \mathbb{Z}} q^{n^2}$. The Fourier expansion of δ starts as follows:

$$\delta(z) = q - 56q^4 + 120q^5 - 240q^8 + 9q^9 + 1440q^{12} - 1320q^{13} + \dots.$$

Computational data for δ is listed in Table 1.

In our second example, we consider the cusp form $G = \eta(z)^2\eta(11z)^2$ of weight 2 and level 11 corresponding to the elliptic curve $X_0(11)$. Here $\eta = q^{1/24}\prod_{n \ge 1}(1 - q^n)$ is the Dedekind eta function. A cusp form of weight $3/2$ and level 44 in the plus space corresponding to G under the Shimura lift is

$$g(z) = \left(\theta(11z)\eta(2z)\eta(22z)\right)|U_4 \in S_{3/2}^+(44, \chi_0),$$

Table 2. *The proportion of positive coefficients of g*

X	10	10^2	10^3	10^4	10^5	10^6
$R_{\text{tot}}^+(g, X)$	0.500	0.500	0.500	0.496042	0.501022	0.499544
$R_{\text{fund}}^+(g, X)$	1.000	0.500	0.503	0.491968	0.500861	0.499589

see [Du] §2. Here U_4 denotes the usual Hecke operator of index 4. The Fourier expansion of g starts as follows:

$$g(z) = q^3 - q^4 - q^{11} - q^{12} + q^{15} + 2q^{16} + q^{20} - q^{23} - q^{27} - q^{31} + q^{44} + q^{55} + \dots$$

Computational data for g is listed in Table 2.

Bibliography

[Br]   J. H. Bruinier, On a theorem of Vigneras, Abh. Math. Sem. Univ. Hamburg **68** (1998), 163–168.

[Ch]   *The Riemann Hypothesis and Hilbert's tenth problem*, Gordon and Breach Science Publishers, New York-London-Paris (1965).

[CS]   J. B. Conrey and K. Soundararajan, Real zeros of quadratic Dirichlet *L*-functions, Invent. Math. **150** (2002), 1–44.

[Du]   N. Dummigan, Congruences of modular forms and Selmer groups, Math. Res. Lett. **8** (2001), 479–494.

[IKS] H. Iwaniec, W. Kohnen, and J. Sengupta, The first negative Hecke eigenvalue, Intern. J. Number Theory **3** (2007), 355–363.

[KKP] M. Knopp, W. Kohnen, and W. Pribitkin, On the signs of Fourier coefficients of cusp forms, Ramanujan J. **7** (2003), 269–277.

[Ko1] W. Kohnen, Fourier coefficients of modular forms of half integral weight, Math. Ann. **271** (1985), 237–268.

[Ko2] W. Kohnen, A simple remark on eigenvalues of Hecke operators on Siegel modular forms, Abh. Math. Sem. Univ. Hamburg **57** (1986), 33–36.

[KoSe] W. Kohnen and J. Sengupta, On the first sign change of Hecke eigenvalues of newforms, Math. Z. **254** (2006), 173–184.

[KSL] W. Kohnen, I. Shparlinski, and Y. K. Lau, On the number of sign changes of Hecke eigenvalues of newforms, J. Australian Math. Soc., to appear.

[KZ]   W. Kohnen and D. Zagier, Values of *L*-series of modular forms at the center of the critical strip, Invent. Math. **64** (1981), 175–198.

[Ni]   S. Niwa, Modular forms of half-integral weight and the integral of certain theta-functions, Nagoya Math. J. **56** (1975), 147–161.

[Sh]   G. Shimura, On modular forms of half integral weight, Ann. of Math. (2) **97** (1973), 440–481.

[Wa]   J.-L. Waldspurger, Sur les coefficients de Fourier des formes modulaires de poids demi-entier, J. Math. Pures et Appl. **60** (1981), 375–484.

Gauss map on the theta divisor
and Green's functions

Robin de Jong*

Abstract

In an earlier paper we constructed a Cartier divisor on the theta divisor of a principally polarised abelian variety whose support is precisely the ramification locus of the Gauss map. In this note we discuss a Green's function associated to this locus. For jacobians we relate this Green's function to the canonical Green's function of the corresponding Riemann surface.

1 Introduction

In [7] we investigated the properties of a certain theta function η defined on the theta divisor of a principally polarised complex abelian variety (ppav for short). Let us recall its definition. Fix a positive integer g and denote by $\mathbb{H}_g$ the complex Siegel upper half space of degree g. On $\mathbb{C}^g \times \mathbb{H}_g$ we have the Riemann theta function

$$\theta = \theta(z, \tau) = \sum_{n \in \mathbb{Z}^g} e^{\pi i\,^t n \tau n + 2\pi i\,^t n z} \,.$$

Here and henceforth, vectors are column vectors and ^{t} denotes transpose. For any fixed τ, the function $\theta = \theta(z)$ on $\mathbb{C}^g$ gives rise to an (ample, symmetric and reduced) divisor Θ on the torus $A = \mathbb{C}^g/(\mathbb{Z}^g + \tau\mathbb{Z}^g)$ which, by this token, acquires the structure of a ppav. The theta function θ can be interpreted as a tautological section of the line bundle $O_A(\Theta)$ on A.

* Supported by a grant from the Netherlands Organisation for Scientific Research (NWO).

Write θ_i for the first order partial derivative $\partial\theta/\partial z_i$ and θ_{ij} for the second order partial derivative $\partial^2\theta/\partial z_i\partial z_j$. Then we define η by

$$\eta = \eta(z,\tau) = \det\begin{pmatrix} \theta_{ij} & \theta_j \\ {}^t\theta_i & 0 \end{pmatrix}.$$

We consider the restriction of η to the vanishing locus of θ on $\mathbb{C}^g \times \mathbb{H}_g$.

In [7] we proved that for any fixed τ the function η gives rise to a global section of the line bundle $O_\Theta(\Theta)^{\otimes g+1} \otimes \lambda^{\otimes 2}$ on Θ in $A = \mathbb{C}^g/(\mathbb{Z}^g + \tau\mathbb{Z}^g)$; here λ is the trivial line bundle $\mathrm{H}^0(A, \omega_A) \otimes_{\mathbb{C}} O_\Theta$, with ω_A the canonical line bundle on A. When viewed as a function of two variables (z, τ) the function η transforms as a theta function of weight $(g+5)/2$ on $\theta^{-1}(0)$. If τ is fixed then the support of η on Θ is exactly the closure in Θ of the ramification locus $\mathrm{R}(\gamma)$ of the Gauss map on the smooth locus Θ^s of Θ. Recall that the Gauss map on Θ^s is the map

$$\gamma : \Theta^s \longrightarrow \mathbb{P}(T_0 A)^\vee$$

sending a point x in Θ^s to the tangent space $T_x\Theta$, translated over x to a subspace of $T_0 A$. It is well-known that the Gauss map on Θ^s is generically finite exactly when (A, Θ) is indecomposable; in particular the section η is non-zero for such ppav's.

It turns out that the form η has a rather nice application in the study of the geometry of certain codimension-2 cycles on the moduli space of ppav's. For this application we refer to the paper [5].

The purpose of the present note is to discuss a certain real-valued variant $\|\eta\| : \Theta \to \mathbb{R}$ of η. In the case that (A, Θ) is the jacobian of a Riemann surface X we will establish a relation between this $\|\eta\|$ and the canonical Green's function of X. In brief, note that in the case of a jacobian of a Riemann surface X we can identify Θ^s with the set of effective divisors of degree $g-1$ on X that do not move in a linear system; thus for such divisors D it makes sense to define $\|\eta\|(D)$. On the other hand, note that Θ^s carries a canonical involution σ coming from the action of -1 on A, and moreover note that sense can be made of evaluating the canonical (exponential) Green's function G of X on pairs of effective divisors of X. The relation that we shall prove is then of the form

$$\|\eta\|(D) = e^{-\zeta(D)} \cdot G(D, \sigma(D));$$

here D runs through the divisors in Θ^s, and ζ is a certain continuous function on Θ^s. The ζ from the above formula is intimately connected with the geometry of intersections $\Theta \cap (\Theta + R - S)$, where R, S are distinct points on X. Amusingly, the limits of such intersections where R and S approach each other are hyperplane sections of the Gauss map corresponding to points on the canonical image of X, so the Gauss map on the theta divisor is connected with the above formula in at least two different ways.

2 Real-valued variant of η

Let $(A = \mathbb{C}^g/(\mathbb{Z}^g + \tau\mathbb{Z}^g)$, $\Theta = \mathrm{div}\,\theta)$ be a ppav as in the introduction. As we said, the function η transforms like a theta function of weight $(g + 5)/2$ and order $g + 1$ on Θ. This implies that if we define

$$\|\eta\| = \|\eta\|(z, \tau) = (\det Y)^{(g+5)/4} \cdot e^{-\pi(g+1)^t y \cdot Y^{-1} \cdot y} \cdot |\eta(z, \tau)|,$$

where $Y = \mathrm{Im}\,\tau$ and $y = \mathrm{Im}\,z$, we obtain a (real-valued) function which is invariant for the action of Igusa's transformation group $\Gamma_{1,2}$ of matrices $\gamma = \left(\begin{smallmatrix} a & b \\ c & d \end{smallmatrix}\right)$ in $\mathrm{Sp}(2g, \mathbb{Z})$ with a, b, c, d square matrices such that the diagonals of both $^t ac$ and $^t bd$ consist of even integers. Recall that $\Gamma_{1,2}$ acts on $\mathbb{C}^g \times \mathbb{H}_g$ via

$$(z, \tau) \mapsto \left({}^t(c\tau + d)^{-1}z, (a\tau + b)(c\tau + d)^{-1} \right).$$

It follows that $\|\eta\|$ is a well-defined function on Θ, equivariant with respect to isomorphisms $(A, \Theta) \overset{\sim}{\longrightarrow} (A', \Theta')$ coming from the symplectic action of $\Gamma_{1,2}$ on $\mathbb{H}_g$. Note that the zero locus of $\|\eta\|$ on Θ coincides with the zero locus of η on Θ. In fact, if (A, Θ) is indecomposable then the function $-\log\|\eta\|$ is a Green's function on Θ associated to the closure of $R(\gamma)$.

The definition of $\|\eta\|$ is a variant upon the definition of the function

$$\|\theta\| = \|\theta\|(z, \tau) = (\det Y)^{1/4} \cdot e^{-\pi^t y \cdot Y^{-1} \cdot y} \cdot |\theta(z, \tau)|$$

that one finds in [4], p. 401. We note that $\|\theta\|$ should be seen as the norm of θ for a canonical hermitian metric $\|\cdot\|_{\mathrm{Th}}$ on $O_A(\Theta)$; we obtain $\|\eta\|$ as the norm of η for the induced metric on $O_\Theta(\Theta)^{\otimes g+1} \otimes \lambda^{\otimes 2}$. Here $\mathrm{H}^0(A, \omega_A)$ has the standard metric given by putting $\|dz_1 \wedge \ldots \wedge dz_g\| = (\det Y)^{1/2}$.

The curvature form of $(O_A(\Theta), \|\cdot\|_{\mathrm{Th}})$ on A is the translation-invariant $(1, 1)$-form

$$\mu = \frac{i}{2} \sum_{k=1}^{g} dz_k \wedge \overline{dz_k}.$$

The (g, g)-form $\frac{1}{g!}\mu^g$ is a Haar measure for A giving A measure 1. As μ represents Θ we have

$$\frac{1}{g!} \int_\Theta \mu^{g-1} = 1.$$

If (A, Θ) is indecomposable then $\log \|\eta\|$ is integrable with respect to μ^{g-1} and the integral

$$\frac{1}{g!} \int_{\Theta} \log \|\eta\| \cdot \mu^{g-1}$$

is a natural real-valued invariant of (A, Θ). We come back to it in Remark 4.6 below.

3 Arakelov theory of Riemann surfaces

The purpose of this section and the next is to investigate the function $\|\eta\|$ in more detail for jacobians. There turns out to be a natural connection with certain real-valued invariants occurring in the Arakelov theory of Riemann surfaces. We begin by recalling the basic notions from this theory [1] [4].

Let X be a compact and connected Riemann surface of positive genus g, fixed from now on. Denote by ω_X its canonical line bundle. On $H^0(X, \omega_X)$ we have a natural inner product $(\omega, \eta) \mapsto \frac{i}{2} \int_X \omega \wedge \overline{\eta}$; we fix an orthonormal basis $(\omega_1, \ldots, \omega_g)$ with respect to this inner product.

We put

$$\nu = \frac{i}{2g} \sum_{k=1}^{g} \omega_k \wedge \overline{\omega_k} .$$

This is a $(1, 1)$-form on X, independent of our choice of $(\omega_1, \ldots, \omega_g)$ and hence canonical. In fact, if one denotes by (J, Θ) the jacobian of X and by $j \colon X \hookrightarrow J$ an embedding of X into J using line integration, then $\nu = \frac{1}{g} j^* \mu$ where μ is the translation-invariant form on J discussed in the previous section. Note that $\int_X \nu = 1$.

The canonical Green's function G of X is the unique non-negative function $X \times X \to \mathbb{R}$ which is non-zero outside the diagonal and satisfies

$$\frac{1}{i\pi} \partial \overline{\partial} \log G(P, \cdot) = \nu(P) - \delta_P , \quad \int_X \log G(P, Q) \nu(Q) = 0$$

for each P on X; here δ denotes Dirac measure. The functions $G(P, \cdot)$ give rise to canonical hermitian metrics on the line bundles $O_X(P)$, with curvature form equal to ν.

From G, a smooth hermitian metric $\| \cdot \|_{\mathrm{Ar}}$ can be put on ω_X by declaring that for each P on X, the residue isomorphism

$$\omega_X(P)[P] = (\omega_X \otimes_{O_X} O_X(P))[P] \xrightarrow{\sim} \mathbb{C}$$

is an isometry. Concretely this means that if $z : U \to \mathbb{C}$ is a local coordinate around P on X then

$$\|dz\|_{\mathrm{Ar}}(P) = \lim_{Q \to P} |z(P) - z(Q)|/G(P, Q).$$

The curvature form of the metric $\| \cdot \|_{\mathrm{Ar}}$ on ω_X is equal to $(2g - 2)\nu$.

We conclude with the delta-invariant of X. Write $J = \mathbb{C}^g/(\mathbb{Z}^g + \tau\mathbb{Z}^g)$ and $\Theta = \mathrm{div}\,\theta$. There is a standard and canonical identification of (J, Θ) with $(\mathrm{Pic}_{g-1} X, \Theta_0)$ where $\mathrm{Pic}_{g-1} X$ is the set of linear equivalence classes of divisors of degree $g - 1$ on X, and where $\Theta_0 \subseteq \mathrm{Pic}_{g-1} X$ is the subset of $\mathrm{Pic}_{g-1} X$ consisting of the classes of effective divisors. By the identification $(J, \Theta) \cong (\mathrm{Pic}_{g-1} X, \Theta_0)$ the function $\|\theta\|$ can be interpreted as a function on $\mathrm{Pic}_{g-1} X$.

Now recall that the curvature form of $(O_J(\Theta), \| \cdot \|_{\mathrm{Th}})$ is equal to μ. This boils down to an equality of currents

$$\frac{1}{i\pi} \partial\bar{\partial} \log \|\theta\| = \mu - \delta_\Theta$$

on J. On the other hand one has for generic $P_1, \ldots, P_g$ on X that $\|\theta\|(P_1 + \cdots + P_g - Q)$ vanishes precisely when Q is one of the points P_k. This implies that on X the equality of currents

$$\frac{1}{i\pi} \partial_Q\bar{\partial}_Q \log \|\theta\|(P_1 + \cdots + P_g - Q) = j^*\mu - \sum_{k=1}^{g} \delta_{P_k} = g\nu - \sum_{k=1}^{g} \delta_{P_k}$$

holds. Since also

$$\frac{1}{i\pi} \partial_Q\bar{\partial}_Q \log \prod_{k=1}^{g} G(P_k, Q) = g\nu - \sum_{k=1}^{g} \delta_{P_k}$$

we may conclude, by compactness of X, that

$$\|\theta\|(P_1 + \cdots + P_g - Q) = c(P_1, \ldots, P_g) \cdot \prod_{k=1}^{g} G(P_k, Q)$$

for some constant $c(P_1, \ldots, P_g)$ depending only on $P_1, \ldots, P_g$. A closer analysis (cf. [4], p. 402) reveals that

$$c(P_1, \ldots, P_g) = e^{-\delta/8} \cdot \frac{\|\det \omega_i(P_j)\|_{\mathrm{Ar}}}{\prod_{k<l} G(P_k, P_l)}$$

for some constant δ which is then by definition the delta-invariant of X. The argument to prove this equality uses certain metrised line bundles and their curvature forms on sufficiently big powers X^r of X. A variant of this argument occurs in the proof of our main result below.

4 Main result

In order to state our result, we need some more notation and facts. We still have our fixed Riemann surface X of positive genus g and its jacobian (J, Θ). The following lemma is well-known.

Lemma 4.1. *Under the identification* $\Theta \cong \Theta_0$, *the smooth locus* Θ^s *of* Θ *corresponds to the subset* Θ_0^s *of* Θ_0 *of divisors that do not move in a linear system. Furthermore, there is a tautological surjection* Σ *from the* $(g-1)$-*fold symmetric power* $X^{(g-1)}$ *of* X *onto* Θ_0. *This map* Σ *is an isomorphism over* Θ_0^s.

The lemma gives rise to identifications $\Theta^s \cong \Theta_0^s \cong \Upsilon$ with Υ a certain open subset of $X^{(g-1)}$. We fix and accept these identifications in all that follows. Note that the set Υ carries a canonical involution σ, coming from the action of -1 on J. For D in Υ the divisor $D + \sigma(D)$ of degree $2g - 2$ is always a canonical divisor.

The next lemma gives a description of the ramification locus of the Gauss map on $\Theta^s \cong \Upsilon$.

Lemma 4.2. *Under the identification* $\Theta^s \cong \Upsilon$ *the ramification locus of the Gauss map on* Θ^s *corresponds to the set of divisors* D *in* Υ *such that* D *and* $\sigma(D)$ *have a point in common.*

Proof. According to [3], p. 691 the ramification locus of the Gauss map is given by the set of divisors $E + P$ with E effective of degree $g - 2$ and P a point such that on the canonical image of X the divisor $E + 2P$ is contained in a hyperplane. But this condition on E and P means that $E + 2P$ is dominated by a canonical divisor, or equivalently, that P is contained in the conjugate $\sigma(E + P)$ of $E + P$. The lemma follows. □

If $D = P_1 + \cdots + P_m$ and $D' = Q_1 + \cdots + Q_n$ are two effective divisors on X we define $G(D, D')$ to be

$$G(D, D') = \prod_{i=1}^{m} \prod_{j=1}^{n} G(P_i, Q_j).$$

Clearly the value $G(D, D')$ is zero if and only if D and D' have a point in common. Applying this to the above lemma, we see that the function $D \mapsto G(D, \sigma(D))$ on Υ vanishes precisely on the ramification locus of the Gauss map. As a consequence $G(D, \sigma(D))$ and $\|\eta\|(D)$ have exactly the same zero locus. It looks therefore as if a relation

$$\|\eta\|(D) = e^{-\zeta(D)} \cdot G(D, \sigma(D))$$

should hold for D on Υ with ζ a suitable continuous function. The aim of the rest of this note is to prove this relation, and to compute ζ explicitly.

We start with

Proposition 4.3. *Let* $Y = \Upsilon \times X \times X$. *The map* $\|\Lambda\| : Y \to \mathbb{R}$ *given by*

$$\|\Lambda\|(D, R, S) = \frac{\|\theta\|(D + R - S)}{G(R, S)G(D, S)G(\sigma(D), R)}$$

is continuous and nowhere vanishing. Furthermore $\|\Lambda\|$ *factors via the projection of* Y *onto* Υ.

Proof. The numerator $\|\theta\|(D + R - S)$ vanishes if and only if $R = S$ or $D = E + S$ for some effective divisor E of degree $g - 2$ or $D + R$ is linearly equivalent to an effective divisor E' of degree g such that $E' = E'' + S$ for some effective divisor E'' of degree $g - 1$. The latter condition is precisely fulfilled when the linear system $|D + R|$ is positive dimensional, or equivalently, by Riemann-Roch, when $D + R$ is dominated by a canonical divisor, i.e. when R is contained in $\sigma(D)$. It follows that the numerator $\|\theta\|(D + R - S)$ and the denominator $G(R, S)G(D, S)G(\sigma(D), R)$ have the same zero locus on Y. Fixing a divisor D in Υ and using what we have said in Section 3 it is seen that the currents

$$\frac{1}{i\pi}\partial\overline{\partial} \log \|\theta\|(D + R - S) \text{ and } \frac{1}{i\pi}\partial\overline{\partial} \log (G(R, S)G(D, S)G(\sigma(D), R))$$

are both the same on $X \times X$. We conclude that $\|\Lambda\|$ is non-zero and continuous and depends only on D. $\qquad\square$

We also write $\|\Lambda\|$ for the induced map on Υ. Our main result is

Theorem 4.4. *Let D be an effective divisor of degree $g - 1$ on X, not moving in a linear system. Then the formula*

$$\|\eta\|(D) = e^{-\delta/4} \cdot \|\Lambda\|(D)^{g-1} \cdot G(D, \sigma(D))$$

holds.

Proof. Fix two distinct points R, S on X. We start by proving that there is a non-zero constant c depending only on X such that

$$(*) \quad \|\eta\|(D) = c \cdot G(D, \sigma(D)) \left(\frac{\|\theta\|(D + R - S)}{G(R, S)G(D, S)G(\sigma(D), R)} \right)^{g-1}$$

for all D varying through Υ. We would be done if we could show that

$$\frac{1}{i\pi}\partial\overline{\partial} \log \|\eta\|(D)$$

and

$$\frac{1}{i\pi}\partial\bar\partial \log\left(G(D,\sigma(D))\left(\frac{\|\theta\|(D+R-S)}{G(R,S)G(D,S)G(\sigma(D),R)}\right)^{g-1}\right)$$

define the same currents on Υ. Indeed, then the function $\phi(D)$ given by

$$\log\|\eta\|(D) - \log\left(G(D,\sigma(D))\left(\frac{\|\theta\|(D+R-S)}{G(R,S)G(D,S)G(\sigma(D),R)}\right)^{g-1}\right)$$

is pluriharmonic on Υ, hence on Θ^s, and since Θ^s is open in Θ with boundary empty or of codimension ≥ 2, and since Θ is normal (cf. [8], Theorem 1') we may conclude that ϕ is constant.

To prove equality of

$$\frac{1}{i\pi}\partial\bar\partial \log\|\eta\|(D)$$

and

$$\frac{1}{i\pi}\partial\bar\partial \log\left(G(D,\sigma(D))\left(\frac{\|\theta\|(D+R-S)}{G(R,S)G(D,S)G(\sigma(D),R)}\right)^{g-1}\right)$$

on Υ it suffices to prove that their pullbacks are equal on $\Upsilon' = p^{-1}(\Upsilon)$ in X^{g-1} under the canonical projection $p : X^{g-1} \to X^{(g-1)}$.

First of all we compute the pullback under p of

$$\frac{1}{i\pi}\partial\bar\partial \log\|\eta\|(D)$$

on Υ'. Let $\pi_i : X^{g-1} \to X$ for $i = 1,\ldots,g-1$ be the projections onto the various factors. We have seen that the curvature form of $O_J(\Theta)$ is μ, hence the curvature form of $O_\Theta(\Theta)^{\otimes g+1}$ is $(g+1)\mu_\Theta$. According to [4], p. 397 the pullback of μ_Θ to X^{g-1} under the canonical surjection $\Sigma : X^{g-1} \to \Theta$ can be written as

$$\frac{i}{2}\sum_{k=1}^{g}\left(\sum_{i=1}^{g-1}\pi_i^*(\omega_k)\right)\wedge\left(\sum_{i=1}^{g-1}\pi_i^*(\overline{\omega_k})\right).$$

Here $(\omega_1,\ldots,\omega_g)$ is an orthonormal basis for $H^0(X,\omega_X)$ which we fix. Let's call the above form ξ. It follows that

$$p^*\frac{1}{i\pi}\partial\bar\partial \log\|\eta\|(D) = (g+1)\xi - \delta_{p^*R(\gamma)}$$

as currents on Υ'. Here $R(\gamma)$ is the ramification locus of the Gauss map on Υ.

Next we consider the pullback under p of

$$\frac{1}{i\pi}\partial\bar{\partial}\log\left(G(D,\sigma(D))\left(\frac{\|\theta\|(D+R-S)}{G(R,S)G(D,S)G(\sigma(D),R)}\right)^{g-1}\right)$$

on Υ'. The factor $\|\theta\|(D+R-S)$ accounts for a contribution equal to ξ, and both of the factors $G(D,S)$ and $G(\sigma(D),R)$ give a contribution $\sum_{i=1}^{g-1}\pi_i^*(\nu)$. We find

$$p^*\frac{1}{i\pi}\partial\bar{\partial}\log\left(\frac{\|\theta\|(D+R-S)}{G(R,S)G(D,S)G(\sigma(D),R)}\right)^{g-1} = (g-1)(\xi-2\sum_{i=1}^{g-1}\pi_i^*(\nu)).$$

We are done if we can prove that

$$p^*\frac{1}{i\pi}\partial\bar{\partial}\log G(D,\sigma(D)) = 2\xi + (2g-2)\sum_{i=1}^{g-1}\pi_i^*(\nu) - \delta_{p^*R(\gamma)}.$$

For this consider the product $\Upsilon'\times\Upsilon'\subseteq X^{g-1}\times X^{g-1}$. For $i,j=1,\ldots,g-1$ denote by $\pi_{ij}:X^{g-1}\times X^{g-1}\to X\times X$ the projection onto the i-th factor of the left X^{g-1}, and onto the j-th factor of the right X^{g-1}. Denoting by Φ the smooth form represented by $\frac{1}{i\pi}\partial\bar{\partial}\log G(P,Q)$ on $X\times X$ it is easily seen that we can write

$$p^*\frac{1}{i\pi}\partial\bar{\partial}\log G(D,\sigma(D)) + \delta_{p^*R(\gamma)} = (\sigma^*\sum_{i,j=1}^{g-1}\pi_{ij}^*\Phi)|_\Delta\,;$$

here $\Delta\cong\Upsilon'$ is the diagonal in $\Upsilon'\times\Upsilon'$ and σ^* is the action on symmetric $(1,1)$-forms on Υ' induced by the automorphism $(x,y)\mapsto(x,\sigma(y))$ of $\Upsilon\times\Upsilon$. Let q_1,q_2 be the projections of $X\times X$ onto the first and second factor, respectively. Then according to [1], Proposition 3.1 we have

$$\Phi_x = \frac{i}{2g}\sum_{k=1}^{g}q_1^*(\omega_k)\wedge q_1^*(\overline{\omega_k}) + \frac{i}{2g}\sum_{k=1}^{g}q_2^*(\omega_k)\wedge q_2^*(\overline{\omega_k})$$

$$- \frac{i}{2}\sum_{k=1}^{g}q_1^*(\omega_k)\wedge q_2^*(\overline{\omega_k}) - \frac{i}{2}\sum_{k=1}^{g}q_2^*(\omega_k)\wedge q_1^*(\overline{\omega_k})\,.$$

Note that $q_1\cdot\pi_{ij}=\pi_i$ and $q_2\cdot\pi_{ij}=\pi_j$; this gives

$$\pi_{ij}^*\Phi = \frac{i}{2g}\sum_{k=1}^{g}\pi_i^*(\omega_k)\wedge\pi_i^*(\overline{\omega_k}) + \frac{i}{2g}\sum_{k=1}^{g}\pi_j^*(\omega_k)\wedge\pi_j^*(\overline{\omega_k})$$

$$- \frac{i}{2}\sum_{k=1}^{g}\pi_i^*(\omega_k)\wedge\pi_j^*(\overline{\omega_k}) - \frac{i}{2}\sum_{k=1}^{g}\pi_j^*(\omega_k)\wedge\pi_i^*(\overline{\omega_k})\,.$$

Next note that σ acts as -1 on $H^0(X, \omega_X)$; this implies, at least formally, that

$$\sigma^* \pi_{ij}^* \Phi = \frac{i}{2g} \sum_{k=1}^{g} \pi_i^*(\omega_k) \wedge \pi_i^*(\overline{\omega_k}) + \frac{i}{2g} \sum_{k=1}^{g} \pi_j^*(\omega_k) \wedge \pi_j^*(\overline{\omega_k})$$

$$+ \frac{i}{2} \sum_{k=1}^{g} \pi_i^*(\omega_k) \wedge \pi_j^*(\overline{\omega_k}) + \frac{i}{2} \sum_{k=1}^{g} \pi_j^*(\omega_k) \wedge \pi_i^*(\overline{\omega_k})$$

$$= \pi_i^*(\nu) + \pi_j^*(\nu) + \frac{i}{2} \sum_{k=1}^{g} \pi_i^*(\omega_k) \wedge \pi_j^*(\overline{\omega_k})$$

$$+ \frac{i}{2} \sum_{k=1}^{g} \pi_j^*(\omega_k) \wedge \pi_i^*(\overline{\omega_k}) .$$

We obtain for $(\sigma^* \sum_{i,j=1}^{g-1} \pi_{ij}^* \Phi)|_\Delta$ the expression

$$\sum_{i,j=1}^{g-1} \pi_i^*(\nu) + \pi_j^*(\nu) + \frac{i}{2} \sum_{k=1}^{g} \sum_{i,j=1}^{g-1} \pi_i^*(\omega_k) \wedge \pi_j^*(\overline{\omega_k}) + \pi_j^*(\omega_k) \wedge \pi_i^*(\overline{\omega_k})$$

$$= (2g-2) \sum_{i=1}^{g-1} \pi_i^*(\nu) + i \sum_{k=1}^{g} \left((\sum_{i=1}^{g-1} \pi_i^*(\omega_k)) \wedge (\sum_{i=1}^{g-1} \pi_i^*(\overline{\omega_k})) \right)$$

$$= (2g-2) \sum_{i=1}^{g-1} \pi_i^*(\nu) + 2\xi ,$$

and this gives us what we want.

It remains to prove that the constant c is equal to $e^{-\delta/4}$. We use the following lemma.

Lemma 4.5. *Let* $\mathrm{Wr}(\omega_1, \ldots, \omega_g)$ *be the Wronskian on* $(\omega_1, \ldots, \omega_g)$, *considered as a global section of* $\omega_X^{\otimes g(g+1)/2}$. *Let* P *be any point on* X. *Then the equality*

$$\|\eta\|((g-1)P) = e^{-(g+1)\delta/8} \cdot \|\mathrm{Wr}(\omega_1, \ldots, \omega_g)\|_{\mathrm{Ar}}(P)^{g-1}$$

holds. Left and right hand side are non-vanishing for generic P.

Proof. Let $\kappa \colon X \to \Theta$ be the map given by sending P on X to the linear equivalence class of $(g-1) \cdot P$. According to [6], Lemma 3.2 we have a canonical isomorphism

$$\kappa^*(O_\Theta(\Theta)) \otimes \omega_X^{\otimes g} \xrightarrow{\sim} \omega_X^{\otimes g(g+1)/2} \otimes \kappa^*(\lambda)^{\otimes -1}$$

of norm $e^{\delta/8}$. It follows that we have a canonical isomorphism

$$\kappa^* \left(O_\Theta(\Theta)^{\otimes g+1} \otimes \lambda^{\otimes 2} \right) \xrightarrow{\sim} \left(\omega_X^{\otimes g(g+1)/2} \otimes \kappa^*(\lambda)^{\otimes -1} \right)^{\otimes g-1}$$

of norm $e^{(g+1)\delta/8}$. Chasing these isomorphisms using [7], Theorem 5.1 one sees that the global section $\kappa^* \eta$ of

$$\kappa^* \left(O_\Theta(\Theta)^{\otimes g+1} \otimes \lambda^{\otimes 2} \right)$$

is sent to the global section

$$\left(\xi_1 \wedge \ldots \wedge \xi_g \mapsto \frac{\xi_1 \wedge \ldots \wedge \xi_g}{\omega_1 \wedge \ldots \wedge \omega_g} \cdot \mathrm{Wr}(\omega_1, \ldots, \omega_g) \right)^{\otimes g-1}$$

of

$$\left(\omega_X^{\otimes g(g+1)/2} \otimes \kappa^*(\lambda)^{\otimes -1} \right)^{\otimes g-1} .$$

The claimed equality follows. The non-vanishing for generic P follows from $\mathrm{Wr}(\omega_1, \ldots, \omega_g)$ being non-zero as a section of $\omega_X^{\otimes g(g+1)/2}$. $\qquad \square$

We can now finish the proof of Theorem 4.4. Using the defining relation

$$\|\theta\|(P_1 + \cdots + P_g - S) = e^{-\delta/8} \frac{\|\det \omega_i(P_j)\|_{\mathrm{Ar}}}{\prod_{k<l} G(P_k, P_l)} \prod_{k=1}^{g} G(P_k, S)$$

mentioned earlier for δ we can rewrite equality (*) as

$$\|\eta\|(D) = c \cdot e^{-(g-1)\delta/8} \frac{G(D, \sigma(D))}{G(R, \sigma(D))^{g-1}} \left(\frac{\|\det \omega_i(P_j)\|_{\mathrm{Ar}}}{\prod_{k<l} G(P_k, P_l)} \right)^{g-1} ;$$

here we have set $D = P_1 + \cdots + P_{g-1}$ and $P_g = R$. Letting the P_j approach R we find, by a similar computation as in [6], proof of Lemma 3.2,

$$\|\eta\|((g-1)R) = c \cdot e^{-(g-1)\delta/8} \cdot \|\mathrm{Wr}(\omega_1, \ldots, \omega_g)\|_{\mathrm{Ar}}(R)^{g-1} .$$

Lemma 4.5 gives $c \cdot e^{-(g-1)\delta/8} = e^{-(g+1)\delta/8}$, in other words $c = e^{-\delta/4}$. $\qquad \square$

Remark 4.6. It was shown by J.-B. Bost [2] that there is an invariant A of X such that for each pair of distinct points R, S on X the formula

$$\log G(R, S) = \frac{1}{g!} \int_{\Theta + R - S} \log \|\theta\| \cdot \mu^{g-1} + A$$

holds. An inspection of the proof as for example given in [9], Section 5 reveals that the integrals

$$\frac{1}{g!} \int_{\Theta^s} \log G(D, S) \cdot \mu(D)^{g-1} \quad \text{and} \quad \frac{1}{g!} \int_{\Theta^s} \log G(\sigma(D), R) \cdot \mu(D)^{g-1}$$

are zero and hence from the definition of $\| \Lambda \|$ we can write

$$A = -\frac{1}{g!} \int_{\Theta^s} \log \| \Lambda \| (D) \cdot \mu(D)^{g-1} .$$

A combination with the formula in Theorem 4.4 yields

$$-\frac{1}{g!} \int_{\Theta} \log \| \eta \| \cdot \mu^{g-1} = \frac{\delta}{4} + (g-1)A - \frac{1}{g!} \int_{\Theta^s} \log G(D, \sigma(D)) \cdot \mu(D)^{g-1} .$$

This formula might be considered interesting since the left hand side is an invariant of ppav's, whereas the right hand side is only defined for Riemann surfaces.

Bibliography

[1] S.Y. Arakelov, An intersection theory for divisors on an arithmetic surface. Izv. Akad. Nauk SSSR Ser. Mat. 38 (1974), 1179–1192.

[2] J.-B. Bost, Fonctions de Green-Arakelov, fonctions thêta et courbes de genre 2. C. R. Acad. Sci. Paris Sér. I Math. 305 (1987), no. 14, 643–646.

[3] O. Debarre, Le lieu des variétés abéliennes dont le diviseur thêta est singulier a deux composantes. Ann. Sci. Ecole Norm. Sup. (4) 25 (1992), no. 6, 687–707.

[4] G. Faltings, Calculus on arithmetic surfaces. Ann. of Math. (2) 119 (1984), no. 2, 387–424.

[5] S. Grushevsky and R. Salvati Manni, Singularities of the theta divisor at points of order two. Int. Math. Res. Not. IMRN 2007, no. 15, Art. ID rmn045.

[6] R. de Jong, Arakelov invariants of Riemann surfaces. Doc. Math. 10 (2005), 311–329.

[7] R. de Jong, Theta functions on the theta divisor. To appear in Rocky Mountain J. Math.

[8] G. Kempf, On the geometry of a theorem of Riemann. Ann. of Math. (2) 98 (1973), 178–185.

[9] R. Wentworth, The asymptotics of the Arakelov-Green's function and Faltings' delta invariant. Comm. Math. Phys. 137 (1991), no. 3, 427–459.

A control theorem for the images of Galois actions on certain infinite families of modular forms

Luis Dieulefait[†]

A letter with the result

November 11, 2006

Dear Colleagues,

After reading a recent preprint by G. Wiese[1] on the images of Galois representations attached to classical modular forms and applications to inverse Galois theory I have been thinking again on the results of Ribet and Serre[2] giving "image as large as possible for almost every prime" in the non-CM case. The result of Wiese is used to realize as Galois group over $\mathbb{Q}$ the group $\mathrm{PGL}_2(\ell^r)$ or $\mathrm{PSL}_2(\ell^r)$ for a fixed ℓ and exponent r larger than any given exponent r_0. I haven't read his proof in detail, but since it uses "good-dihedral" primes and the result is for fixed ℓ, I can imagine that both the good-dihedral prime and the modular form are constructed ad hoc to realize only the desired linear group in this specific characteristic (and large-but-unknown exponent). In any case, since his method works for every ℓ, the result that he obtains is quite interesting.

On the other hand, of course, you and I believe that a stronger result is true, namely, a uniform result: so, assume for a moment that I just consider, for a fixed small prime t and every exponent n, a modular form f_n of level $u_n \cdot t^n$ and weight 2 (and trivial character) (*), where u_n is prime to t, without CM. Assume also that we know somehow that the images of the ℓ-adic and mod ℓ Galois representations attached to f_n are "as large as possible" for every prime ℓ and for every f_n. In other words, the special family f_n has image

† Research supported by project MTM2006-04895, MECD, Spain
[1] see [Wi]
[2] see [Se1], [Ri1] and [Ri2]

79

large *for every prime* instead of just for almost every prime (which is what Ribet's result shows in general for any modular form without CM). If such a family f_n exists, the well-known relation between the conductor and the minimal field of definition for a Galois representation with values in a finite field (as explained in Serre's Duke 1987 paper on modularity conjectures[3] and exploited by Brumer[4] to show that for a large power t^n in the level the field of coefficients of the projective representations attached to f_n must contain the real part of a cyclotomic field of t^m-roots of unity, with m going to infinity as n does) implies that the family f_n gives another proof of Wiese's Galois realizations result, and in a uniform way: for any given ℓ and exponent r_0 we know that taking any element f_n with n sufficiently large in our family we will be realizing not only the desired projective linear group in characteristic ℓ and with exponent greater than r_0 but also a similar linear group for a set of primes of density as close to 1 as desired (but smaller than 1), always with exponent larger than r_0. So, instead of realizing the desired "linear group over a finite field with large exponent" for an isolated prime our modular forms f_n will do the job for a large density set of primes.

As far as I know, it is not yet known how to construct an infinite family of modular forms with growing level f_n as the one described above, having all of them large image for every prime. But I can construct a family with a slightly weaker property, that is still good enough to derive the above conclusion regarding realizations of linear groups as Galois groups over $\mathbb{Q}$. In particular the family f_n that I have found with levels as in (*) has, of course, the property that the degree of the corresponding field of coefficients goes to infinity with n (because of Brumer's result and the factor t^n in the level), and concerning the images of the corresponding Galois representations it has the property that for each f_n with $n \geq 4$ we can give an upper bound to the set of exceptional primes computed as a function only of the level of f_n (i.e., all the information we need is the value of the level, not a single eigenvalue is needed) and in particular we can easily show that for any given prime $\ell > 3$ there is a value n_0 such that ℓ is not exceptional for f_n for any exponent $n > n_0$, where here "exceptional" means dihedral, reducible or (for $\ell = 5$) some of the other cases of small image in Dickson's result[5].

Let us show one example of such a family f_n: for any $n \geq 4$ take f_n to be *any* modular form of level $2 \cdot 3^n$, weight 2, trivial character. Because of semistability at 2 none of them has CM. Since the large ramification at 3 for n

---

[3] see [Se2]

[4] see [Br]

[5] i.e., cases of maximal proper subgroups of $PGL_2(\mathbb{F}_{p^r})$ as given by Dickson's classification, cf. for example [Se1]

sufficiently large makes easy to see (again, using the ideas of Serre and Brumer on conductors) that the small special groups in Dickson's list can not occur for $\ell = 5$ and it is well known that these groups cannot occur for larger ℓ in weight 2, we can concentrate on the two problems that have to be solved to control the images of the representations attached to f_n for any n: to control dihedral primes and to control reducible primes. In both cases we will use the large ramification at 3 and the semistability at 2 to do so.

Dihedral primes: Let $n \geq 4$ and $\ell > 3$ be a dihedral prime for f_n. Using the arguments created by Serre and Ribet, we see that the only possibility is that the mod ℓ representation is induced from the quadratic number field ramifying[6] only at 3. Also, since a dihedral image does not contain unipotent elements, this mod ℓ representation is unramified at 2. Since 2 is a non-square mod 3 this implies that the trace a_2 of the image of Frob 2 in this mod ℓ representation has to be 0 (i.e., as usual in the residually CM case, half of the traces have to be 0, and a_2 is in that half). On the other hand, the ℓ-adic representation attached to f has semistable ramification at 2, so we are in a case of raising the level (or lowering the level, depending on the perspective), and as you know very well this can only happen if $a_2 = \pm 3$ (these are numbers which only exist mod ℓ). Putting the two things together we conclude that 0 and ± 3 are the same mod ℓ, and since $\ell > 3$ this gives us a contradiction.

Reducible primes: this time assume $\ell > 3$ is a reducible prime for $f_n, n \geq 4$. We anticipate that now such a prime can exist (for example 7 is reducible for some newforms of level 162) but we just want to bound the set of reducible primes in terms of the level $2 \cdot 3^n$. Again, we will use the local information at 2 and 3 to do so. For simplicity of the exposition, we assume that $n = 2 \cdot u$ is even. Since the mod ℓ representation is reducible (we semisimplify if necessary, so assume it is semisimple) and using the value of the level of f_n (and, because $\ell > 3$, it is well-known[7] that residually the conductor at 3 will be exactly 3^n) we know that it is just the direct sum $\chi \cdot \psi \oplus \psi^{-1}$, where χ is the mod ℓ cyclotomic character and ψ is a character of conductor exactly 3^u (remember that u is half of n, so it is at least 2). Computing the trace of the image of Frob 2 for the mod ℓ representation this time we obtain $a_2 = 2 \cdot \psi(2) + \psi^{-1}(2)$. Here the important thing to observe is that the order of ψ is $\phi(3^u) = 2 \cdot 3^{u-1}$ or 3^{u-1} (where ϕ is Euler's function), and that 2 is primitive modulo 3^u, so the order of the element $\psi(2)$ is also $2 \cdot 3^{u-1}$ or 3^{u-1}. On the other hand, using again raising the level since the ℓ-adic representation is semistable at 2 we must have $a_2 = \pm 3$. Comparing the two formulas for a_2 the first observation

---

[6] Because the weight is 2, for $\ell > 3$ the quadratic field cannot ramify at ℓ, and due to semistability at 2 it cannot ramify at 2, see [Ri4] for the proof of both claims

[7] see for example [Ri3]

is that the roots of the characteristic polynomial of the image of Frob 2 are 1, 2 or -1, -2, in particular they belong to the prime field $\mathbb{F}_\ell$, so $\psi(2)$ must be in this field, and looking at the order of this element this means that 3^{u-1} divides $\ell - 1$ (call this condition (@)). This already shows that for n sufficiently large any prime ℓ given a priori will not be reducible (thus, will not be exceptional), because the maximal power of 3 dividing $\ell - 1$ is finite.

Just for fun, let us bound the set of possibly reducible primes for f_n: Comparing the two formulas for a_2 (comparing the roots of the polynomials deduced from both formulas) and using the information on the order of $\psi(2)$ we conclude that any reducible ℓ must satisfy, in addition to (@), the condition: ℓ divides $2^{2 \cdot 3^{(u-1)}} - 1$. So, this is a bound for the set of reducible primes for f_n. For example for $n = 4$ (thus $u = 2$) we conclude that ℓ has to be congruent to 1 mod 3 and divides $2^6 - 1 = 63$, thus the prime 7 may be reducible (and it is so for some newforms of level $2 \cdot 3^4 = 162$), but it is the only possible reducible prime $\ell > 3$ (computing reducible primes for all newforms of this level using the method in my thesis confirms this fact).

Conclusion: For any newform in the family f_n described above, if $n = 2 \cdot u$ (we assume it is even for simplicity) the residual image is "large" for any prime which is not congruent to 1 modulo 3^{u-1}. This, together with the fact proved by Serre and Brumer that as n (the exponent of the 3-part of the conductor) goes to infinity the exponents of the fields of coefficients of the projective residual representations also go to infinity, has as a corollary that with our family f_n we are realizing, for any prime $\ell > 3$, projective linear groups over the field of ℓ^r elements for r arbitrarily large as Galois groups over $\mathbb{Q}$, and we are realizing these groups in a uniform way (i.e., for sufficiently large n we obtain these groups not only for a given ℓ but also for large density sets of primes, all with large exponent). Each of them is realized as an extension unramified outside $6 \cdot \ell$.

Of course we can construct other similar examples taking other suitable pairs of primes instead of 2 and 3, we can also take more general levels having semistable ramification at more primes, and other variations. The main point is that we can bound the set of exceptional primes for *all* modular forms in an infinite family of increasing conductor, which is an interesting result that of course can not be obtained using just the computational method explained in my thesis years ago. It is interesting to observe how a very simple ramification condition (one semistable prime and other dividing the conductor with a large power) was enough to obtain "uniformly large" images, to explain that only primes that are very split can be exceptional, thus to generate a lot of large exponent linear groups as Galois groups. Maybe other combinations of ramification conditions can lead to similar, or even stronger, results.

The idea used to control dihedral primes is an idea I had in Paris in 2002, when considering dihedral primes for the case of Q-curves coming from diophantine equations. The new idea is the idea to control reducible primes, which I had in Berkeley last week during the modularity conference (but I knew since I saw Wiese's paper months ago that the arguments of Serre and Brumer should be the key: obtaining linear groups over fields with large exponents as Galois groups using these results is something I wanted to do already when starting my thesis).

Best regards,

Luis Dieulefait

Bibliography

[Br]   Brumer, A., *The rank of $J_0(N)$*, S.M.F. Astérisque **228** (1995), 41–68.

[Ri1] Ribet, K.A., *On ℓ-adic representations attached to modular forms*, Invent. Math. **28** (1975), 245–275.

[Ri2] ———, *On ℓ-adic representations attached to modular forms II*, Glasgow Math. J. **27** (1985), 185–194.

[Ri3] ———, *Report on mod ℓ representations of* $\mathrm{Gal}(\bar{\mathbb{Q}}/\mathbb{Q})$, Motives (Seattle, WA, 1991), Proc. Sympos. Pure Math. **55** Part 2, Amer. Math. Soc., (1994), 639–676.

[Ri4] ———, *Images of semistable Galois representations*, Pacific J. of Math. **181** (1997), 277–297.

[Se1] Serre, J.-P., *Propriétés galoisiennes des points d'ordre fini des courbes elliptiques*, Invent. Math. **15** (1972), 259–331.

[Se2] ———, *Sur les représentations modulaires de degré* 2 *de* $\mathrm{Gal}(\bar{\mathbb{Q}}/\mathbb{Q})$, Duke Math. J. **54** (1987), 179–230.

[Wi]   Wiese, G., *On projective linear groups over finite fields as Galois groups over the rational numbers*, this volume, pp. 343–350.

Galois realizations of families of Projective Linear Groups via cusp forms

Luis Dieulefait*

1 Introduction

Let S_k be the space of cusp forms of weight k for $\mathrm{SL}_2(\mathbb{Z})$ and write $S_2(N)$ for the space of cusp forms of weight 2 for $\Gamma_0(N)$.

We are going to consider the Galois representations attached to eigenforms in these spaces, whose images have been determined by Ribet and Momose (see [Ri 75] for S_k and [Mo 81], [Ri 85] for $S_2(N)$).

Our purpose is to use these representations to realize as Galois groups over $\mathbb{Q}$ some linear groups of the following form: $\mathrm{PSL}_2(p^r)$ if r is even and $\mathrm{PGL}_2(p^r)$ if r is odd. In order to ease the notation, we will call both these families of linear groups $\mathrm{PXL}_2(p^r)$, so that PXL stands for PSL if r is even and PGL if r is odd.

Extending the results in [Re-Vi], where it is shown that for $r \leq 10$ these groups are Galois groups over $\mathbb{Q}$ for infinitely many primes p, we will cover the cases $r = 11, 13, 17$ and 19, using the representations attached to eigenforms in S_k and again the cases 11 and 17 using the ones coming from $S_2(N)$.

We will give the explicit criterion for the case $r = 3$: for every prime $p > 3$ such that $p \equiv 2, 3, 4, 5 \mod 7$ the group $\mathrm{PGL}_2(p^3)$ is a Galois group over $\mathbb{Q}$.

Assuming the following conjecture (known as Maeda's conjecture): "The characteristic polynomial $P_{2,k}$ of the Hecke operator T_2 acting on S_k is irreducible over $\mathbb{Q}$, for all k", we will prove that for every prime exponent $r \geq 3$, $\mathrm{PGL}_2(p^r)$ is a Galois group over $\mathbb{Q}$ for infinitely many primes p.

* Research supported by project MTM2006-04895, MECD, Spain

Finally, applying results of [Br 96] we will prove that there exist infinitely many exponents r for which $\mathrm{PXL}_2(p^r)$ are Galois groups over $\mathbb{Q}$ for infinitely many primes p.

Remark: This article was written in 1998 and it corresponds to a Research Project advised by Nuria Vila that the author did as part of his PhD at the Universitat de Barcelona, previous to his thesis.

2 Galois representations attached to eigenforms in S_k

Generalizing the result of [Ri 75] for $r = 2$, in [Re-Vi] sufficient conditions are given for $\mathrm{PXL}_2(p^r)$ to be a Galois group over $\mathbb{Q}$. They are the following:

Criterion 2.1. *Let k be such that $\dim_{\mathbb{C}} S_k = r$.*

Let $P_{2,k}$ be the characteristic polynomial of the Hecke operator T_2 acting on S_k and let λ be one of its roots.

Let p be a prime such that $p \notin \Sigma_{k,\lambda}$, where $\Sigma_{k,\lambda}$ is a finite set of primes that can be computed in terms of k and λ.

Then if $P_{2,k}$ is irreducible modulo p (which implies, in particular, that it is irreducible over $\mathbb{Q}$), $\mathrm{PXL}_2(p^r)$ is a Galois group over $\mathbb{Q}$.

Remark 2.2. The condition that $P_{2,k}$ is irreducible mod p implies that there are infinitely many inert primes in $\mathbb{Q}(\lambda)$ (besides, $\mathbb{Q}(\lambda) = \mathbb{Q}_f$ for some eigenform f). From this, $\mathrm{PXL}_2(q^r)$ is realized as a Galois group over $\mathbb{Q}$ for infinitely many primes q ([Re-Vi]).

Corollary 2.3. *Suppose that there is a prime p_0 such that $P_{2,k}$ is irreducible mod p_0. Then there are infinitely many primes p not in $\Sigma_{k,\lambda}$ satisfying this and for all of them $\mathrm{PXL}_2(p^r)$ is a Galois group over $\mathbb{Q}$, where $r = \dim_{\mathbb{C}} S_k$. The corresponding Galois number field ramifies only at p.*

Remark 2.4. The existence of such a prime for $r = 2, 3, 4, \ldots., 10$ is verified in [Re-Vi], thus 2.3 applies to these exponents.

In [Bu 96] it is proved that for $r = 11, 13, 17, 19$ the polynomial $P_{2,12r}$ is irreducible modulo 479, 353, 263, 251 respectively. Then applying 2.3 we obtain:

Corollary 2.5. $\mathrm{PGL}_2(p^r)$ *is a Galois group over $\mathbb{Q}$ for $r = 11, 13, 17, 19$, for infinitely many primes p in each case, with ramification only at p.*

The following conjecture is due to Maeda and is widely believed:

Conjecture 2.6. *For every k, the characteristic polynomial $P_{2,k}$ of the Hecke operator T_2 acting on S_k is irreducible over $\mathbb{Q}$.*

Even assuming 2.6 we are not in condition of applying 2.3 for other values of r. However, in case r is prime we can use the following :

Lemma 2.7. *Let K be a number field of prime degree p over $\mathbb{Q}$. Then there exist infinitely many rational primes inert in K.*

Proof. Let N be the normal closure of K and $G = \text{Gal}(N/\mathbb{Q})$. It is clear that $\#G = [N : \mathbb{Q}]$ satisfies

$$p \mid \#G, \qquad \text{and} \quad \#G \mid p! \,. \tag{2.1}$$

Let H be a p–Sylow subgroup of G, whose order is p, and let L be its fixed field, so that $H = \text{Gal}(N/L)$. N/L being a cyclic extension of degree p, we can apply class field theory ([Ne], pag. 85) to see that there are infinitely many primes Q of L inert in N/L. Applying this fact, together with the multiplicativity of the residual degree and (2.1) we see that there are infinitely many inert primes q in K. $\qquad \square$

Theorem 2.8. *Assume the truth of 2.6. Then for every prime exponent r, there exist infinitely many primes p such that $\text{PGL}_2(p^r)$ is a Galois group over $\mathbb{Q}$, with ramification only at p.*

Proof. Let $k = 12r$. If $P_{2,k}$ is irreducible over $\mathbb{Q}$, calling λ one of its roots we have: $\mathbb{Q}(\lambda) = \mathbb{Q}_f$, for some eigenform f and $\dim_{\mathbb{C}} S_k = r = [\mathbb{Q}_f : \mathbb{Q}]$. The previous lemma implies that there are infinitely many inert primes in $\mathbb{Q}_f/\mathbb{Q}$ and we can apply 2.3. $\qquad \square$

3 Galois representations attached to newforms in $S_2(N)$

Let f be a newform of weight 2 in $\Gamma_0(N)$. We apply the following theorem, ([Ri 85], [Re 95]):

Theorem 3.1. *Let N be squarefree and P_2 be the characteristic polynomial of T_2 acting on $S_2(N)$. Let λ be a simple root of P_2 such that*

$$\text{there exists a newform } f \in S_2(N) \text{ verifying } \mathbb{Q}(\lambda) = \mathbb{Q}_f \tag{*}$$

Then for every rational prime p outside a finite set $\Sigma_{N,\lambda}$ inert in $\mathbb{Q}_f$, $\text{PXL}_2(p^r)$ is a Galois group over $\mathbb{Q}$ where $r = [\mathbb{Q}_f : \mathbb{Q}]$, with corresponding number field unramified outside $p \cdot N$.

Remark 3.2. The fact that the nebentypus $\varepsilon = 1$ and N is squarefree implies that f neither has complex multiplication nor inner twists. This is used to obtain the surjectivity of the Galois representations.

In [Wa 73], a table of P_2 polynomials, we see that for $N = 229, 239$ there are simple factors of degree $11, 17$ respectively. It can be verified with further computations that condition (*) holds in these two examples. Invoking again 2.7 we can apply 3.1 and conclude:

Corollary 3.3. $PGL_2(p^r)$ *is a Galois group over* $\mathbb{Q}$ *for* $r = 11, 17$ *and infinitely many primes* p *in both cases.*

In the case of prime level N the exceptional set $\Sigma_{N,\lambda}$ can be 'removed' by using the following new result [Ri 97]:

Proposition 3.4. *Let* $\mathbf{T}$ *be the Hecke ring, the ring of endomorphisms over* $\mathbb{Q}$ *of* $J_0(N)$, *for prime* N. *Let* $\mathfrak{R}$ *be a maximal ideal of* $\mathbf{T}$ *of residual characteristic* $p \geq 5$, *and with* $\mathbf{T}/\mathfrak{R} = \mathbb{F}_{p^r}$. *Then if* $\mathfrak{R}$ *is not Eisenstein,* $PXL_2(p^r)$ *is Galois over* $\mathbb{Q}$ *with corresponding number field unramified outside* $p \cdot N$.

Remark 3.5. (i) In the prime level case there are no oldforms, and we have the identification:

$$\mathbf{T} \otimes \mathbb{Q} \cong \prod_{f \in \Sigma} \mathbb{Q}_f$$

where Σ is a set of representatives of all newforms modulo the action of $\mathrm{Gal}(\overline{\mathbb{Q}}/\mathbb{Q})$.

(ii) The Eisenstein ideal $\mathfrak{I} \subseteq \mathbf{T}$ is the one generated by the elements: $1 + l - \mathbf{T}_l$ ($l \neq N$), $1 + \omega$, where l takes prime values and ω is the Atkin-Lehner involution. An Eisenstein prime is a prime ideal $\beta \subseteq \mathbf{T}$ in the support of $\mathfrak{I}$. Let $n = \mathrm{num}(\frac{N-1}{12})$. We have a one-to-one correspondence between Eisenstein primes β and prime factors p of n, given by:

$$\text{Prime factors of } n \quad \longleftrightarrow \quad \text{Eisenstein primes}$$
$$p \quad \longleftrightarrow \quad (\mathfrak{I}, p)$$

Besides, for every Eisenstein prime β, $\mathbf{T}/\beta = \mathbb{F}_p$. In particular, whenever the inertia at some prime above p in $\mathbb{Q}_f$ is nontrivial (and these are the cases we are interested in) the involved prime will not be Eisenstein.

Corollary 3.6. $PSL_2(p^2)$ *is Galois over* $\mathbb{Q}$, *for every*

$$p \equiv \pm 2 \quad \mathrm{mod}\ 5, \quad p \geq 7.$$

Proof. Consider level $N = 23$. In this case $\mathbf{T} \otimes \mathbb{Q} \cong \mathbb{Q}(\sqrt{5})$, $n = 11$. If $\mathfrak{R}$ is Eisenstein, $\mathbf{T}/\mathfrak{R} = \mathbb{F}_{11}$. In $\mathbb{Q}(\sqrt{5})$ the inert primes are the $p \equiv \pm 2 \mod 5$. The result follows from 3.4. $\square$

Remark 3.7. The same result is obtained in [Me 88] by a different approach.

Corollary 3.8. $\mathrm{PSL}_2(p^2)$ *is a Galois group over* $\mathbb{Q}$, *for every*

$$p \equiv \pm 3 \mod 8, \quad p \geq 5.$$

Proof. Consider level $N = 29$. Here $\mathbf{T} \otimes \mathbb{Q} \cong \mathbb{Q}(\sqrt{2})$, $n = 7$. If $\mathfrak{R}$ is Eisenstein, $\mathbf{T}/\mathfrak{R} = \mathbb{F}_7$. The inert primes are the $p \equiv \pm 3 \mod 8$. Apply 3.4. $\square$

Remark 3.9. This same result is obtained in [Re 95] using 3.1 where the exceptional set is explicitated and proved to be disjoint from the set of inert primes.

Corollary 3.10. $\mathrm{PGL}_2(p^3)$ *is a Galois group over* $\mathbb{Q}$, *for every*

$$p \equiv 2, 3, 4, 5 \mod 7, \quad p \geq 5.$$

Proof. Consider level $N = 97$. We found in [Wa 73] that for this level there is a newform f with $\mathbb{Q}_f$ equal to the splitting field of the polynomial:

$$x^3 + 4x^2 + 3x - 1$$

This field is the real cyclotomic field: $\mathbb{Q}(\zeta_7 + \zeta_7^{-1})$. The inert primes in this field are the primes that when reduced mod 7 give a generator of the group $(\mathbb{Z}/7\mathbb{Z})^*$ or the square of such a generator, corresponding to the cases of residual degree 6 and 3 in $\mathbb{Q}(\zeta_7)$, respectively. These are the following: $p \equiv 2, 3, 4, 5 \mod 7$. Applying 3.4 and 3.5-1 (and once again the fact that the coefficients of Eisenstein forms are in the prime field) we obtain the desired result. $\square$

We now consider the case of arbitrary level N, again with trivial nebentypus. We still need the assumption that f is a newform without CM (complex multiplication). In this general case, we can apply the surjectivity result of [Ri 85], after replacing $\mathbb{Q}_f$ by the field $F_f \subseteq \mathbb{Q}_f$ defined as follows (we give three equivalent definitions, see [Ri 80]):

Definition 3.11. Let A_f be the abelian variety associated to f, and let $E = \mathrm{End}\, A_f \otimes \mathbb{Q}$ be its algebra of endomorphisms. We define F_f to be the centre of E. Equivalently, if Γ is the set of all immersions $\gamma : \mathbb{Q}_f \to \mathbb{C}$ such that there exists a Dirichlet character χ with: $\gamma(a_p) = \chi(p)a_p$ for almost every p, where $\sum a_n q^n = f$; then $F_f = \mathbb{Q}_f^\Gamma$, the fixed field of Γ. This coincides with the field generated over $\mathbb{Q}$ by the a_p^2, p ranging through almost every prime (for example, through all p not dividing N).

The following theorem can be deduced from the results in [Ri 85]:

Theorem 3.12. *Let f be a newform in $S_2(N)$ without CM. Let p be a rational prime outside a finite set $\Sigma_{N,f}$ and let i be the residual degree in $F_f / \mathbb{Q}$ of some $P \mid p$. Then $\mathrm{PXL}_2(p^i)$ is a Galois group over $\mathbb{Q}$ with corresponding extension of $\mathbb{Q}$ unramified outside $p \cdot N$.*

In order to obtain Galois realizations, we need some information about the fields F_f. The best result is the following ([Br 96]):

Theorem 3.13. *Let f be as in 3.12. Suppose that $p^{r_p} \parallel N$. Let $s_p = \left\lceil \frac{r_p}{2} - 1 - \frac{1}{p-1} \right\rceil$ and ζ a primitive p^{s_p}-root of unity. Then $F_f \supseteq \mathbb{Q}(\zeta + \zeta^{-1})$ if $p > 2$ (resp. $\mathbb{Q}(\zeta^2 + \zeta^{-2})$ if $p = 2$).*

As a particular case, let $p > 3$, with $p^3 \parallel N$. Then $s_p = \left\lceil \frac{1}{2} - \frac{1}{p-1} \right\rceil = 1$. If f does not have CM, $F_f \supseteq \mathbb{Q}(\zeta_p + \zeta_p^{-1})$. If q is a rational prime inert in $\mathbb{Q}(\zeta_p)$, that is to say: $q \mod p$ generates the multiplicative group $\mathbb{F}_p^*$, and if Q is a prime over q in O_{F_f}, the ring of integers of F_f, we have: $\frac{p-1}{2} \mid f_Q(F_f : \mathbb{Q})$, where $f_Q(F_f : \mathbb{Q})$ denotes the residual degree of the prime Q.

All these residual degrees are bounded by $[F_f : \mathbb{Q}]$, then among the infinitely many primes congruent to generators of $\mathbb{F}_p^*$ we can pick out an infinite subset of primes $\{q_i\}_{i \in \mathbb{N}}$ such that for all of them there exists a prime Q_i in O_{F_f} over q_i with:

$$ f_{Q_i}(F_f : \mathbb{Q}) = l \cdot \frac{p-1}{2}, \qquad l \text{ independent of } i $$

Combining this with 3.12 we get:

Theorem 3.14. *Let $p > 3$ be a prime and let N be such that $p^3 \parallel N$. Then if there is a newform $f \in S_2(N)$ without CM, there exists a number $l = l(p)$ such that for infinitely many primes q : $\mathrm{PXL}_2(q^{l(p-1)/2})$ is a Galois group over $\mathbb{Q}$. The primes q can all be chosen congruent $\mod p$ to generators of $\mathbb{F}_p^*$.*

Remark 3.15. (i) There is always an exceptional set $\Sigma_{N,f}$ that has to be excluded in order to apply 3.12.

(ii) The values of $l = l(p)$ are bounded by:

$$ l \leq [F_f : \mathbb{Q}(\zeta_p + \zeta_p^{-1})] < [\mathbb{Q}_f : \mathbb{Q}] \leq \dim S_2^{\mathrm{new}}(N) . $$

Remark 3.16. In order to apply this result we need to ensure that: "For every odd prime p there exists a positive integer N with $p^3 \parallel N$ and such that no newform $f \in S_2(N)$ has complex multiplication" (in fact, we would have enough with the existence of one f without CM of such level). Taking $N = p^3 \cdot t$, t odd prime, $t \neq p$, it is well-known that this holds, in fact if a newform f of this level had CM the corresponding abelian variety A_f, factor of $J_0(N)$, would also have CM, contradicting the fact that it has multiplicative reduction at t, as can be seen looking at its Néron model.

We see from the remark above that 3.14 applies for every prime $p > 3$ (take $N = p^3 \cdot t$) so that we have:

Corollary 3.17. *Let $p > 3$ be a prime. Then there exists a positive integer $l = l(p)$ such that for infinitely many primes q : $\mathrm{PXL}_2(q^{l(p-1)/2})$ is a Galois group over $\mathbb{Q}$.*

The fact that 3.17 holds for every prime $p > 3$ implies the following:

Corollary 3.18. *There exist infinitely many positive integers n such that for every one of them there are infinitely many primes q with $\mathrm{PXL}_2(q^n)$ being a Galois group over $\mathbb{Q}$. Moreover, an infinite number of these exponents n is even.*

Bibliography

[Br 96]   Brumer, A., *The rank of $J_0(N)$*, S.M.F. Astérisque **228** (1995), 41–68.

[Bu 96]   Buzzard, K., *On the eigenvalues of the Hecke operator T_2*, J. of Number Theory **57** (1996), 130–132.

[Me 88]   Mestre, J.F., *Courbes hyperelliptiques à multiplications réelles*, C.R. Acad. Sci. Paris **307** (1988), 721–724.

[Mo 81]   Momose, F., *On the l-adic representations attached to modular forms*, J. Fac. Sci. Univ. Tokyo, Sect. IA Math. **28**:1 (1981), 89–109.

[Ne]   Neukirch, J.,*Class Field Theory*, Springer Verlag (1986).

[Re 95]   Reverter, A.,*Construccions arithmètico-geomètriques de grups de Galois*, (thesis) Universitat de Barcelona (1995).

[Re-Vi]   Reverter, A. and Vila, N., *Some projective linear groups over finite fields as Galois groups over $\mathbb{Q}$*, Contemporary Math. **186** (1995), 51–63.

[Ri 75]   Ribet, K.A., *On l-adic representations attached to modular forms*, Invent. Math. **28** (1975), 245–275.

[Ri 77]   ———, *Galois representations attached to eigenforms with nebentypus*, LNM 601 (1977), Springer-Verlag.

[Ri 80]   ———, *Twists of modular forms and endomorphisms of Abelian Varieties*, Math. Ann. **253** (1980), 43–62.

[Ri 85]   ———, *On l-adic representations attached to modular forms II*, Glasgow
            Math. J. **27** (1985), 185–194.
[Ri 97]   ———, *Images of semistable Galois representations*, Pacific J. of Math.
            181, 3 (1997), 277–297.
[Wa 73]   Wada, H., *A table of Hecke operators II*, Proc. Japan Acad. **49** (1973),
            380–384.

A strong symmetry property of Eisenstein series

Bernhard Heim

1 Introduction and statement of results

Eisenstein series play a critical role in number theory. For two hundred years they have been an essential tool in the analysis of automorphic L-functions and in studying properties of quadratic forms in one and several variables. The construction is clear and straightforward, while their properties are sometimes very surprising. The arithmetic of their Fourier coefficients, and their analytic properties are still not completely understood. There are many connections with the Riemann hypothesis and other famous unsolved problems in number theory.

Eisenstein series are named after Ferdinand Gotthold Eisenstein (1823 - 1852). Let k be an even integer larger than 2 and let τ be in the upper complex half-space. One of the simplest Eisenstein series is defined by

$$E_k(\tau) := \frac{1}{2} \sum_{m,n \in \mathbb{Z},\, (m,n)=1} (m\tau + n)^{-k}. \tag{1.1}$$

It has the transformation property

$$E_k \left(\frac{a\tau + b}{c\tau + d} \right) = (c\tau + d)^k E_k(\tau) \tag{1.2}$$

for $\left(\begin{smallmatrix} a & b \\ c & d \end{smallmatrix} \right) \in \mathrm{SL}_2(\mathbb{Z})$. It has a Fourier expansion with rational Fourier coefficients with bounded denominators, involving divisor functions and Bernoulli numbers, and is connected with special values of the Riemann zeta function.

To understand special values of more general types of L-functions, this simple version of Eisenstein series has been extended in many directions. Siegel and Klingen studied Eisenstein series attached to the symplectic group, in order to study, for example, quadratic forms and the structure of Siegel modular forms in several variables. This culminated in the Siegel-Weil formula [We65] and the structure theorem. Later Klingen introduced the Eisenstein

93

series now called *Klingen type* [Kl90]. In another direction, Maass, Roelcke, and Selberg [Se56] studied real analytic Eisenstein series in the context of differential operators and spectral theory. Langlands [La76] succeeded in showing remarkable general analytic properties, i.e., meromorphic continuation to the whole complex plane and functional equation, for a wide range of reductive groups. This has applications in the Rankin-Selberg and the Shahidi methods to study analytic and arithmetic properties of automorphic L-functions. The arithmetic properties of the Fourier coefficients play a fundamental role in the study of the arithmetic of the special values. Garrett's integral representation of the triple L-function [Ga87] was an unexpected example of a different sort.

Yet another direction appears in the brilliant work of the late H. Maass [Ma79], who found a new relation satisfied by the Fourier coefficients of holomorphic Eisenstein series of Siegel type of degree 2. Automorphic functions with this property he called the *Spezialschar*. His beautiful work on this subject made it possible to understand and prove the main part of the Saito-Kurokawa conjecture [Za80]. Recently Skinner used results of Shimura on delicate properties of Eisenstein series to attack the Iwasawa conjecture (see also [Br07]). This brief review suggests that that new features of Eisenstein series should be fruitful.

In this paper we present a new method to study Fourier coefficients of holomorphic and non-holomorphic Eisenstein series simultaneously. This leads to a fundamental identity we state now. We mainly focus on the real analytic Eisenstein series on Siegel upper half-space $\mathbb{H}_2$ of degree 2 to make our method clear and to not burden the discussion with other technical considerations.

Let $E_k^{(2)}(Z, s)$ be the real analytic Eisenstein series of weight k and $Z \in \mathbb{H}_2$ with respect to the Siegel modular group $\mathrm{Sp}_2(\mathbb{Z})$ and $s \in \mathbb{C}$ with $2\,\mathrm{Re}(s) + k > 3$. For details we refer to section 3. This function is not holomorphic as a function of Z on $\mathbb{H}_2$, but does satisfy the transformation rule of a modular form. Since it is periodic with respect to the real part X of Z it has a Fourier expansion:

$$E_k^{(2)}(X + iY, s) = \sum_N A(N, Y; s)\, e^{2\pi i\, \mathrm{tr}\,(NX)}. \tag{1.3}$$

where Y is the imaginary part of Z and N is summed over half-integral matrices $N = \begin{pmatrix} n & r/2 \\ r/2 & m \end{pmatrix}$.

Then the following identity holds between the Fourier coefficients $A(N, Y; s)$. Let $G[H] := H^t G H$ for appropriate matrices G and H. We have for all prime numbers p and for all half-integral $N = \begin{pmatrix} n & r/2 \\ r/2 & m \end{pmatrix}$ the formula

$$
p^{k-1} A\left(\left(\begin{array}{cc} \frac{n}{p} & \frac{r}{2p} \\ \frac{r}{2p} & m \end{array}\right), p\, Y; s\right) - A\left(\left(\begin{array}{cc} n & \frac{r}{2} \\ \frac{r}{2} & p\,m \end{array}\right), Y; s\right) \tag{1.4}
$$

$$
= p^{k-1} A\left(\left(\begin{array}{cc} n & \frac{r}{2p} \\ \frac{r}{2p} & \frac{m}{p} \end{array}\right), Y\left[\left(\begin{array}{cc} 1 & 0 \\ 0 & p \end{array}\right)\right]; s\right)
$$

$$
- A\left(\left(\begin{array}{cc} pn & \frac{r}{2} \\ \frac{r}{2} & m \end{array}\right), Y\left[\left(\begin{array}{cc} p^{-\frac{1}{2}} & 0 \\ 0 & p^{\frac{1}{2}} \end{array}\right)\right]; s\right).
$$

Here we put $A(N, Y, s) = 0$ if N is not half-integral.

The nature of the Fourier coefficients $A(N, Y; s)$ is complicated, involving special values of Dirichlet L-series (Siegel series) and Bessel functions of higher order. One has to distinguish the various cases of the rank of N. Nevertheless, our method works without any explicit knowledge of these formulas, and is completely explicit. Moreover it also works inthe case of Hecke summation.

This paper is organized in the following way. In §2 we recall some basic aspects of Shimura's approach to the theory of Hecke. This will be used to define a new kind of operators, which do not act on the space of modular forms, but nevertheless inherit interesting properties. In §3 we prove a decomposition of the real-analytic Eisenstein series, essentially based on consideration of two subseries $A_k(Z, s)$ and $B_k(Z, s)$, concerning which we prove several properties. In §4 we present the main result of this paper, namely, we show that real-analytic Eisenstein series satisfy the strong symmetry property

$$
E_k^{(2)}(Z, s)| \bowtie T = 0 \tag{1.5}
$$

for all Hecke operators T, which will be explained in that paragraph in detail. We also give an example of a family of modular forms which do not have this property. Finally, we give applications, for example, the fundamental identity among the Fourier coefficients.

2 Hecke Theory à la Shimura

For $k \in \mathbb{N}$ be even let M_k be the space of elliptic modular forms of weight k with respect to the full modular group $\Gamma = \mathrm{SL}_2(\mathbb{Z})$. Let $f \in M_k$. Hecke introduced the operators $T_n, n \in \mathbb{N}$ given by

$$
T_n(f)(\tau) := n^{k-1} \sum_{d|n} d^{-k} \sum_{b=0}^{d-1} f\left(\frac{n\tau + bd}{d^2}\right), \tag{2.1}
$$

which map modular forms to modular forms. These operators commute with each other. They are multiplicative and self-adjoint with respect to the Petersson scalar product on the space of cusp forms. The vector space M_k has a basis of simultaneous eigenforms. The eigenvalues $\lambda_n(f)$ are totally real integers and are proportional to the n-th Fourier coefficients of the eigenform. Shimura [Sh71] studied systematically the underlying Hecke algebra. The realization of this Hecke algebra on the space of modular forms gives then the Hecke operators above.

We start with some basic constructions [Sh71]. Let (R, S) be a Hecke pair, meaning that R is a subgroup of the group S and for each $s \in S$ the coset space $R \backslash RsR$ is finite. For P be a principal ideal domain, R acts on the right on the P-module $L_P(R, S)$ of formal finite sums $X = \sum_j a_j Rs_j$ with $a_j \in P, s_j \in S$. The subset $H_P(R, S)$ of elements invariant under this action forms a ring with the multiplication

$$\left(\sum_i a_i Rg_i \right) \circ \left(\sum_j b_j Rh_j \right) := \sum_{i,j} a_i b_j Rg_i h_j. \qquad (2.2)$$

This ring is called the associated Hecke ring or algebra. It is convenient to identify the left coset decomposition of the double cosets $RsR = \bigsqcup_j Rs_j$ with $\sum_j Rs_j \in H_P(R, S)$ which form a basis of the P-module $H_P(R, S)$. Hence double cosets are identified with a full system of representatives of the R-left coset decomposition of the double coset.

Now we apply this construction to our situation. For $l \in \mathbb{N}$ put

$$M(l) \quad := \quad \bigsqcup_{d|l,\ d|\frac{l}{d}} \Gamma \begin{pmatrix} d & 0 \\ 0 & l/d \end{pmatrix} \Gamma. \qquad (2.3)$$

Then we set $M_\infty := \bigsqcup_{l=1, n=1}^\infty \begin{pmatrix} n^{-1} & 0 \\ 0 & n^{-1} \end{pmatrix} M(l)$. The following property is well-known.

Lemma 2.1. *We have that* (Γ, M_∞) *is a Hecke pair.*

Let $\mathcal{H}$ be the corresponding Hecke algebra of the Hecke pair (Γ, M_∞) over $\mathbb{Q}$. Then we have the Hecke pair

$$\left(\Gamma, \ \cup_{l \in \mathbb{Z}} M(p^l) \right)$$

for all prime p with corresponding Hecke algebra $\mathcal{H}_p$. By the elementary divisor theorem

$$\mathcal{H} = \otimes_p \mathcal{H}_p. \qquad (2.4)$$

Let $\mathbb{T}_l = \Gamma \backslash M(l)$. Then the Hecke algebra $\mathcal{H}_p$ is generated by the $\mathbb{T}_p$, the special double cosets $\Gamma \begin{pmatrix} 1 & 0 \\ 0 & p \end{pmatrix} \Gamma$, and $\Gamma \begin{pmatrix} p^{-1} & 0 \\ 0 & p^{-1} \end{pmatrix} \Gamma$. Here

$$\Gamma \begin{pmatrix} 1 & 0 \\ 0 & p \end{pmatrix} \Gamma = \Gamma \begin{pmatrix} p & 0 \\ 0 & 1 \end{pmatrix} + \sum_{a=0}^{p-1} \Gamma \begin{pmatrix} 1 & a \\ 0 & p \end{pmatrix}. \tag{2.5}$$

Let $GL_2^+(\mathbb{R})$ the set of $\mathbb{R}$-valued 2×2 matrices with positive determinant. Let $M \in GL_2^+(\mathbb{R})$. Define $\widetilde{M} := \det(M)^{-\frac{1}{2}} M$.

Definition 2.2. *The action of the Hecke algebra $\mathcal{H}$ on M_k is induced by double cosets. Let $g \in GL_2^+(\mathbb{Q})$ and $f \in M_k$. Then*

$$f|_k[\Gamma g \Gamma] := \sum_{A \in \Gamma \backslash \Gamma g \Gamma} f|_k \widetilde{A}. \tag{2.6}$$

Here $|_k$ is the Petersson slash operator. In particular, the normalized Hecke operators are defined by

$$\mathbb{T}_n(f) := n^{\frac{k}{2}-1} \sum_{A \in \Gamma \backslash M(n)} f|_k \widetilde{A}. \tag{2.7}$$

Remark. The Hecke operators $\mathbb{T}_n$ coincide with the classical Hecke operators T_n on the space M_k. For f be a primitive form, the eigenvalue of T_n is the n-th Fourier coefficient of f.

Shimura's approach to Hecke theory can be generalized to introduce new operators related to classical Hecke operators, and which coincide in certain special situations.

Let $A = \begin{pmatrix} a & b \\ c & d \end{pmatrix}$ and $B = \begin{pmatrix} e & f \\ g & h \end{pmatrix}$. Then

$$A \times B := \begin{pmatrix} a & 0 & b & 0 \\ 0 & e & 0 & f \\ c & 0 & d & 0 \\ 0 & g & 0 & h \end{pmatrix} \tag{2.8}$$

gives an embedding of $SL_2(\mathbb{R}) \times SL_2(\mathbb{R})$ into the symplectic group $Sp_2(\mathbb{R})$ of degree 2. Let $A \in GL_2(\mathbb{R})$ with $\det(A) = l > 1$. We put

$$\widetilde{A}^\bullet := \begin{pmatrix} l^{-1/2}a & l^{-1/2}b \\ l^{-1/2}c & l^{-1/2}d \end{pmatrix} \times \begin{pmatrix} 1 & 0 \\ 0 & 1 \end{pmatrix}, \tag{2.9}$$

and similarly define $\widetilde{A}_\bullet$. Let $F : \mathbb{H}_2 \longrightarrow \mathbb{C}$ with $F|_k g^\bullet = F$ for all $g \in \Gamma$. Let $A \in GL_2^+(\mathbb{Q})$. Define the Hecke operator

$$F|_k \widetilde{[\Gamma A \Gamma]}^\bullet := \sum_{\gamma \in \Gamma \backslash \Gamma A \Gamma} F|_k \widetilde{\gamma}^\bullet, \tag{2.10}$$

and similarly $F|_k \widetilde{[\Gamma A \Gamma]}_\bullet$. For simplicity put $|_k T^\bullet$ and $|_k T_\bullet$ for $T \in \mathcal{H}$.

3 Eisenstein series decompositon

In this section we state and prove a decomposition formula for $E_k^{(2)}(Z, s)$. It is essentially constructed from two functions. The symplectic group $\mathrm{Sp}_n(\mathbb{R})$ acts on the Siegel upper-half space $\mathbb{H}_n$ of degree n via

$$\left(\begin{smallmatrix} A & B \\ C & D \end{smallmatrix}\right)(Z) := (AZ + B)(CZ + D)^{-1}.$$

We put $j\left(\left(\begin{smallmatrix} A & B \\ C & D \end{smallmatrix}\right), Z\right) := \det(CZ + D)$. Let $\Gamma_n := \mathrm{Sp}_n(\mathbb{Z})$ be the Siegel modular group and let $\Gamma_{n,0}$ be the subgroup of all elements with $C = 0$.

Definition 3.1. Let k be an even integer and let $n \in \mathbb{N}$. Define the real analytic Eisenstein series of weight k and genus n on $\mathbb{H}_n \times \mathcal{D}_k^n$, where

$$\mathcal{D}_k^n := \{s \in \mathbb{C} \mid 2\,\mathrm{Re}(s) + k > n + 1\},$$

by

$$E_k^{(n)}(Z, s) := \sum_{g \in \Gamma_{n,0} \backslash \Gamma_n} j(g, Z)^{-k}\, \delta\,(g(Z))^s. \qquad (3.1)$$

Here $\delta\,(Z) := \det(\,\mathrm{Im}(Z))$.

The infinite sum in (3.1) converges absolutely and uniformly on compacts on the set $\mathbb{H}_n \times \mathcal{D}_k^n$. From Langlands' theory [La76], $E_k^{(n)}(Z, s)$ has a meromorphic continuation in s to the whole complex plane, and satisfies a functional equation. In particular, let k be an even positive integer, let $\xi(s) := \pi^{\frac{s}{2}}\Gamma(\frac{s}{2})\zeta(s)$ and $\Gamma_n(s) := \prod_{j=1}^n \Gamma(s - \frac{j-1}{2})$. Here $\Gamma(s)$ is the Gamma function and $\zeta(s)$ the Riemann zeta function. Then the function

$$\mathbb{E}_k^{(n)}(Z, s) := \frac{\Gamma_n(s + \frac{k}{2})}{\Gamma_n(s)} \cdot \xi(2s) \prod_{i=1}^{[n/2]} \xi(4s - 2i)\, E_k^{(n)}\left(Z, s - \frac{k}{2}\right) \qquad (3.2)$$

is invariant under $s \mapsto \frac{n+1}{2} - s$ and is an entire function in s (see [Mi91]). Here $[x]$ is the largest integer smaller or equal to x. When $n = 1$ the function

$$\mathbb{E}_k(\tau, s) = \Gamma\left(s + \frac{k}{2}\right)\zeta(2s)\,\pi^{-s}\, E_k\left(\tau, s - \frac{k}{2}\right) \qquad (3.3)$$

is entire and is invariant under $s \mapsto 1 - s$. Moreover, for $n = 2$ the function

$$\mathbb{E}_k^{(2)}(Z, s) \;=\; \Gamma(s)\,\Gamma\left(s + \frac{k}{2}\right)\Gamma\left(s + \frac{k-1}{2}\right)2^{2s-2}\,\pi^{-s-\frac{1}{2}}$$

$$\zeta(2s)\,\zeta(4s - 2)\, E_k^{(2)}\left(Z, s - \frac{k}{2}\right)$$

entire and invariant under $s \mapsto \frac{3}{2} - s$.

For a positive even integer k with $k > n + 1$ the function $E_k^{(n)}(Z) := E_k^{(n)}(Z, 0)$ is the holomorphic Siegel Eisenstein series. It has a Fourier expansion with rational coefficients. Moreover the denominators are bounded. In the real analytic case the situation is somehow different. The Fourier coefficient depend on the imaginary part of Z and involve confluent hypergeometric functions. Moreover, one has to study Hecke summation if one is interested in the case $k = n + 1$ and $s = 0$, for example. Let k be an even integer. Then $\mathcal{D}_k := \{s \in \mathbb{C} \,|\, 2\, Re(s) + k > 3\}$. It is well known that $E_2^{(2)}(Z, 0)$ is finite. But we do not want to go into this topic further. We parametrize $Z \in \mathbb{H}_2$ by $\left(\begin{smallmatrix} \tau & z \\ z & \tilde{\tau} \end{smallmatrix}\right)$ and define $\varphi_k(Z) := \tau + 2z + \tilde{\tau}$,. For simplicity, put $\chi_{k,s}(g, Z) := j(g, Z)^{-k} |j(g, Z)|^{-2s}$ and $\Phi_{k,s} := \varphi_k(Z)^{-k} |\varphi_k(Z)|^{-2s}$ for $g \in \mathrm{Sp}_2(\mathbb{R})$. Also let $\Gamma_\infty = \Gamma_{1,0}$ and $\mathbb{H} = \mathbb{H}_1$. Let $|_k$ be the Petersson slash operator. We drop the symbol for the weight k if it is clear from the context.

Definition 3.2. For $k \in \mathbb{Z}$ be even we define two $\mathbb{C}$-valued functions A_k (resp. B_k) on $\mathbb{H}_2 \times \mathcal{D}_k$ by

$$(Z, s) \mapsto \delta(Z)^s \sum_{g, h \in \Gamma_\infty \backslash \Gamma} \chi_{k,s}\left(g^\bullet h_\bullet, Z\right) \quad and$$

$$(Z, s) \mapsto \delta(Z)^s \sum_{g \in \Gamma} \Phi_{k,s}\left(g_\bullet(Z)\right) \chi_{k,s}\left(g_\bullet, Z\right).$$

These functions turn out to be subseries of the real analytic Eisenstein series of degree two, with similiar convergence properties.

Theorem 3.3. Let k be an even integer. Let $Z \in \mathbb{H}_2$ and $s \in \mathcal{D}_k$. Then

$$E_k^{(2)}(Z, s) = A_k(Z, s) + \sum_{m=1}^{\infty} B_k \bigg| \left(\Gamma \begin{pmatrix} m & 0 \\ 0 & m^{-1} \end{pmatrix} \Gamma\right)^\bullet (Z, s) \; m^{-2s-k}. \quad (3.4)$$

Proof. From Garrett [Ga84], [Ga87] we know how to study coset systems of the type

$$\Gamma_{2n,0} \backslash \Gamma_{2n} / \Gamma_n \times \Gamma_n$$

in the context of the doubling method. Similarly, we obtain a useful $\Gamma_{2,0}$-left coset decomposition of Γ_2 given by $R_0 \bigsqcup R_1$ with

$$R_0 = \Gamma_\infty \backslash \Gamma \times \Gamma_\infty \backslash \Gamma \text{ and } R_1 = \bigsqcup_{m=1}^{\infty} g_m \left(\Gamma \times \Gamma(m) \backslash \Gamma\right). \quad (3.5)$$

Here $\Gamma(m) := \{g \in \Gamma | \begin{pmatrix} 0 & 1/m \\ m & 0 \end{pmatrix} g \begin{pmatrix} 0 & 1/m \\ m & 0 \end{pmatrix} \in \Gamma\}$ and

$$g_m := \begin{pmatrix} 0 & 0 & -1 & 0 \\ 0 & 1 & 0 & 0 \\ 1 & m & 0 & 0 \\ 0 & 0 & -m & 1 \end{pmatrix}. \tag{3.6}$$

The subseries related to the representatives g_m $(\Gamma \times \Gamma(m)\backslash\Gamma)$ is

$$\delta(Z)^s \sum_{g \in \Gamma,\, h \in \Gamma(m)\backslash\Gamma} \chi_{k,s} (g_m (g \times h), Z). \tag{3.7}$$

Let $\mathbb{M}_m$ be the diagonal 4×4 matrix with $(1, m, 1, m^{-1})$ on the diagonal. Then $j(g_m, Z) = j(g_1, \mathbb{M}_m(Z))$. Hence we obtain, for (3.7):

$$\delta(Z)^s \sum_{g \in \Gamma,\, g \in \Gamma(m)\backslash\Gamma} \Phi_{k,s} (\mathbb{M}_m (g \times h)(Z)) \chi_{k,s} (g \times h, Z).$$

Let # be the automorphism of $SL_2(\mathbb{R})$ given by $\begin{pmatrix} a & b \\ c & d \end{pmatrix}^{\#} := \begin{pmatrix} d & b \\ c & a \end{pmatrix}$ of $SL_2(\mathbb{R})$. Then we can prove in a straightforward manner the symmetric relation

$$\Phi_{k,s} \left(g^{\bullet}(Z)\right) \chi_{k,s}(g^{\bullet}, Z) = \Phi_{k,s} \left(g_{\bullet}^{\#}(Z)\right) \chi_{k,s}(g_{\bullet}^{\#}, Z). \tag{3.8}$$

By the elementary divisor theorem we obtain for our subseries the expression

$$\delta(Z)^s m^{(k+2s)} \sum_{\gamma \in \Gamma \begin{pmatrix} m & 0 \\ 0 & m^{-1} \end{pmatrix}\Gamma} \Phi_{k,s} (1_2 \times \gamma)(Z)) \chi_{k,s} (1_2 \times \gamma, Z). \tag{3.9}$$

Now we can apply again the symmetry relation and obtain the formula (3.4) in our theorem. □

Corollary 3.4. Let k be an even integer. Let $Z \in \mathbb{H}_2$ and $s \in \mathcal{D}_k$. Then

$$E_k^{(2)} (Z, s) = A_k(Z, s) + \sum_{m=1}^{\infty} \left(B_k| \left(\Gamma \begin{pmatrix} m & 0 \\ 0 & m^{-1} \end{pmatrix} \Gamma\right)_{\bullet} (Z, s)\right) m^{-2s-k}. \tag{3.10}$$

Let F be a complex-valued function on $\mathbb{H}_2$. Let $k \in \mathbb{N}_0$ be even. Then we say that F is Γ-modular of weight k if $F|_k\gamma^{\bullet} = F_k|\gamma_{\bullet} = F$ for all $\gamma \in \Gamma$.

Corollary 3.5. The functions $A_k(Z, s)$ and $B_k(Z, s)$ are Γ-modular.

4 Strong symmetry of Eisenstein series

Let F be a complex valued C^{∞} function on the Siegel upper half-space of degree 2 with the transformation property of a modular form of even weight k with respect to $Sp_2(\mathbb{Z})$. Let $f(\tau, \tilde{\tau}) := F \begin{pmatrix} \tau & 0 \\ 0 & \tilde{\tau} \end{pmatrix}$. Then we have the symmetry

$$f(\tau, \tilde{\tau}) = f(\tilde{\tau}, \tau), \tag{4.1}$$

since $F|U = F$ with

$$U := \begin{pmatrix} 0 & 1 & 0 & 0 \\ 1 & 0 & 0 & 0 \\ 0 & 0 & 0 & 1 \\ 0 & 0 & 1 & 0 \end{pmatrix}.$$

It is worth noting that this does *not* imply that if we apply Hecke operators $T \in \mathcal{H}$ on f by fixing one of the variables that such a symmetry still holds. Let for example F be the holomorphic Klingen Eisenstein series of degree 2 and weight 12 attached to the Ramanujan Δ-function. Then it can be shown that

$$f(\tau, \tilde{\tau}) = E_{12}(\tau)\Delta(\tilde{\tau}) + E_{12}(\tilde{\tau})\Delta(\tau) + \alpha\Delta(\tau)\Delta(\tilde{\tau}), \tag{4.2}$$

with $\alpha \in \mathbb{C}$. Since infinitely many Hecke eigenvalues of the Eisenstein series and the Δ function are different, it is obvious that

$$f|\widetilde{T_p}^{\bullet} - f|\widetilde{T_p}_{\bullet} \neq 0 \tag{4.3}$$

for (at least) one prime number p.

The real analytic Eisenstein series $E_k^{(2)}(Z, s)$ of degree two has an important symmetry which had not been discovered before. Let T be an element of the Hecke algebra $\mathcal{H}$. We will show in this section that, if we apply T as an operator on the Eisenstein series to the two embeddings $T^{\bullet}$ and $T_{\bullet}$ we get the same new function, i.e.,

$$\left(E_k^{(2)}|\widetilde{T}^{\bullet}\right)(Z, s) = \left(E_k^{(2)}|\widetilde{T}_{\bullet}\right)(Z, s). \tag{4.4}$$

From the viewpoint of physics this can been seen as a scattering experiment with an object X, in which we hit the object from outside with T_p for different prime numbers and look at the reaction. For example, if we knew in advance that the object were a holomorphic Eisenstein series, then we could conclude that it is of Siegel type.

Actually we show that the subseries $A_k(Z, s)$ and

$$B_k^m(Z, s) := B_k|\left(\Gamma\begin{pmatrix} m & 0 \\ 0 & m^{-1} \end{pmatrix}\Gamma\right)_{\bullet}(Z, s) \tag{4.5}$$

already have the strong symmetry property. Furthermore, the function $A_k(Z, s)$ turns out to be an eigenfunction.

Proposition 4.1. *Let k be an even integer and $s \in \mathcal{D}_k$. For $T \in \mathcal{H}$ we have*

$$\left(A_k|\widetilde{T}^{\bullet}\right)(Z, s) = \left(A_k|\widetilde{T}_{\bullet}\right)(Z, s) = \lambda(T)A_k(T, s), \tag{4.6}$$

with $\lambda(T) \in \mathbb{C}$.

Proof. We have that

$$A_k(Z, s) = \sum_{g, h \in \Gamma_\infty \backslash \Gamma} j(g^\bullet h_\bullet, Z)^{-k} \delta \left(g^\bullet h_\bullet(Z) \right)^s.$$

At this point we note that

$$g^\bullet h_\bullet = h_\bullet g^\bullet \quad \text{and} \quad j(g^\bullet h_\bullet, Z) = j(g^\bullet, h_\bullet(Z)) j(h_\bullet, Z).$$

Since the series convergences absolutely and uniformly on compacts in $\mathbb{H}_2 \times \mathcal{D}_k$ we can interchange summation to obtain

$$A_k(Z, s) = \sum_{h \in \Gamma_\infty \backslash \Gamma} j(h_\bullet, Z)^{-k} \sum_{g \in \Gamma_\infty \backslash \Gamma} j(g^\bullet, h_\bullet(Z))^{-k} \delta(g^\bullet(h_\bullet(Z))^s$$

$$= \sum_{h \in \Gamma_\infty \backslash \Gamma} E_k \left((h_\bullet(Z))^*, s \right) j(h, Z_*)^{-k}.$$

Let $Z = \left(\begin{smallmatrix} \tau & z \\ z & \tilde{\tau} \end{smallmatrix} \right)$. Here $Z^* := \tau$ and $Z_* := \tilde{\tau}$. By the same procedure we obtain

$$A_k(Z, s) = \sum_{g \in \Gamma_\infty \backslash \Gamma} E_k \left((g^\bullet(Z))_*, s \right) j(g, Z^*)^{-k}.$$

Now let $T \in \mathcal{H}$ and $T = \sum_j a_j \Gamma t_j$. Then we have

$$\left(A_k | \tilde{T}^\bullet \right)(Z, s)$$

$$= \sum_j a_j \sum_{h \in \Gamma_\infty \backslash \Gamma} E_k \left((h_\bullet \tilde{t}_j^\bullet(Z))^*, s \right) j(h, \tilde{t}_j(Z)_*)^{-k} j(\tilde{t}_j^\bullet, \tilde{\tau})^{-k}.$$

Hence,

$$\left(A_k | \tilde{T}^\bullet \right)(Z, s)$$

$$= \sum_j a_j \sum_{h \in \Gamma_\infty \backslash \Gamma} E_k \left((\tilde{t}_j^\bullet h_\bullet(Z))^*, s \right) j \left(\tilde{t}_j^\bullet, h_\bullet(Z)^* \right)^{-k} j(h_\bullet, Z)^{-k}$$

$$= \sum_{h \in \Gamma_\infty \backslash \Gamma} \left(E_k | \tilde{T}^\bullet \right) \left(h_\bullet(Z)^*, s \right) j(h_\bullet, Z)^{-k}. \tag{4.7}$$

It is well known that $E_k(\tau, s)$ with $\tau \in \mathbb{H}$ is a Hecke eigenform. This leads to $\left(A_k | \tilde{T}^\bullet \right)(Z, s) = \lambda(\tilde{T}) A_k(Z, s)$. The same argument works for $\left(A_k | \tilde{T}_\bullet \right)(Z, s)$ with the same eigenvalue. This proves the proposition. $\qquad \square$

Proposition 4.2. *Let k be an even integer. Let $m \in \mathbb{N}$ and let $T \in \mathcal{H}$. Then we have*

$$\left(B_k^m | \tilde{T}^\bullet \right)(Z, s) = \left(B_k^m | \tilde{T}_\bullet \right)(Z, s) \tag{4.8}$$

for all $(Z, s) \in \mathbb{H}_2 \times \mathcal{D}_k$.

Proof. Let $T = \sum_j a_j \Gamma g_j$ with $a_j \in \mathbb{C}$ and $g_j \in \text{Gl}_2^+(\mathbb{Q})$. Then we have

$$\left(B_k^m | \widetilde{T}^\bullet\right)(Z, s) = \sum_j a_j \, B_k | \left(\Gamma \begin{pmatrix} m & 0 \\ 0 & m^{-1} \end{pmatrix} \Gamma\right)_\bullet \widetilde{g_j}_\bullet(Z, s)$$

$$= \sum_j a_j \, B_k | \widetilde{g_j}^\bullet \left(\Gamma \begin{pmatrix} m & 0 \\ 0 & m^{-1} \end{pmatrix} \Gamma\right)_\bullet (Z, s)$$

since the Hecke algebra $\mathcal{H}$ is commutative. Hence we can reduce our calculations to the case $m = 1$. Then we have for $\left(B_k | \widetilde{T}^\bullet\right)(Z, s)$ the expression

$$\sum_j a_j \, \delta\left(\widetilde{g_j}^\bullet(Z)\right)^s \sum_{g \in \Gamma} \Phi_{k,s}\left((g_\bullet \widetilde{g_j}^\bullet)(Z)\right) \chi_{k,s}\left(g_\bullet, \widetilde{g_j}^\bullet(Z)\right) \, j(g_j^\bullet, Z)^{-k}.$$

To proceed further we use the cocycle property

$$\chi_{k,s}\left(g_\bullet \widetilde{g_j}^\bullet, Z\right) = \chi_{k,s}\left(g_\bullet, \widetilde{g_j}^\bullet(Z)\right) \chi_{k,s}\left(\widetilde{g_j}^\bullet, Z\right)$$

and the transformation property $\delta\left(\widetilde{g_j}^\bullet(Z)\right)^s = \delta(Z)^s | j(\widetilde{g_j}^\bullet, Z|^{-2s}$. Hence $\left(B_k | \widetilde{T}^\bullet\right)(Z, s)$ is equal to

$$\sum_j a_j \delta(Z)^s \sum_{g \in \Gamma} \Phi_{k,s}\left((g_\bullet \widetilde{g_j}^\bullet)(Z)\right) \chi_{k,s}\left((g_\bullet \widetilde{g_j}^\bullet), Z\right).$$

Now we apply the symmetry relation and note that T is invariant with respect to the automorphism #. Then we obtain

$$\sum_j \delta(Z)^s \sum_{g \in \Gamma} \Phi_{k,s}\left((\widetilde{g_j}_\bullet g_\bullet)(Z)\right) \chi_{k,s}\left((\widetilde{g_j}_\bullet g_\bullet), Z\right).$$

Finally we use the Γ-invariance property of $\Phi_{k,s}$ and $\chi_{k,s}$. This leads to

$$\sum_j \delta(Z)^s \sum_{g \in \Gamma} \Phi_{k,s}\left((g_\bullet \widetilde{g_j}_\bullet)(Z)\right) \chi_{k,s}\left((g_\bullet \widetilde{g_j}_\bullet), Z\right). \tag{4.9}$$

This gives the proposition. $\qquad\square$

For $T \in \mathcal{H}$ and even integer k let $|_k \bowtie T$ be the operator $|_k \widetilde{T}^\bullet - |_k \widetilde{T}_\bullet$. If a Γ-modular function is annihilated by this operator, we say that is satisfies the strong symmetry property. This makes sense since this property turns out to classify certain subspaces and gives a fundamental identity between Fourier coefficients. Summarizing our results, we have

Theorem 4.3. *Let k be an even integer. Let T be an element of the Hecke algebra $\mathcal{H}$. Let $(Z, s) \in \mathbb{H}_2 \times \mathcal{D}_k$. Then we have*

$$E_k^{(2)} | \bowtie T(Z, s) = 0. \tag{4.10}$$

Corollary 4.4. *The strong symmetry (4.10) of the Eisenstein series is also preserved under meromorphic continuation.*

It would be interesting to study the implication of this property for the residues in relation with the Siegel-Weil formula.

5 Applications of the strong symmetry property

In [He06] we have shown that a Siegel modular form F of degree 2 with respect to the Siegel modular group $Sp_2(\mathbb{Z})$ F is a Saito-Kurokawa lift if and only if F has the strong symmetry property. Moreover, this can be used to study the non-vanishing of certain special values predicted by the Gross-Prasad conjecture and in the context of the Maass-Spezialschar results recently proven by Ichino. Our proof in the holomorphic case was based on the interplay between Taylor coefficients and certain differential operators. In this paper in the setting of real analytic Eisenstein series the proof does not work. That was the reason why we gave a new one and which works just because of the definition of an Eisenstein series via certain left cosets.

Theorem 5.1. Let k be an even integer. Let $F : \mathbb{H}_2 \longrightarrow \mathbb{C}$ be a $\mathbb{C}^\infty$-function which satisfies the transformation law $F|_k\gamma = F$ for all $\gamma \in \Gamma_2$. Then the following conditions are equivalent:

(i) $F|\bowtie_T = 0$ *for all $T \in \mathcal{H}$;*
(ii) $F|\bowtie_{T_p} = 0$ *for all prime numbers p;*
(iii)

$$
p^{k-1} F \begin{pmatrix} p\tau & pz \\ pz & \tilde{\tau} \end{pmatrix} + \frac{1}{p} \sum_{\lambda \ (\mathrm{mod}\ p)} F \begin{pmatrix} \frac{\tau+\lambda}{p} & z \\ z & \tilde{\tau} \end{pmatrix}
$$
$$
= p^{k-1} F \begin{pmatrix} \tau & pz \\ pz & p\tilde{\tau} \end{pmatrix} + \frac{1}{p} \sum_{\mu \ (\mathrm{mod}\ p)} F \begin{pmatrix} \tau & z \\ z & \frac{\tilde{\tau}+\mu}{p} \end{pmatrix} .
$$
(5.1)

Proof. We first show that (i) $\Longleftrightarrow$ (ii). The direction from left to right is clear since it is a specialization. The other direction follows from the fact that the Hecke algebra $\mathcal{H}$ is the infinite restricted tensor product of all local Hecke algebras $\mathcal{H}_p$. Here p runs through the set of all primes. Hence it is sufficient to focus on the generators of $\mathcal{H}_p$. Here one has to be careful. This conclusion works only because everything is compatible with sums of operators and the underlying Hecke algebras are commutative. Now, since the local Hecke algebras are essentially generated by T_p we are done.

Next we show that (ii) $\iff$ (iii). We have seen that

$$T_p = \Gamma \begin{pmatrix} p & 0 \\ 0 & 1 \end{pmatrix} + \sum_{\lambda \pmod p} \Gamma \begin{pmatrix} 1 & \lambda \\ 0 & p \end{pmatrix}.$$

We use this explicit description to calculate $F|\widetilde{T_p}^{\bullet}$ and $F|\widetilde{T_p}_{\bullet}$. Finally we make a change of variable $z \mapsto p^{\frac{1}{2}} z$. $\qquad\square$

We parametrize $Z \in \mathbb{H}_2$ with $Z = \begin{pmatrix} \tau & z \\ z & \tilde{\tau} \end{pmatrix}$. Let $X = \begin{pmatrix} \tau_x & z_x \\ z_x & \tilde{\tau}_x \end{pmatrix}$ be the real part of Z and let $Y = \begin{pmatrix} \tau_y & z_y \\ z_y & \tilde{\tau}_y \end{pmatrix}$ be the imaginary part of Z. Comparing Fourier coefficients in (5.1) we deduce the following result:

Theorem 5.2. Let $k \in \mathbb{N}_0$ be even and let $F : \mathbb{H}_2 \longrightarrow \mathbb{C}$ be a Γ-modular function of weight k. Assume that F has Fourier expansion of the form

$$F(Z) = \sum_N A(N, Y) \, e\{N X\}, \tag{5.2}$$

summing over all half-integral symmetric 2×2 matrices. Then $F|_k \bowtie_T = 0$ for all Hecke operators $T \in \mathcal{H}$ if and only if the Fourier coefficients of F satisfy for all prime numbers p the identity

$$p^{k-1} A \left(\begin{pmatrix} \frac{n}{p} & \frac{r}{2p} \\ \frac{r}{2p} & m \end{pmatrix} \begin{pmatrix} p\tau_y & pz_y \\ pz_y & \tilde{\tau}_y \end{pmatrix} \right) + A \left(\begin{pmatrix} pn & \frac{r}{2} \\ \frac{r}{2} & m \end{pmatrix}, \begin{pmatrix} \frac{\tau_y}{p} & z_y \\ z_y & \tilde{\tau}_y \end{pmatrix} \right) \tag{5.3}$$

$$= p^{k-1} A \left(\begin{pmatrix} n & \frac{r}{2p} \\ \frac{r}{2p} & \frac{m}{p} \end{pmatrix}, \begin{pmatrix} \tau_y & pz_y \\ pz_y & p\tilde{\tau}_y \end{pmatrix} \right) + A \left(\begin{pmatrix} n & \frac{r}{2} \\ \frac{r}{2} & pm \end{pmatrix}, \begin{pmatrix} \tau_y & z_y \\ z_y & \frac{\tilde{\tau}_y}{p} \end{pmatrix} \right).$$

Bibliography

[Bo84] S. Böcherer: *Über die Fourierkoeffizienten der Siegelschen Eisensteinreihen.* Manuscripta Math. **45** (1984), 273–288.

[Bo85] S. Böcherer: Über die Funktionalgleichung automorpher L-Funktionen zur Siegelschen Modulgruppe. J. reine angew. Math. **362** (1985), 146-168.

[Br07] J. Brown: *Saito-Kurokawa lifts and applications to the Bloch-Kato conjecture.* Compositio Math. **143** part 2. (2007), 290–322.

[Ga84] P. Garrett: *Pullbacks of Eisenstein series; applications.* Automorphic forms of several variables (Katata, 1983), 114–137, Progr. Math., **46** Birkhäuser Boston, Boston, MA, 1984.

[Ga87] P. Garrett: *Decomposition of Eisenstein series: triple product L-functions.* Ann. Math. **125** (1987), 209–235.

[GP92] Gross and Prasad: *On the decomposition of a representation of SO_n when restricted to SO_{n-1}.* Canad. J. Math. **44** (1992), 974–1002.

[Ha97] A. Haruki: Explicit formulae of Siegel Eisenstein series. manuscripta math. **92** (1997), 107–134.

[He06]   B. Heim: *On the Spezialschar of Maass*. ArXiv: 0801.1804v1 [math.NT].

[Ich05] A. Ichino: Pullbacks of Saito-Kurokawa Lifts. Invent. Math. **162** (2005), 551–647.

[Ik01]   T. Ikeda: On the lifting of elliptic cusp forms to Siegel cusp forms of degree 2*n*. Ann. of Math. **154** no. 3 (2001), 641–681.

[Ka59]   G. Kaufhold: Dirichletsche Reihe mit Funktionalgleichung in der Theorie der Modulfunktionen 2. Grades. Math. Ann **137** (1959), 454–476.

[KK05]   W. Kohnen, H. Kojima: A Maass space in higher genus. Compos. Math. **141** No. 2 (2005), 313–322.

[Kl90]   H. Klingen: Introductory lectures on Siegel modular forms. Cambridge Studies in Advanced Mathematics, 20. Cambridge University Press, Cambridge, 1990.

[La76]   R. Langlands: On the functional equations satisfied by Eisenstein series. Lect. Notes Math. 544, Berlin-Heidelberg-New York **544** (1976).

[Ma64]   H. Maass: *Die Fourierkoeffizienten der Eisensteinreihen zweiten Grades*. Mat.-Fys. Medd. Danske Vid. Selsk. **34** (1964).

[Ma71]   H. Maass: Siegel's modular forms and Dirichlet series. Lect. Notes Math. 216, Berlin-Heidelberg-New York **544** (1971).

[Ma79]   H. Maass: Über eine Spezialschar von Modulformen zweiten Grades I,II,III. Invent. Math. **52, 53, 53** (1979), 95–104, 249–253, 255–265.

[Mi91]   S. Mizumoto: Poles and residues of standard L-functions attached to Siegel modular forms. Math. Ann. **289** (1991), 589–612.

[Se56]   A. Selberg: Harmonic analysis and discontinuous groups in weakly symmetric Riemannian spaces with applications to Dirichlet series. J. Ind. Math. Soc. **20** (1956), 47–50.

[Sh71]   G. Shimura: *Introduction to the Arithmetical Theory of Automorphic Functions*. Princeton, Iwanami Shoten and Princeton Univ. Press, (1971).

[Sh83]   G. Shimura: On Eisenstein series. Duke Math. J. **50** (1983), 417–476.

[Si39]   C. L. Siegel: *Einführung in die Theorie der Modulformen n-ten Grades*. Math. Ann. **116** (1939), 617–657.

[Si64]   C. L. Siegel: *Über die Fourierschen Koeffizienten der Eisensteinschen Reihen*. Math. Fys. Medd. Danske Vid. Selsk. **34** (1964).

[We65]   A. Weil: *Sur la formule de Siegel dans la theorie des groupes classiques*. Acta Math. **113** (1965), 1–87.

[Za77]   D. Zagier: *Modular forms whose Fourier coefficients involve zeta-functions of quadratic fields*. Modular functions of one variable, VI (Proc. Second Internat. Conf., Univ. Bonn, Bonn, 1976), pp. 105–169. Lecture Notes in Math., Vol. 627, Springer, Berlin, 1977.

[Za80]   D. Zagier, Sur la conjecture de Saito-Kurokawa (d'après H. Maass), *Sém. Delange-Pisot-Poitou 1979/1980, Progress in Math. 12* (1980), 371–394.

A conjecture on a Shimura type correspondence for Siegel modular forms, and Harder's conjecture on congruences

Tomoyoshi Ibukiyama

For modular forms of one variable there is the famous correspondence of Shimura between modular forms of integral weight and half integral weight (cf. [15]). In this paper, we propose a similar conjecture for vector valued Siegel modular forms of degree two and provide numerical evidence and conjectural dimensional equality (Section 1, Main Conjecture 1.1; a short announcement was made in [11].) We also propose a half-integral version of Harder's conjecture in [4]. Our version is deduced in a natural way from our Main Conjecture. While the original conjecture deals with congruences between eigenvalues of Siegel modular forms and modular forms of one variable our version is stated as a congruence between L-functions of a Siegel cusp form and a Klingen type Eisenstein series.

We give here a rough indication of the content of our Main Conjecture. This is restricted to the case of level one, but stated as a precise bijective correspondence as follows.

Conjecture *For any natural number $k \geq 3$ and any even integer $j \geq 0$, there is a linear isomorphism*

$$S^+_{\det^{k-1/2} \operatorname{Sym}(j)}\left(\Gamma_0(4), (\frac{-4}{*})\right) \cong S_{\det^{j+3} \operatorname{Sym}(2k-6)}(\operatorname{Sp}(2, \mathbb{Z}))$$

which preserves L-functions.

Here the superscript $+$ means a certain subspace of new forms or a "level one" part. The details of the notation and our definitions of L-functions will be explained in section 1.

In section 1, after reviewing the definitions of Siegel modular forms of integral and half-integral weight and their L-functions, we give a precise statement of our Main Conjecture and a half-integral version of Harder's conjecture. In section 2 we compare dimensions and also give a supplementary conjecture on dimensions. In section 3 we give numerical examples which support our conjecture. In section 4 we explain how to calculate the numerical examples. In section 5 we review correspondence between Jacobi forms and Siegel modular forms of half-integral and define vector valued Klingen type Eisenstein series which is used in the half-integral version of Harder's conjecture. In the appendix we give tables of Fourier coefficients which we used.

The author would like to thank Professor Tsushima for showing him a conjectural dimension formula of vector valued Jacobi forms, which gave the author the motivation to start this research.

1 Main Conjecture

In this section, after reviewing the definitions quickly, we give our main conjecture.

1.1 Vector valued Siegel modular forms of integral weight

We denote by H_n the Siegel upper half space of degree n.

$$H_n = \left\{ Z = X + iY \in M_n(\mathbb{C}) \mid X = {}^t X, Y = {}^t Y \in M_n(\mathbb{R}), Y > 0 \right\},$$

where $Y > 0$ means that Y is positive definite. For any natural number N, we put

$$\Gamma_0^{(n)}(N) = \left\{ g = \left(\begin{smallmatrix} A & B \\ C & D \end{smallmatrix} \right) \in M_{2n}(\mathbb{Z}) \mid {}^t g J g = J, C \equiv 0 \bmod N \right\}$$

where $J = \left(\begin{smallmatrix} 0_n & 1_n \\ -1_n & 0_n \end{smallmatrix} \right)$ and 1_n or 0_n is the $n \times n$ unit or the zero matrix. When $n = 2$, we sometimes write $\Gamma_0^{(2)}(N) = \Gamma_0(N)$. We also write $\Gamma_n = \Gamma_0^{(n)}(1) = \mathrm{Sp}(n, \mathbb{Z})$.

Now we define vector valued Siegel modular forms of degree $n = 2$ of integral weight and their spinor L-functions. First we recall the irreducible representations of $\mathrm{GL}_2(\mathbb{C})$. For variables u_1, u_2 and $g \in \mathrm{GL}_2(\mathbb{C})$, we put $(v_1, v_2) = (u_1, u_2)g$. We define the $(j + 1) \times (j + 1)$ matrix $\mathrm{Sym}_j(g)$ by

$$(v_1^j, v_1^{j-1} v_2, \ldots, v_2^j) = (u_1^j, u_1^{j-1} u_2, \ldots, u_2^j) \mathrm{Sym}_j(g).$$

Then Sym_j gives the symmetric tensor representation of degree j of $\mathrm{GL}_2(\mathbb{C})$. We denote by $V_j \cong \mathbb{C}^{j+1}$ the representation space of Sym_j. The space V_j can be identified with the space of polynomials $P(u, v)$ in two variables u, v of homogeneous degree j, where the action is given by $P((u, v)g)$ for $g \in \mathrm{GL}_2$. If ρ is a rational irreducible representation of $\mathrm{GL}_2(\mathbb{C})$, then there exist an integer k and a positive integer j such that $\rho = \det^k \mathrm{Sym}_j$. We denote this representation by $\rho_{k,j}$. Any V_j-valued holomorphic function $F(Z)$ of H_2 is said to be a Siegel modular form of weight $\rho_{k,j}$ belonging to Γ_2 if we have

$$F(\gamma Z) = \rho_{k,j}(CZ + D)F(Z)$$

for any $\gamma = \begin{pmatrix} A & B \\ C & D \end{pmatrix} \in \Gamma_2$. We denote by $A_{k,j}(\Gamma_2)$ the linear space over $\mathbb{C}$ of these functions. If j is odd, then -1_4 acts as multiplication by -1 and we have $A_{k,j}(\Gamma_2) = 0$. We define the Siegel Φ-operator by

$$(\Phi F)(\tau_1) = \lim_{\lambda \to \infty} F\begin{pmatrix} \tau_1 & 0 \\ 0 & i\lambda \end{pmatrix},$$

where $\tau_1 \in H_1$. It is well-known that all the components of the vector $\Phi(F)$ except for the first one always vanish and the first component is in $S_{k+j}(\Gamma_1)$ (e.g. [1]). If $\Phi(F) = 0$ for $F \in A_{k,j}(\Gamma_2)$, we say that F is a cusp form. We denote by $S_{k,j}(\Gamma_2)$ the space of cusp forms. If k is odd (and j is even), then since $S_{k+j}(\Gamma_1) = 0$, we have $A_{k,j}(\Gamma_2) = S_{k,j}(\Gamma_2)$.

Now we define Hecke operators and the spinor L functions. For any natural number m, we put

$$T(m) = \{\delta \in M_4(\mathbb{Z}); \ ^t\delta J \delta = mJ\}.$$

The action of $T(m)$ on $F \in M_{j,k}$ is defined by

$$F|_{(k,j)}T(m) = m^{2k+j-3} \sum_{M \in \Gamma_2 \backslash T(m)} F|_{(k,j)}[M]$$

where we put

$$F|_{(k,j)}M = \rho_{k,j}(CZ + D)^{-1}F(MZ)$$

for $M = \begin{pmatrix} A & B \\ C & D \end{pmatrix} \in T(m)$. For a common Hecke eigenform $F \in S_{k,j}(\Gamma_2)$ we denote by $\lambda(p^\nu)$ the eigenvalue of $T(p^\nu)$, i.e., we put $T(p^\nu)F = \lambda(p^\mu)F$. The spinor L-function $L(s, F)$ of the common Hecke eigenform $F \in S_{k,j}(\Gamma_2)$ is defined to be

$$L(s, F, Sp) = \prod_{p:\text{ prime}} L_p(s, F)$$

where $L_p(s, F)$ equals

$$\left(1 - \lambda(p)p^{-s} + (\lambda(p)^2 - \lambda(p^2) - p^{\mu-1})p^{-2s} - \lambda(p)p^{\mu-3s} + p^{2\mu-4s}\right)^{-1}$$

and $\mu = 2k + j - 3$ (cf. e.g. [1]).

1.2 Vector valued Siegel modular forms of half-integral weight

First we define Siegel modular forms of half integral weight with or without character. We denote by ψ the Dirichlet character modulo 4 defined by $\psi(a) = \left(\frac{-4}{a}\right)$ for any odd a. We define a character of $\Gamma_0(4)$ by $\psi(\det(D))$ for any $g = \left(\begin{smallmatrix} A & B \\ C & D \end{smallmatrix}\right) \in \Gamma_0(4)$ and denote this character also by ψ. To fix an automorphy factor of half-integral weight, we define a theta function on H_2 by

$$\theta(Z) = \sum_{p \in \mathbb{Z}^2} e({}^t p Z p),$$

where $e(x) = \exp(2\pi i x)$. A vector valued Siegel modular form F of weight $\det^{k-1/2} \mathrm{Sym}_j$ belonging to $\Gamma_0(4)$ with character ψ^l ($l = 0$ or 1) is defined to be a V_j-valued holomorphic function $F(Z)$ of H_2 such that

$$F(\gamma Z) = \psi(\gamma)^l \left(\frac{\theta(\gamma Z)}{\theta(Z)}\right)^{2k-1} \mathrm{Sym}_j(CZ + D)F(Z)$$

for any $\gamma = \left(\begin{smallmatrix} A & B \\ C & D \end{smallmatrix}\right) \in \Gamma_0(4)$. Let $A_{k-1/2, j}(\Gamma_0(4), \psi^l)$ denote the space of these functions. We note that if j is odd, then $A_{k-1/2, j}(\Gamma_0(4), \psi^l) = 0$ since -1_4 acts as multiplication by -1. We also see that any $F \in A_{k-1/2, j}(\Gamma_0(4), \psi)$ (here $l = 1$) is a cusp form since there are no half-integral modular forms of $\Gamma_0^{(1)}(4)$ with character ψ. For modular forms of half integral weight of one variable, Kohnen introduced the "plus" subspace to pick up a "level one" part and has shown that it is isomorphic to the space of modular forms of integral weight of level one (see [14]). Later it was shown that that this space is also isomorphic to the space of Jacobi forms of index one in Eichler-Zagier [3]. This notion of plus space was generalized to general degree and used in the comparison with holomorphic and skew holomorphic Jacobi forms of general degree (cf. [9], [5],[7]). We review this "plus" subspace for our case. We write the Fourier expansion of $F \in S_{k-1/2, j}(\Gamma_0(4), \psi)$ by

$$F(Z) = \sum_T a(T)e(\mathrm{Tr}(TZ))$$

where T runs over half-integral positive definite symmetric matrices. The subspace of $S_{k-1/2, j}(\Gamma_0(4), \psi^l)$ consisting of those F such that $a(T) = 0$ unless $T \equiv (-1)^{k+l-1} \mu^t \mu \bmod 4$ for some column vector $\mu \in \mathbb{Z}^2$ is called a plus subspace and denoted by $S_{k-1/2, j}^+(\Gamma_0(4), \psi^l)$. This is a higher dimensional analogue of the Kohnen plus space and should be regarded as the level one part of $S_{k-1/2, j}(\Gamma_0(4), \psi^l)$. In section 5, we review an isomorphism of this space to the space of holomorphic or skew holomorphic Jacobi forms.

The theory of Hecke operators on Siegel modular forms of half-integral weight was developped by Zhuravlev [21] [22]. (See also Ibukiyama [8] in case of vector valued forms of degree two.) We review it here. (Our normalization is slighly different from his original definition.) We define $\widetilde{\mathrm{GSp}}^+(2, \mathbb{R})$ as the set of elements $(g, \phi(Z))$ where

$$g \in \mathrm{GSp}^+(2, \mathbb{R}) = \left\{ g \in \mathrm{GL}_4(\mathbb{R}) \mid {}^t g J g = n(g) J, n(g) \in \mathbb{R}_+^\times \right\}$$

and $\phi(Z)$ is a holomorphic function such that $|\phi(Z)| = |\det(CZ + D)|^{1/2}$. $\widetilde{\mathrm{GSp}}^+(2, \mathbb{R})$ becomes a group via the product

$$(g_1, \phi_1(Z))(g_2, \phi_2(Z)) = (g_1 g_2, \phi_1(g_2 Z)\phi_2(Z)).$$

We can identify $\Gamma_0(4)$ with a subgroup $\tilde{\Gamma}_0(4)$ of $\widetilde{\mathrm{GSp}}^+(2, \mathbb{R})$ by embedding $\gamma \mapsto (\gamma, \theta(\gamma Z)/\gamma(Z))$. For any element $(g, \phi(Z)) \in \widetilde{\mathrm{GSp}}^+(2, \mathbb{R})$ with ${}^t g J g = m^2 J$, we put

$$g' = m^{-1}g = \begin{pmatrix} A_1 & B_1 \\ C_1 & D_1 \end{pmatrix}.$$

We define an action of $\mathrm{GSp}^+(2, \mathbb{R})$ on V_j-valued functions F on H_2 by

$$F|_{k-1/2, j}[(g, \phi(Z))] = \mathrm{Sym}_j (C_1 Z + D_1)^{-1} \phi(Z)^{-2k+1} F(gZ).$$

For any prime number p, we put

$$K_1(p^2) = \left(\begin{pmatrix} 1 & 0 & 0 & 0 \\ 0 & p & 0 & 0 \\ 0 & 0 & p^2 & 0 \\ 0 & 0 & 0 & p \end{pmatrix}, p^{1/2} \right) \qquad K_2(p^2) = \left(\begin{pmatrix} 1 & 0 & 0 & 0 \\ 0 & 1 & 0 & 0 \\ 0 & 0 & p^2 & 0 \\ 0 & 0 & 0 & p^2 \end{pmatrix}, p \right)$$

For the $\tilde{\Gamma}_0(4)$ double cosets

$$T_i(p) = \tilde{\Gamma}_0(4) K_i(p^2) \tilde{\Gamma}_0(4) = \cup_\nu \tilde{\Gamma}_0(4) \tilde{g}_\nu$$

we define

$$F|_{k-1/2, j} T_i(p) = p^{i(k+j-7/2)} \sum_\nu F|_{k-1/2, \rho}[\tilde{g}_\nu] \psi(\det(D))$$

where D is right lower 2×2-matrix of $n(g_\nu)^{-1/2} g_\nu \in \mathrm{Sp}(2, \mathbb{R})$ and g_ν is the projection of $\tilde{g}_\nu$ on its first argument. For any odd prime p and any $F \in A_{k-1/2, j}(\Gamma_0(4), \psi^l)$, assume that $F|T_1(p) = \lambda(p)F$ and $F|T_2(p) = \omega(p)F$. We put $\lambda^*(p) = \psi(p)^l \lambda(p)$, where, as before, $\psi(p) = \left(\frac{-1}{p} \right)$. Then the Euler p-factor of the L-function of F is defined to be

$$\left(1 - \lambda^*(p)p^{-s} + (p\omega(p) + p^{\nu-2}(1+p^2))p^{-2s} - \lambda^*(p)p^{\nu-3s} + p^{2\nu-4s} \right)^{-1},$$

where $\nu = 2k + 2j - 3$. We remark that when $p = 2$, we can also define an Euler 2-factor for $F \in S_{k-1/2, j}^+(\Gamma_0(4), \psi)$ in the same way as in [7]. Indeed, we can similarly define $T_i^*(2)$ as in [9] and [5] by the pull back of the Hecke operators on holomorphic or skew holomorphic Jacobi forms. Denoting by $\lambda^*(2)$ and $\omega^*(2) = \omega(2)$ the eigenvalues of these operators, we can then define an Euler 2-factor as above. For details, see [7] or section 4.2 and 5 of this paper.

1.3 Main Conjecture

We propose the following conjecture.

Conjecture 1.1. *For any integer $k \geq 3$ and even integer $j \geq 0$, there exists a linear isomorphism ϕ of $S_{k-1/2, j}^+(\Gamma_0(4), \psi)$ onto $S_{j+3, 2k-6}(\Gamma_2)$ such that*

$$L(s, F) = L(s, \phi(F), \mathrm{Sp}).$$

In the above, the scalar valued case occurs only when $j = 0$ on the left or $k = 3$ on the right, respectively and when both sides are scalar valued, they are zero. So it is essential to treat the vector valued forms. The above conjecture is false when j is odd, since the left hand side is zero and the right hand side is not zero in general in this case. It is not clear which kind of modification is necessary for odd j.

This conjecture can be proved in principle by Selberg trace formula, but no concrete trace formula is known at the moment except for dimension formulas. To use trace formulas, we need a conjecture comparing not only $T(p^\delta)$ or $T_i(p)$ but all the Hecke operators. So we would like to describe this. Let a, b, c and d be natural numbers such that $a \leq b \leq d \leq c$ with $a + c = b + d = 2\delta$ for some natural integer δ. We denote by $T(p^a, p^b, p^c, p^d)$ the Hecke operator obtained by the Γ_2 double coset determined by the diagonal matrix whose diagonal components are (p^a, p^b, p^c, p^d). If we take sums of two of a, b, c, d, respectively, we have six sums but only two of them equal δ and there are four other terms. Our guess for comparing the Hecke operators is

$$T(p^a, p^b, p^c, p^d) \to \psi(p)^\delta T_{\mathrm{half}}(p^{a+b}, p^{a+d}, p^{c+d}, p^{b+c})$$

up to normalizing factors, where T_{half} is the Hecke operator defined by the $\widetilde{\Gamma}_0(4)$ double coset containing the diagonal matrix in the parenthesis.

We explain several reasons why we believe our Main Conjecture.

(1) The above weight correspondence is explained as follows. By the Langlands conjectures, Siegel modular forms of $\mathrm{Sp}(2, \mathbb{Q})$ should correspond to automorphic forms belonging to the compact twist whose real form is $\mathrm{Sp}(2) = \{g \in M_2(\mathbb{H}); g^t g = 1_2\}$, where $\mathbb{H}$ denotes the Hamilton quaternions. It was observed by Y. Ihara (cf. [12]) that the holomorphic discrete series representation of $\mathrm{Sp}(2, \mathbb{R})$ corresponding to the weight $\det^k \mathrm{Sym}_j$ should correspond

to the irreducible representation of Sp(2) corresponding to the Young diagram $(k + j - 3, k - 3)$ by comparing the character of the representations. On the other hand, Sp(2) is isogenous to SO(5) and starting from automorphic forms belonging to SO(5), by the theta correspondence we can construct Siegel modular forms of the double cover of $Sp(2, \mathbb{R})$ of weight $\det^{(j+5)/2} \mathrm{Sym}_{k-3}$ (cf. [8]). In other words, the weight $\det^{k-1/2} \mathrm{Sym}(j)$ should correspond the weight $\det^{j+3} \mathrm{Sym}_{2k-6}$ as stated as above and we have no other choice. By the way, we cannot prove our conjecture by this theta correspondence, since in our case the level equals one, while the compact Sp(2) has always a level greater than one.

(2) The dimension of $S_{k,j}(\Gamma_2)$ is known by Igusa for $j = 0$ and by Tsushima for $j > 0$ under the condition that $k \geq 5$. On the other hand, the dimension of half-integral Siegel modular forms are known by Tsushima. Furthermore, the dimension of holomorphic Jacobi forms and skew holomorphic Jacobi forms are known also by Tsushima as far as we assume a standard vanishing theorem of cohomology which has not been proved in the non-scalar valued case. So, we have a proven dimension formula for $S_{k,j}(\Gamma_2)$ and conjectural dimensions for $S^+_{k-1/2,j}(\Gamma_0(4), \psi)$ for $k > 4$. We can compare these two and we can show that they coincide.

These proven and conjectural dimensions are given in the table in section 2.

(3) Several numerical examples of Euler factors which support the conjecture will be given for spaces of small dimensions in section 3.

1.4 A half-integral version of Harder's conjecture

In his paper [4], Harder proposed a conjecture on certain congruences between eigenvalues of Siegel modular forms and modular forms of one variables. I understand that it can be stated as follows. Let $f \in S_{2k+j-2}(\mathrm{SL}_2(\mathbb{Z}))$ be a common Hecke eigenform of weight $2k + j - 2$ of one variable with p-th eigenvalue $c(p)$. Then there exists a Siegel modular form $F \in S_{k,j}(\Gamma_2)$ which is a Hecke common eigenform with eigenvalues $\lambda(p^\delta)$ with respect to $T(p^\delta)$ such that the following condition is satisfied.

$$1 - \lambda(p)T + \left(\lambda(p)^2 - \lambda(p^2) - p^{2k+j-4}\right)T^2$$
$$- \lambda(p)p^{2k+j-3}T^3 + p^{4k+2j-6}T^4$$
$$\equiv \left(1 - p^{k-2}T\right)\left(1 - p^{k+j-1}T\right)\left(1 - c(p)T + p^{2k+j-3}T^2\right) \bmod \mathfrak{l},$$

where T is a variable and $\mathfrak{l}$ is a certain prime ideal which divides a certain critical value of $L(s, f)$. (For a deeper explanation, see [4].) The left hand side is the Euler p-factor of F if we put $T = p^{-s}$. But, as far as we can see from examples, there are no Siegel modular forms having the right hand side as the Euler p-factor.

Now since we have a conjectural correspondence between $S_{k,j}(\Gamma_2)$ and $S^+_{(j+5)/2,k-3}(\Gamma_0(4), \psi)$ for odd k, we can give a half-integral version of Harder's conjectures, and in this case we can say more. We assume that $k \geq 3$ and $j \geq 0$ is even. We take $f \in S_{2k+j-2}(\mathrm{SL}_2(\mathbb{Z}))$ as above. Then there exists $g \in S^+_{k+j/2-1/2}(\Gamma_0^{(1)}(4))$ which corresponds to f by Shimura correspondence (cf. Kohnen [14]). As we shall see in section 5, when $j + 3 > 5$, associates to g there exists a Klingen type Eisenstein series $E_{((j+5)/2,k-3)}(Z, g) \in S^+_{(j+5)/2,k-3}(\Gamma_0(4))$ (*without character*) such that

$$L(s, E_{(j+5)/2,k-3}(Z, g)) = \zeta(s - k + 2)\zeta(s - k - j + 1)L(s, f)$$

Hence we propose

Conjecture 1.2. *Assume that $k > 5$. For any Klingen type Eisenstein series $E(Z, g) \in S^+_{k-1/2,j}(\Gamma_0(4))$ as above (with $g \in S^+_{k+j-1/2}(\Gamma_0^{(1)}(4)) \cong S_{2k+2j-2}(\mathrm{SL}_2(\mathbb{Z}))$), there exists a Hecke eigen cusp form*

$$F \in S^+_{k-1/2,j}(\Gamma_0(4), \psi)$$

such that the Hecke eigenvalues are congruent to those of $E(Z, g)$ modulo the above ideal $\mathfrak{l}$.

This type of congruence between cusp forms and Eisenstein series are well-known for the one variable case, so it seems interesting to state Harder's conjecture in this way.

2 Dimension formulas

We review here Tsushima's formula for $\dim S_{k,j}(\Gamma_2)$ for $k \geq 5$ in [18]. Tsushima also gave a conjectural dimension formulas for vector valued holomorphic or skew holomorphic Jacobi forms of any index under the assumption that $k \geq 4$ and assuming a standard conjecture on the vanishing of obstruction cohomology, which is satisfied when $j = 0$. By the isomorphism we shall define in section 5 this implies also conjectural dimension formulas for the plus space $S^+_{k-1/2,j}(\Gamma_0(4), \psi^l)$. He stated his results in the form of polynomials in k and j defined accordingly to the residue classes of k, j modulo certain natural numbers. Here we restate these results using generating functions and find:

Theorem 2.1. *For $k \geq 4$ and even $j \geq 2$, $\dim S_{j+3,2k-6}(\Gamma_2)$ is equal to the conjectural formula of $\dim S^+_{k-1/2,j}(\Gamma_0(4), \psi)$.*

For small k or j, examples for the dimensions in question are given as follows.

$$\sum_{j=0}^{\infty} \dim S_{5,j}(\Gamma_2)s^j = \frac{s^{18} + s^{20} + s^{24}}{(1 - s^6)(1 - s^8)(1 - s^{10})(1 - s^{12})}$$

$$\sum_{j=0}^{\infty} \dim S_{7,j}(\Gamma_2)s^j = \frac{s^{12} + s^{14} + s^{16} + s^{18} + s^{20}}{(1 - s^6)(1 - s^8)(1 - s^{10})(1 - s^{12})}$$

$$\sum_{k=1}^{\infty} \dim S^+_{k-1/2,0}(\Gamma_0(4), \psi)t^k = \frac{t^{21}}{(1 - t^3)(1 - t^4)(1 - t^5)(1 - t^6)}$$

$$\sum_{k=1}^{\infty} \text{cdim} S^+_{k-1/2,2}(\Gamma_0(4), \psi)t^k = \frac{t^{12}(1 + t + t^3)}{(1 - t^3)(1 - t^4)(1 - t^5)(1 - t^6)}$$

$$\sum_{k=1}^{\infty} \text{cdim} S^+_{k-1/2,4}(\Gamma_0(4), \psi)t^k = \frac{t^9(1 + t + t^2 + t^3 + t^4)}{(1 - t^3)(1 - t^4)(1 - t^5)(1 - t^6)}$$

where cdim means the conjectured dimension.

For the reader's convenience, we quote here Tsushima's formula for $\dim S_{k,j}(\Gamma_2)$ for any odd $k \geq 5$ and even j using generating functions. The values $\dim S_{3,j}(\Gamma_2)$ are not known, but in Table 1 below, we give them as our conjecture. (See below.)

We have

$$\sum_{j=0}^{\infty} \sum_{k=3}^{\infty} \dim S_{2j+3,2k-6}(\Gamma_2)t^k s^j$$

$$= \frac{f(t, s)}{(1 - s^2)(1 - s^3)(1 - s^5)(1 - s^6)(1 - t^3)(1 - t^4)(1 - t^5)(1 - t^6)}$$

where $f(t, s)$ is given below and $\dim S_{3,2k-6}(\Gamma_2)$ (the case $j = 0$) are conjectural values.

The coefficients of $t^k s^j$ of $f(t, s)$ are given in Table 1. A part of our main conjecture says that we should have

$$S^+_{5/2,j}(\Gamma_0(4), \psi) \cong S_{j+3,0}(\Gamma_2)$$

and

$$S_{3,2k-6}(\Gamma_2) \cong S^+_{k-1/2,0}(\Gamma_0(4), \psi).$$

In each case, while the dimension of the right hand side is known for all k or j, the one for the left hand side is not known. So, assuming these isomorphisms, we are naturally led to the following conjecture on dimension.

Table 1.

$k\backslash j$	0	1	2	3	4	5	6	7	8	9	10	11	12	13	14	15	16
0	0	0	0	0	0	0	0	0	0	0	0	0	0	0	0	0	0
1	0	0	0	0	0	0	0	0	0	0	0	0	0	0	0	0	0
2	0	0	0	0	0	0	0	0	0	0	0	0	0	0	0	0	0
3	0	0	0	0	0	0	0	0	0	0	0	0	0	0	0	0	1
4	0	0	0	0	0	0	0	0	0	1	1	0	1	1	0	-1	0
5	0	0	0	0	0	0	1	1	1	1	1	0	0	0	0	0	0
6	0	0	0	0	1	1	1	1	1	1	1	0	0	0	0	0	-1
7	0	0	0	1	1	1	2	2	1	-1	0	0	-1	-1	0	1	-1
8	0	0	0	1	1	1	2	2	1	-1	-1	0	-2	-2	0	1	-1
9	0	0	1	2	2	1	0	0	-1	-3	-3	0	-1	-1	0	1	-1
10	0	0	1	1	1	1	0	-1	-2	-3	-4	-1	-2	-1	0	1	1
11	0	0	1	1	1	1	-2	-3	-4	-3	-4	-2	0	1	1	0	1
12	0	1	1	0	-1	-1	-3	-4	-3	-2	-2	-1	1	1	0	0	2
13	0	1	1	-1	-1	-1	-3	-5	-4	0	0	-1	2	3	1	-1	1
14	0	0	0	0	-1	-2	-2	-2	-2	-1	1	0	2	2	1	0	1
15	0	1	0	-1	-2	-2	-1	-2	-1	2	4	1	2	2	1	-1	-1
16	0	0	0	0	-1	-1	0	0	0	1	3	1	1	1	1	0	-1
17	0	0	0	0	-1	-1	1	1	1	1	3	2	0	0	0	0	-1
18	0	0	0	0	0	0	1	1	1	1	1	1	0	0	0	0	-1
19	0	0	0	0	0	0	0	1	2	1	0	1	0	-1	-1	0	0
20	0	0	0	0	1	1	0	0	0	0	-1	0	0	0	0	0	0
21	1	0	-1	-1	0	0	-1	1	2	1	-1	0	0	-1	-1	0	1

Conjecture 2.2. *We have*

$$\sum_{j=0}^{\infty} \dim S_{5/2,j}^{+}(\Gamma_0(4), \psi)t^j \;=\; \frac{t^{32}}{(1 - t^4)(1 - t^6)(1 - t^{10})(1 - t^{12})}$$

$$\sum_{j=0}^{\infty} \dim S_{3,j}(\Gamma_2)s^j \;=\; \frac{s^{36}}{(1 - s^6)(1 - s^8)(1 - s^{10})(1 - s^{12})}.$$

Actually, if $j > 0$, these conjectured dimensions are equal to those obtained by putting $k = 3$ in the general formula of $\dim S_{k-1/2,j}^{+}(\Gamma_0(4), \psi)$ or by putting $k = 3$ in the general formula of $\dim S_{k,j}(\Gamma_2)$.

3 Numerical examples

We have the following table of dimensions of $S_{k,j}(\Gamma_2)$ due to Tsushima (cf. [18]).

(k,j)	(5,18)	(5,20)	(5,22)	(5,24)	(5,26)
dim $S_{k,j}(\Gamma_2)$	1	1	0	2	2
(k,j)	(7,10)	(7,12)	(7,14)	(7,16)	(7,18)
dim $S_{k,j}(\Gamma_2)$	0	1	1	1	2

We also have the following table of dimensions of $S_{k-1/2,j}(\Gamma_2)$ and conjectural dimensions of $S^+_{k-1/2,j}(\Gamma_0(4), \psi)$ due to Tsushima (cf. [19], [20]).

(k,j)	(12,2)	(13,2)	(14,2)	(15,2)	(16,2)
dim $S_{k-1/2,j}(\Gamma_0(4), \psi)$	32	45	58	77	96
dim $S^+_{k-1/2,j}(\Gamma_0(4), \psi)$	1	1	0	2	2
(k,j)	(8,4)	(9,4)	(10,4)	(11,4)	(12,4)
dim $S_{k-1/2,j}(\Gamma_0(4), \psi)$	20	32	45	65	86
dim $S^+_{k-1/2,j}(\Gamma_0(4), \psi)$	0	1	1	1	2

We give below the basis of the above spaces and their Euler 2 and 3 factors excluding the case $S_{7,18}(\Gamma_2)$ and $S_{23/2,4}(\Gamma_0(4))$. The Fourier coefficients we used will be given in the Appendix.

3.1 Eigenforms of integral weight

We construct elements in $S_{k,j}(\Gamma_2)$ by theta functions with harmonic polynomials. For any $x = (x_i)$, $y = (y_i) \in \mathbb{C}^8$, we put $(x, y) = {}^t xy$. Let $a, b \in \mathbb{C}^8$ be vectors such that $(a, a) = (a, b) = (b, b) = 0$. We use the lattice E_8 defined by

$$E_8 = \left\{ x = (x_i) \in \mathbb{Q}^8 \;\middle|\; 2x_i \in \mathbb{Z}, x_i - x_j \in \mathbb{Z}, \sum_{i=1}^{8} x_i \in 2\mathbb{Z} \right\}.$$

This is the unique unimodular lattice of rank 8 up to isomorphism. For a variable $Z \in H_2$, we write

$$Z = \begin{pmatrix} \tau & z \\ z & \omega \end{pmatrix}$$

and for integers $k \geq 4$ and $j \in \mathbb{Z}_{\geq 0}$, we define $\vartheta_{k,j,a,b}(Z)$ to be the sum

$$\sum_{x,y \in E_8} \begin{vmatrix} (x,a) & (x,b) \\ (y,a) & (y,b) \end{vmatrix}^{k-4} (xu + yv, a)^j e\left(\frac{1}{2}((x,x)\tau + 2(x,y)z + (y,y)\omega)\right),$$

where we write $e(x) = \exp(2\pi i x)$. Then identifying V_j with homogeneous polynomials in u and v, we have $\theta_{k,j,a,b} \in A_{k,j}(\Gamma_2)$. (This is more or less forklore and we omit the proof.) Now a nuisance here is that this theta function often vanishes identically and we must choose a and b carefully to get non-zero forms. Here we put

$$a_1 = (2, 1, i, i, i, i, i, 0) \qquad b_1 = (1, -1, i, i, 1, -1, -i, i)$$

or

$$a_2 = (3, 2i, i, i, i, i, i, 0) \qquad b_2 = (1, i, -1, i, 1, i, -i, 1).$$

We define

$$F_{5,18} = \theta_{5,18,a_1,b_1}$$

$$F_{5,20} = \theta_{5,20,a_1,b_1}$$

$$f_{5,24a} = \theta_{5,24,a_1,b_1}/(2^{25} \cdot 3 \cdot 5^3 \cdot 13)$$

$$f_{5,24b} = \theta_{5,24,a_2,b_2}/(2^{25} \cdot 3^7 \cdot 5^4 \cdot 13)$$

$$f_{5,26a} = \theta_{5,26,a_1,b_1}/(2^{28} \cdot 3^2 \cdot 5^4 \cdot 13)$$

$$f_{5,26b} = \theta_{5,26,a_2,b_2}/(2^{28} \cdot 3^6 \cdot 5^4 \cdot 13)$$

$$F_{7,12} = \theta_{7,12,a_1,b_1}/(2^{22} \cdot 3^2 \cdot 5^5 \cdot 7^2)$$

$$F_{7,14} = \theta_{7,14,a_1,b_1}/(2^{23} \cdot 3^4 \cdot 5^2 \cdot 11 \cdot 181)$$

$$F_{7,16} = \theta_{7,16,a_1,b_1}/(2^{27} \cdot 3^3 \cdot 5^3 \cdot 11^2 \cdot 19)$$

Then these forms are non-zero, $F_{k,j} \in S_{k,j}(\Gamma_2)$ and $f_{k,ja}$, $f_{k,jb} \in S_{k,j}(\Gamma_2)$ for all the above forms. Moreover, the $F_{k,j}$ are common Hecke eigenforms. We also put

$$F_{5,24a} = -28741829 f_{5,24a} + (-966968929 + 821420\sqrt{4657}) f_{5,24b}$$

$$F_{5,24b} = 28741829 f_{5,24b} + (966968929 + 821420\sqrt{4657}) f_{5,24b}$$

$$F_{5,26a} = (-171241458523 + 631327288\sqrt{99661}) f_{5,26a}$$
$$\qquad - 5095416151 f_{5,26b}$$

$$F_{5,26b} = (171241458523 + 631327288\sqrt{99661}) f_{5,26a}$$
$$\qquad + 5095416151 f_{5,26b}.$$

Then $F_{5,24a}$, $F_{5,24b}$, $F_{5,26a}$, $F_{5,26b}$ are also common Hecke eigenforms.

3.2 Structure of half-integral weight

Since it seems to be difficult to compute the plus space directly, we first give basis of $S_{k-1/2,j}(\Gamma_0(4), \psi)$ and then, calculating Fourier coefficients we find elements in the plus space. We consider the graded ring $A = \sum_{k=0}^{\infty} A_{2k}(\Gamma_0(4), \psi^k)$ of scalar valued Siegel modular forms of even weight belonging to $\Gamma_0(4)$ with character ψ^k for weight k. For each j, the module $\oplus_{k=1}^{\infty} S_{k-1/2,j}(\Gamma_0(4), \psi)$ is an A-module. The explicit structure of A was given in [7]. It is a weighted polynomial ring $A = \mathbb{C}[f_1, g_2, x_2, f_3]$ generated by the following algebraically independent four forms

$$
\begin{aligned}
f_1 &= (\theta_{0000}(2Z))^2 = \theta^2, \\
x_2 &= (\theta_{0000}(2Z)^4 + \theta_{0001}(2Z)^4 + \theta_{0010}(2Z)^4 + \theta_{0011}(2Z)^4)/4, \\
g_2 &= (\theta_{0000}(2Z)^4 + \theta_{0100}(2Z)^4 + \theta_{1000}(2Z)^4 + \theta_{1100}(2Z)^4), \\
f_3 &= (\theta_{0001}(2Z)\theta_{0010}(2Z)\theta_{0011}(2Z))^2.
\end{aligned}
$$

where, for any $m = (m', m'') \in \mathbb{Z}^4$, we define theta constants $\theta_m(Z)$ as usual by

$$
\theta_m(Z) = \sum_{p \in \mathbb{Z}^2} e\left(\frac{1}{2}{}^t(p + \frac{m'}{2})Z(p + \frac{m'}{2}) + {}^t(p + \frac{m'}{2})\frac{m''}{2}\right).
$$

For $j=2$ or $j=4$, the explicit structure of $\oplus_{k=1}^{\infty} S_{k-1/2,j}(\Gamma_0(4), \psi)$ as A-module is given in [10]. When $j=2$, this is a free A-module of rank 3 generated by $F_{11/2,3} \in S_{11/2,3}(\Gamma_0(4), \psi)$ and $G_{13/2,2}, H_{13/2,2} \in S_{13/2,2}(\Gamma_0(4), \psi)$. When $j=4$, the module $\oplus_{k=1}^{\infty} S_{k-1/2,4}(\Gamma_0(4), \psi)$ is a non-free A-module generated $F_{9/2,4a}, F_{9/2,4b}, F_{9/2,4c} \in S_{9/2,4}(\Gamma_0(4), \psi)$ and $F_{11/2,4a}, F_{11/2,4b}, F_{11/2,4c} \in S_{11/2,4}(\Gamma_0(4), \psi)$. The fundamental relation of the generators of the A-module is given by

$$
\begin{aligned}
&(18f_1^2 - 6g_2 - 12x_2)F_{11/2,4a} + (576f_1^2 - 96g_2 - 768x_2)F_{11/2,4c} \\
&\quad + (-30f_1^3 + 27f_3 + 13f_1g_2 + 8f_1x_2)F_{9/2,4a} \\
&\quad + (-24f_1^3 + 4f_1g_2 + f_1x_2)F_{9/2,4b} + (-6f_1^3 + 9f_3 + 3f_1g_2)F_{9/2,4c} = 0
\end{aligned}
$$

in the space $M_{15/2,4}(\Gamma_0(4), \psi)$. All these modular forms are constructed by Rankin-Cohen type differential operators starting from θ, g_2, x_2, f_3. For details, we refer to [10] and we omit them here.

3.3 Eigenforms of half-integral weight in the plus space

By calculating enough Fourier coefficients, we compute basis for various plus spaces (assuming the conjectured dimension formulas hold true). Then by calculating the action of Hecke operators $T_i(3)$, we can give common eigenforms. We define $F_{k-1/2,j}$ or $F_{k-1/2,j\epsilon}$ ($\epsilon = a$ or b) as below. For each k, j, this is a common eigenforms of all the Hecke operators $T_i(p)$ ($i = 1, 2$) belonging to $S^+_{k-1/2,j}(\Gamma_0(4), \psi)$.

The form $F_{23/2,2}$ is defined to be $1/2717908992$ times

$$(1134 f_1^5 - 1728 f_1^3 x_2 - 378 f_1^3 g_2 - 162 f_1^2 f_3 + 648 f_1 x_2^2 + 324 f_1 g_2 x_2 + 108 x_2 f_3$$
$$+ 54 g_2 f_3)G_{13/2,2} + (-4680 f_1^5 + 8640 f_1^3 x_2 - 1188 f_1^3 g_2 + 12636 f_1^2 f_3$$
$$- 13968 f_1 x_2^2 + 72 f_1 g_2^2 + 936 f_1 g_2 x_2 - 2376 x_2 f_3 + 108 g_2 f_3)H_{13/2,2} + (297 f_1^6$$
$$+ 54 f_1^4 x_2 - 81 f_1^4 g_2 - 540 f_1^3 f_3 - 180 f_1^2 x_2^2 - 108 f_1^2 g_2 x_2 - 9 f_1^2 g_2^2 + 360 f_1 x_2 f_3$$
$$+ 180 f_1 g_2 f_3 + 8 x_2^3 + 12 g_2 x_2^2 + 6 g_2^2 x_2 + g_2^3)F_{11/2,2}.$$

The form $F_{25/2,2}$ is defined to be $1/3623878656$ times

$$(1053 f_1^6 - 2826 f_1^4 x_2 - 333 f_1^4 g_2 - 648 f_1^3 f_3 + 828 f_1^2 x_2^2 + 684 f_1 g_2 x_2 - 9 f_1^2 g_2^2$$
$$+ 432 f_1 x_2 f_3 + 216 f_1 g_2 f_3 + 392 x_2^3 + 204 g_2 x_2^2 + 6 g_2^2 x_2 + g_2^3)G_{13/2,2}$$
$$+ (-4374 f_1^6 + 828 f_1^4 x_2 + 3654 f_1^4 g_2 - 1296 f_1^3 f_3 + 1656 f_1^2 x_2^2 - 72 f_1^2 g_2 x_2$$
$$- 738 f_1^2 g_2^2 + 864 f_1 x_2 f_3 + 432 f_1 g_2 f_3 - 176 x_2^3 - 552 g_2 x_2^2 - 228 g_2^2 x_2$$
$$+ 2 g_2^3)H_{13/2,2} + (54 f_1^7 - 54 f_1^5 g_2 + 612 f_1^5 x_2 + 3456 f_1^4 f_3 + 18 f_1^3 g_2^2$$
$$- 168 f_1^3 g_2 x_2 - 696 f_1^3 x_2^2 - 1584 f_1^2 x_2 f_3 - 1152 f_1^2 g_2 f_3 + 176 f_1 x_2^3 + 72 f_1 g_2 x_2^2$$
$$- 12 f_1 g_2^2 x_2 - 2 f_1 g_2^3 - 240 g_2 x_2 f_3 - 480 x_2^2 f_3)F_{11/2,2}.$$

The form $f_{29/2,2a}$ is defined to be $1/14495514624$ times

$$G_{13/2,2}(2430 f_1^8 - 49086 f_1^6 x_2 - 1134 f_1^6 g_2 + 3888 f_1^5 f_3 + 117612 f_1^4 x_2^2$$
$$+ 16362 f_1^4 g_2 x_2 + 162 f_1^4 g_2^2 + 23328 f_1^3 x_2 f_3 - 1296 f_1^3 g_2 f_3 - 65448 f_1^2 x_2^3$$
$$- 28944 f_1^2 g_2 x_2^2 - 162 f_1^2 g_2^2 x_2 - 18 f_1^2 g_2^3 + 1944 f_1^2 f_3^2 - 8640 f_1 g_2 x_2 f_3$$
$$- 17280 f_1 x_2^2 f_3 + 6 g_2^3 x_2 + 36 g_2^2 x_2^2 + 2760 g_2 x_2^3 + 5424 x_2^4 - 648 g_2 f_3^2$$
$$- 1296 x_2 f_3^2) + H_{13/2,2}(-3645 f_1^8 + 73548 f_1^6 x_2 + 28512 f_1^6 g_2 + 153576 f_1^5 f_3$$
$$- 46656 f_1^4 x_2^2 - 154116 f_1^4 g_2 x_2 - 4266 f_1^4 g_2^2 - 173664 f_1^3 x_2 f_3 - 47952 f_1^3 g_2 f_3$$
$$+ 72144 f_1^2 x_2^3 + 72432 f_1^2 g_2 x_2^2 + 26676 f_1^2 g_2^2 x_2 - 216 f_1^2 g_2^3 + 42768 f_1^2 f_3^2$$
$$+ 7560 f_1 g_2^2 f_3 + 12960 f_1 g_2 x_2 f_3 - 56160 f_1 x_2^2 f_3 + 15 g_2^4 + 132 g_2^3 x_2 - 2928 g_2^2 x_2^2$$
$$- 7440 g_2 x_2^3 - 2352 x_2^4 - 1296 g_2 f_3^2 + 10368 x_2 f_3^2) + F_{11/2,2}(-1620 f_1^9$$
$$- 972 f_1^7 g_2 + 15012 f_1^7 x_2 - 20736 f_1^6 f_3 + 756 f_1^5 g_2^2 - 3564 f_1^5 g_2 x_2 - 15624 f_1^5 x_2^2$$

$$- 89856 f_1^4 x_2 f_3 + 6912 f_1^4 g_2 f_3 + 816 f_1^3 x_2^3 + 2688 f_1^3 g_2 x_2^2 - 36 f_1^3 g_2^2 x_2 - 84 f_1^3 g_2^3$$
$$- 16848 f_1^3 f_3^2 + 34560 f_1^2 g_2 x_2 f_3 + 79200 f_1^2 x_2^2 f_3 + 5616 f_1 g_2 f_3^2 + 11232 f_1 x_2 f_3^2$$
$$- 52 f_1 g_2^3 x_2 - 312 f_1 g_2^2 x_2^2 + 720 f_1 g_2 x_2^3 + 2272 f_1 x_2^4 - 3360 g_2 x_2^2 f_3 - 6720 x_2^3 f_3).$$

Next, the form $f_{29/2,2b}$ is defined as $1/28991029248$ times

$$G_{13/2,2}(+486 f_1^8 - 12636 f_1^6 x_2 - 162 f_1^6 g_2 + 972 f_1^5 f_3 + 30888 f_1^4 x_2 x_2$$
$$+ 4104 f_1^4 g_2 x_2 + 7776 f_1^3 x_2 f_3 - 324 f_1^3 g_2 f_3 - 17424 f_1^2 x_2^3 - 7560 f_1^2 g_2 x_2^2$$
$$+ 486 f_1^2 f_3^2 - 2808 f_1 g_2 x_2 f_3 - 5616 f_1 x_2^2 f_3 + 768 g_2 x_2^3 + 1536 x_2^4 - 162 g_2 f_3^2$$
$$- 324 x_2 f_3^2) + H_{13/2,2}(-486 f_1^8 + 23976 f_1^6 x_2 + 5508 f_1^6 g_2 + 38880 f_1^5 f_3$$
$$- 17280 f_1^4 x_2^2 - 38880 f_1^4 g_2 x_2 - 756 f_1^4 g_2^2 - 38880 f_1^3 x_2 f_3 - 14904 f_1^3 g_2 f_3$$
$$+ 15840 f_1^2 x_2^3 + 19152 f_1^2 g_2 x_2^2 + 6912 f_1^2 g_2 g_2 x_2 - 72 f_1^2 g_2^3 + 8748 f_1^2 f_3^2$$
$$+ 2376 f_1 g_2^2 f_3 + 3888 f_1 g_2 x_2 f_3 - 12096 f_1 x_2^2 f_3 + 6 g_2^4 + 48 g_2^3 x_2 - 816 g_2^2 x_2^2$$
$$- 2112 g_2 x_2^3 - 672 x_2^4 - 324 g_2 f_3^2 + 1944 x_2 f_3^2) + F_{11/2,2}(-405 f_1^9 - 243 f_1^7 g_2$$
$$+ 3780 f_1^7 x_2 - 7209 f_1^6 f_3 + 189 f_1^5 g_2^2 - 918 f_1^5 g_2 x_2 - 3600 f_1^5 x_2^2 - 22302 f_1^4 x_2 f_3$$
$$+ 2457 f_1^4 g_2 f_3 - 144 f_1^3 x_2^3 + 588 f_1^3 g_2 x_2^2 - 21 f_1^3 g_2^3 - 6156 f_1^3 f_3^2 - 27 f_1^2 g_2^2 f_3$$
$$+ 8964 f_1^2 g_2 x_2 f_3 + 20916 f_1^2 x_2^2 f_3 + 2052 f_1 g_2 f_3^2 + 4104 f_1 x_2 f_3^2 - 14 f_1 g_2^3 x_2$$
$$- 84 f_1 g_2^2 x_2^2 + 216 f_1 g_2 x_2^3 + 656 f_1 x_2^4 + 3 g_2^3 f_3 + 18 g_2^2 x_2 f_3 - 924 g_2 x_2^2 f_3$$
$$- 1896 x_2^3 f_3).$$

We go on defining

$$F_{29/2,2a} = 442 f_{29/2,2a} + (-4207 + 15\sqrt{4657}) f_{29/2,2b}$$
$$F_{29/2,2b} = 442 f_{29/2,2a} + (-4207 - 15\sqrt{4657}) f_{29/2,2b}.$$

We define $f_{31/2,2a}$ as $1/1391569403904$ times

$$F_{11/2}(+162 f_1^{10} - 216 f_1^8 x_2 - 162 f_1^8 g_2 + 162 f_1^7 f_3 1152 f_1^6 x_2^2 - 180 f_1^6 g_2 x_2$$
$$+ 54 f_1^6 g_2^2 - 1188 f_1^5 x_2 f_3 - 162 f_1^5 g_2 f_3 - 2208 f_1^4 x_2^3 + 72 f_1^4 g_2 x_2^2 + 144 f_1^4 g_2^2 x_2$$
$$- 6 f_1^4 g_2^3 + 54 f_1^3 g_2^2 f_3 + 504 f_1^3 g_2 x_2 f_3 + 2520 f_1^3 x_2^2 f_3 - 20 f_1^2 g_2^3 x_2 - 120 f_1^2 g_2^2 x_2^2$$
$$+ 336 f_1^2 g_2 x_2^3 + 992 f_1^2 x_2^4 - 432 f_1^2 x_2 f_3^2 - 6 f_1 g_2^3 f_3 - 36 f_1 g_2^2 x_2 f_3 - 648 f_1 g_2 x_2^2 f_3$$
$$- 1200 f_1 x_2^3 f_3 + 144 g_2 x_2 f_3^2 + 288 x_2^2 f_3^2) + G_{13/2}(567 f_1^9 - 999 f_1^7 g_2 + 756 f_1^7 x_2$$
$$- 1377 f_1^6 f_3 + 405 f_1^5 g_2^2 + 594 f_1^5 g_2 x_2 - 3024 f_1^5 x_2^2 + 3618 f_1^4 x_2 f_3 + 513 f_1^4 g_2 f_3$$
$$+ 1584 f_1^3 x_2^3 + 540 f_1^3 g_2 x_2^2 - 216 f_1^3 g_2^2 x_2 - 45 f_1^3 g_2^3 - 648 f_1^3 f_3^2 - 27 f_1^2 g_2^2 f_3$$
$$- 972 f_1^2 g_2 x_2 f_3 - 1836 f_1^2 x_2^2 f_3 + 216 f_1 g_2 f_3^2 + 432 f_1 x_2 f_3^2 - 6 f_1 g_2^3 x_2$$
$$- 36 f_1 g_2^2 x_2^2 - 72 f_1 g_2 x_2^3 - 48 f_1 x_2^4 + 3 g_2^3 f_3 + 18 g_2^2 x_2 f_3 + 36 g_2 x_2^2 f_3 + 24 x_2^3 f_3)$$
$$+ H_{13/2}(-2754 f_1^9 + 594 f_1^7 g_2 + 13176 f_1^7 x_2 - 6642 f_1^6 f_3 + 378 f_1^5 g_2^2$$

$$- 4860 f_1^5 g_2 x_2 - 13824 f_1^5 x_2^2 + 11124 f_1^4 x_2 f_3 + 3618 f_1^4 g_2 f_3 + 2592 f_1^3 x_2^3$$

$$+ 2232 f_1^3 g_2 x_2^2 + 288 f_1^3 g_2^2 x_2 - 90 f_1^3 g_2^3 - 1296 f_1^3 f_3^2 - 486 f_1^2 g_2^2 f_3 - 2808 f_1^2 g_2 x_2 f_3$$

$$- 3672 f_1^2 x_2^2 f_3 + 432 f_1 g_2 f_3^2 + 864 f_1 x_2 f_3^2 - 12 f_1 g_2^3 x_2 + 216 f_1 g_2^2 x_2^2$$

$$+ 1008 f_1 g_2 x_2^3 + 1056 f_1 x_2^4 + 6 g_2^3 f_3 - 108 g_2^2 x_2 f_3 - 504 g_2 x_2^2 f_3 - 528 x_2^3 f_3).$$

Next we define $f_{31/2,2b}$ as $1/521838526464$ times

$$F_{11/2}(2673 f_1^{10} - 3078 f_1^8 x_2 - 2511 f_1^8 g_2 - 47628 f_1^7 f_3 - 15336 f_1^6 x_2^2 + 864 f_1^6 g_2 x_2$$

$$+ 702 f_1^6 g_2^2 + 1944 f_1^5 x_2 f_3 1 + 16524 f_1^5 g_2 f_3 - 144 f_1^4 x_2^3 + 6264 f_1^4 g_2 x_2^2$$

$$+ 540 f_1^4 g_2^2 x_2 - 18 f_1^4 g_2^3 + 77760 f_1^4 f_3^2 - 324 f_1^3 g_2^2 f_3 + 9072 f_1^3 g_2 x_2 f_3$$

$$+ 45360 f_1^3 x_2^2 f_3 - 15 f_1^2 g_2^4 - 192 f_1^2 g_2^3 x_2 - 360 f_1^2 g_2^2 x_2^2 + 3840 f_1^2 g_2 x_2^3$$

$$+ 7824 f_1^2 x_2^2 - 25920 f_1^2 g_2 f_3^2 - 51840 f_1^2 x_2 f_3^2 + 36 f_1 g_2^3 f_3 + 216 f_1 g_2^2 x_2 f_3$$

$$- 8208 f_1 g_2 x_2^2 f_3 - 16992 f_1 x_2^3 f_3 + 1 g_2^5 + 10 g_2^4 x_2 - 8 g_2^3 x_2^2 - 208 g_2^2 x_2^3 - 496 g_2 x_2^2$$

$$- 352 x_2^5) + G_{13/2}(10206 f_1^9 - 10206 f_1^7 g_2 - 29160 f_1^7 x_2 - 164754 f_1^6 f_3$$

$$+ 3402 f_1^5 g_2^2 + 22356 f_1^5 g_2 x_2 - 23328 f_1^5 x_2^2 + 251748 f_1^4 x_2 f_3 + 55890 f_1^4 g_2 f_3$$

$$+ 68256 f_1^3 x_2^3 + 1944 f_1^3 g_2 x_2^2 - 5184 f_1^3 g_2^2 x_2 - 378 f_1^3 g_2^3 + 23328 f_1^3 f_3^2$$

$$- 486 f_1^2 g_2^2 f_3 - 48600 f_1^2 g_2 x_2 f_3 - 87480 f_1^2 x_2^2 f_3 - 7776 f_1 g_2 f_3^2 - 15552 f_1 x_2 f_3^2$$

$$+ 324 f_1 g_2^3 x_2 + 1944 f_1 g_2^2 x_2^2 - 11664 f_1 g_2 x_2^3 - 28512 f_1 x_2^4 + 54 g_2^3 f_3 + 324 g_2^2 x_2 f_3$$

$$- 1944 g_2 x_2^2 f_3 - 4752 x_2^3 f_3) + H_{13/2}(-43740 f_1^9 + 18468 f_1^7 g_2 + 136080 f_1^7 x_2$$

$$+ 813564 f_1^6 f_3 + 2916 f_1^5 g_2^2 - 48600 f_1^5 g_2 x_2 - 15552 f_1^5 x_2^2 - 1417176 f_1^4 x_2 f_3$$

$$+ 96228 f_1^4 g_2 f_3 - 212544 f_1^3 x_2^3 + 159408 f_1^3 g_2 x_2^2 - 2592 f_1^3 g_2^2 x_2 - 1620 f_1^3 g_2^3$$

$$- 1819584 f_1^3 f_3^2 + 1620 f_1^2 g_2^2 f_3 - 71280 f_1^2 g_2 x_2 f_3 1483920 f_1^2 x_2^2 f_3 - 15552 f_1 g_2 f_3^2$$

$$+ 342144 f_1 x_2 f_3^2 + 72 f_1 g_2^4 + 1224 f_1 g_2^3 x_2 - 13392 f_1 g_2^2 x_2^2 - 97056 f_1 g_2 x_2^3$$

$$+ 614592 f_1 x_2^4 + 108 g_2^3 f_3 - 1944 g_2^2 x_2 f_3 - 14256 g_2 x_2^2 f_3 + 104544 x_2^3 f_3).$$

We define

$$F_{31/2,2a} = (-144 + 48\sqrt{99661}) f_{31/2,2a} + f_{31/2,2b}$$

$$F_{31/2,2b} = (-144 - 48\sqrt{99661}) f_{31/2,2a} + f_{31/2,2b}$$

We define $F_{17/2,4}$ to be $1/884736$ times

$$(F_{11/2,4a}(36 f_1^3 - 12 f_1 g_2 - 24 f_1 x_2) + F_{11/2,4b}(-144 f_1^3 + 48 f_1 g_2 + 96 f_1 x_2)$$

$$+ F_{11/2,4c}(1728 f_1^3 - 576 f_1 g_2 - 1152 f_1 x_2 + 1728 f_3) + F_{9/2,4a}(27 f_1^4 - 270 f_1 f_3$$

$$- 72 f_1^2 g_2 + 84 f_1^2 x_2 + 3 g_2^2 + 44 g_2 x_2 + 76 x_2^2) + F_{9/2,4b}(-342 f_1^4 - 72 f_1 f_3$$

$$+ 156 f_1^2 g_2 + 216 f_1^2 x_2 - 14 g_2^2 - 24 g_2 x_2 + 8 x_2^2) + F_{9/2,4c}(-63 f_1^4 - 90 f_1 f_3$$

$$+ 108 f_1^2 x_2 + g_2^2 + 4 g_2 x_2 + 4 x_2^2))$$

We set $F_{19/2,4} := f_{19/2,4}/589824$, where $f_{19/2,4}$ is

$$F_{11/2,4c}(-1296f_1^4 + 96f_1^2g_2 - 2112f_1^2x_2 - 16g_2^2 - 64g_2x_2 - 64x_2^2 + 3456f_1f_3)$$
$$+ F_{9/2,4a}(-15f_1^5 - 70f_1^3g_2364f_1^3x_2 + 9f_1g_2^2 + 36f_1g_2x_2 - 396f_1x_2^2 + 96f_1^2f_3$$
$$+ 16g_2f_3 - 40x_2f_3) + F_{9/2,4b}(+150f_1^5 + 44f_1^3g_2 - 440f_1^3x_2 - 10f_1g_2^2$$
$$- 40f_1g_2x_2 + 248f_1x_2^2 + 336f_1^2f_3 + 16g_2f_3 - 304x_2f_3) + F_{9/2,4c}(-21f_1^5$$
$$- 18f_1^3g_2 + 132f_1^3x_2 + 3f_1g_2^2 + 12f_1g_2x_2 - 132f_1x_2^2 + 48f_1^2f_3 - 24x_2f_3)$$

and finally $F_{21/2,4} := f_{21/2,4}/10616832$, with $f_{21/2,4}$ given by

$$F_{11/2,4b}(-162f_1^5 + 432f_1^3x_2 + 54f_1^3g_2 - 162f_1^2f_3 - 216f_1x_2^2 - 108f_1g_2x_2$$
$$+ 108x_2f_3 + 54g_2f_3) + F_{11/2,4c}(+7776f_1^5 - 22464f_1^3x_2 + 432f_1^3g_2 + 9072f_1^2f_3$$
$$+ 11520f_1x_2^2 + 288f_1g_2x_2 - 144f_1g_2^2 - 6048x_2f_3 - 432g_2f_3)F_{9/2,4a}(+459f_1^6$$
$$+ 486f_1^4x_2 - 729f_1^4g_2 - 810f_1^3f_3 - 972f_1^2x_2^2 + 540f_1^2g_2x_2 + 108f_1^2g_2^2$$
$$+ 1188f_1x_2f_3 - 54f_1g_2f_3 + 8x_2^3 + 12g_2x_2^2 + 6g_2^2x_2 + 1g_2^3 - 243f_3^2)$$
$$+ F_{9/2,4b}(-1269f_1^6 - 882f_1^4x_2 + 1395f_1^4g_2 + 1944f_1^3f_3 + 3444f_1^2x_2^2$$
$$- 1092f_1^2g_2x_2 - 219f_1^2g_2^2 - 4752f_1x_2f_3 + 108f_1g_2f_3 + 8x_2^3 + 12g_2x_2^2 + 6g_2^2x_2$$
$$+ g_2^3 + 1296f_3^2) + F_{9/2,4c}(+108f_1^6 + 288f_1^4x_2 - 234f_1^4g_2 - 324f_1^3f_3 - 384f_1^2x_2^2$$
$$+ 156f_1^2g_2x_2 + 39f_1^2g_2^2 + 432f_1x_2f_3 - 81f_3^2)$$

3.4 Numerical Examples of L-functions

In this section, we give our results on the Euler 2-factors and 3-factors of the common eigenforms listed above. We verified that they all support our conjecture. We shall explain how to calculate these examples in section 4.

$S_{5,18}(\Gamma_2)$ and $S_{23/2,2}^+(\Gamma_0(4), \psi)$.

$$H_2(s, F_{23/2,2}) = H_2(s, F_{5,18})$$
$$= 1 + 2880T - 26378240T^2 + 2880 \cdot 2^{25}T^3 + 2^{50}T^4$$
$$H_3(s, F_{23/2,2}) = H_3(s, F_{5,18})$$
$$= 1 + 538970T + 2046223302870T^2 + 539870 \cdot 3^{25}T^3 + 3^{50}T^4$$

$S_{5,20}(\Gamma_2)$ and $S_{25/2,2}^+(\Gamma_0(4), \psi)$.

$$H_2(s, F_{25/2,2}) = H_2(s, F_{5,20})$$
$$= 1 + 240T - 29204480T^2 + 240 \cdot 2^{27}T^3 + 2^{54}T^4$$
$$H_3(s, F_{25/2,2}) = H_3(s, F_{5,20})$$
$$= 1 - 1645560T - 2281745279610T^2 - 1645560 \cdot 3^{27}T^3 + 3^{54}T^4$$

$S_{5,24}(\Gamma_2)$ and $S^+_{29/2,2}(\Gamma_0(4), \psi)$.

$$H_2(s, F_{29/2,a}) = 1 - (-8040 + 600\sqrt{4657})T + (742973440 - 1843200\sqrt{4657})T^2$$
$$- (-8040 + 600\sqrt{4657})2^{31}T^3 + 2^{62}T^4$$

$$H_2(s, F_{29/2,b}) = 1 - (-8040 - 600\sqrt{4657})T + (742973440 + 1843200\sqrt{4657})T^2$$
$$- (-8040 - 600\sqrt{4657})2^{31}T^3 + 2^{62}T^4$$

$$H_3(s, F_{29/2,a}) = 1 - (4187160 - 194400\sqrt{4657})T + 196830(65242301$$
$$+ 4016320\sqrt{4657})T^2 - (4187160 - 194400\sqrt{4657})3^{31}T^3$$
$$+ 3^{62}T^4$$

$$H_3(s, F_{29/2,b}) = 1 - (4187160 + 194400\sqrt{4657})T + 196830(65242301$$
$$- 4016320\sqrt{4657})T^2 - (4187160 + 194400\sqrt{4657})3^{31}T^3$$
$$+ 3^{62}T^4$$

$S_{5,26}(\Gamma_2)$ and $S^+_{31/2,2}(\Gamma_0(4), \psi)$.

$$H_2(s, F_{31/2,a}) = 1 - (27072 + 192\sqrt{99661})T + (4836327424$$
$$- 9732096\sqrt{99661})T^2 - (27072 + 192\sqrt{99661}) \cdot 2^{33}T^3$$
$$+ 2^{66}T^4$$

$$H_2(s, F_{31/2,b}) = 1 - (27072 - 192\sqrt{99661})T + (4836327424$$
$$+ 9732096\sqrt{99661})T^2 - (27072 - 192\sqrt{99661}) \cdot 2^{33}T^3$$
$$+ 2^{66}T^4$$

$$H_3(s, F_{31/2,a}) = H_3(s, F_{5,26a})$$
$$= 1 - (-9567144 - 59904\sqrt{99661})T + (-268954900275114$$
$$+ 16134754093056\sqrt{99661})T^2 - (-9567144$$
$$- 59904\sqrt{99661}) \cdot 3^{33}T^3 + 3^{66}T^4$$

$$H_3(s, F_{31/2,b}) = H_3(s, F_{5,26b})$$
$$= 1 - (-9567144 + 59904\sqrt{99661})T + (-268954900275114$$
$$- 16134754093056\sqrt{99661})T^2 - (-9567144$$
$$+ 59904\sqrt{99661}) \cdot 3^{33}T^3 + 3^{66}T^4$$

$S_{7,12}(\Gamma_2)$ and $S^+_{17/2,4}(\Gamma_0(4), \psi)$.

$$H_2(s, F_{17/2,4}) = H_2(s, f_{7,12})$$
$$= 1 + 480T + 5754880T^2 + 480 \cdot 3^{26}T^3 + 3^{52}T^4$$

$$H_3(s, F_{17/2,4}) = H_3(s, f_{7,12})$$
$$= 1 + 73080T - 97880212890T^2 + 73080 \cdot 3^{23}T^3 + 3^{46}T^4$$

$S_{7,14}(\Gamma_2)$ and $S_{19/2,4}^+(\Gamma_0(4), \psi)$.

$$H_2(s, F_{19/2,4}) = 1 + 3696T + 18116608T^2 + 3696 \cdot 2^{25}T^3 + 2^{50}T^4$$

$$H_3(s, F_{19/2,4}) = 1 - 511272T + 377292286422T^2 - 511272 \cdot 3^{25}T^3 \cdot 3^{50}T^4$$

$S_{7,16}(\Gamma_2)$ and $S_{21/2,4}^+(\Gamma_0(4), \psi)$.

$$H_2(s, F_{21/2,4}) = 1 - 13440T + 166912000T^2 - 13440 \cdot 2^{27}T^3 + 2^{54}T^4$$

$$H_3(s, F_{21/2,4}) = 1 + 1487160T - 2487701893050T^2 + 1487160 \cdot 3^{27}T^3 + 3^{54}T^4$$

4 How to calculate eigenvalues

In this section, we would like to show how to calculate the Euler factors of section 3 from the Fourier coefficients of the corresponding Siegel modular forms.

4.1 Integral weight

For the theory of vector valued forms of integral weight, we refer to Arakawa [1]. For a common eigenform $F \in A_{k,j}(\Gamma_2)$, we write the Fourier expansion as

$$F(Z) = \sum_T A(T)e(\mathrm{Tr}(TZ))$$

where T runs over 2×2 half-integral positive semi-definite symmetric matrices and $A(T) \in \mathbb{C}^{j+1}$. By automorphy, we have $A(UT^tU) = \rho_{k,j}(U)A(T)$ for any $U \in GL_2(\mathbb{Z})$. For the Hecke operator $T(p^\delta)$, denote by $A(p^\delta; T)$ the Fourier coefficient at T of $T(p^\delta)F$. For a fixed $T = \begin{pmatrix} a & b/2 \\ b/2 & c \end{pmatrix}$, and for any $U \in GL_2(\mathbb{Z})$, we define a_U, b_U, c_U by the relation ${}^tUTU = \begin{pmatrix} a_U & b_U/2 \\ b_U/2 & c_U \end{pmatrix}$, and for any non-negative integers α, β, we put

$$d_{\alpha,\beta} = \begin{pmatrix} p^\alpha & 0 \\ 0 & p^{\alpha+\beta} \end{pmatrix}.$$

Let $R(p^\beta)$ denote a complete set of representatives of $SL_2(\mathbb{Z})/{}^t\Gamma_0^{(1)}(p^\beta)$. For example, we can take

$$R(p) = \left\{ \begin{pmatrix} 1 & x \\ 0 & 1 \end{pmatrix}, \begin{pmatrix} 0 & 1 \\ -1 & 0 \end{pmatrix}; \ x \in \mathbb{Z}/p\mathbb{Z} \right\},$$

$$R(p^2) = \left\{ \begin{pmatrix} 1 & x \\ 0 & 1 \end{pmatrix}, \begin{pmatrix} py & 1 \\ -1 & 0 \end{pmatrix}; \ x \in \mathbb{Z}/p^2\mathbb{Z}, \ y \in \mathbb{Z}/p\mathbb{Z} \right\}$$

Here, for simplicity, we write $\rho_j = \mathrm{Sym}_j$. Then we have

$$A(p^\delta; T) = \sum_{\alpha+\beta+\gamma=\delta} p^{\beta(k+j-2)+\gamma(2k+j-3)}$$

$$\sum_{\substack{U \in R(p^\beta), \\ a_U \equiv 0 \, p^{\beta+\gamma}, \\ b_U \equiv c_U \equiv 0 \bmod p^\gamma}} \rho_j(d_{0,\beta}U)^{-1} A\left(p^\alpha \begin{pmatrix} a_U \, p^{-\beta-\gamma} & b_U \, p^{-\gamma}/2 \\ b_U \, p^{-\gamma}/2 & c_U \, p^{\beta-\gamma} \end{pmatrix}\right)$$

For practical use, we need more explicit examples which are given below. Here we write $A(p^\delta, T) = A(p^\delta, (a, c, b))$ for $T = \begin{pmatrix} a & b/2 \\ b/2 & c \end{pmatrix}$ for simplicity.

$A(2, (1, 1, 1)) = A(2, 2, 2)$,

$A(4, (1, 1, 1)) = A(4, 4, 4)$,

$A(2, (2, 2, 2)) = A(4, 4, 4) + 2^{k+j-2}\left(\rho_j \begin{pmatrix} 1 & 0 \\ 0 & 2^{-1} \end{pmatrix} A(1, 4, 2) + \rho_j \begin{pmatrix} 1 & 1 \\ 0 & 2 \end{pmatrix}^{-1} A(3, 4, 6)\right.$

$\left. + \rho_j \begin{pmatrix} 0 & 1 \\ -2 & 0 \end{pmatrix}^{-1} A(1, 4, -2)\right) + 2^{2k+j-3} A(1, 1, 1)$

$= A(4, 4, 4) + 2^{k-2}\left(\rho_j \begin{pmatrix} 1 & 1 \\ -1 & 1 \end{pmatrix} A(1, 3, 0) + \rho_j \begin{pmatrix} 2 & 0 \\ 1 & 1 \end{pmatrix} A(1, 3, 0)\right.$

$\left. + \rho_j \begin{pmatrix} 1 & -1 \\ 2 & 0 \end{pmatrix} A(1, 3, 0)\right) + 2^{2k+j-3} A(1, 1, 1)$,

$A(2, (1, 1, 0)) = A(2, 2, 0) + 2^{k+j-2}\rho_j \begin{pmatrix} 1 & 1 \\ 0 & 2 \end{pmatrix}^{-1} A(1, 2, 2)$

$= A(2, 2, 0) + 2^{k-2}\rho_j \begin{pmatrix} 1 & -1 \\ 1 & 1 \end{pmatrix} A(1, 1, 0)$,

$A(4, (1, 1, 0)) = A(4, 4, 0) + 2^{k-2}\rho_j \begin{pmatrix} 1 & -1 \\ 1 & 1 \end{pmatrix} A(2, 2, 0)$,

$A(2, (2, 2, 0)) = A(4, 4, 0) + 2^{k-2}\rho_j \begin{pmatrix} 1 & -1 \\ 1 & 1 \end{pmatrix} A(2, 2, 0) + 2^{k-2}\left(\rho_j \begin{pmatrix} 2 & 0 \\ 0 & 1 \end{pmatrix} A(1, 4, 0)\right.$

$\left. + \rho_j \begin{pmatrix} 0 & -1 \\ 2 & 0 \end{pmatrix} A(1, 4, 0)\right) + 2^{2k+j-3} A(1, 1, 0)$,

$A(3, (1, 1, 0)) = A(3, 3, 0)$,

$A(9, (1, 1, 0)) = A(9, 9, 0)$,

$A(3, (3, 3, 0)) = A(9, 9, 0) + 3^{k-2}\left(\rho_j \begin{pmatrix} 3 & 0 \\ 0 & 1 \end{pmatrix} A(1, 9, 0) + \rho_j \begin{pmatrix} 2 & -1 \\ 1 & 1 \end{pmatrix} A(2, 5, 2)\right.$

$\left. + (-1)^k \rho_j \begin{pmatrix} -2 & 1 \\ 1 & 1 \end{pmatrix} A(2, 5, 2) + \rho_j \begin{pmatrix} 0 & -1 \\ 3 & 0 \end{pmatrix} A(1, 9, 0)\right)$

$+ 3^{2k+j-3} A(1, 1, 0)$,

$A(3, (1, 1, 1)) = A(3, 3, 3) + 3^{k+j-2}\rho_j \begin{pmatrix} 1 & -1/3 \\ 0 & 1/3 \end{pmatrix} A(1, 3, 3)$

$= A(3, 3, 3) + 3^{k-2}\rho_j \begin{pmatrix} 2 & -1 \\ 1 & 1 \end{pmatrix} A(1, 1, 1)$,

$A(9, (1, 1, 1)) = A(9, 9, 9) + 3^{k-2}\rho_j \begin{pmatrix} 2 & -1 \\ 1 & 1 \end{pmatrix} A(3, 3, 3)$,

$$A(3, (3, 3, 3)) = A(9, 9, 9) + 3^{k-2}\rho_j \begin{pmatrix} 2 & -1 \\ 1 & 1 \end{pmatrix} A(3, 3, 3) + 3^{k-2}\Big(\rho_j \begin{pmatrix} 3 & 0 \\ 1 & 1 \end{pmatrix} A(1, 7, 1)$$

$$+ (-1)^k \rho_j \begin{pmatrix} -2 & 1 \\ 1 & 1 \end{pmatrix} A(1, 7, 1) + \rho_j \begin{pmatrix} 1 & -1 \\ 3 & 0 \end{pmatrix} A(1, 7, -1)\Big)$$

$$+ 3^{2k+j-3} A(1, 1, 1).$$

4.2 Half-integral weight

This section is essentially due to Zhuravlev [21], [22]. The scalar valued case is explained in detail in [7]. For $h \in S^+_{k-1/2,j}(\Gamma_0(4), \psi)$, we write the Fourier expansion as

$$h(Z) = \sum_T A(T)e(\mathrm{Tr}(TZ)).$$

When p is odd, we use the notation $A(T_i(p); (a, c, b)) = A(T_i(p), T)$ for the Fourier coefficients of $T_i(p)h$ for $i = 1, 2$. Then we have

$$A(T_1(p); T) = \alpha_{1,1,0}(T) + \alpha_{1,1,1}(T) + \alpha_{1,2,0}(T),$$

where

$$\alpha_{1,1,0}(T) = p^{2k+j-4}\psi(p) \sum_{U \in R(p)} \rho_j(U)^{-1}\rho_j \begin{pmatrix} p & 0 \\ 0 & 1 \end{pmatrix} A\left(\begin{pmatrix} p^{-1} & 0 \\ 0 & 1 \end{pmatrix} U T^t U \begin{pmatrix} p^{-1} & 0 \\ 0 & 1 \end{pmatrix}\right),$$

and

$$\alpha_{1,1,1}(T) = \psi(p) p^j \rho_j(U^{-1})\rho_j \begin{pmatrix} 1 & 0 \\ 0 & p^{-1} \end{pmatrix} \sum_{U \in R(p)} A\left(\begin{pmatrix} 1 & 0 \\ 0 & p \end{pmatrix} U T^t U \begin{pmatrix} 1 & 0 \\ 0 & p \end{pmatrix}\right)$$

and

$$\alpha_{1,2,0}(T) = \begin{cases} \left(\dfrac{(-1)^{k+1}a}{p}\right) p^{k+j-2}A(T) & \text{if } p \nmid a \text{ and } p \mid \det(2T) \\[2mm] \left(\dfrac{(-1)^{k+1}c}{p}\right) p^{k+j-2}A(T) & \text{if } p \mid a \text{ and } p \mid \det(2T) \\[2mm] 0 & \text{otherwise.} \end{cases}$$

Moreover, we have

$$A(T_2(p); T) = \sum_{0 \le i+j \le 2} \alpha_{2,i,j}(T),$$

where

$$\alpha_{2,0,0}(T) = p^{4k+3j-8}A(p^{-2}T)$$

$$\alpha_{2,0,1}(T) = p^{2k+2j-5} \sum_{U \in R(p^2)} \rho_j(U^{-1})\rho_j \begin{pmatrix} p & 0 \\ 0 & p^{-1} \end{pmatrix} A\left(\begin{pmatrix} p^{-1} & 0 \\ 0 & p \end{pmatrix} U T^t U \begin{pmatrix} p^{-1} & 0 \\ 0 & p \end{pmatrix}\right)$$

$$\alpha_{2,0,2}(T) = p^j A(p^2 T)$$

$$\alpha_{2,1,0}(T) = p^{3k+2j-7}\psi(p)^k \sum_{U \in R(p)} \left(\frac{m}{p}\right) \rho_j(U^{-1})\left(\left(\begin{smallmatrix} p & 0 \\ 0 & 1 \end{smallmatrix}\right)\right)$$
$$\times A\left(\left(\begin{smallmatrix} p^{-1} & 0 \\ 0 & 1 \end{smallmatrix}\right)\right)UT^tU\left(\left(\begin{smallmatrix} p^{-1} & 0 \\ 0 & 1 \end{smallmatrix}\right)\right),$$

where $UT^tU = \left(\begin{smallmatrix} * & * \\ * & m \end{smallmatrix}\right)$. Further,

$$\alpha_{2,1,1}(T) = p^{k+2j-3}\psi(p)^k \sum_{U \in R(p)} \left(\frac{m}{p}\right) \rho_j(U^{-1})\left(\left(\begin{smallmatrix} 1 & 0 \\ 0 & p^{-1} \end{smallmatrix}\right)\right)$$
$$\times A\left(\left(\begin{smallmatrix} 1 & 0 \\ 0 & p \end{smallmatrix}\right)\right)UT^tU\left(\left(\begin{smallmatrix} 1 & 0 \\ 0 & p \end{smallmatrix}\right)\right),$$

where $UT^tU = \left(\begin{smallmatrix} m & * \\ * & * \end{smallmatrix}\right)$, and

$$\alpha_{2,2,0}(T) = \begin{cases} -p^{2k+2j-6}A(T) & \text{if } p \nmid \det(2T) \\ (p-1)p^{2k+2j-6}A(T) & \text{if } p \mid \det(2T). \end{cases}$$

When $p = 2$, we denote by $T_i^*(2)$ $(i = 1, 2)$ the Hecke operator obtained as the pullback of the Hecke operators on Jacobi forms at 2 (see section 5). This amounts to take $\psi(p)T_i(p)$ for odd p. We must modify the above formula of Fourier coefficients $A(T_i^*(2), T)$ of $T_i^*(p)F$, but this can be done in the same way as in [7] p.516–517. To be more precise, we omit $\psi(p)$ in $\alpha_{1,1,0}(T)$ and $\alpha_{1,1,1}(T)$, replace $k + 1$ by k in $\left(\frac{(-1)^{k+1}a}{p}\right)$ and $\left(\frac{(-1)^{k+1}c}{p}\right)$ in $\alpha_{1,2,0}(T)$, and replace everywhere the condition $p \mid \det(2T)$ or $p \nmid \det(2T)$ by $8 \mid \det(T)$ or $8 \nmid \det(T)$, respectively, and interpret the symbol $\left(\frac{x}{p}\right)$ as 0, 1 or -1 for $x \equiv 0$ mod 4, 1 mod 8, or 5 mod 8, respectively.

We have the following formulas.

$A(T_1(3), (3, 3, 2))$
$$= \psi(3)\left(\rho_j \left(\begin{smallmatrix} 0 & -1 \\ 3 & 0 \end{smallmatrix}\right) A(3, 27, -6) + \rho_j \left(\begin{smallmatrix} 3 & 0 \\ 0 & 1 \end{smallmatrix}\right) A(3, 27, 6)\right.$$
$$\left.+3^j \rho_j \left(\begin{smallmatrix} 1 & -1/3 \\ 0 & 1/3 \end{smallmatrix}\right) A(8, 27, 24) + 3^j \rho_j \left(\begin{smallmatrix} 1 & -2/3 \\ 0 & 1/3 \end{smallmatrix}\right) A(19, 27, 42)\right)$$
$$= \psi(3)\left(\rho_j \left(\begin{smallmatrix} 2 & -1 \\ 1 & 1 \end{smallmatrix}\right) A(8, 11, 8) + \rho_j \left(\begin{smallmatrix} 3 & 0 \\ 0 & 1 \end{smallmatrix}\right) A(3, 27, 6)\right.$$
$$\left.+\rho_j \left(\begin{smallmatrix} 0 & -1 \\ 3 & 0 \end{smallmatrix}\right) A(3, 27, -6) + \rho_j \left(\begin{smallmatrix} 1 & -2 \\ 1 & 1 \end{smallmatrix}\right) A(19, 4, 4)\right)$$

$A(T_2(3), (3, 3, 2))$
$$= 3^j A(27, 27, 18) + 3^{2k+j-5}\left(\rho_j \left(\begin{smallmatrix} 0 & -1 \\ 9 & 3 \end{smallmatrix}\right) A(4, 27, -20)\right.$$
$$\left.+\rho_j \left(\begin{smallmatrix} 9 & -3 \\ 0 & 1 \end{smallmatrix}\right) A(4, 27, 20)\right) + 3^{k+j-3}\psi(3)^k\left(-\rho_j \left(\begin{smallmatrix} 3 & -1 \\ 0 & 1 \end{smallmatrix}\right) A(8, 27, 24)\right.$$
$$\left.+\rho_j \left(\begin{smallmatrix} 3 & -2 \\ 0 & 1 \end{smallmatrix}\right) A(19, 27, 21)\right) - 3^{2k+2j-6}A(3, 3, 2)$$

$$= 3^j A(27, 27, 18) + 3^{2k+j-5} \left(\rho_j \left(\begin{smallmatrix} 3 & -3 \\ 2 & 1 \end{smallmatrix} \right) A(4, 3, 4) \right.$$
$$+ \rho_j \left(\begin{smallmatrix} 2 & -1 \\ 3 & 3 \end{smallmatrix} \right) A(4, 3, -4) \right) + 3^{k+j-3} \psi(3)^k \left(-\rho_j \left(\begin{smallmatrix} 2 & -1 \\ 1 & 1 \end{smallmatrix} \right) A(8, 11, 8) \right.$$
$$+ \rho_j \left(\begin{smallmatrix} 1 & -2 \\ 1 & 1 \end{smallmatrix} \right) A(19, 4, 4) \right) - 3^{2k+2j-6} A(3, 3, 2)$$

$A(T_1(3), (3, 4, 0))$

$$= \psi(3) \left(\rho \left(\begin{smallmatrix} 3 & 0 \\ 0 & 1 \end{smallmatrix} \right) A(3, 36, 0) + \rho_j \left(\begin{smallmatrix} 3 & -1 \\ 0 & 1 \end{smallmatrix} \right) A(7, 36, 24) \right.$$
$$+ \rho_j \left(\begin{smallmatrix} 3 & -2 \\ 0 & 1 \end{smallmatrix} \right) A(19, 36, 48) + \rho_j \left(\begin{smallmatrix} 0 & -1 \\ 3 & 0 \end{smallmatrix} \right) A(4, 27, 0) \right)$$
$$+ 3^{k+j-2} A(3, 4, 0)$$

$A(T_2(3), (3, 4, 0))$

$$= 3^j A(27, 36, 0) + 3^{k+j-3} \psi(3)^k \left(\rho_j \left(\begin{smallmatrix} 3 & -1 \\ 0 & 1 \end{smallmatrix} \right) A(7, 36, 24) \right.$$
$$+ \rho_j \left(\begin{smallmatrix} 3 & -2 \\ 0 & 1 \end{smallmatrix} \right) A(19, 36, 48) + \rho_j \left(\begin{smallmatrix} 0 & -1 \\ 0 & 3 \end{smallmatrix} \right) A(4, 27, 0) \right)$$
$$+ 2 \cdot 3^{2k+2j-6} A(3, 4, 0)$$

$A(T_1(3), (1, 4, 0))$

$$= \rho_j \left(\begin{smallmatrix} 3 & 0 \\ 0 & 1 \end{smallmatrix} \right) A(1, 36, 0) + 3^j \rho_j \left(\begin{smallmatrix} 1 & -1/3 \\ 0 & 1/3 \end{smallmatrix} \right) A(5, 36, 24)$$
$$+ 3^j \rho_j \left(\begin{smallmatrix} 1 & -2/3 \\ 0 & 1/3 \end{smallmatrix} \right) A(17, 36, 48) + 3^j \rho_j \left(\begin{smallmatrix} 0 & -1/3 \\ 1 & 0 \end{smallmatrix} \right) A(4, 9, 0)$$
$$= \rho_j \left(\begin{smallmatrix} 3 & 0 \\ 0 & 1 \end{smallmatrix} \right) A(1, 36, 0) + \rho_j \left(\begin{smallmatrix} 2 & -1 \\ 1 & 1 \end{smallmatrix} \right) A(5, 17, 14)$$
$$+ \rho_j \left(\begin{smallmatrix} 1 & -2 \\ 1 & 1 \end{smallmatrix} \right) A(17, 5, 14) + \rho_j \left(\begin{smallmatrix} 0 & -1 \\ 3 & 0 \end{smallmatrix} \right) A(4, 9, 0)$$

$A(T_2(3), (1, 4, 0))$

$$= 3^j A(9, 36, 0) + 3^{k+j-3} \psi(3)^k \left(\rho_j \left(\begin{smallmatrix} 3 & 0 \\ 0 & 1 \end{smallmatrix} \right) A(1, 36, 0) \right.$$
$$- \rho_j \left(\begin{smallmatrix} 3 & -1 \\ 0 & 1 \end{smallmatrix} \right) A(5, 36, 24) - \rho_j \left(\begin{smallmatrix} 3 & -2 \\ 0 & 1 \end{smallmatrix} \right) A(17, 36, 48)$$
$$+ \rho_j \left(\begin{smallmatrix} 0 & -1 \\ 3 & 0 \end{smallmatrix} \right) A(4, 9, 0) \right) - 3^{2k+2j-6} A(1, 4, 0)$$

$A(T_1^*(2), (1, 4, 0))$

$$= \rho_j \left(\begin{smallmatrix} 2 & 0 \\ 0 & 1 \end{smallmatrix} \right) A(1, 16, 0) + \rho_j \left(\begin{smallmatrix} 2 & -1 \\ 0 & 1 \end{smallmatrix} \right) A(5, 16, 16)$$
$$+ \rho_j \left(\begin{smallmatrix} 0 & -1 \\ 2 & 0 \end{smallmatrix} \right) A(4, 4, 0) + 2^{2k+j-4} \rho_j \left(\begin{smallmatrix} 0 & -1 \\ 2 & 0 \end{smallmatrix} \right) A(1, 1, 0)$$

$A(T_1^*(2), (4, 5, 0))$

$$= \rho_j \left(\begin{smallmatrix} 2 & 0 \\ 0 & 1 \end{smallmatrix} \right) A(4, 20, 0) + \rho_j \left(\begin{smallmatrix} 2 & -1 \\ 0 & 1 \end{smallmatrix} \right) A(9, 20, 20)$$
$$+ \rho_j \left(\begin{smallmatrix} 0 & -1 \\ 2 & 0 \end{smallmatrix} \right) A(5, 16, 0) + 2^{2k+j-4} \rho_j \left(\begin{smallmatrix} 2 & 0 \\ 0 & 1 \end{smallmatrix} \right) A(1, 5, 0)$$

$A(T_2^*(2), (1, 4, 0))$

$$= 2^{2k+j-5} \left(\rho_j \left(\begin{smallmatrix} 0 & -1 \\ 4 & 0 \end{smallmatrix} \right) A(1, 4, 0) + \rho_j \left(\begin{smallmatrix} 1 & -1 \\ 2 & 2 \end{smallmatrix} \right) A(2, 2, 0) \right) + 2^j A(4, 16, 0)$$

$$+2^{3k+2j-7}e\big(((-1)^{k+1}+1)/8\big)\rho_j \left(\begin{smallmatrix}0 & -1\\2 & 0\end{smallmatrix}\right) A(1,1,0)$$

$$+2^{k+j-3}\Big(e\big(((-1)^{k+1}+1)/8\big)\rho_j \left(\begin{smallmatrix}2 & 0\\0 & 1\end{smallmatrix}\right) A(1,16,0)$$

$$+e\big((5(-1)^{k+1}+1)/8\big)\rho_j \left(\begin{smallmatrix}2 & -1\\0 & 1\end{smallmatrix}\right) A(5,16,16)\Big) - 2^{2k+2j-6}A(1,4,0)$$

$A(T_2^*(2),(4,5,0))$

$$= \quad 2^{2k+j-5}\Big(\rho_j \left(\begin{smallmatrix}4 & 0\\0 & 1\end{smallmatrix}\right) A(1,20,0) + \rho_j \left(\begin{smallmatrix}4 & -2\\0 & 1\end{smallmatrix}\right) A(6,20,20)\Big)$$

$$+2^j A(16,20,0) + 2^{3k+2j-7}e\big(((-1)^{k+1}5+1)/8\big)\rho_j \left(\begin{smallmatrix}2 & 0\\0 & 1\end{smallmatrix}\right) A(1,5,0)$$

$$+2^{k+j-3}(e\big(((-1)^{k+1}9+1)/8\big)\rho_j \left(\begin{smallmatrix}2 & -1\\0 & 1\end{smallmatrix}\right) A(9,20,20)$$

$$+\rho_j \left(\begin{smallmatrix}0 & -1\\2 & 0\end{smallmatrix}\right) A(5,16,0)) - 2^{2k+2j-6}A(4,5,0)$$

In the above, for $F \in S_{k-1/2}^+(\Gamma_0(4).\psi)$, by definition of the plus space, we have $A(1,4,0)=0$ if k is odd. So we can always assume that k is even.

$A(T_1^*(2),(3,4,0))$

$$= \quad 2^{2k+j-4}\rho_j \left(\begin{smallmatrix}0 & -1\\2 & 0\end{smallmatrix}\right) A(1,3,0) + \rho_j \left(\begin{smallmatrix}2 & 0\\0 & 1\end{smallmatrix}\right) A(3,16,0)$$

$$+\rho_j \left(\begin{smallmatrix}2 & -1\\0 & 1\end{smallmatrix}\right) A(7,16,16) + \rho_j \left(\begin{smallmatrix}0 & -1\\2 & 0\end{smallmatrix}\right) A(4,12,0)$$

$A(T_2^*(2),(3,4,0))$

$$= \quad 2^{2k+j-5}\Big(\rho_j \left(\begin{smallmatrix}0 & -1\\4 & 0\end{smallmatrix}\right) A(1,12,0) + (-1)\rho_j \left(\begin{smallmatrix}0 & 1\\4 & -2\end{smallmatrix}\right) A(4,12,12)\Big)$$

$$+2^j A(12,16,0) + 2^{3k+2j-7}e\big(((-1)^{k+1}3+1)/8\big)\rho_j \left(\begin{smallmatrix}0 & -1\\2 & 0\end{smallmatrix}\right) A(1,3,0)$$

$$+2^{k+j-3}\Big(e\big(((-1)^{k+1}3+1)/8\big)\rho_j \left(\begin{smallmatrix}2 & 0\\0 & 1\end{smallmatrix}\right) A(3,16,0)$$

$$+e\big(((-1)^{k+1}7+1)/8\big)\rho_j \left(\begin{smallmatrix}2 & -1\\0 & 1\end{smallmatrix}\right) A(7,16,16)\Big) - 2^{2k+2j-6}A(3,4,0)$$

$A(T_1^*(2),(3,3,2))$

$$= \quad 2^{2k+j-4}\rho_j \left(\begin{smallmatrix}2 & -1\\0 & 1\end{smallmatrix}\right) A(2,3,4) + \rho_j \left(\begin{smallmatrix}2 & 0\\0 & 1\end{smallmatrix}\right) A(3,12,4)$$

$$+\rho_j \left(\begin{smallmatrix}2 & -1\\0 & 1\end{smallmatrix}\right) A(8,12,16) + (-1)\rho_j \left(\begin{smallmatrix}0 & 1\\2 & 0\end{smallmatrix}\right) A(3,12,4)$$

$$-2^{k+j-2}A(3,3,2)$$

$A(T_2^*(2),(3,3,2))$

$$= \quad 2^{2k+j-5}\Big(\rho_j \left(\begin{smallmatrix}4 & -1\\0 & 1\end{smallmatrix}\right) A(2,12,8) + \rho_j \left(\begin{smallmatrix}4 & -3\\0 & 1\end{smallmatrix}\right) A(9,12,20)\Big)$$

$$+2^j A(12,12,8) + 2^{3k+2j-7}e\big(((-1)^{k+1}3+1)/8\big)\rho_j \left(\begin{smallmatrix}2 & -1\\0 & 1\end{smallmatrix}\right) A(2,3,4)$$

$$+2^{k+j-3}e\big(((-1)^{k+1}3+1)/8\big)$$

$$\times \left(\rho_j \begin{pmatrix} 2 & 0 \\ 0 & 1 \end{pmatrix} A(3, 12, 4) + (-1)\rho_j \begin{pmatrix} 0 & 1 \\ 2 & 0 \end{pmatrix} A(3, 12, 4) \right)$$
$$+2^{2k+2j-6} e \big(((-1)^{k+1}3 + 1)/4\big) A(3, 3, 2)$$

In the above formula, if k is even, then by the very definition of the plus space, we have $A(3, 3, 2) = 0$, so we can always assume that k is odd.

5 Jacobi forms and Klingen type Eisenstein series

Here, for the reader's convenience, we explain the isomorphism between vector valued holomorphic or skew holomorphic Jacobi forms of index one and the plus space of vector valued Siegel modular forms of half integral weight of general degree. We also define Klingen type Jacobi Eisenstein series, and hence Klingen type Eisenstein series of half integral weight in the plus space. For simplicity, we assume here that $n = 2$ though the results can be easily generalized. Most of the materials in this section are in [3], [17], [2], [9], [7], [5], [13], and [23], [6]. A definition of vector valued Klingen Eisenstein series has not been treated in the above papers and we sketch it here but it is almost the same as in the known cases.

5.1 Definition of holomorphic and skew holomorphic Jacobi forms

The symplectic group $\mathrm{Sp}(n, \mathbb{R})$ acts on $H_n \times \mathbb{C}^n$ by

$$M(\tau, z) = ((a\tau + b)(c\tau + d)^{-1}, {}^t(c\tau + d)^{-1}\tau)$$

for

$$(\tau, z) \in H_n \times \mathbb{C}^n, \quad M = \begin{pmatrix} a & b \\ c & d \end{pmatrix} \in \mathrm{Sp}(n, \mathbb{R}) \quad (a, b, c, d \in M_n(\mathbb{R})) .$$

The Jacobi group $\mathrm{Sp}(2, \mathbb{R})^J$ is defined by

$$\{(M, ([\mu, \nu], \kappa)); M \in \mathrm{Sp}(2, \mathbb{R}), \ \mu, \nu \in \mathbb{R}^2, \kappa \in \mathbb{R}\}$$

as a set, with product given by

$$(M, ([0, 0], 0))(1_{2n}, ([\lambda, \mu], \kappa)) = (M, ([\lambda, \mu], \kappa))$$
$$(1_{2n}, ([\lambda, \mu], \kappa))(M, ([0, 0], 0)) = (M, ([{}^t a\lambda + {}^t c\mu, {}^t b\lambda + {}^t c\mu], \kappa))$$
$$(M, ([0, 0], 0))(M', [(0, 0), 0)) = (MM', ([0, 0], 0))$$

and

$$(1_{2n}, ([\lambda, \mu], \kappa))(1_{2n}, ([\lambda', \mu'], \kappa'))$$
$$= (1_{2n}, ([\lambda + \lambda', \mu + \mu'], \kappa + \kappa' + {}^t\lambda\mu' - {}^t\mu\lambda')) .$$

We identify $\mathrm{Sp}(n, \mathbb{R})$ and the Heisenberg group $H(\mathbb{Z}) = \{([\lambda, \mu], \kappa)\}$ with the corresponding subgroup of $\mathrm{Sp}(n, \mathbb{R})^J$, respectively. We define Γ_n^J the subgroup of $Sp(n, \mathbb{R})^J$ such that $M \in \Gamma_n$, λ, $\mu \in \mathbb{Z}^n$, $\kappa \in \mathbb{Z}$. For any irreducible polynomial representation ρ of $\mathrm{GL}_n(\mathbb{C})$ and for any function $F(\tau, z)$ on $H_n \times \mathbb{C}^n$, we define two kinds of group actions $|\gamma$ or $|^{sk}\gamma$ of $\mathrm{Sp}(n, \mathbb{R})^J$ (of index 1). One is

$$
\begin{aligned}
F|_1([\lambda, \mu], \kappa) &= e({}^t\lambda\tau\lambda + 2{}^t\lambda z + {}^t\lambda\mu + \kappa)F(\tau, z + \tau\lambda + \mu) \\
F|_{\rho,1}M &= e(-{}^tz(c\tau + d)^{-1}cz)\rho(c\tau + d)^{-1}F(M\tau, {}^t(c\tau + d)z)
\end{aligned}
$$

($M \in \mathrm{Sp}(n, \mathbb{R})$, λ, $\mu \in \mathbb{R}^n$, $\kappa \in \mathbb{R}$) and the other one is given by

$$
F|_1^{sk}([\lambda, \mu], \kappa) = F|_1([\lambda, \mu], \kappa)
$$

and

$$
\begin{aligned}
F|_{\rho,1}^{sk}M = e(-{}^tz(c\tau + d)^{-1}cz) \\
\times \left(\frac{|\det(c\tau + d)|}{\det(c\tau + d)}\right)\overline{\rho(c\tau + d)^{-1}}F(M\tau, {}^t(c\tau + d)z),
\end{aligned}
$$

where $\overline{*}$ means complex conjugation. We say that F is a holomorphic Jacobi form of weight ρ of index 1 if
(0) F is holomorphic on $H_n \times \mathbb{C}^n$.
(1) $F|_{\rho,1}\gamma = F$ for any $\gamma \in \Gamma_n^J$ and
(2) $F(\tau, z)$ has the Fourier expansion of the following form,

$$
F(\tau, z) = \sum_{N \in L_n^*, r \in \mathbb{Z}^n} A(N, r)e(\mathrm{tr}(N\tau) + {}^trz),
$$

where we denote by L_n^* the set of all half integral symmetric matrices and (N, r) runs over all elements in $L_n^* \times \mathbb{Z}^n$ such that $4N - r^tr \geq 0$ (positive semi-definite). The space of such functions is denoted by $J_{\rho,1}$. For $\mu \in \mathbb{Z}^n$, we define

$$
\vartheta_\mu(\tau, z) = \sum_{p \in \mathbb{Z}^n} e\left({}^t(p + \mu/2))\tau(p + \mu/2) + 2{}^t(p + \mu/2)z\right).
$$

Then for any $F \in J_{\rho,1}$, we can write

$$
F(\tau, z) = \sum_{\mu \in (\mathbb{Z}/2\mathbb{Z})^n} h_\mu(\tau)\vartheta_\mu(\tau, z),
$$

where the $h_\mu(\tau)$ are holomorphic and uniquely determined by F. We define

$$
\sigma(F) = \sum_{\mu \in (\mathbb{Z}/2\mathbb{Z})^n} h_\mu(4\tau).
$$

The definition of skew holomorphic Jacobi forms was introduced by Skoruppa for $n = 1$ and by Arakawa for general n. We say that $F(\tau, z)$ is a skew holomorphic Jacobi form of weight ρ of index 1 if

(0) F is holomorphic with respect to z and real analytic with respect to the real and the imaginary part of τ.

(1) $F|^{sk}_{\rho,1} \gamma = F$ for any $\gamma \in \Gamma^J_n$. and

(2) F has a Fourier expansion of the following form.

$$F(\tau, z) = \sum_{N \in L^*_n, r \in \mathbb{Z}^n} A(N, r) e(tr(N\tau - \frac{1}{2}i(4N - r^t r)y)) e(^t rz) ,$$

where y is the imaginary part of τ and (N, r) runs over $L^*_n \times \mathbb{Z}^n$ such that $r^t r - 4N \geq 0$. The space of such functions is denoted by $J^{skew}_{\rho,1}$. For any $F \in J_{\rho,1}$, we have

$$F(\tau, z) = \sum_{\mu \in (\mathbb{Z}/2\mathbb{Z})^n} h_\mu(\tau) \vartheta_\mu(\tau, z)$$

where the $h_\mu(\tau)$ are uniquely determined functions. We define

$$\sigma(F) = \sum_{\mu \in (\mathbb{Z}/2\mathbb{Z})^n} h_\mu(-4\bar\tau).$$

The definition of the Hecke operators $T_i(p)$ $(0 \leq i \leq n)$ associated with the diagonal matrix $K_i(p^2) = (1_{n-i}, p1_i, p^2 1_{n-i}, p1_i)$ for holomorphic or skew holomorphic Jacobi forms is given in [9] or [5] for the scalar valued case. The vector valued case is obtained by replacing the automorphy factor $\det^k$ by ρ. For any ρ, we write $\rho = \det^k \rho_0$ where k is the largest integer such that ρ_0 is a polynomial representation. Then Siegel modular forms of half-integral weight $\det^{k-1/2} \rho_0$ are defined similarly by taking as automorphy factor $(\theta(\gamma\tau)/\theta(\tau))^{2k-1} \rho_0(c\tau + d)$. The definition of the plus space is the same as in the scalar valued case, parity depending only on k and the character. The definition of Hecke operators $T_{half,i}(p)$ on half-integral weight associated with the $\widetilde\Gamma_0(4)$ double coset containing $(K_i(p^2), p^{(n-i)/2})$ is similar.

Theorem 5.1 (cf. [9],[7], [5],[13]). *For any irreducible representation $\rho = \det^k \rho_0$ as above, the linear map σ gives an isomorphism*

$$J_{\rho,1}(\Gamma_2) \cong A^+_{\det^{k-1/2} \rho_0}(\Gamma_0(4), \psi^k).$$

and

$$J^{skew}_{\rho,1}(\Gamma_2) \cong A^+_{\det^{k-1/2} \rho_0}(\Gamma_0(4), \psi^{k-1}).$$

For odd primes p, this isomorphism σ commutes with Hecke operators, i.e.,

$$T_i(p^2)F = p^{(3n+i)/2}\left(\frac{-1}{p}\right)^{(k+\delta)i} T_{\text{half},i}(p)(\sigma(F))$$

where $\delta = 0$ or 1 for holomorphic or skew holomorphic Jacobi forms, respectively.

For simplicity we assume now that $n = 2$ and denote by $\rho_{k,j}$ the representation $\det^k \text{Sym}_j$ as before. The Φ operator on Siegel modular forms of half-integral weight is defined as usual. As for Jacobi forms, for any function $F(\tau, z)$ on $H_2 \times \mathbb{C}^2$, the Siegel Φ-operator is defined by

$$(\Phi F)(\tau_1, z_1) = \lim_{\lambda \to \infty} F\left(\begin{pmatrix} \tau_1 & 0 \\ 0 & i\lambda \end{pmatrix}, \begin{pmatrix} z_1 \\ 0 \end{pmatrix}\right),$$

where $(\tau_1, z_1) \in H_1 \times \mathbb{C}$ (cf. [23]). If $F \in J_{\rho_{k,j},1}$ or $J_{\rho_{k,j},1}^{\text{skew}}$, then we have $\Phi(F) = \phi e_1$ where $e_1 = {}^t(1, 0, \dots, 0)$ and ϕ is a Jacobi form of degree one belonging to $J_{k+j,1}$ or $J_{k+j,1}^{\text{skew}}$, respectively. It is well-known and easy to see that $J_{k+j,1} = 0$ or $J_{k+j,1}^{\text{skew}} = 0$ if $k + j$ is odd or even respectively. We sometimes identify ϕe_1 with ϕ. We would like to define F such that $\Phi(F) = \phi$ for a given ϕ. This can be done using Klingen type Jacobi Eisenstein series similarly as in [23] and [6]. To define these we need some notation. We put

$$P_1(\mathbb{Z}) = \left\{\begin{pmatrix} \mathbb{Z} & 0 & \mathbb{Z} & \mathbb{Z} \\ \mathbb{Z} & \mathbb{Z} & \mathbb{Z} & \mathbb{Z} \\ \mathbb{Z} & 0 & \mathbb{Z} & \mathbb{Z} \\ 0 & 0 & 0 & \mathbb{Z} \end{pmatrix}\right\} \cap \Gamma_2$$

and

$$P_1(\mathbb{Z})^J = \left\{(M, ([\lambda, \mu], \kappa)) \mid M \in P_1(\mathbb{Z}); \lambda = \begin{pmatrix} \lambda_1 \\ 0 \end{pmatrix} \in \mathbb{Z}^2, \mu \in \mathbb{Z}^2, \kappa \in \mathbb{Z}\right\}.$$

We denote by j an even natural number. First of all, we assume that k is even. We take a holomorphic Jacobi form $\phi(\tau_1, z_1) \in J_{k+j,1}$ of weight $k + j$ and of index 1 where $(\tau_1, z_1) \in H \times \mathbb{C}$. For $\tau = \begin{pmatrix} \tau_1 & z_0 \\ z_0 & \tau_2 \end{pmatrix} \in H_2$ and $z = \begin{pmatrix} z_1 \\ z_2 \end{pmatrix} \in \mathbb{C}^2$, we define f_ϕ by $f_\phi(\tau, z) = \phi(\tau, z)$. We put

$$E_{(k,j)}(\tau, z, \phi) = \sum_{\gamma \in P_1(\mathbb{Z})^J \backslash \Gamma_2^J} (f_\phi|_{\rho_{k,j},1}\gamma)(\tau, z).$$

This converges when $k > 5$. We have $E_{(k,j)} \in J_{\rho_{k,j},1}$ and $\Phi(F) = \phi e_1$. Next we assume that k is odd. For $\phi \in J_{k+j,1}^{\text{skew}}$, we put

$$E_{(k,j)}^{\text{skew}}(\tau, z, \phi) = \sum_{\gamma \in P_1(\mathbb{Z})^J \backslash \Gamma_2^J} (f_\phi|_{\rho_{k,j},1}^{sk}\gamma)(\tau, z).$$

This converges also when $k > 5$. We have $E^{\text{skew}}_{(k,j)} \in J^{\text{skew}}_{\rho_{k,j},1}$ and $\Phi(F) = \phi e_1$. We can show that $E_{(k,j)}(\tau, z, \phi)$ or $E^{\text{skew}}_{(k,j)}(\tau, z, \phi)$ is a Hecke eigenform if and only if ϕ is so. For any $F \in J_{\rho_{k,j},1}$ or $J^{\text{skew}}_{\rho_{k,j},1}$, it is easy to see that we have $\Phi(\sigma(F)) = \sigma(\Phi(F))$. We have $\sigma(E_{(k,j)}) \in A^+_{k-1/2,j}(\Gamma_0(4))$ and $\sigma(E_{(k,j)}) \in A^+_{k-1/2,j}(\Gamma_0(4))$. We call these functions also the Klingen type Eisenstein series of half-integral weight. Now by Eichler-Zagier [3] or Skoruppa [16], the above ϕ corresponds to a modular form g of half-integral weight $(k + j) - 1/2$ of one variable in the plus space. By Kohnen [14], this g corresponds to a modular form f of integral weight $2k + 2j - 2$. If ϕ is a Hecke eigen form, then so is f. It is proved in the same way as in [7] pp.517–518 that $L(s, \sigma(E_{(k,j)}))$ or $L(s, \sigma(E^{\text{skew}}_{(k,j)}))$ is equal to

$$\zeta(s - j - 1)\zeta(s - 2k - j + 4)L(s, f).$$

Appendix: Table of Fourier coefficients

We give here some Fourier coefficients which are needed to calculate Euler 3-factors.

A.1 Integral weight

For the sake of simplicity, in the tables below, we give $\binom{j}{\nu}^{-1}$ times the ν-th component of the Fourier coefficients.

Fourier coefficients of $F_{5,18}$

$(1, 1, 1)$	$(0, 0, 0, 0, 0, -13, -39, -56, -42, 0, 42, 56, 39, 13, 0, 0, 0, 0, 0)$
$(3, 3, 3)$	$(0, -72176832, -72176832, -35823060, 530712, 22121775,$ $28950129, 24964254, 14112630, 0, -14112630, -24964254,$ $-28950129, -22121775, -530712, 35823060, 72176832,$ $72176832, 0)$
$(1, 7, 1)$	$(0, 0, 0, 0, 0, 55692, 167076, -312984, -2031624, -752976,$ $12641832, 29628144, 1511244, -186309396, -743285088,$ $-545257440, 7212022272, 66466842624, 363771233280)$

Fourier coefficients of $F_{5,20}$

$(1, 1, 1)$	$(0, 0, 0, 0, 0, -10, -30, -47, -48, -30, 0, 30, 48, 47, 30, 10,$ $0, 0, 0, 0, 0)$
$(3, 3, 3)$	$(0, -271257984, -271257984, -152285184, -33312384,$ $10991250, -19374282, -71372997, -91968912, -62764362,$ $0, 62764362, 91968912, 71372997, 19374282, -10991250,$ $33312384, 152285184, 271257984, 271257984, 0)$
$(1, 7, 1)$	$(0, 0, 0, 0, 0, 42840, 128520, 264708, 459072, -1507896,$ $-10082880, -8142120, -8142120, 66806208, 194800572,$ $-6014376, -1196526600, -3357573120, -17302150656,$ $-115867874304, -1848572928, 3544437657600)$

Fourier coefficients of $f_{5,24a}$ **and** $f_{5,24b}$. We denote by (a, c, b, v) the v-th component of the Fourier coefficients at (a, c, b). If $a = c$, then the v-th component is easily obtained from the $(j - v)$-th component by the relation

$$(-1)^k \rho_j \begin{pmatrix} 0 & 1 \\ 1 & 0 \end{pmatrix} \begin{pmatrix} a & b/2 \\ b/2 & a \end{pmatrix} = \begin{pmatrix} a & b/2 \\ b/2 & a \end{pmatrix}.$$

So we omit half of the components.

Fourier coefficients of $f_{5,24a}$ and $f_{5,24b}$

Matrix	$f_{5,24a}$	$f_{5,24b}$
$(1, 1, 1, 0)$	0	0
$(1, 1, 1, 1)$	0	0
$(1, 1, 1, 2)$	0	0
$(1, 1, 1, 3)$	0	0
$(1, 1, 1, 4)$	0	0
$(1, 1, 1, 5)$	-1065725580	32022220
$(1, 1, 1, 6)$	-3197176740	96066660
$(1, 1, 1, 7)$	-4633759802	139577242
$(1, 1, 1, 8)$	-3614881088	109997888
$(1, 1, 1, 9)$	-490246470	17470950
$(1, 1, 1, 10)$	2280138630	-65162790
$(1, 1, 1, 11)$	2398195767	-69714887
$(1, 1, 1, 12)$	0	0

(3, 3, 3, 0)	0	0
(3, 3, 3, 1)	−57680814147751872	1965334455387072
(3, 3, 3, 2)	−57680814147751872	1965334455387072
(3, 3, 3, 3)	−79892317549985472	2530636395440832
(3, 3, 3, 4)	−102103820952219072	3095938335494592
(3, 3, 3, 5)	−88644356379770868	2616377386560948
(3, 3, 3, 6)	−39513923832640860	1091953548639900
(3, 3, 3, 7)	13841245593223650	−520288602797250
(3, 3, 3, 8)	3997492080187536	−1263304492279200
(3, 3, 3, 9)	32753921485717278	−972251330000958
(3, 3, 3, 10)	11358118125502722	−274488111823842
(3, 3, 3, 11)	−1274040532609515	98637117316155
(3, 3, 3, 12)	0	0
(1, 7, 1, 0)	0	0
(1, 7, 1, 1)	0	0
(1, 7, 1, 2)	0	0
(1, 7, 1, 3)	0	0
(1, 7, 1, 4)	0	0
(1, 7, 1, 5)	4565568384720	−137183190480
(1, 7, 1, 6)	13696705154160	−411549571440
(1, 7, 1, 7)	−20125272546792	560249344872
(1, 7, 1, 8)	−144419047573248	4161562046208
(1, 7, 1, 9)	−40199266068696	1269696712536
(1, 7, 1, 10)	978023462535000	−27744078783960
(1, 7, 1, 11)	1416905286935292	−40456367945532
(1, 7, 1, 12)	−4941068164623360	140166523284480
(1, 7, 1, 13)	−14069719692603036	396569445933276
(1, 7, 1, 14)	20781880935902856	−614348042111496
(1, 7, 1, 15)	115846315862908680	−3315904201378440
(1, 7, 1, 16)	−45366806527725312	1588017523099392
(1, 7, 1, 17)	−905887869644583000	26592097111617240
(1, 7, 1, 18)	−722781032781399984	18020442954662064
(1, 7, 1, 19)	5659154519219412624	−176346931613794704
(1, 7, 1, 20)	13313426015445373440	−382512025094976000
(1, 7, 1, 21)	−31897628338076621568	1006879261703793408
(1, 7, 1, 22)	−274312629286921032192	7649076209441812992
(1, 7, 1, 23)	775542991272035728896	−27722461484811050496
(1, 7, 1, 24)	18282369435943326658560	−563378797820507074560

Fourier coefficients of $f_{5,26a}$ **and** $f_{5,26b}$

Matrix	$f_{5,26a}$	$f_{5,26b}$
(1, 1, 1, 0)	0	0
(1, 1, 1, 1)	0	0
(1, 1, 1, 2)	0	0
(1, 1, 1, 3)	0	0
(1, 1, 1, 4)	0	0
(1, 1, 1, 5)	−65362127	−698150645
(1, 1, 1, 6)	−196086381	−2094451935
(1, 1, 1, 7)	−293270485	−3083495479
(1, 1, 1, 8)	−258012162	−2559872886
(1, 1, 1, 9)	−92619264	−581765568
(1, 1, 1, 10)	99390228	1629055260
(1, 1, 1, 11)	197894213	2672858375
(1, 1, 1, 12)	150772479	1957582341
(1, 1, 1, 13)	0	0
(3, 3, 3, 0)	0	0
(3, 3, 3, 1)	8828682282969120	267842148592629600
(3, 3, 3, 2)	8828682282969120	267842148592629600
(3, 3, 3, 3)	−5270746587314076	47392227548316684
(3, 3, 3, 4)	−19370175457597272	−173057693495996232
(3, 3, 3, 5)	−22713697964516067	−234166624356665745
(3, 3, 3, 6)	−15301314108070461	−135934565033691855
(3, 3, 3, 7)	−4293834304435281	7922247106615029
(3, 3, 3, 8)	3147931030214646	83687574697944498
(3, 3, 3, 9)	4514603375393028	59187990031424172
(3, 3, 3, 10)	1948236106302108	−16207124944380300
(3, 3, 3, 11)	−793744931583747	−69036654729809745
(3, 3, 3, 12)	−1374600847934457	−59198826380455971
(3, 3, 3, 13)	0	0
(1, 7, 1, 0)	0	0
(1, 7, 1, 1)	0	0
(1, 7, 1, 2)	0	0
(1, 7, 1, 3)	0	0
(1, 7, 1, 4)	0	0
(1, 7, 1, 5)	280011352068	2990877363180
(1, 7, 1, 6)	840034056204	8972632089540
(1, 7, 1, 7)	−64803927852	−7006354420644

(1, 7, 1, 8)	−4179374640360	−69897700767096
(1, 7, 1, 9)	−5957294015376	−58447436865840
(1, 7, 1, 10)	7439078119248	294803996050800
(1, 7, 1, 11)	87684101757252	770654941762860
(1, 7, 1, 12)	356834320429356	30375725391396
(1, 7, 1, 13)	100273026367800	−3441731807036952
(1, 7, 1, 14)	−4078946072879964	−8820348405789492
(1, 7, 1, 15)	−9910281374768412	−3591595424266740
(1, 7, 1, 16)	20223662622618096	57878193051642960
(1, 7, 1, 17)	108510171590526048	109772358846004512
(1, 7, 1, 18)	−84232703324738808	−697171710871429416
(1, 7, 1, 19)	−1093102324323664140	−2574974409947325252
(1, 7, 1, 20)	−10267763962648572	71814735496516966620
(1, 7, 1, 21)	11129160276266097060	55901776861279992780
(1, 7, 1, 22)	8613076555184317920	49190764056017351328
(1, 7, 1, 23)	−121153026606762050208	−472827529196854612704
(1, 7, 1, 24)	−282688453093682845440	−1514111683998966877440
(1, 7, 1, 25)	1122572073240871925760	4629136865780229580800
(1, 7, 1, 26)	8479594535903665574400	6404196822211303352000

Fourier coefficients of $F_{7,12}$, $F_{7,14}$ and $F_{7,16}$

Matrix	$F_{7,12}$	$F_{7,14}$	$F_{7,16}$
(1, 1, 1, 0)	0	0	0
(1, 1, 1, 1)	0	0	0
(1, 1, 1, 2)	0	0	0
(1, 1, 1, 3)	−3	−11	13
(1, 1, 1, 4)	−6	−22	26
(1, 1, 1, 5)	−5	−23	30
(1, 1, 1, 6)	0	−14	25
(1, 1, 1, 7)	5	0	14
(1, 1, 1, 8)	6	14	0
(1, 7, 1, 0)	0	0	0
(1, 7, 1, 1)	0	0	0
(1, 7, 1, 2)	0	0	0
(1, 7, 1, 3)	12852	47124	−55692
(1, 7, 1, 4)	25704	94248	−111384

Fourier coefficients of $F_{7,12}$, $F_{7,14}$ and $F_{7,16}$ (cont.)

Matrix	$F_{7,12}$	$F_{7,14}$	$F_{7,16}$
$(1, 7, 1, 5)$	-183060	-69948	-280440
$(1, 7, 1, 6)$	-613440	-445464	-562860
$(1, 7, 1, 7)$	139860	-617904	839664
$(1, 7, 1, 8)$	3482136	-172872	5725440
$(1, 7, 1, 9)$	-9955764	7715484	12379752
$(1, 7, 1, 10)$	-80317440	36284472	15574860
$(1, 7, 1, 11)$	-97843680	90293148	-20239920
$(1, 7, 1, 12)$	367804800	151132608	-187748136
$(1, 7, 1, 13)$	$*$	-237714048	-298479636
$(1, 7, 1, 14)$	$*$	-2784862080	820572480
$(1, 7, 1, 15)$	$*$	$*$	-21346022880
$(1, 7, 1, 16)$	$*$	$*$	-198792921600
$(3, 3, 3, 0)$	0	0	0
$(3, 3, 3, 1)$	-1443420	-22517352	107469180
$(3, 3, 3, 2)$	-1443420	-22517352	107469180
$(3, 3, 3, 3)$	-312201	-17099181	44380791
$(3, 3, 3, 4)$	819018	-11681010	-18707598
$(3, 3, 3, 5)$	923085	-7094385	-50519700
$(3, 3, 3, 6)$	0	-3339306	-51055515
$(3, 3, 3, 7)$	-923085	0	-30740472
$(3, 3, 3, 8)$	-819018	3339306	0

A.2 half-integral weight

The coefficients in the next tables are given as row vectors, though they are regarded as column vectors in the main text.

Fourier coefficients of $F_{23/2,2}$

T	coefficient
$(4, 9, 0)$	$(0, 47115, 0)$
$(1, 4, 0)$	$(0, 1, 0)$
$(1, 36, 0)$	$(-12729, -62577, -79704)$
$(9, 36, 0)$	$(0, -15424819875, 0)$
$(5, 17, 14)$	$(-12729, -37119, -29856)$

Fourier coefficients of $F_{25/2,2}$

T	coefficient
(3, 3, 2)	(−1, 0, 1)
(3, 4, 4)	(1, 2, 0)
(11, 8, 8)	(−88659, −177318, 0)
(27, 27, 18)	(186439477563, 0, −186439477563)
(19, 4, 4)	(291057, 582114, 0)
(3, 27, 6)	(0, 506412, 506412)

Fourier coefficients of $f_{29/2,a}$

T	coefficient
(3, 4, 0)	(0, 148, 0)
(3, 36, 0)	(0, −639309348, 0)
(4, 27, 0)	(0, −313727592, 0)
(7, 36, 24)	(182700984, 954103848, 565559136)
(19, 36, 48)	(205843728, −177014424, −565559136)
(27, 36, 0)	(0, −2064912998383584, 0))

Fourier coefficients of $f_{29/2,b}$

T	coefficient
(3, 4, 0)	(0, 16, 0)
(3, 36, 0)	(0, −85729104, 0)
(4, 27, 0)	(0, −33480480, 0)
(7, 36, 24)	(23092896, 117705696, 57084768)
(19, 36, 48)	(37528032, 3536160, −57084768)
(27, 36, 0)	(0, −176777487932544, 0)

Fourier coefficients of $f_{31/2,2a}$ and $f_{31/2,2b}$

T	$f_{31/2,2a}$	$f_{31/2,2b}$
(1, 4, 0)	(0, 0, 0)	(0, 1, 0)
(1, 36, 0)	(0, 0, 0)	(0, −3139803, 0)
(4, 9, 0)	(0, −408, 0)	(0, 114939, 0)

Fourier coefficients of $f_{31/2,2a}$ and $f_{31/2,2b}$ (cont.)

T	$f_{31/2,2a}$	$f_{31/2,2b}$
$(5, 17, 14)$	$(-232, 68, 188)$	$(-5433, -42063, -36240)$
$(5, 36, 24)$	$(-232, -396, 24)$	$(-5433, -52929, -83736)$
$(9, 36, 0)$	$(0, 6557038128, 0)$	$(0, -198464650763139, 0)$
$(17, 5, 14)$	$(-188, -68, 232)$	$(36240, 42063, 5433)$
$(17, 36, 48)$	$(-188, -444, -24)$	$(36240, 114543, 83736)$

Fourier coefficients of $F_{17/2,4}$

T	coefficient
$(3, 4, 0)$	$(0, -2, 0, 16, 0)$
$(7, 36, 24)$	$(22596, 249036, 738720, 569952, -211104)$
$(3, 36, 0)$	$(0, -3078, 0, -867024, 0)$
$(19, 36, 48)$	$(268776, 1325868, 2237760, 1414368, 211104)$
$(4, 27, 0)$	$(0, 31200, 0, -357660, 0)$
$(27, 36, 0)$	$(0, 1747574352, 0, -11599739136, 0)$

Fourier coefficients of $F_{19/2,4}$

T	coefficient
$(1, 4, 0)$	$(0, 0, 0, -1, 0)$
$(5, 7, 14)$	$(-6279, -35121, -74448, -68520, -23640)$
$(1, 36, 0)$	$(0, 0, 0, 113643, 0)$
$(17, 36, 48)$	$(23640, 163080, 421848, 484137, 208008)$
$(4, 9, 0)$	$(0, 15885, 0, -7128, 0)$
$(9, 36, 0)$	$(0, -93533616, 0, 1200752667, 0)$

Fourier coefficients of $F_{21/2,4}$

T	coefficient
$(3, 3, 2)$	$(0, 1, 0, -1, 0)$
$(8, 11, 8)$	$(0, -39798, -59697, -44307, -12204)$
$(4, 3, 4)$	$(0, -2, -3, -1, 0)$

(3, 27, 6)   (0, −115182, −345546, −634230, −403866)
(19, 4, 4)   (141102, 304965, 68283, 45522, 0)
(27, 27, 18)   (−8571080448, −53384619699, 0, 53384619699,
                                                        8571080448)

Bibliography

[1] T. Arakawa, Vector valued Siegel's modular forms of degree two and the associated Andrianov L-functions, Manuscripta. Math. 44(1983), 155-185.

[2] T. Arakawa, Siegel's formula for Jacobi forms. Internat. J. Math. 4 (1993), no. 5, 689–719.

[3] M. Eichler and D. Zagier, The theory of Jacobi forms. Progress in Mathematics, 55. Birkhäuser Boston, Inc., Boston, MA, 1985. v+148 pp.

[4] G. Harder, A congruence between a Siegel and an elliptic modular form, Colloquium Bonn, February 7, 2003. In: Bruinier, van der Geer, Harder, Zagier, The 1-2-3 of Modular Forms. Universitext, Springer Verlag 2008.

[5] S. Hayashida, Skew holomorphic Jacobi forms of index 1 and Siegel modular forms of half integral weights, J. Number Theory 106(2004), 200–218.

[6] S. Hayashida, Klingen type Eisenstein series of skew holomorphic Jacobi forms, Comm. Math. Univ. St. Pauli, 52 no.2(2003). 219–228.

[7] S. Hayashida and T. Ibukiyama, Siegel modular forms of half integral weight and a lifting conjecture. J. Math. Kyoto Univ. 45 (2005), no. 3, 489–530.

[8] T. Ibukiyama, Construction of half integral weight Siegel modular forms of $Sp(2, R)$ from automorphic forms of the compact twist $Sp(2)$. J. Reine Angew. Math. 359 (1985), 188–220.

[9] T. Ibukiyama, On Jacobi Forms and Siegel modular forms of half integral weights, Comment. Math. Univ. St. Paul 41 No. 2 (1992), 109-124.

[10] T. Ibukiyama, Vector valued Siegel modular forms of half-integral weight, preprint.

[11] T. Ibukiyama, Siegel modular forms of weight three and conjectural correspondence of Shimura type and Langlands type, *The conference on L-functions*, Ed. L. Weng and M. Kaneko, World Scientific (2007), 55–69.

[12] Y. Ihara, On certain arithmetical Dirichlet series. J. Math. Soc. Japan 16 1964 214–225, and his Master Thesis Univ. Tokyo 1963.

[13] S. Kimura, On vector valued Siegel modular forms of half integral weight and Jacobi Forms, Master thesis in Osaka University(2005) pp. 82.

[14] W. Kohnen, Modular forms of half-integral weight on $\Gamma_0(4)$. Math. Ann. 248 (1980), no. 3, 249–266.

[15] G. Shimura, On modular forms of half integral weight. Ann. of Math. (2) 97 (1973), 440–481.

[16] N-. P. Skoruppa,Über den Zusammenhang zwischen Jacobiformen und Modulformen halbganzen Gewichts, pp. 163, Thesis, Bonn 1984.

[17] N-. P. Skoruppa, Developments in the theory of Jacobi forms. Automorphic functions and their applications (Khabarovsk, 1988), 167–185, Acad. Sci. USSR, Inst. Appl. Math., Khabarovsk, 1990.

[18] R. Tsushima, An explicit dimension formula for the spaces of generalized automorphic forms with respect to Sp(2, Z). Proc. Japan Acad. Ser. A Math. Sci. 59 (1983), no. 4, 139–142.

[19] R. Tsushima, Dimension formula for the spaces of Siegel cusp forms of half integral weight and degree two. Comment. Math. Univ. St. Pauli 52 (2003), no. 1, 69–115.

[20] R. Tsushima, On the dimension formula for the spaces of Jacobi forms of degree two. Automorphic forms and L-functions (Kyoto, 1999). Sūrikaisekikenkyūsho Kōkyūroku No. 1103 (1999), 96–110.

[21] V. G. Zhuravlev, Hecke rings for a covering of the symplectic group (Russian) Mat. Sb. (N.S.) 121(163) (1983), no. 3, 381–402.

[22] V. G. Zhuravlev, Euler expansions of theta-transformations of Siegel modular forms of half integer weight and their analytic properties. (Russian) Mat. Sb. (N.S.) 123(165) (1984), no. 2, 174–194.

[23] C. Ziegler, Jacobi forms of higher degree. Abh. Math. Sem. Univ. Hamburg 59 (1989), 191–224.

Petersson's trace formula and the Hecke eigenvalues of Hilbert modular forms

Andrew Knightly* and Charles Li

Abstract

Using an explicit relative trace formula, we obtain a Petersson trace formula for holomorphic Hilbert modular forms. Our main result expresses a sum (over a Hecke eigenbasis) of products of Fourier coefficients and Hecke eigenvalues in terms of generalized Kloosterman sums and Bessel functions. As an application we show that the normalized Hecke eigenvalues for a fixed prime p have an asymptotic weighted equidistribution relative to a polynomial times the Sato-Tate measure, as the norm of the level goes to ∞.

Mathematics Subject Classification: 11F70, 11F72, 11F30, 11L05

1 Introduction

Let $h \in S_k(\Gamma_0(N))$ be a Hecke eigenform (k even), and for a prime $p \nmid N$ define the normalized Hecke eigenvalue v_p^h by

$$p^{-\frac{k-1}{2}} T_p h = v_p^h h.$$

The Ramanujan-Petersson conjecture asserts that $|v_p^h| \leq 2$. This is a theorem of Deligne. Because we have assumed trivial central character, the operator T_p is self-adjoint, so its eigenvalues are real numbers, and thus

$$v_p^h \in [-2, 2].$$

* The first author was supported by NSA grant H98230-06-1-0039.

For a fixed non-CM newform h, the Sato-Tate conjecture predicts that the set $\{v_p^h : p \nmid N\}$ is equidistributed in $[-2, 2]$ relative to the Sato-Tate measure

$$d\mu_\infty(x) = \begin{cases} \frac{1}{\pi}\sqrt{1 - \frac{x^2}{4}}dx & \text{when } x \in [-2, 2], \\ 0 & \text{otherwise.} \end{cases} \qquad (1.1)$$

Taylor has recently proven this in many cases when $k = 2$ [Ta].

When the prime p is fixed, the normalized eigenvalues of T_p on the space $S_k(\Gamma_0(N))$ are asymptotically equidistributed relative to the measure

$$d\mu_p(x) = \frac{p+1}{(p^{1/2} + p^{-1/2})^2 - x^2}d\mu_\infty(x)$$

as $k + N \to \infty$. This was proven in the late 1990's by Serre [Se], and independently (for $N = 1$) by Conrey, Duke and Farmer [CDF]. An extension of the result (in the level aspect only) to Hilbert modular forms is given in [Li2]. These results have an antecedent in work of Bruggeman in 1978, who proved that the eigenvalues of T_p on Maass forms of level 1 exhibit the same distribution ([Br], Sect. 4; see also Sarnak [Sa]). Even earlier, in 1968 Birch proved a similar result for sizes of elliptic curves over $\mathbf{F}_p$ [Bi].

In this paper we prove a different result on the asymptotic distribution of Hecke eigenvalues, valid over a totally real number field F. The following is a special case of a more general result (cf. Theorem 6.6). Notation and terminology will be defined precisely later on.

Theorem 1.1. *Let F be a totally real number field, and let m be a totally positive element of the inverse different $\mathfrak{d}^{-1} \subset F$. For a cusp form φ on $\mathrm{GL}_2(F)\backslash \mathrm{GL}_2(\mathbf{A}_F)$ with trivial central character, let W_m^φ denote its mth Fourier coefficient (see (3.7)). Define a weight*

$$w_\varphi = \frac{|W_m^\varphi(1)|^2}{\|\varphi\|^2}.$$

Fix a prime ideal $\mathfrak{p} \nmid m\mathfrak{d}$. Let $\{\varphi\}$ be an orthogonal basis of eigenforms of prime-to-$\mathfrak{p}$ level $\mathfrak{N}$ and weight $(k_1, \ldots, k_r)$, with all $k_j > 2$ and even. For such φ, let

$$v_\mathfrak{p}^\varphi \in [-2, 2]$$

denote the associated normalized eigenvalue of the Hecke operator $T_\mathfrak{p}$. Then the w_φ-weighted distribution of the eigenvalues $v_\mathfrak{p}^\varphi$ is asymptotically uniform relative to the Sato-Tate measure as the norm of $\mathfrak{N}$ goes to ∞. This means that for any continuous function $f : \mathbf{R} \longrightarrow \mathbf{C}$,

$$\lim_{\mathrm{N}(\mathfrak{N})\to\infty} \frac{\sum_\varphi f(v_\mathfrak{p}^\varphi)w_\varphi}{\sum_\varphi w_\varphi} = \int_\mathbf{R} f(x)d\mu_\infty(x).$$

Noteworthy is the fact that here the measure and the weights are independent of $\mathfrak{p}$, in contrast to Serre's (unweighted) result. The Sato-Tate measure $d\mu_\infty$ is nonzero on any subinterval of $[-2, 2]$, so in particular the above illustrates the density of the Hecke eigenvalues in $[-2, 2]$.

The proof of the above theorem involves a trace formula which may be of independent interest. In a previous paper [KL1], we detailed the way in which the classical 1932 Petersson trace formula can be realized as an explicit relative trace formula, as a conceptual alternative to the usual method using Poincaré series. Here we extend the technique and result to cusp forms on $GL_2(\mathbf{A}_F)$, where F is an arbitrary totally real number field. We work with spaces $A_k(\mathfrak{N}, \omega)$ of holomorphic Hilbert cusp forms of weight $k = (k_1, \ldots, k_r)$, all strictly greater than 2, on the Hecke congruence subgroups $\Gamma_0(\mathfrak{N})$. Our main result is Theorem 5.11 in which a sum (over a Hecke eigenbasis) of terms involving the associated Fourier coefficients, Petersson norms and eigenvalues of a Hecke operator $T_\mathfrak{n}$, is expressed in terms of generalized Kloosterman sums and Bessel functions.

The incorporation of Hecke eigenvalues is a novel feature of this generalized sum formula. In Section 6 we prove a more general version of the above weighted distribution theorem by an argument based on bounding the terms of the sum formula in terms of $\mathbb{N}(\mathfrak{N})$, following the technique for the $F = \mathbf{Q}$ case from [Li1]. This is analogous to the way in which Serre's result follows from bounding the terms of the Eichler-Selberg formula for $\mathrm{tr}(T_{p^m})$.

In Section 7 we give some variants of the generalized Petersson formula, including a simplified version for the case where F has narrow class number 1. In Corollary 7.3, setting $\mathfrak{n} = 1$ for the trivial Hecke operator, we recover the classical extension of Petersson's formula to Hilbert modular forms, as follows from a 1954 paper of Gundlach, [Gu]. His results are valid for cusp forms of uniform weight $k \geq 2$. Refer to §2 of [Lu] for an overview of the classical derivation.

In 1980 Kuznetsov gave an analog of Petersson's formula involving Maass forms and Eisenstein series on the spectral side, [Ku]. As an application, he used his formula to give estimates for sums of Kloosterman sums. His work was reformulated in a representation-theoretic setting by various authors, see [CPS] and its references, notably [MW] where the general rank-1 case is treated. More recently Bruggeman, Miatello and Pacharoni gave a general Kuznetsov trace formula for automorphic forms on $SL_2(F_\infty)$ of uniform even weight ([BMP], Theorem 2.7.1). An important application of their formula is an estimate for sums of Kloosterman sums.

In contrast to the above results, here we are concerned only with holomorphic cusp forms, i.e. those whose infinity types are discrete series. For our test function, we take discrete series matrix coefficients at the infinite components. This serves to isolate the holomorphic part of the cuspidal spectrum. At the finite places we take Hecke operators, which introduces Hecke eigenvalues into the final formula.

Thanks: We would like to thank Don Blasius for a helpful discussion about the Ramanujan conjecture. We also thank the referee for many insightful comments.

2 General setting

We recall the general setting of Jacquet's relative trace formula [Ja]. Full details for the discussion below are given in §2 of [KL1]. Let F be a number field, with adele ring $\mathbf{A}$. Let G be a reductive algebraic group defined over F. Let H be an F-subgroup of $G \times G$, with $H(\mathbf{A})$ unimodular. We assume that $H(F) \backslash H(\mathbf{A})$ is compact. Define a right action of H on G by $g(x, y) = x^{-1} g y$. For $g \in G$, let H_g be the stabilizer of g, i.e.

$$H_g = \{(x, y) \in H \mid x^{-1} g y = g\}.$$

For $\delta \in G(F)$, let $[\delta]$ be the $H(F)$-orbit of δ in $G(F)$, i.e.

$$[\delta] = \{x^{-1} \delta y \mid (x, y) \in H(F)\}.$$

Each element of $[\delta]$ can be expressed uniquely in the form $u^{-1} \delta v$ for some $(u, v) \in H_\delta(F) \backslash H(F)$.

Let f be a continuous function on $G(\mathbf{A})$, and let

$$K(x, y) = \sum_{\gamma \in G(F)} f(x^{-1} \gamma y) \quad (x, y \in G(\mathbf{A})) \tag{2.1}$$

be the associated kernel function. We assume that the above sum is uniformly absolutely convergent on compact subsets of $H(\mathbf{A})$. In particular, $K(x, y)$ is a continuous function on the compact set $H(F) \backslash H(\mathbf{A})$. Let $\chi(x, y)$ be a character of $H(\mathbf{A})$, invariant under $H(F)$. Consider the expression

$$\int_{H(F) \backslash H(\mathbf{A})} K(x, y) \chi(x, y) d(x, y), \tag{2.2}$$

where $d(x, y)$ is a Haar measure on $H(\mathbf{A})$. A relative trace formula results from computing this integral using spectral and geometric expressions for $K(x, y)$. Using the geometric expression (2.1), it is straightforward to show that the integral (2.2) is equal to $\sum_{[\delta]} I_\delta(f)$, where

$$I_\delta(f) = \int_{H_\delta(F) \backslash H(\mathbf{A})} f(x^{-1} \delta y) \chi(x, y) d(x, y).$$

We see that

$$I_\delta(f) = \int_{H_\delta(\mathbf{A})\backslash H(\mathbf{A})} f(x^{-1}\delta y) \left[\int_{H_\delta(F)\backslash H_\delta(\mathbf{A})} \chi(rx, sy) d(r, s) \right] d(x, y).$$

An orbit $[\delta]$ is **relevant** if χ is trivial on $H_\delta(\mathbf{A})$. From the above, we see that $I_\delta = 0$ whenever $[\delta]$ is not relevant. Indeed the expression in the brackets is equal to $\chi(x_0, y_0) \int_{H_\delta(F)\backslash H_\delta(\mathbf{A})} \chi(r, s) d(r, s)$, where $(x_0, y_0) \in H(\mathbf{A})$ is any representative for (x, y), and this integral vanishes unless δ is relevant.

3 Preliminaries

3.1 Notation

Throughout this paper we work over a totally real number field $F \neq \mathbf{Q}$. Let

$$r = [F : \mathbf{Q}],$$

and let $\sigma_1, \ldots, \sigma_r$ be the distinct embeddings $F \hookrightarrow \mathbf{R}$. Let $\infty_1, \ldots, \infty_r$ denote the corresponding archimedean valuations. Let $\mathcal{O}$ be the ring of integers of F. We will generally use Gothic letters $\mathfrak{a}$, $\mathfrak{b}$ etc. to denote fractional ideals of F. We reserve $\mathfrak{p}$ for prime ideals. Let $v = v_\mathfrak{p}$ be the discrete valuation corresponding to $\mathfrak{p}$. Let $\mathcal{O}_v$ be the ring of integers in the local field F_v, and let $\varpi_v \in \mathcal{O}_v$ be a generator of the maximal ideal $\mathfrak{p}_v = \mathfrak{p}\mathcal{O}_v$.

Let $\mathrm{N} : F \to \mathbf{Q}$ denote the norm map. For a nonzero ideal $\mathfrak{a} \subset \mathcal{O}$, let $\mathbb{N}(\mathfrak{a}) = |\mathcal{O}/\mathfrak{a}|$ denote the absolute norm. This extends by multiplicativity to the group of nonzero fractional ideals of F. For $\alpha \in F^*$, we define

$$\mathbb{N}(\alpha) = \mathbb{N}(\alpha\mathcal{O}) = |\mathrm{N}(\alpha)|.$$

We also use the above norms in the local setting with the analogous meanings. We normalize the absolute value on F_v by $|\varpi_v|_v = \mathbb{N}(\mathfrak{p})^{-1}$.

By Dirichlet's unit theorem, the unit group of F is

$$\mathcal{O}^* \cong (\mathbf{Z}/2\mathbf{Z}) \times \mathbf{Z}^{r-1}.$$

Letting $\mathcal{O}^{*2}$ denote the subgroup consisting of squares of units, we see that $\mathcal{O}^*/\mathcal{O}^{*2} \cong (\mathbf{Z}/2\mathbf{Z})^r$ is a finite group. Let

$$U = \{u_1, \ldots, u_{2^r}\} \subset \mathcal{O}^* \tag{3.1}$$

be a fixed set of representatives for $\mathcal{O}^*/\mathcal{O}^{*2}$.

For a fractional ideal $\mathfrak{a} \subset F$, let $\mathrm{ord}_v \mathfrak{a}$ (or $\mathrm{ord}_\mathfrak{p} \mathfrak{a}$) denote the order. Let $\mathfrak{a}_v$ (or $\mathfrak{a}_\mathfrak{p}$) denote its localization, so $\mathfrak{a}_v = \varpi_v^{\mathrm{ord}_v(\mathfrak{a})}\mathcal{O}_v$. Let $\mathbf{A}$ denote the adele ring of F, with finite adeles $\mathbf{A}_{\mathrm{fin}}$, so that

$$\mathbf{A} = F_\infty \times \mathbf{A}_{\mathrm{fin}},$$

where $F_\infty = F \otimes \mathbf{R} \cong \mathbf{R}^r$. Let $\widehat{\mathcal{O}} = \prod_{v<\infty} \mathcal{O}_v$. Generally if $\mathfrak{a}$ is a fractional ideal of F, we write $\widehat{\mathfrak{a}} = \mathfrak{a}\widehat{\mathcal{O}} = \prod_{v<\infty} \mathfrak{a}_v \subset \mathbf{A}_{\text{fin}}$. Let $\text{Cl}(F)$ be the class group of F, of cardinality $h(F)$. For a fractional ideal $\mathfrak{a}$, let $[\mathfrak{a}]$ represent its image in the class group.

Let F^+ denote the set of totally positive elements of F. We let F_∞^+ denote the subset of F_∞ of vectors whose entries are all positive. Let

$$\mathfrak{d}^{-1} = \{x \in F | \, \text{tr}_{\mathbf{Q}}^F(x\mathcal{O}) \subset \mathbf{Z}\}$$

denote the inverse different. We set $\mathfrak{d}_+^{-1} = \mathfrak{d}^{-1} \cap F^+$.

We use the `mathtt` font to represent finite ideles. Thus we write

$$\mathtt{a} \in \mathbf{A}_{\text{fin}}^*, \quad \widehat{\mathfrak{a}} = \mathtt{a}\widehat{\mathcal{O}}, \quad \mathfrak{a} = \mathcal{O} \cap \widehat{\mathfrak{a}}. \tag{3.2}$$

In the other direction, given a fractional ideal $\mathfrak{a} \subset F$, there is an element $\mathtt{a} \in \mathbf{A}_{\text{fin}}^*$ such that (3.2) holds. Explicitly, we can take $\mathtt{a}_v = \varpi_v^{\text{ord}_v \, \mathfrak{a}}$. The element $\mathtt{a}$ is unique up to $\widehat{\mathcal{O}}^*$. We define norms of ideles by taking the products of the local norms. For example, in the situation of (3.2), we have

$$|\mathtt{a}|_{\text{fin}} = \prod_{v<\infty} |\mathtt{a}_v|_v = \prod_{v<\infty} \mathbb{N}(\mathtt{a}_v)^{-1} = \mathbb{N}(\mathtt{a})^{-1} = \mathbb{N}(\mathfrak{a})^{-1}.$$

We use Roman or Greek letters for rational elements $a, \alpha \in F$.

3.2 Haar measure

We use Lebesgue measure on $\mathbf{R}$, and take the product measure on $F_\infty \cong \mathbf{R}^r$. We normalize Haar measure on each non-archimedean completion F_v by taking $\text{meas}(\mathcal{O}_v) = 1$. This choice induces a Haar measure on $\mathbf{A}_{\text{fin}}$ with $\text{meas}(\widehat{\mathcal{O}}) = 1$, and because $\mathbf{A} = F + F_\infty \times \widehat{\mathcal{O}}$,

$$\text{meas}(F\backslash\mathbf{A}) = \text{meas}((F \cap \widehat{\mathcal{O}})\backslash(F_\infty \times \widehat{\mathcal{O}})) = \text{meas}((\mathcal{O}\backslash F_\infty) \times \widehat{\mathcal{O}})$$

$$= \text{meas}(\mathcal{O}\backslash F_\infty) = d_F^{1/2}, \tag{3.3}$$

where d_F is the discriminant of F. The resulting measure on $\mathbf{A}$ is not self-dual.

Let $G = \text{GL}_2$. We normalize Haar measure on $G(\mathbf{R})$ as follows. Use Lebesgue measure dx to define a measure dn on the unipotent subgroup $N(\mathbf{R}) \cong \mathbf{R}$. On $\mathbf{R}^*$ we use the measure $dx/|x|$. We take the product measure dm on the diagonal subgroup $M(\mathbf{R}) \cong \mathbf{R}^* \times \mathbf{R}^*$, and normalize dk on $\text{SO}_2(\mathbf{R})$ to have total measure 1. On $G(\mathbf{R}) = M(\mathbf{R})N(\mathbf{R})\text{SO}_2(\mathbf{R})$ we take $dg = dm \, dn \, dk$.

Let

$$K_{\text{fin}} = \prod_{v<\infty} K_v = \prod_{v<\infty} G(\mathcal{O}_v) = G(\widehat{\mathcal{O}})$$

be the standard maximal compact subgroup of $G(\mathbf{A}_{\text{fin}})$. We normalize Haar measure on $G(\mathbf{A}_{\text{fin}})$ by taking $\text{meas}(K_v) = 1$ for all $v < \infty$. We let Z denote the center of G, and write

$$\overline{G} = G/Z.$$

We normalize Haar measure on $\overline{G}(F_v)$ by taking $\text{meas}(\overline{K}_v) = 1$.

3.3 Hilbert modular forms

Let $\mathfrak{N} \subset \mathcal{O}$ be an integral ideal of F. Let $\mathbf{k} = (\mathbf{k}_1, \ldots, \mathbf{k}_r)$ be an r-tuple of positive integers, each greater than or equal to 2. Let

$$\omega : F^* \backslash \mathbf{A}^* \to \mathbf{C}^*$$

be a unitary Hecke character. Write $\omega = \prod_v \omega_v$. We assume that:

(i) the conductor of ω divides $\mathfrak{N}$

(ii) $\omega_{\infty_j}(x) = \text{sgn}(x)^{\mathbf{k}_j}$ for all $1 \le j \le r$.

The first condition means that ω_v is trivial on $1+\mathfrak{N}_v$ for all $v|\mathfrak{N}$, and unramified for all $v \nmid \mathfrak{N}$. In other words, ω_{fin} is trivial on the open set

$$\widehat{\mathcal{O}}_{\mathfrak{N}}^* \overset{\text{def}}{=} \widehat{\mathcal{O}}^* \cap (1 + \mathfrak{N}\widehat{\mathcal{O}}) = \prod_{v|\mathfrak{N}}(1 + \mathfrak{N}_v) \prod_{v \nmid \mathfrak{N}} \mathcal{O}_v^*.$$

Thus ω can be viewed as a character of the ray class group mod $\mathfrak{N}$

$$\text{Cl}(\mathfrak{N}) \cong \mathbf{A}^* / [F^* (F_\infty^+ \times \widehat{\mathcal{O}}_{\mathfrak{N}}^*)].$$

We freely identify $Z(\mathbf{A})$ with $\mathbf{A}^*$ throughout this paper. For example if $z \in Z(\mathbf{A})$ we write $\omega(z)$. Let $G = \text{GL}_2$ and let

$$L^2(\omega) = L^2(\overline{G}(F)\backslash\overline{G}(\mathbf{A}), \omega)$$

be the space of left $G(F)$-invariant functions on $G(\mathbf{A})$ which transform by ω under the center and which are square integrable over $\overline{G}(F)\backslash\overline{G}(\mathbf{A})$. Let R denote the right regular representation of $G(\mathbf{A})$ on $L^2(\omega)$. Let $L_0^2(\omega)$ be the subspace of cuspidal functions. We know that the restriction of R to $L_0^2(\omega)$ decomposes as a discrete sum of irreducible representations. These are the cuspidal representations π. Every such π factorizes as a restricted tensor product of admissible local representations $\otimes \pi_v$.

Define groups

$$K_0(\mathfrak{N}) = \{ \begin{pmatrix} a & b \\ c & d \end{pmatrix} \in K_{\text{fin}} | c \in \mathfrak{N}\widehat{\mathcal{O}} \}$$

and

$$K_1(\mathfrak{N}) = \{\begin{pmatrix} a & b \\ c & d \end{pmatrix} \in K_0(\mathfrak{N}) \mid d \in 1 + \mathfrak{N}\widehat{\mathcal{O}}\}.$$

These are open compact subgroups of $G(\mathbf{A}_{\mathrm{fin}})$. Let

$$H_{\mathrm{k}}(\mathfrak{N}, \omega) = \bigoplus \pi, \tag{3.4}$$

where π runs through all cuspidal representations in $L_0^2(\omega)$ for which:

(i) $\pi_{\mathrm{fin}} = \otimes_{v<\infty}\pi_v$ contains a nonzero $K_1(\mathfrak{N})$-fixed vector
(ii) $\pi_{\infty_j} = \pi_{\mathrm{k}_j}$ is the discrete series representation of $G(\mathbf{R})$ of weight k_j, for $j = 1, \ldots, r$.

The central character of π_{k_j} is given by

$$\chi_{\pi_{\mathrm{k}_j}}(x) = \mathrm{sgn}(x)^{\mathrm{k}_j} = \omega_{\infty_j}(x).$$

For a discrete series representation π_{∞_j}, let $v_{\pi_{\infty_j}}$ be a lowest weight vector, unique up to nonzero multiples. Define the subspace

$$A_{\mathrm{k}}(\mathfrak{N}, \omega) = \bigoplus_{\pi} \mathbf{C}v_{\pi_{\infty_1}} \otimes \cdots \otimes v_{\pi_{\infty_r}} \otimes \pi_{\mathrm{fin}}^{K_1(\mathfrak{N})}, \tag{3.5}$$

where π runs through all irreducible summands of $H_{\mathrm{k}}(\mathfrak{N}, \omega)$. Here $\pi_{\mathrm{fin}}^{K_1(\mathfrak{N})}$ is the subspace of $K_1(\mathfrak{N})$-fixed vectors in the space of π_{fin}.

Proposition 3.1. *The space $A_{\mathrm{k}}(\mathfrak{N}, \omega)$ defined above is equal to the set of $\varphi \in L_0^2(\omega)$ satisfying:*

(i) $\varphi(gk_{\mathrm{fin}}) = \varphi(g)$ *for all $k_{\mathrm{fin}} \in K_1(\mathfrak{N})$*
(ii) $\varphi(gk_{\infty}) = \prod_{j=1}^{r} e^{i\mathrm{k}_j\theta_j}\varphi(g)$ *for all $k_{\infty} = \prod_j k_{\theta_j} \in K_{\infty} = \mathrm{SO}_2(\mathbf{R})^r$*
(iii) *For any fixed $x \in G(\mathbf{A})$ and $1 \le j \le r$, the function $g_{\infty_j} \mapsto \varphi(xg_{\infty_j})$ is annihilated by $R(E^-)$, where $E^- = \begin{pmatrix} 1 & -i \\ -i & -1 \end{pmatrix} \in \mathfrak{gl}_2(\mathbf{C})$ and R denotes the right regular action of $G(F_{\infty_j})$.*

The elements of $A_{\mathrm{k}}(\mathfrak{N}, \omega)$ are continuous functions on $G(\mathbf{A})$.

Proof. The proof is the same as for the case $F = \mathbf{Q}$ given in Theorem 12.6 of [KL2]. For the continuity, see Lemma 3.3 of [Li2]. □

By a general theorem of Harish-Chandra, the space $A_{\mathrm{k}}(\mathfrak{N}, \omega)$ is finite-dimensional (cf. [HC], [BJ]). In particular the set of π in (3.4) is finite.

3.4 Fourier coefficients

Let $\theta : \mathbf{A} \longrightarrow \mathbf{C}^*$ be the standard character of $\mathbf{A}$. Explicitly,

(i) $\theta_\infty(x) = e^{-2\pi i (x_1 + \cdots + x_r)}$ for $x = (x_1, \ldots, x_r) \in F_\infty$

(ii) For $v < \infty$, θ_v is the composition

$$\theta_v : F_v \xrightarrow{\mathrm{tr}^{F_v}_{\mathbf{Q}_p}} \mathbf{Q}_p \to \mathbf{Q}_p/\mathbf{Z}_p \hookrightarrow \mathbf{Q}/\mathbf{Z} \xrightarrow{e^{(2\pi i \cdot)}} \mathbf{C}^*.$$

Note that θ_v is trivial precisely on the local inverse different $\mathfrak{d}_v^{-1} = \mathfrak{d}^{-1}\mathcal{O}_v = \{x \in F_v \mid \mathrm{tr}^{F_v}_{\mathbf{Q}_p}(x) \in \mathbf{Z}_p\}$. Recall that θ is trivial on F, and that every character on $F\backslash \mathbf{A}$ is of the form

$$\theta_m(x) = \theta(-mx)$$

for some $m \in F$. This identifies the discrete group F with the dual group $\widehat{F\backslash \mathbf{A}}$.

Consider the unipotent subgroup $N = \{\left(\begin{smallmatrix} 1 & * \\ & 1 \end{smallmatrix}\right)\}$ of G. As topological groups, $N(\mathbf{A}) \cong \mathbf{A}$, so characters of the two can be identified. For $m \in F$, we identify θ_m with a character on $N(F)\backslash N(\mathbf{A})$ in the obvious way.

For any smooth cusp form φ on $G(\mathbf{A})$ and $g \in G(\mathbf{A})$, the map $n \mapsto \varphi(ng)$ is a continuous function on $N(F)\backslash N(\mathbf{A})$, with a Fourier expansion

$$\varphi\left(\begin{pmatrix} 1 & x \\ & 1 \end{pmatrix} g\right) = \frac{1}{d_F^{1/2}} \sum_{m \in F} W_m^\varphi(g)\theta_m(x). \tag{3.6}$$

The coefficients are Whittaker functions defined by

$$W_m^\varphi(g) = \int_{F\backslash \mathbf{A}} \varphi\left(\begin{pmatrix} 1 & x \\ & 1 \end{pmatrix} g\right) \theta(mx)\,dx. \tag{3.7}$$

The purpose of the factor of $d_F^{-1/2}$ in (3.6) is to balance the fact that by our choice, $\mathrm{meas}(F\backslash \mathbf{A}) = d_F^{1/2}$. This non-selfdual measure is convenient for the calculations later on.

Let $y_1, \ldots, y_h \in \mathbf{A}_{\mathrm{fin}}^*$ be representatives for $\mathbf{A}_{\mathrm{fin}}^*/F^*\widehat{\mathcal{O}}^*$, the class group of F. Then

$$G(\mathbf{A}) = \bigcup_{i=1}^h G(F)\left[B(F_\infty)^+ K_\infty \times \begin{pmatrix} y_i & \\ & 1 \end{pmatrix} K_1(\mathfrak{N})\right]$$

([Hi], §9.1). Here B denotes the subgroup of invertible upper triangular matrices. It follows that an element $\varphi \in A_k(\mathfrak{N}, \omega)$ is determined by the values

$$\varphi\left(\begin{pmatrix} 1 & x \\ & 1 \end{pmatrix}_\infty \begin{pmatrix} y & \\ & 1 \end{pmatrix}_\infty \begin{pmatrix} y_i & \\ & 1 \end{pmatrix}_{\mathrm{fin}}\right)$$

for $y \in F_\infty^+$ and $x \in F_\infty$. We define for $y \in \mathbf{A}^*$ the notation

$$W_m^\varphi(y) = W_m^\varphi\left(\begin{pmatrix} y & \\ & 1 \end{pmatrix}\right). \tag{3.8}$$

By the above remarks, φ is determined by the coefficients $W_m^\varphi(y)$ for $y \in \mathbf{A}^*$.

Proposition 3.2. *For $\varphi \in A_k(\mathfrak{N}, \omega)$,*

- $W_m^\varphi(g) = W_1^\varphi((^m{}_{1})g)$ *for all $m \in F^*$ and $g \in G(\mathbf{A})$*
- *For any $u \in \widehat{\mathcal{O}}^*$ and $y \in \mathbf{A}^*$, $W_m^\varphi(y) = W_m^\varphi(uy)$ for all $m \in F^*$*
- *For $y \in \mathbf{A}_{\mathrm{fin}}^*$, $W_1^\varphi(y) \neq 0$ only if $y \in \widehat{\mathfrak{d}}^{-1}$.*

(Here and throughout, we identify $\mathbf{A}_{\mathrm{fin}}^*$ with $\{1_\infty\} \times \mathbf{A}_{\mathrm{fin}}^* \subset \mathbf{A}^*$.)

Proof. These are standard facts. The first follows by a change of variables in (3.7) plus the left $G(F)$-invariance of φ. The $K_1(\mathfrak{N})$-invariance of φ gives the second, and also implies that for $u \in \widehat{\mathcal{O}}$

$$\varphi((^1{}_{1}^{\,x})(^y{}_{1})) = \varphi((^1{}_{1}^{\,x})(^y{}_{1})(^1{}_{1}^{\,u})) = \varphi((^1{}_{1}^{\,x+uy})(^y{}_{1})).$$

This means that $W_m^\varphi(y) = W_m^\varphi(y)\theta(-muy)$ for all $u \in \widehat{\mathcal{O}}$, which gives the third when $m = 1$. (See also §9.1 of [Hi].) $\qquad\square$

Proposition 3.3. *For any $\varphi \in A_k(\mathfrak{N}, \omega)$ and any $y \in F_\infty^+$, $W_m^\varphi(y) = 0$ unless $m \in \mathfrak{d}_+^{-1} = \mathfrak{d}^{-1} \cap F^+$.*

Proof. By the definition of cuspidality, the constant term $\varphi_N(g)$ vanishes for a.e. $g \in G(\mathbf{A})$. Because φ is actually continuous, it follows that $\varphi_N(g) = 0$ for all g. Therefore when $m = 0$,

$$W_0^\varphi(g) = \varphi_N(g) = 0. \tag{3.9}$$

To φ we can attach a holomorphic function on $\mathbf{H}^r$ by

$$h(x + iy) = \varphi((^y{}_{1}^{\,x})_\infty \times 1_{\mathrm{fin}}) \prod_{j=1}^r y_j^{-k_j/2}. \tag{3.10}$$

Then h is a Hilbert modular form for the group $\Gamma_1(\mathfrak{N}) = \mathrm{SL}_2(F) \cap K_1(\mathfrak{N})$, so it has a Fourier expansion of the form

$$h(x + iy) = a_0(h) + \sum_{m \in \mathfrak{d}_+^{-1}} a_m(h)e^{-2\pi \, \mathrm{tr}(my)}\theta_{\infty, m}(x),$$

where $\mathrm{tr}(my) = \sum_{j=1}^r \sigma_j(m)y_j$. For any $y \in F_\infty^+$, the Fourier coefficients $a_m(h)$ and $W_m^\varphi(y)$ are related by

$$W_m^\varphi(y) = d_F^{1/2}(\prod_{j=1}^r y_j^{k_j/2})e^{-2\pi \, \mathrm{tr}(my)}a_m(h). \tag{3.11}$$

This follows immediately by equating the classical and adelic Fourier expansions of $\varphi((^{y_\infty}_{0}{}^{\,x_\infty}_{1}))$. Using (3.9), this implies that $W_m^\varphi(y) = 0$ unless $m \in \mathfrak{d}_+^{-1}$. $\qquad\square$

4 Construction of the test function

The right regular action of an element $f \in L^1(G(\mathbf{A}), \omega^{-1})$ on $L^2(\omega)$ is given by

$$R(f)\phi(x) = \int_{\overline{G}(\mathbf{A})} f(g)\phi(xg)dg. \tag{4.1}$$

In this section we will construct a continuous integrable function f such that $R(f)$ has finite rank (vanishing on $A_k(\mathfrak{N}, \omega)^{\perp}$), and acts like a Hecke operator on $A_k(\mathfrak{N}, \omega)$. The function will be defined locally:

$$f = f_\infty f_{\text{fin}} = \prod_{j=1}^{r} f_{\infty j} \prod_{v<\infty} f_v.$$

4.1 Archimedean test functions

For $j = 1, \ldots, r$ let v_{k_j} be a lowest weight unit vector for π_{k_j} and let d_{k_j} be the formal degree of π_{k_j} relative to the measure on $G(\mathbf{R})$ fixed in Sect. 3.2. We take $f_{\infty j}(g)$ to be the normalized matrix coefficient $d_{k_j}\overline{\langle \pi_{k_j}(g)v_{k_j}, v_{k_j}\rangle}$. Explicitly, if $g = \left(\begin{smallmatrix} a & b \\ c & d \end{smallmatrix}\right)$, then (cf. [KL2], Theorem 14.5)

$$f_{\infty j}(g) = \begin{cases} \dfrac{(k_j - 1)}{4\pi} \dfrac{\det(g)^{k_j/2}(2i)^{k_j}}{(-b + c + (a+d)i)^{k_j}} & \text{if } \det(g) > 0 \\[2ex] 0 & \text{otherwise.} \end{cases} \tag{4.2}$$

This function is integrable over $\overline{G}(\mathbf{R})$ if and only if $k_j > 2$. Therefore in order for (4.1) to converge we must assume henceforth that

$$k_j > 2 \quad (j = 1, \ldots, r).$$

Proposition 4.1. *Let* $f_\infty = \prod_j f_{\infty j}$ *be as above, and suppose* f_{fin} *is a bi-$K_1(\mathfrak{N})$-invariant function on* $G(\mathbf{A}_{\text{fin}})$ *satisfying*

$$f_{\text{fin}}(zg) = \omega_{\text{fin}}(z)^{-1} f_{\text{fin}}(g)$$

and whose support is compact mod $Z(\mathbf{A}_{\text{fin}})$. Then $R(f)$ vanishes on $A_k(\mathfrak{N}, \omega)^{\perp}$ (the orthogonal complement in $L^2(\omega)$) and its image is a subspace of $A_k(\mathfrak{N}, \omega)$.

Proof. The case $F = \mathbf{Q}$ is proven in Corollary 13.13 of [KL2], and the general case is no different. The main point is that $\int_{N(\mathbf{R})} f_{\infty j}(g_1 n g_2) dn = 0$ for each j. Using this, one shows that $R(f)\phi$ is cuspidal for each $\phi \in L^2(\omega)$. By the left $K_1(\mathfrak{N})$-invariance of f_{fin}, it is easy to see that $R(f)\phi$ is $K_1(\mathfrak{N})$-invariant, while the matrix coefficients project onto the span of the lowest weight vectors of the discrete series of the appropriate weight. Thus $R(f)\phi \in A_k(\mathfrak{N}, \omega)$. (See also [Li2], Prop. 2.2). $\qquad\square$

4.2 Non-Archimedean test functions

We now specify the local factors of f more precisely. Fix a discrete valuation v of F, and let $\mathfrak{p}$ be the corresponding prime ideal of $\mathcal{O}$. Let $\mathfrak{N} \subset \mathcal{O}$ be the ideal fixed earlier, and let $\mathfrak{N}_v = \mathfrak{N}\mathcal{O}_v$ be its localization.

Let $G_v = G(F_v)$, and similarly for its subgroups $Z_v = Z(F_v)$, etc. Suppose $f_v : G_v \to \mathbf{C}$ is a locally constant function whose support is compact modulo Z_v. Then the Hecke operator $R(f_v)$ is defined by

$$R(f_v)\phi(x) = \int_{\overline{G}_v} f_v(g)\phi(xg)dg$$

for any continuous function ϕ on G_v. Note that the integrand is not always well-defined modulo Z_v. In our situation, ϕ will be a function satisfying $\phi(zg) = \omega_v(z)\phi(g)$ for all $z \in Z_v$ and $g \in G_v$. Therefore we must require f_v to transform under the center by ω_v^{-1}.

The Hecke algebra of bi-$K_1(\mathfrak{N})_v$-invariant functions is generated by functions supported on sets of the form

$$Z_v K_1(\mathfrak{N})_v\, x_v\, K_1(\mathfrak{N})_v$$

for $x_v \in G_v$.

Fix an integral ideal $\mathfrak{n}_v$ in $\mathcal{O}_v$. We assume that $\mathfrak{n}_v$ is coprime to $\mathfrak{N}_v$, i.e., that either $\mathfrak{n}_v$ or $\mathfrak{N}_v$ is equal to $\mathcal{O}_v$. Define a set

$$M(\mathfrak{n}_v, \mathfrak{N}_v) = \{\begin{pmatrix} a & b \\ c & d \end{pmatrix} \in M_2(\mathcal{O}_v) |\, c \in \mathfrak{N}_v,\ (ad - bc)\mathcal{O}_v = \mathfrak{n}_v\}.$$

The determinant condition is equivalent to $\det g \in \varpi_v^{\mathrm{ord}_v(\mathfrak{n}_v)}\mathcal{O}_v^*$. By the Cartan decomposition of G_v we have

$$M(\mathfrak{n}_v, \mathfrak{N}_v) = \begin{cases} \displaystyle\bigcup_{\substack{i+j=\mathrm{ord}_v(\mathfrak{n}_v) \\ i \geq j \geq 0}} K_v \begin{pmatrix} \varpi_v^i & \\ & \varpi_v^j \end{pmatrix} K_v & \text{if } v \nmid \mathfrak{N} \\[2em] K_0(\mathfrak{N})_v & \text{if } v | \mathfrak{N}. \end{cases}$$

Clearly this is a compact set. We need to define a bi-$K_1(\mathfrak{N})_v$-invariant function $f_{\mathfrak{n}_v}$, supported on

$$Z_v M(\mathfrak{n}_v, \mathfrak{N}_v),$$

and with central character ω_v^{-1}. If $v|\mathfrak{N}$, for $k = \left(\begin{smallmatrix} a & b \\ c & d \end{smallmatrix}\right) \in K_0(\mathfrak{N})_v$ define

$$\omega_v(k) = \omega_v(d).$$

Because $c \in \mathfrak{N}_v$, one easily sees that this is a character of $K_0(\mathfrak{N})_v$. Now for $z \in Z_v$ and $m \in M(\mathfrak{n}_v, \mathfrak{N}_v)$, define

$$f_{\mathfrak{n}_v}(zm) = \begin{cases} \omega_v(z)^{-1} & \text{if } v \nmid \mathfrak{N} \\ \psi(\mathfrak{N}_v)\omega_v(z)^{-1}\omega_v(m)^{-1} & \text{if } v|\mathfrak{N}. \end{cases} \tag{4.3}$$

Here, when $\mathfrak{p}|\mathfrak{N}$,

$$\psi(\mathfrak{N}_v) = \text{meas}(\overline{K_1(\mathfrak{N})_v})^{-1} = [K_v : K_0(\mathfrak{N})_v] = \mathbb{N}(\mathfrak{p})^{\text{ord}_v(\mathfrak{N})}(1 + \mathbb{N}(\mathfrak{p})^{-1}).$$

It is straightforward to show that $f_{\mathfrak{n}_v}$ is well-defined.

Lemma 4.2. *Suppose* $g = \left(\begin{smallmatrix} a & b \\ c & d \end{smallmatrix}\right) \in G(F_v)$ *and that* $(\det g) = \mathfrak{n}_v$. *Then* $f_{\mathfrak{n}_v}(g) \neq 0$ *if and only if* $g \in M_2(\mathcal{O}_v)$ *and* $c \in \mathfrak{N}_v$.

Proof. Note that $f_{\mathfrak{n}_v}(g) \neq 0$ if and only if $g = zm$, with $z \in Z(F_v)$, $m \in M(\mathfrak{n}_v, \mathfrak{N}_v)$. Taking determinants we see that z is a unit in $\mathcal{O}_v$ (identifying Z_v with F_v^*). Thus z can be absorbed into m, so in fact $g \in M(\mathfrak{n}_v, \mathfrak{N}_v)$ as required. $\square$

Proposition 4.3. *The adjoint of the operator* $R(f_{\mathfrak{p}_v^\ell})$ *on* $L^2(G_v, \omega_v)$ *is given by*

$$R(f_{\mathfrak{p}_v^\ell})^* = \omega_v(\varpi_v^\ell)^{-1} R(f_{\mathfrak{p}_v^\ell}).$$

Proof. We have $R(f_{\mathfrak{p}_v^\ell})^* = R(f_{\mathfrak{p}_v^\ell}^*)$ where $f_{\mathfrak{p}_v^\ell}^*(g) = \overline{f_{\mathfrak{p}_v^\ell}(g^{-1})}$. If $g = zm$, then

$$g^{-1} = z^{-1}m^{-1} = (z^{-1}\varpi_v^{-\ell})(\varpi_v^\ell m^{-1}) \in Z_v M(\mathfrak{n}_v, \mathfrak{N}_v).$$

The proposition follows easily from this. (Note that $\ell = 0$ if $v|\mathfrak{N}$.) $\square$

When $v \nmid \mathfrak{N}$, the functions $f_{\mathfrak{n}_v}$ defined above linearly span the spherical Hecke algebra of bi-K_v-invariant complex-valued functions on G_v with central character ω_v^{-1}.

Now suppose $v \nmid \mathfrak{N}$ and write $\mathfrak{n}_v = \mathfrak{p}_v^\ell$ for $\ell > 0$. Let $\chi\left(\left(\begin{smallmatrix} a & b \\ c & d \end{smallmatrix}\right)\right) = \chi_1(a)\chi_2(d)$ be an unramified character of the Borel subgroup $B(F_v)$, and let (π, V_χ) be the representation of G_v obtained from χ by normalized induction. We assume that

$$\chi_1(z)\chi_2(z) = \omega_v(z)$$

for all $z \in Z_v$. Define a function $\phi_0 \in V_\chi$ by

$$\phi_0(\left(\begin{smallmatrix} a & b \\ 0 & d \end{smallmatrix}\right)k) = |a/d|_v^{1/2}\chi_1(a)\chi_2(d). \tag{4.4}$$

Then ϕ_0 spans the 1-dimensional space of K_v-fixed vectors in V_χ.

Proposition 4.4. *Let* $q_v = \mathbb{N}(\mathfrak{p}_v)$. *Then* $\pi(f_{\mathfrak{p}_v^\ell})\phi_0 = \lambda_{\mathfrak{p}_v^\ell}\phi_0$, *where*

$$\lambda_{\mathfrak{p}_v^\ell} = q_v^{\ell/2}\sum_{j=0}^{\ell}\chi_1(\varpi_v)^j\,\chi_2(\varpi_v)^{\ell-j}.$$

Proof. Because $f_{\mathfrak{p}_v^\ell}$ is left K_v-invariant, $R(f_{\mathfrak{p}_v^\ell})\phi_0$ is again fixed by K_v, and hence $R(f_{\mathfrak{p}_v^\ell})\phi_0 = \lambda\phi_0$ for some $\lambda \in \mathbf{C}$. The action of π is the same as the action of R, so it suffices to compute $\lambda = R(f_{\mathfrak{p}_v^\ell})\phi_0(1)$. Using the well-known left coset decomposition ([KL2], Lemma 13.4)

$$M(\mathfrak{p}_v^\ell, \mathfrak{N}_v) = \bigcup_{j=0}^{\ell}\bigcup_{a\in\mathcal{O}_v/\mathfrak{p}_v^j}\begin{pmatrix}\varpi_v^j & a \\ & \varpi_v^{\ell-j}\end{pmatrix}K_v,$$

we see that

$$\begin{aligned}
\lambda &= \int_{\overline{G}_v}f_{\mathfrak{p}_v^\ell}(g)\phi_0(g)dg \\
&= \int_{M(\mathfrak{p}_v^\ell, \mathfrak{N}_v)}\phi_0(g)dg \\
&= \sum_{j=0}^{\ell}\sum_{a\in\mathcal{O}_v/\mathfrak{p}_v^j}\phi_0(\left(\begin{smallmatrix}\varpi_v^j & a \\ & \varpi_v^{\ell-j}\end{smallmatrix}\right)) \qquad (\text{meas}(\overline{K}_v) = 1) \\
&= \sum_{j=0}^{\ell}q_v^j\left|\varpi_v^j/\varpi_v^{\ell-j}\right|_v^{1/2}\chi_1(\varpi_v)^j\chi_2(\varpi_v)^{\ell-j} \\
&= q_v^{\ell/2}\sum_{j=0}^{\ell}\chi_1(\varpi_v)^j\,\chi_2(\varpi_v)^{\ell-j},
\end{aligned}$$

as claimed. $\square$

Proposition 4.5. *With notation as above,*

$$q_v^{-\ell/2}\omega_v(\varpi_v)^{-\ell/2}\lambda_{\mathfrak{p}_v^\ell} = X_\ell(q_v^{-1/2}\omega_v(\varpi_v)^{-1/2}\lambda_{\mathfrak{p}_v}) \tag{4.5}$$

where

$$X_\ell(2\cos\theta) = \frac{\sin(\ell+1)\theta}{\sin\theta} = e^{i\ell\theta} + e^{i(\ell-2)\theta} + \cdots + e^{-i\ell\theta}$$

is the Chebyshev polynomial of degree ℓ.

Proof. Let $\alpha_{\mathfrak{p}} = \omega_v(\varpi_v)^{-1/2}\chi_1(\varpi_v)$ and $\beta_{\mathfrak{p}} = \omega_v(\varpi_v)^{-1/2}\chi_2(\varpi_v)$. Note that $\alpha_{\mathfrak{p}}\beta_{\mathfrak{p}} = 1$. Hence we may write $\alpha_{\mathfrak{p}} = e^{i\theta}$, $\beta_{\mathfrak{p}} = e^{-i\theta}$, and $\alpha_{\mathfrak{p}} + \beta_{\mathfrak{p}} = 2\cos\theta$ for some $\theta \in \mathbf{C}$. By the previous proposition, the left-hand side of (4.5) is

$$\sum_{j=0}^{\ell} \alpha_{\mathfrak{p}}^j \beta_{\mathfrak{p}}^{\ell-j} = \sum_{j=0}^{\ell} e^{ij\theta} e^{-i(\ell-j)\theta} = \sum_{j=0}^{\ell} e^{i(2j-\ell)\theta} = X_\ell(2\cos\theta).$$

This proves the result since $q_v^{-1/2}\omega(\varpi_v)^{-1/2}\lambda_{\mathfrak{p}_v} = \alpha_{\mathfrak{p}} + \beta_{\mathfrak{p}} = 2\cos\theta$. $\qquad\square$

4.3 Global Hecke operators

Finally we define the global Hecke operator. Fix an ideal $\mathfrak{n}$ in $\mathcal{O}$, relatively prime to $\mathfrak{N}$. Define a function on $\mathbf{A}_{\text{fin}}$ by

$$f_{\mathfrak{n}} = \prod_v f_{\mathfrak{n}_v},$$

where $f_{\mathfrak{n}_v}$ is defined as in the previous subsection. Then $f_{\mathfrak{n}}$ is bi-$K_1(\mathfrak{N})$-invariant, and supported on $Z(\mathbf{A}_{\text{fin}})M(\mathfrak{n}, \mathfrak{N})$, where

$$M(\mathfrak{n}, \mathfrak{N}) = \prod_{v<\infty} M(\mathfrak{n}_v, \mathfrak{N}_v)$$

$$= \left\{ \begin{pmatrix} a & b \\ c & d \end{pmatrix} \in M_2(\widehat{\mathcal{O}}) \,\bigg|\, c \in \widehat{\mathfrak{N}}, \ (ad-bc)\widehat{\mathcal{O}} = \widehat{\mathfrak{n}} \right\}.$$

It is also clear that

$$f_{\mathfrak{n}}(zg) = \omega_{\text{fin}}(z)^{-1} f_{\mathfrak{n}}(g) \quad (z \in Z(\mathbf{A}_{\text{fin}}), \ g \in G(\mathbf{A}_{\text{fin}})).$$

We define the operator

$$T_{\mathfrak{n}} = R(f_\infty \times f_{\mathfrak{n}})$$

on $L_0^2(\omega)$, which we can view as an operator on $A_{\text{k}}(\mathfrak{N}, \omega)$ by Proposition 4.1. The family of operators $T_{\mathfrak{n}}$ for $(\mathfrak{n}, \mathfrak{N}) = 1$ is simultaneously diagonalizable (see Lemma 6.3 below).

The following proposition and its corollaries spell out the connection between Hecke eigenvalues and Fourier coefficients.

Proposition 4.6. *Given* $\mathfrak{n}$, *choose* $n, d \in \mathbf{A}_{\text{fin}}^*$ *such that* $n\widehat{\mathcal{O}} = \widehat{\mathfrak{n}}$ *and* $d\widehat{\mathcal{O}} = \widehat{\mathfrak{d}}$. *Then for any* $\varphi \in A_{\text{k}}(\mathfrak{N}, \omega)$,

$$W_1^{T_{\mathfrak{n}}\varphi}(1/d) = \mathbb{N}(\mathfrak{n})W_1^{\varphi}(n/d).$$

Proof. We use the left coset decomposition

$$M(\mathfrak{n}, \mathfrak{N}) = \bigcup_{\substack{r,s \in \hat{\mathcal{O}}/\hat{\mathcal{O}}^* \\ rs\hat{\mathcal{O}}=\hat{\mathfrak{n}}}} \bigcup_{t \in \hat{\mathcal{O}}/r\hat{\mathcal{O}}} \begin{pmatrix} r & t \\ 0 & s \end{pmatrix} K_0(\mathfrak{N}),$$

which is proven as in [KL2], Lemma 13.5. We see that

$$T_{\mathfrak{n}}\varphi(\begin{pmatrix} y & x \\ & 1 \end{pmatrix}) = \int_{\overline{G}(\mathbf{A}_{\text{fin}})} f_{\mathfrak{n}}(g)\varphi(\begin{pmatrix} y & x \\ & 1 \end{pmatrix} g)dg$$

$$= \sum_{r,s} \sum_{t} \varphi(\begin{pmatrix} y & x \\ & 1 \end{pmatrix} \begin{pmatrix} r & t \\ & s \end{pmatrix}).$$

Note that

$$\varphi(\begin{pmatrix} y & x \\ & 1 \end{pmatrix} \begin{pmatrix} r & t \\ & s \end{pmatrix}) = \varphi(\begin{pmatrix} yr & yt+xs \\ & s \end{pmatrix}) = \omega_{\text{fin}}(s)\varphi(\begin{pmatrix} \frac{yr}{s} & \frac{yt}{s}+x \\ & 1 \end{pmatrix})$$

$$= \frac{\omega_{\text{fin}}(s)}{d_F^{1/2}}\left[W_0^\varphi(\frac{yr}{s}) + \sum_{m \in F^*} W_1^\varphi(\frac{myr}{s})\theta_m(\frac{yt}{s}+x) \right].$$

Therefore

$$W_1^{T_{\mathfrak{n}}\varphi}(y) = d_F^{1/2} \cdot [\text{coeff. of } \theta_1(x)] = \sum_{r,s} \omega_{\text{fin}}(s)W_1^\varphi(\frac{yr}{s}) \sum_t \theta(-\frac{yt}{s}).$$

Now suppose $y_\infty = 1$ and identify y with y_{fin}. Then we can assume that $yr/s \in \hat{\mathfrak{d}}^{-1}$ since otherwise $W_1^\varphi(\frac{yr}{s}) = 0$. Then $\theta(-yt/s)$ is a well-defined character of $t \in \hat{\mathcal{O}}/r\hat{\mathcal{O}}$, and

$$\sum_{t \in \hat{\mathcal{O}}/r\hat{\mathcal{O}}} \theta(-yt/s) = \begin{cases} \mathbb{N}(r) & \text{if } -y/s \in \hat{\mathfrak{d}}^{-1} \\ 0 & \text{otherwise.} \end{cases}$$

Thus

$$W_1^{T_{\mathfrak{n}}(\varphi)}(y) = \sum_{\substack{r,s \\ y/s\in\hat{\mathfrak{d}}^{-1}}} \omega_{\text{fin}}(s)\mathbb{N}(r)W_1^\varphi(\frac{yr}{s}) \qquad (y \in \mathbf{A}_{\text{fin}}^*). \qquad (4.6)$$

Now take $y = 1/d$. Then $y/s \in \hat{\mathfrak{d}}^{-1}$ only if $s = 1 \in \hat{\mathcal{O}}/\hat{\mathcal{O}}^*$. We can therefore take $r = \mathfrak{n}$ and $s = 1$, so

$$W_1^{T_{\mathfrak{n}}\varphi}(1/d) = \mathbb{N}(\mathfrak{n})W_1^\varphi(\mathfrak{n}/d),$$

as claimed. $\qquad \square$

Using the above, we can express Hecke eigenvalues in terms of Fourier coefficients and vice versa.

Corollary 4.7. *Suppose φ is an eigenvector of $T_\mathfrak{n}$ with eigenvalue $\lambda_\mathfrak{n}$. Then if $W_1^\varphi(1/\mathfrak{d}) \neq 0$,*

$$\lambda_\mathfrak{n} = \frac{\mathbb{N}(\mathfrak{n}) W_1^\varphi(\mathfrak{n}/\mathfrak{d})}{W_1^\varphi(1/\mathfrak{d})}.$$

Corollary 4.8. *If $(m\mathfrak{d}, \mathfrak{N}) = 1$, then for any $T_{m\mathfrak{d}}$-eigenfunction $\varphi \in A_k(\mathfrak{N}, \omega)$ with $W_1^\varphi(1/\mathfrak{d}) = 1$ and $T_{m\mathfrak{d}}\varphi = \lambda_{m\mathfrak{d}}\varphi$, we have*

$$W_m^\varphi(1) = \frac{e^{2\pi r} \prod_{j=1}^{r} \sigma_j(m)^{k_j/2-1}}{d_F e^{2\pi \operatorname{tr}(m)}} \lambda_{m\mathfrak{d}}.$$

Proof. Apply Cor. 4.7 with $\mathfrak{n} = m\mathfrak{d}$ and $\mathfrak{n} = m\mathfrak{d}$. We get

$$\lambda_{m\mathfrak{d}} = \mathbb{N}(m\mathfrak{d}) W_1^\varphi(1_\infty \times m_{\operatorname{fin}}) = \mathbb{N}(m\mathfrak{d}) W_m^\varphi(m_\infty^{-1} \times 1_{\operatorname{fin}}). \tag{4.7}$$

Here $m_\infty = (\sigma_1(m), \ldots, \sigma_r(m)) \in F_\infty^+$. Using (3.11), it is straightforward to show that

$$W_m^\varphi(m_\infty^{-1}) = (\prod_{j=1}^{r} \sigma_j(m)^{-k_j/2}) e^{2\pi \operatorname{tr}(m)} e^{-2\pi \operatorname{tr}(1)} W_m^\varphi(1).$$

Substituting this into (4.7) and using $\mathbb{N}(m\mathfrak{d}) = d_F \mathbb{N}(m)$ gives the result. $\quad\square$

5 A Hilbert modular Petersson trace formula

Let $f = f_\infty \times f_\mathfrak{n}$ for an ideal $(\mathfrak{n}, \mathfrak{N}) = 1$, and recall that $T_\mathfrak{n} = R(f)$ is the associated Hecke operator. Let $\mathcal{F}$ be any orthogonal basis for $A_k(\mathfrak{N}, \omega)$. (Later we will require $\mathcal{F}$ to consist of eigenvectors of $T_\mathfrak{n}$.) Then the kernel of $R(f)$ is given by

$$K(x, y) = \sum_{\varphi \in \mathcal{F}} \frac{R(f)\varphi(x)\overline{\varphi(y)}}{\|\varphi\|^2} = \sum_{\gamma \in \overline{G}(F)} f(x^{-1}\gamma y),$$

The first expression is the spectral expansion of the kernel, and the second is the geometric expansion. The equality of the two hinges on the continuity (in x, y) of the geometric expansion. The proof of this continuity given in Prop 3.2. of [Li2] carries over easily to the case of nontrivial central character we consider here.

In this section we apply the technique of Section 2 to the kernel function given above, taking $H = N \times N$. We need to fix a character on $N(F)\backslash N(\mathbf{A}) \times N(F)\backslash N(\mathbf{A})$, which amounts to choosing two characters on $F\backslash \mathbf{A}$. As discussed earlier, every character on $F\backslash \mathbf{A}$ is of the form

$$\theta_m(x) = \theta(-mx)$$

for some $m \in F$. Fix $m_1, m_2 \in F$. Our goal is to obtain a trace formula by computing

$$\int_{N(F)\backslash N(\mathbf{A})} \int_{N(F)\backslash N(\mathbf{A})} K(n_1, n_2)\overline{\theta_{m_1}(n_1)}\theta_{m_2}(n_2)dn_1 dn_2 \qquad (5.1)$$

with the two expressions for the kernel.

5.1 The spectral side

Using the spectral expansion of the kernel, expression (5.1) is easily computed in terms of Hecke eigenvalues and Fourier coefficients of cusp forms. Suppose the basis $\mathcal{F}$ consists of eigenfunctions of T_n. Then for $\varphi \in \mathcal{F}$ we have $R(f)\varphi = \lambda_n^\varphi \varphi$ for some scalar $\lambda_n^\varphi \in \mathbf{C}$. Hence (5.1) is

$$= \sum_{\varphi \in \mathcal{F}} \frac{\lambda_n^\varphi}{\|\varphi\|^2} \int_{N(F)\backslash N(\mathbf{A})} \varphi(n_1)\overline{\theta_{m_1}(n_1)}dn_1 \int_{N(F)\backslash N(\mathbf{A})} \overline{\varphi(n_2)}\theta_{m_2}(n_2)dn_2$$

$$= \sum_{\varphi \in \mathcal{F}} \frac{\lambda_n^\varphi W_{m_1}^\varphi(1)\overline{W_{m_2}^\varphi(1)}}{\|\varphi\|^2}, \qquad (5.2)$$

as in (3.7). By Proposition 3.3, the above expression is nonzero only if

$$m_1, m_2 \in \mathfrak{d}_+^{-1}.$$

We may assume that this holds from now on.

5.2 The geometric side

Here we use the method of Section 2 to compute (5.1) using the geometric expansion of the kernel. This gives a sum $\sum_{[\delta]} I_\delta(f)$, where

$$I_\delta(f) = \int_{H_\delta(F)\backslash H(\mathbf{A})} f(n_1^{-1}\delta n_2)\overline{\theta_{m_1}(n_1)}\theta_{m_2}(n_2)dn_1 dn_2.$$

The orbits $[\delta]$ are in one-to-one correspondence with the double cosets

$$N(F)\backslash \overline{G}(F)/N(F).$$

Let M be the group of invertible diagonal matrices. The Bruhat decomposition is the following partition of $G(F)$ into two cells:

$$G(F) = N(F)M(F) \cup N(F)M(F) \begin{pmatrix} 0 & 1 \\ 1 & 0 \end{pmatrix} N(F).$$

We call these the first and second **Bruhat cells** respectively. This gives

$$N(F)\backslash\overline{G}(F)/N(F) = \{[(\begin{smallmatrix} \gamma & 0 \\ 0 & 1 \end{smallmatrix})]\big|\ \gamma \in F^*\}\bigcup\{[(\begin{smallmatrix} 0 & \mu \\ 1 & 0 \end{smallmatrix})]\big|\ \mu \in F^*\}.$$

We need to determine which of these orbits are relevant in the sense of §2. First let $\delta = (\begin{smallmatrix} \gamma & 0 \\ 0 & 1 \end{smallmatrix}) \in G(F)$. If $((\begin{smallmatrix} 1 & t_1 \\ & 1 \end{smallmatrix}), (\begin{smallmatrix} 1 & t_2 \\ & 1 \end{smallmatrix})) \in H_\delta(\mathbf{A})$, then

$$\begin{pmatrix} 1 & -t_1 \\ & 1 \end{pmatrix}\begin{pmatrix} \gamma & 0 \\ 0 & 1 \end{pmatrix}\begin{pmatrix} 1 & t_2 \\ & 1 \end{pmatrix} = z \begin{pmatrix} \gamma & 0 \\ 0 & 1 \end{pmatrix},$$

for some $z \in Z(F)$. A simple calculation shows that $z = 1$ and $t_1 = \gamma t_2$, so

$$H_\delta(\mathbf{A}) = \left\{(\begin{pmatrix} 1 & \gamma t \\ 0 & 1 \end{pmatrix}, \begin{pmatrix} 1 & t \\ 0 & 1 \end{pmatrix})|\ t \in \mathbf{A}\right\}. \tag{5.3}$$

Thus δ is relevant if and only if

$$\theta((m_1\gamma - m_2)t) = 1$$

for all $t \in \mathbf{A}$, or equivalently, $\gamma = m_2/m_1$ (since $m_1 \in \mathfrak{d}_+^{-1}$ is nonzero).

On the other hand, if $\delta = \begin{pmatrix} 0 & \mu \\ 1 & 0 \end{pmatrix} \in G(F)$, one sees easily that

$$H_\delta(\mathbf{A}) = \{(e, e)\},$$

so all of these matrices are relevant.

5.2.1 *Computation of the first type of I_δ.*

Here we take $m_1, m_2 \in \mathfrak{d}_+^{-1}$, and $\delta = (\begin{smallmatrix} \gamma & \\ & 1 \end{smallmatrix})$ where $\gamma = m_2/m_1$. By (5.3),

$$I_\delta(f) = \int_{\{(\gamma t, t) \in F^2\}\backslash \mathbf{A} \times \mathbf{A}} f((\begin{smallmatrix} 1 & -t_1 \\ & 1 \end{smallmatrix})(\begin{smallmatrix} \gamma & 0 \\ 0 & 1 \end{smallmatrix})(\begin{smallmatrix} 1 & t_2 \\ & 1 \end{smallmatrix}))\theta(m_1 t_1 - m_2 t_2)dt_1 dt_2$$

$$= \int_{\{(\gamma t, t) \in F^2\}\backslash \mathbf{A} \times \mathbf{A}} f(\begin{pmatrix} \gamma & \gamma t_2 - t_1 \\ 0 & 1 \end{pmatrix})\theta(m_1 t_1 - m_2 t_2)\, dt_1 dt_2.$$

Let $t_1' = \gamma t_2 - t_1$ and $t_2' = t_2$. Then because $m_1 \gamma = m_2$, we have $m_1 t_1 - m_2 t_2 = -m_1 t_1'$, so

$$I_\delta = \int_{\{0\} \times F \backslash (\mathbf{A} \times \mathbf{A})} f\left(\begin{pmatrix} \gamma & t_1' \\ 0 & 1 \end{pmatrix}\right) \theta(-m_1 t_1') dt_1' dt_2'$$

$$= \operatorname{meas}(F \backslash \mathbf{A}) \int_{\mathbf{A}} f\left(\begin{pmatrix} m_2/m_1 & t \\ 0 & 1 \end{pmatrix}\right) \theta(-m_1 t) dt$$

$$= d_F^{1/2} \int_{\mathbf{A}} f\left(\begin{pmatrix} m_2 & m_1 t \\ & m_1 \end{pmatrix}\right) \theta(-m_1 t) dt$$

$$= d_F^{1/2} \int_{\mathbf{A}} f\left(\begin{pmatrix} m_2 & t \\ 0 & m_1 \end{pmatrix}\right) \theta(-t) dt.$$

Here we used (3.3) and the fact that $f(zg) = f(g)$ for $z \in Z(F)$.

We factorize the above integral into $(I_\delta)_\infty (I_\delta)_{\mathrm{fin}}$, and incorporate the coefficient $d_F^{1/2}$ into $(I_\delta)_\infty$. First we consider

$$(I_\delta)_{\mathrm{fin}} = \int_{\mathbf{A}_{\mathrm{fin}}} f_{\mathfrak{n}}\left(\begin{pmatrix} m_2 & t \\ & m_1 \end{pmatrix}\right) \theta_{\mathrm{fin}}(-t) dt.$$

We shall compute this locally, as we may since $f_{\mathfrak{n}_v}\left(\begin{pmatrix} m_2 & t \\ & m_1 \end{pmatrix}\right)\theta_v(-t)\big|_{t \in \mathcal{O}_v} \equiv 1$ for almost all v. For any finite place v we have

$$(I_\delta)_v = \int_{F_v} f_{\mathfrak{n}_v}\left(\begin{pmatrix} m_2 & t \\ & m_1 \end{pmatrix}\right) \theta_v(-t) dt.$$

The integrand is nonzero only if

$$\operatorname{ord}_v\left(\frac{m_1 m_2}{s_v^2}\right) = \operatorname{ord}_v(\mathfrak{n}_v)$$

for some $s_v \in F_v^*$. Supposing this is the case,

$$(I_\delta)_v = \int_{F_v} f_{\mathfrak{n}_v}\left(\begin{pmatrix} s_v & \\ & s_v \end{pmatrix}\begin{pmatrix} \frac{m_2}{s_v} & \frac{t}{s_v} \\ & \frac{m_1}{s_v} \end{pmatrix}\right) \theta_v(-t) dt$$

is nonzero only if $\frac{m_2}{s_v}, \frac{m_1}{s_v}, \frac{t}{s_v} \in \mathcal{O}_v$ by Lemma 4.2. Assuming this, we have

$$(I_\delta)_v = \int_{s_v \mathcal{O}_v} \omega_v(s_v)^{-1} f_{\mathfrak{n}_v}\left(\begin{pmatrix} \frac{m_2}{s_v} & \frac{t}{s_v} \\ & \frac{m_1}{s_v} \end{pmatrix}\right) \theta_v(-t) dt.$$

Suppose $v | \mathfrak{N}$. Then by the definition of $f_{\mathfrak{n}_v}$ (see (4.3)),

$$(I_\delta)_v = \omega_v(s_v)^{-1} \omega_v\left(\frac{m_1}{s_v}\right)^{-1} \psi(\mathfrak{N}_v) \int_{s_v \mathcal{O}_v} \theta_v(-t) dt.$$

The integral vanishes unless $s_v \in \mathfrak{d}_v^{-1}$, in which case its value is $|s_v|_v$. Hence

$$(I_\delta)_v = |s_v|_v \, \omega_v(s_v)^{-1} \omega_v(\frac{m_1}{s_v})^{-1} \psi(\mathfrak{N}_v), \quad v|\mathfrak{N}.$$

Now suppose $v \nmid \mathfrak{N}$. Then

$$(I_\delta)_v = \omega_v(s_v)^{-1} \int_{s_v \mathcal{O}_v} \theta_v(-t)dt,$$

which as before vanishes unless $s_v \in \mathfrak{d}_v^{-1}$. So in this case,

$$(I_\delta)_v = |s_v|_v \, \omega_v(s_v)^{-1}, \quad v \nmid \mathfrak{N}.$$

Define

$$\omega_{\mathfrak{N},v} = \begin{cases} \omega_v & \text{if } v|\mathfrak{N} \\ 1 & \text{if } v \nmid \mathfrak{N}, \end{cases} \tag{5.4}$$

and let $\omega_{\mathfrak{N}} = \prod \omega_{\mathfrak{N},v} = \prod_{v|\mathfrak{N}} \omega_v$. Notice that $\omega_{\mathfrak{N}}$ is a character on $\mathbf{A}_{\text{fin}}^*$. Multiplying the above local results together, we obtain the following.

Proposition 5.1. *Suppose* $\delta = \begin{pmatrix} \frac{m_2}{m_1} & 0 \\ 0 & 1 \end{pmatrix} \in G(F)$. *The integral* $(I_\delta)_{\text{fin}}$ *is nonzero if and only if there exists a finite idele* $s \in \mathbf{A}_{\text{fin}}^*$ *such that*

- $\frac{m_1}{s}, \frac{m_2}{s} \in \widehat{\mathcal{O}}$
- $\text{ord}_v(\frac{m_1 m_2}{s_v^2}) = \text{ord}_v(\mathfrak{n})$ *for all* $v < \infty$
- $s \in \widehat{\mathfrak{d}}^{-1}$.

Under these conditions,

$$(I_\delta)_{\text{fin}} = |s|_{\text{fin}} \omega_{\text{fin}}(s)^{-1} \psi(\mathfrak{N}) \omega_{\mathfrak{N}}(m_1/s)^{-1}$$

where $\psi(\mathfrak{N}) = [K_{\text{fin}} : K_0(\mathfrak{N})]$.

Remark: The expression is independent of the choice of s. Indeed if $s' = us$ for $u \in \widehat{\mathcal{O}}^*$, then

$$\omega_{\text{fin}}(s')^{-1} \omega_{\mathfrak{N}}(s') = \prod_{v \nmid \mathfrak{N}} \omega_v(s'_v) = \prod_{v \nmid \mathfrak{N}} \omega_v(s_v) = \omega_{\text{fin}}(s)^{-1} \omega_{\mathfrak{N}}(s)$$

since ω_v is unramified for $v \nmid \mathfrak{N}$.

For purposes of computation, the following lemma is helpful and easily verified.

Lemma 5.2. *With notation as above,* $(I_\delta)_{\text{fin}}$ *is nonzero if and only if for every finite valuation* v *of* F,

- $\xi_v \overset{\text{def}}{=} \frac{1}{2}\,\mathrm{ord}_v(m_1 m_2 \mathfrak{n}^{-1}) \in \mathbf{Z}$
- $\mathrm{ord}_v(m_i) \geq \xi_v \geq -\,\mathrm{ord}_v(\mathfrak{d}_v)$ *for* $i = 1, 2$.

If these conditions hold, then we can define s *by taking* $\mathsf{s}_v = \varpi_v^{\xi_v}$ *for all* v.

For the infinite part, we have the following.

Proposition 5.3. *Let* $\delta = \begin{pmatrix} \frac{m_2}{m_1} & 0 \\ 0 & 1 \end{pmatrix}$ *with* $m_1, m_2 \in F^*$. *Then* $(I_\delta)_\infty$ *is nonzero if and only if* $m_1, m_2 \in F^+$. *Under this condition,*

$$(I_\delta)_\infty = \frac{d_F^{1/2}}{e^{2\pi\,\mathrm{tr}_{\mathbf{Q}}^F(m_1+m_2)}} \prod_{j=1}^{r} \frac{(4\pi)^{k_j-1}\sigma_j(m_1 m_2)^{k_j/2}}{(k_j - 2)!}.$$

Proof. We have

$$(I_\delta)_\infty = d_F^{1/2} \prod_{j=1}^{r} \int_{-\infty}^{\infty} f_{\infty j}\!\left(\begin{pmatrix} \sigma_j(m_2) & t \\ & \sigma_j(m_1) \end{pmatrix} \right) \theta_{\infty j}(-t)\,dt.$$

The integrand is nonzero only if $\sigma_j(m_1 m_2) > 0$. By the formula for $f_{\infty j}$,

$$\int_{-\infty}^{\infty} f_{\infty j}\!\left(\begin{pmatrix} \sigma_j(m_2) & t \\ & \sigma_j(m_1) \end{pmatrix} \right) \theta_{\infty j}(-t)\,dt$$

$$= \frac{k_j - 1}{4\pi}\sigma_j(m_1 m_2)^{k_j/2}(2i)^{k_j} \int_{-\infty}^{\infty} \frac{e^{2\pi i t}}{(-t + i\sigma_j(m_1 + m_2))^{k_j}}\,dt.$$

Using a complex contour integral, this is nonzero only if $\sigma_j(m_1 + m_2) > 0$, in which case it equals

$$\frac{(4\pi)^{k_j-1}}{(k_j - 2)!}\sigma_j(m_1 m_2)^{k_j/2}e^{-2\pi\sigma_j(m_1+m_2)}.$$

Full details are given in [KL1], Proposition 3.4. Because $\sigma_j(m_1)$ and $\sigma_j(m_2)$ have the same sign, the condition $\sigma_j(m_1 + m_2) > 0$ is equivalent to $\sigma_j(m_1), \sigma_j(m_2) > 0$. $\qquad\square$

Proposition 5.4. *Let* $\delta = \begin{pmatrix} \frac{m_2}{m_1} & 0 \\ 0 & 1 \end{pmatrix}$ *for* $m_1, m_2 \in \mathfrak{d}_+^{-1}$. *Then* $I_\delta(f)$ *is nonzero if and only if* $\frac{m_1 m_2}{\mathsf{s}^2}\widehat{\mathcal{O}} = \widehat{\mathfrak{n}}$ *and* $\frac{m_1}{\mathsf{s}}, \frac{m_2}{\mathsf{s}} \in \widehat{\mathcal{O}}$ *for some* $\mathsf{s} \in \widehat{\mathfrak{d}}^{-1}$. *Under this condition,*

$$I_\delta = \left[\prod_{j=1}^{r} \frac{(4\pi\sqrt{\sigma_j(m_1 m_2)})^{k_j-1}}{(k_j - 2)!} \right] \frac{d_F^{1/2}\mathrm{N}(\mathfrak{n})^{1/2}\psi(\mathfrak{N})}{e^{2\pi\,\mathrm{tr}_{\mathbf{Q}}^F(m_1+m_2)}\omega_{\mathfrak{N}}(m_1/\mathsf{s})\omega_{\mathrm{fin}}(\mathsf{s})}.$$

Proof. By Lemma 5.2, we can take $s_v = \varpi_v^{(\operatorname{ord}_v(m_1m_2)-\operatorname{ord}_v(\mathfrak{n}))/2}$. Choose $\mathfrak{n} \in A_{\text{fin}}^*$ such that $\mathfrak{n}\widehat{\mathcal{O}} = \widehat{\mathfrak{n}}$. Then

$$|s|_{\text{fin}} = \left(\frac{|m_1m_2|_{\text{fin}}}{|\mathfrak{n}|_{\text{fin}}}\right)^{1/2}.$$

Note that by the product formula and the fact that $m_1, m_2 \in F^+$,

$$|m_1m_2|_{\text{fin}} = \prod_{j=1}^{r} \sigma_j(m_1m_2)^{-1}.$$

Hence

$$|s|_{\text{fin}} = \mathbb{N}(\mathfrak{n})^{1/2} \prod_{j=1}^{r} \sigma_j(m_1m_2)^{-1/2}.$$

The proposition now follows immediately upon multiplying $(I_\delta)_{\text{fin}}$ and $(I_\delta)_\infty$ together. $\qquad\square$

5.2.2 Computation of the second type of I_δ.

Let $v < \infty$ be a finite valuation of F. Let $\mathfrak{n}_v \in \mathcal{O}_v - \{0\}$ and let $m_{1v}, m_{2v} \in \mathfrak{d}_v^{-1}$. Recall that the conductor of ω_v divides $\mathfrak{N}_v$. For $c_v \in \mathfrak{N}_v - \{0\}$, define a generalized (local) Kloosterman sum by

$$S_{\omega_v}(m_{1v}, m_{2v}; \mathfrak{n}_v; c_v) = \sum_{\substack{s_1,s_2 \in \mathcal{O}_v/c_v\mathcal{O}_v \\ s_1s_2 \equiv \mathfrak{n}_v \bmod c_v\mathcal{O}_v}} \theta_v\left(\frac{m_{1v}s_1 + m_{2v}s_2}{c_v}\right)\omega_v(s_2)^{-1}.$$

The value of the sum is 1 if $c_v \in \mathcal{O}_v^*$. (The above notation is not quite consistent with [KL1], which has $\omega_v(s_1)^{-1}$.)

For $\mathfrak{n} \in \widehat{\mathcal{O}} \cap A_{\text{fin}}^*$, $c \in \widehat{\mathfrak{N}} \cap A_{\text{fin}}^*$, $m_1, m_2 \in \widehat{\mathfrak{d}}^{-1}$, and $\omega_{\mathfrak{N}}$ as in (5.4), we define

$$S_{\omega_{\mathfrak{N}}}(m_1, m_2; \mathfrak{n}; c) = \sum_{\substack{s_1,s_2 \in \widehat{\mathcal{O}}/c\widehat{\mathcal{O}} \\ s_1s_2 \equiv \mathfrak{n} \bmod c\widehat{\mathcal{O}}}} \theta_{\text{fin}}\left(\frac{m_1s_1 + m_2s_2}{c}\right)\omega_{\mathfrak{N}}(s_2)^{-1}.$$

Then

$$S_{\omega_{\mathfrak{N}}}(m_1, m_2; \mathfrak{n}; c) = \prod_{v<\infty} S_{\omega_{\mathfrak{N},v}}(m_{1v}, m_{2v}; \mathfrak{n}_v; c_v).$$

When $\delta = \begin{pmatrix} 0 & \mu \\ 1 & 0 \end{pmatrix}$, we have seen that $H_\delta = \{(e, e)\}$, so

$$I_\delta(f) = \iint_{N(\mathbf{A}) \times N(\mathbf{A})} f(n_1^{-1} \begin{pmatrix} 0 & \mu \\ 1 & 0 \end{pmatrix} n_2) \overline{\theta_{m_1}(n_1)} \theta_{m_2}(n_2) dn_1 dn_2.$$

We will compute this locally. Write $n_i = \begin{pmatrix} 1 & t_i \\ & 1 \end{pmatrix}$, $i = 1, 2$. Then

$$n_1^{-1} \begin{pmatrix} 0 & \mu \\ 1 & 0 \end{pmatrix} n_2 = \begin{pmatrix} -t_1 & \mu - t_1 t_2 \\ 1 & t_2 \end{pmatrix}. \tag{5.5}$$

Proposition 5.5. *Let $m_1, m_2 \in \mathfrak{d}^{-1}$. If $\delta = \begin{pmatrix} & \mu \\ 1 & \end{pmatrix}$, then $(I_\delta)_{\mathrm{fin}}$ is nonzero only if $c^2 \mu \widehat{\mathcal{O}} = \widehat{\mathfrak{n}}$ for some $c \in \widehat{\mathfrak{N}}$. Under this condition, let $n = -c^2 \mu$ be a generator of $\widehat{\mathfrak{n}}$ (the negative sign is for convenience). Then*

$$(I_\delta)_{\mathrm{fin}} = \left[\prod_{j=1}^{r} (-1)^{k_j} \right] \psi(\mathfrak{N}) \omega_{\mathrm{fin}}(c) S_{\omega \mathfrak{N}} (m_1, m_2; n; c).$$

Proof. For a discrete place v,

$$(I_\delta)_v = \iint_{F_v \times F_v} f_{\mathfrak{n}_v} \left(\begin{pmatrix} -t_1 & \mu - t_1 t_2 \\ 1 & t_2 \end{pmatrix} \right) \theta_v (m_1 t_1 - m_2 t_2) dt_1 dt_2.$$

The integrand is nonzero only if there exists $c_v \in F_v^*$ such that

$$\begin{pmatrix} c_v & \\ & c_v \end{pmatrix} \begin{pmatrix} -t_1 & \mu - t_1 t_2 \\ 1 & t_2 \end{pmatrix} \in M(\mathfrak{n}_v, \mathfrak{N}_v).$$

This means

(i) $c_v \in \mathfrak{N}_v$

(ii) $c_v t_1, c_v t_2 \in \mathcal{O}_v$

(iii) $\mathrm{ord}_v(c_v^2 \mu) = \mathrm{ord}_v(\mathfrak{n}_v)$.

Note that the first and third conditions determine c_v up to units. By the third condition, $c_v^2 \mu = -n_v$ for some generator $n_v \in \mathfrak{n}_v$. Now make substitutions in the integral by replacing t_1 and t_2 by $\frac{1}{c_v} t_1$ and $\frac{1}{c_v} t_2$ respectively. Then

$$(I_\delta)_v =$$

$$|c_v|_v^{-2} \iint_{\mathcal{O}_v \times \mathcal{O}_v} f_{\mathfrak{n}_v} (c_v^{-1} \begin{pmatrix} -t_1 & \frac{-n_v - t_1 t_2}{c_v} \\ c_v & t_2 \end{pmatrix}) \theta_v (\frac{m_1 t_1 - m_2 t_2}{c_v}) dt_1 dt_2.$$

The integrand is nonzero if and only if

(i) $c_v \in \mathfrak{N}_v$

(ii) $t_1 t_2 \equiv -n_v \mod c_v \mathcal{O}_v$.

Assuming these hold, the value

$$
f_{\mathfrak{n}_v}\left(c_v^{-1}\begin{pmatrix} -t_1 & \frac{-n_v-t_1t_2}{c_v} \\ c_v & t_2 \end{pmatrix}\right) = \begin{cases} \omega_v(c_v)\psi(\mathfrak{N}_v)\omega_v(t_2)^{-1} & \text{if } v|\mathfrak{N} \\ \\ \omega_v(c_v) & \text{otherwise} \end{cases}
$$

depends only on the residue class of t_2 modulo $\mathfrak{N}_v$. Furthermore, because θ_v is trivial on $\mathfrak{d}_v^{-1}$ and $m_1, m_2 \in \mathfrak{d}^{-1}$, the value $\theta_v(\frac{m_1t_1-m_2t_2}{c_v})$ depends only on the cosets $t_1 + c_v\mathcal{O}_v$ and $t_2 + c_v\mathcal{O}_v$. Thus the entire integrand is constant on cosets of $c_v\mathcal{O}_v$. Each of these cosets has measure $|c_v|_v$, so these measures for t_1 and t_2 will cancel the coefficient $|c_v|_v^{-2}$ in the integral. Therefore

$$
(I_\delta)_v = \psi(\mathfrak{N}_v)\omega_v(c_v) \sum_{\substack{s_1,s_2\in\mathcal{O}_v/c_v\mathcal{O}_v \\ s_1s_2\equiv-n_v \bmod c_v\mathcal{O}_v}} \omega_v(s_2)^{-1}\theta_v(\frac{m_1s_1-m_2s_2}{c_v})
$$

if $v|\mathfrak{N}$, and

$$
(I_\delta)_v = \omega_v(c_v) \sum_{\substack{s_1,s_2\in\mathcal{O}_v/c_v\mathcal{O}_v \\ s_1s_2\equiv-n_v \bmod c_v\mathcal{O}_v}} \theta_v(\frac{m_1s_1-m_2s_2}{c_v})
$$

if $v \nmid \mathfrak{N}$. Replacing s_2 by $-s_2$, in either case we see that

$$
(I_\delta)_v = \psi(\mathfrak{N}_v)\omega_{\mathfrak{N},v}(-1)\omega_v(c_v)S_{\omega_{\mathfrak{N},v}}(m_1, m_2; n_v; c_v).
$$

Multiplying the local results together, we obtain

$$
(I_\delta)_{\text{fin}} = \psi(\mathfrak{N})\omega_{\mathfrak{N}}(-1)\omega_{\text{fin}}(c)S_{\omega_{\mathfrak{N}}}(m_1, m_2; n; c).
$$

The final point is that because $\omega(-1) = 1$ and ω_v is unramified for $v \nmid \mathfrak{N}$,

$$
\omega_{\mathfrak{N}}(-1) = \prod_{v|\mathfrak{N}} \omega_v(-1) = \prod_{v<\infty} \omega_v(-1) = \omega_\infty(-1)^{-1} = \prod_{j=1}^{r}(-1)^{k_j}.
$$

$\square$

Proposition 5.6. *Let* $\delta = \begin{pmatrix} 1 & \mu \\ & 1 \end{pmatrix} \in G(F)$. *Then*

$$
(I_\delta)_\infty = \iint_{F_\infty \times F_\infty} f_\infty\left(\begin{pmatrix} -t_1 & \mu - t_1t_2 \\ 1 & t_2 \end{pmatrix}\right)\theta_\infty(m_1t_1 - m_2t_2)dt_1dt_2 \quad (5.6)
$$

is nonzero only if $m_1, m_2, -\mu$ are all totally positive. Under these conditions,

$$(I_\delta)_\infty = \frac{\mathbb{N}(-\mu)^{1/2}}{e^{2\pi \, \mathrm{tr}_{\mathbb{Q}}^F(m_1+m_2)}}$$

$$\cdot \prod_{j=1}^{r} \frac{(4\pi i)^{k_j} \sqrt{\sigma_j(m_1m_2)}^{-k_j-1} J_{k_j-1}(4\pi \sqrt{-\sigma_j(\mu m_1 m_2)})}{2(k_j-2)!}$$

where J_k is the Bessel J-function.

Proof. The computation of $(I_\delta)_{\infty_j}$ is given in [KL1] Proposition 3.6, and the above is the product of these local computations. Note that since $-\mu \in F^+$, $\mathbb{N}(-\mu) = \mathrm{N}(-\mu) = \prod_{j=1}^{r} \sigma_j(-\mu)$. $\qquad \square$

Multiplying the above results together, we obtain the following.

Proposition 5.7. *Let $\delta = \begin{pmatrix} & \mu \\ 1 & \end{pmatrix} \in G(F)$, and $m_1, m_2 \in \mathfrak{d}_+^{-1}$. Then I_δ is nonzero only if*

    (1) $-\mu$ is totally positive
    (2) $c^2\mu\widehat{\mathcal{O}} = \widehat{\mathfrak{n}}$ for some $c \in \widehat{\mathfrak{N}} \cap \mathbf{A}_{\mathrm{fin}}^*$.

Under these conditions, letting $\mathrm{n} = -c^2\mu$, we have

$$I_\delta = \psi(\mathfrak{N}) \, \omega_{\mathrm{fin}}(c) \, S_{\omega\mathfrak{N}}(m_1, m_2; \mathrm{n}; c) \frac{\mathbb{N}(-\mu)^{1/2}}{2^r e^{2\pi \, \mathrm{tr}_{\mathbb{Q}}^F(m_1+m_2)}}$$

$$\times \prod_{j=1}^{r} \frac{(-4\pi i)^{k_j} \sqrt{\sigma_j(m_1m_2)}^{-k_j-1}}{(k_j-2)!} J_{k_j-1}(4\pi \sqrt{-\sigma_j(\mu m_1 m_2)}).$$

5.2.3 Total contribution of the second Bruhat cell

The total contribution of the second type of δ is a summation of I_δ over all $\mu \in F^*$ satisfying the conditions in Proposition 5.7. For this we need to give a more systematic description of the set of such μ.

Let $c \in \mathbf{A}_{\mathrm{fin}}^*$, and let $\mathfrak{c} = c\widehat{\mathcal{O}} \cap F$ be the associated fractional ideal of F. Then condition (2) of Proposition 5.7 is equivalent to

$$\mathfrak{c}^2(\mu) = \mathfrak{n} \tag{5.7}$$

and

$$\mathfrak{N}|\mathfrak{c}. \tag{5.8}$$

We need to determine the elements μ which satisfy (5.7) for some $\mathfrak{c}$ as in (5.8).

Consider the following equation in the ideal class group

$$1 = [\mathfrak{b}]^2[\mathfrak{n}]. \tag{5.9}$$

If no solution $\mathfrak{b}$ exists, then the contribution of this Bruhat cell is 0. Some examples are given at the end of this section. Otherwise, let

$$[\mathfrak{b}_1], [\mathfrak{b}_2], \ldots, [\mathfrak{b}_t]$$

be the distinct solutions of (5.9). We take the $\mathfrak{b}_i$ to be integral ideals. Because $\mathfrak{b}_i^2\mathfrak{n}$ is principal, there exists a nonzero element $\eta_i \in \mathcal{O}$ such that

$$(\eta_i) = \mathfrak{b}_i^2\mathfrak{n}. \tag{5.10}$$

We fix such generators $\eta_1, \ldots, \eta_t$ once and for all.

Suppose (5.7) holds. Then $[\mathfrak{c}]^{-1} = [\mathfrak{b}_i]$ for some i, so $\mathfrak{c} = s\mathfrak{b}_i^{-1}$ for some $s \in F^*$. Substituting $\mathfrak{c} = s\mathfrak{b}_i^{-1}$ into (5.7), we see that $(\mu) = (s^{-2}\eta_i)$, i.e. $\mu = \frac{-\eta_i u}{s^2}$ for some $u \in \mathcal{O}^*$. Conversely if $\mu = \frac{-\eta_i u}{s^2}$ for some $s \in F^*$, then it satisfies (5.7) with $\mathfrak{c} = s\mathfrak{b}_i^{-1}$.

For each $i = 1, \ldots, t$ fix a generator $b_i \in \widehat{\mathfrak{b}_i}$. We obtain the following lemma:

Lemma 5.8. *An element* $\mu \in F^*$ *satisfies condition* (2) *of Proposition 5.7 if and only if*

$$\mu = \frac{-\eta_i u}{s^2} \tag{5.11}$$

for some $i \in \{1, \ldots, t\}$, $u \in \mathcal{O}^*$ *and* $s \in \mathfrak{b}_i\mathfrak{N}$. *If this condition holds, then we can take* $\mathfrak{c} = s\mathfrak{b}_i^{-1}$ *and* $\mathfrak{n} = \eta_i u \mathfrak{b}_i^{-2}$ *in Proposition 5.7.*

Proof. The above discussion shows that μ satisfies (5.7) if and only if it is given by (5.11) where $\mathfrak{c} = s\mathfrak{b}_i^{-1}$. Then it is easy to check that $\mathfrak{N}|\mathfrak{c}$ if and only if $s \in \mathfrak{b}_i\mathfrak{N}$. The last part of the lemma follows by a simple calculation. $\square$

Lemma 5.9. *In the notation above, if*

$$\frac{-\eta_i u}{s^2} = \frac{-\eta_{i'} u'}{s'^2},$$

then $i = i'$ *and* $u\mathcal{O}^{*2} = u'\mathcal{O}^{*2}$. *Furthermore if we assume* $u, u' \in U$ *as in* (3.1), *then* $u = u'$ *and* $s = \pm s'$.

Proof. By (5.10), we have $\mathfrak{b}_i^2\mathfrak{n}s^{-2} = \mathfrak{b}_{i'}^2\mathfrak{n}s'^{-2}$. Therefore $(\mathfrak{b}_i\mathfrak{b}_{i'}^{-1})^2 = (ss'^{-1})^2$. By unique factorization of ideals, we have $\mathfrak{b}_i\mathfrak{b}_{i'}^{-1} = (ss'^{-1})$. Therefore $[\mathfrak{b}_i] = [\mathfrak{b}_{i'}]$, and hence $i = i'$. With $i = i'$, we have $uu'^{-1} = (ss'^{-1})^2$. This implies that $uu'^{-1} \in \mathcal{O}^* \cap F^{*2} = \mathcal{O}^{*2}$. Then if $u, u' \in U$, it is immediate that $u = u'$, and hence $s = \pm s'$. $\square$

Proposition 5.10. *The total contribution of the second Bruhat cell is*

$$\frac{\psi(\mathfrak{N})}{2^r e^{2\pi \, \mathrm{tr}_{\mathbf{Q}}^F (m_1 + m_2)}} \prod_{j=1}^r \frac{(-4\pi i)^{k_j} \sqrt{\sigma_j (m_1 m_2)}^{-k_j - 1}}{(k_j - 2)!}$$

$$\times \sum_{i=1}^t \sum_{\substack{u \in U \\ \eta_i u \in F^+}} \sum_{\substack{s \in \mathfrak{b}_i \mathfrak{N}/\pm \\ s \neq 0}} \left\{ \omega_{\mathrm{fin}}(s\mathfrak{b}_i^{-1}) S_{\omega \mathfrak{N}} (m_1, m_2; \eta_i u \mathfrak{b}_i^{-2}; s\mathfrak{b}_i^{-1}) \frac{\mathbb{N}(\eta_i u)^{1/2}}{\mathbb{N}(s)} \right.$$

$$\left. \times \prod_{j=1}^r J_{k_j - 1} (4\pi \frac{\sqrt{\sigma_j (\eta_i u m_1 m_2)}}{|\sigma_j(s)|}) \right\}.$$

$$(5.12)$$

5.3 Main result

For $\varphi \in A_k(\mathfrak{N}, \omega)$ and $m \in \mathfrak{d}_+^{-1}$, let W_m^φ denote the Fourier coefficient of φ as in §3.4. If φ is an eigenfunction of the Hecke operator $T_\mathfrak{n} = R(f)$, we write

$$T_\mathfrak{n} \varphi = \lambda_\mathfrak{n}^\varphi \varphi.$$

Equating the geometric and spectral computations of the previous sections, we obtain the following upon multiplying both sides by

$$\frac{e^{2\pi \, \mathrm{tr}_{\mathbf{Q}}^F (m_1 + m_2)}}{\psi(\mathfrak{N})} \prod_{j=1}^r \frac{(k_j - 2)!}{(4\pi \sqrt{\sigma_j (m_1 m_2)})^{k_j - 1}}.$$

Theorem 5.11. *Let* $\mathfrak{n}$ *and* $\mathfrak{N}$ *be integral ideals with* $(\mathfrak{n}, \mathfrak{N}) = 1$. *Let* $k = (k_1, \ldots, k_r)$ *with all* $k_j > 2$. *Let* $\mathcal{F}$ *be an orthogonal basis for* $A_k(\mathfrak{N}, \omega)$ *consisting of eigenfunctions for the Hecke operator* $T_\mathfrak{n}$. *Then for any* $m_1, m_2 \in \mathfrak{d}_+^{-1}$,

$$\frac{e^{2\pi \, \mathrm{tr}_{\mathbf{Q}}^F (m_1 + m_2)}}{\psi(\mathfrak{N})} \left[\prod_{j=1}^r \frac{(k_j - 2)!}{(4\pi \sqrt{\sigma_j (m_1 m_2)})^{k_j - 1}} \right] \sum_{\varphi \in \mathcal{F}} \frac{\lambda_\mathfrak{n}^\varphi W_{m_1}^\varphi (1) \overline{W_{m_2}^\varphi (1)}}{\|\varphi\|^2}$$

$$= \widehat{T}(m_1, m_2, \mathfrak{n}) \frac{d_F^{1/2} \mathbb{N}(\mathfrak{n})^{1/2}}{\omega_\mathfrak{N}(m_1/s) \omega_{\mathrm{fin}}(s)}$$

$$+ \sum_{i=1}^t \sum_{\substack{u \in U \\ \eta_i u \in F^+}} \sum_{\substack{s \in \mathfrak{b}_i \mathfrak{N}/\pm \\ s \neq 0}} \left\{ \omega_{\mathrm{fin}}(s\mathfrak{b}_i^{-1}) S_{\omega \mathfrak{N}} (m_1, m_2; \eta_i u \mathfrak{b}_i^{-2}; s\mathfrak{b}_i^{-1}) \right.$$

$$\left. \times \frac{\mathbb{N}(\eta_i u)^{1/2}}{\mathbb{N}(s)} \prod_{j=1}^r \frac{2\pi}{(\sqrt{-1})^{k_j}} J_{k_j - 1} (4\pi \frac{\sqrt{\sigma_j (\eta_i u m_1 m_2)}}{|\sigma_j(s)|}) \right\},$$

where:

- $\widehat{T}(m_1, m_2, \mathfrak{n}) \in \{0, 1\}$ *is nonzero if and only if there exists* $\mathsf{s} \in \widehat{\mathfrak{d}}^{-1}$ *such that* $m_1, m_2 \in \mathsf{s}\widehat{\mathcal{O}}$ *and* $m_1 m_2 \widehat{\mathcal{O}} = \mathsf{s}^2 \widehat{\mathfrak{n}}$ *(see also Lemma 5.2)*
- U *is a set of representatives for* $\mathcal{O}^* / \mathcal{O}^{*2}$
- $\mathfrak{b}_i \widehat{\mathcal{O}} = \widehat{\mathfrak{b}}_i$ *for* $\mathfrak{b}_i$ *as in (5.9) (for* $i = 1, \ldots, t$)
- $\eta_i \in F$ *generates the principal ideal* $\mathfrak{b}_i^2 \mathfrak{n}$
- $\omega_{\mathfrak{N}} = \displaystyle\prod_{v \mid \mathfrak{N}} \omega_v \times \prod_{v \nmid \mathfrak{N}} 1.$

Remarks: (1) The above represents only part of a larger picture because we have treated just those Fourier coefficients coming from the identity component of $\overline{G}(F) \backslash \overline{G}(\mathbf{A}) / K_\infty K_0(\mathfrak{N})$. The general case would result from integrating $K\left(\left(\begin{smallmatrix} y_j & \\ & 1 \end{smallmatrix}\right) n_1, \left(\begin{smallmatrix} y_i & \\ & 1 \end{smallmatrix}\right) n_2\right)$, where $y_1, \ldots, y_h \in \mathbf{A}_{\text{fin}}^*$ are representatives for the class group of F.

(2) We can choose the basis so that each $\varphi \in V_\pi$ for some cuspidal representation π. Then the eigenvalue $\lambda_{\mathfrak{n}}^\varphi$ depends only on π, and not on φ. This can be seen from (6.5) and (6.6) below.

Example 5.12. *Suppose* $F = \mathbf{Q}[\sqrt{d}]$ *is a real quadratic field with narrow class number 1 (e.g.* $d = 2, 5$).

For such fields, the fundamental unit ε has negative norm. We can take $U = \{\pm 1, \pm \varepsilon\}$, and $\mathfrak{n} = (\eta)$, with $\eta \in F^+$. See Theorem 7.2 below.

Example 5.13. *Let* $F = \mathbf{Q}[\sqrt{d}]$ *be a real quadratic field with class number 1 and narrow class number 2 (e.g.* $d = 3$).

In this case, the fundamental unit ε has positive norm. If $\mathfrak{n}$ does not have a totally positive generator (i.e. if $\mathfrak{n} = (\eta)$ is nontrivial in the narrow class group), then there is no $u \in U$ for which $\eta u \in F^+$. Thus for such $\mathfrak{n}$, the entire Kloosterman term in the above theorem vanishes.

For general F, η_i can only make a nontrivial contribution to the Kloosterman term if it satisfies the sign condition

$$(\text{sgn } \eta_i^{\sigma_1}, \ldots, \text{sgn } \eta_i^{\sigma_r}) = \pm(\text{sgn } \varepsilon^{\sigma_1}, \ldots, \text{sgn } \varepsilon^{\sigma_r})$$

for some unit ε in a fundamental unit system. We now give some examples to illustrate various possibilities for the ideals $\mathfrak{b}_i$.

Example 5.14. *F has class number 2.*

In the class group $\text{Cl}(F)$, the equation $[\mathfrak{b}]^2[\mathfrak{n}] = 1$ has solutions if and only if $[\mathfrak{n}]$ is trivial (since $x^2 = 1$ for all $x \in \text{Cl}(F)$). Hence the Kloosterman term is nonzero only if $\mathfrak{n} = (\eta)$ is a principal ideal. In this case, we can take $\mathfrak{b}_1$ to be any non-principal ideal and $\mathfrak{b}_2 = \mathcal{O}$.

Example 5.15. *F has class number* 3.

Now we have $x^2 = x^{-1}$ for all $x \in \text{Cl}(F)$. Therefore the equation $[\mathfrak{b}]^2[\mathfrak{n}] = 1$ has a unique solution in $\text{Cl}(F)$, and we can take $\mathfrak{b} = \mathfrak{n}$.

Example 5.16. *F has odd class number.*

Suppose $h = |\text{Cl}(F)|$ is odd. Then the equation $[\mathfrak{b}]^2[\mathfrak{n}] = 1$ has a unique solution. For example we can take $\mathfrak{b} = \mathfrak{n}^{(h-1)/2}$.

6 Weighted distribution of Hecke eigenvalues

As an application of Theorem 5.11, we will show that relative to a certain measure, the (normalized) eigenvalues of the Hecke operator $T_{\mathfrak{p}}$ have a weighted equidistribution in the interval $[-2, 2]$ as $\mathbb{N}(\mathfrak{N}) \to \infty$.

6.1 Estimates

As a function of $\mathfrak{N}$, the contribution of the first Bruhat cell (given in Proposition 5.4) has order

$$\psi(\mathfrak{N}) = [K_{\text{fin}} : K_0(\mathfrak{N})] = \mathbb{N}(\mathfrak{N}) \prod_{\mathfrak{p}|\mathfrak{N}} (1 + \frac{1}{\mathbb{N}\mathfrak{p}}).$$

Here we will show that this is the dominant term in the Petersson trace formula as $\mathbb{N}(\mathfrak{N}) \to \infty$. For this, we need to show that the contribution (5.12) of the second Bruhat cell is small in comparison.

We start with the following naive estimate (essentially the triangle inequality) for the Kloosterman sums.

Lemma 6.1. *For any $m_1, m_2 \in \mathfrak{d}^{-1}$, nonzero $\mathfrak{n} \in \widehat{\mathcal{O}}$ and $c \in \widehat{\mathfrak{N}} \cap \mathbf{A}_{\text{fin}}^*$,*

$$|S_{\omega\mathfrak{N}}(m_1, m_2; \mathfrak{n}; c)| \leq \mathbb{N}(\mathfrak{n})\mathbb{N}(c).$$

Proof. It suffices to prove the lemma locally. Note that

$$\mathbb{N}(c) = \prod_{v < \infty} \mathbb{N}(c_v) = \prod_{v < \infty} |\mathcal{O}_v / c_v \mathcal{O}_v|.$$

We have

$$|S_{\omega_{\mathfrak{N},v}}(m_1, m_2; n_v; c_v)|$$

$$\leq \sum_{\substack{s_1, s_2 \in \mathcal{O}_v/c_v\mathcal{O}_v \\ s_1 s_2 \equiv n_v \bmod c_v\mathcal{O}_v}} |\theta_v(\frac{m_1 s_1 + m_2 s_2}{c_v})\omega_{\mathfrak{N},v}(s_2)^{-1}|$$

$$= \sum_{\substack{s_1, s_2 \in \mathcal{O}_v/c_v\mathcal{O}_v \\ s_1 s_2 \equiv n_v \bmod c_v\mathcal{O}_v}} 1.$$

In the case where $\mathrm{ord}_v(n_v) > \mathrm{ord}_v(c_v)$, we can replace n_v by c_v without loss of generality. Thus assuming $0 \leq \mathrm{ord}_v(n_v) \leq \mathrm{ord}_v(c_v)$, we need to count the elements of the following set:

$$\{(s_1, s_2) \in (\mathcal{O}_v/c_v\mathcal{O}_v)^2 \,|\, s_1 s_2 \equiv n_v \bmod c_v\mathcal{O}_v\}. \tag{6.1}$$

For a fixed $s_1 \in \mathcal{O}_v/c_v\mathcal{O}_v$ (we can assume that $0 \leq \mathrm{ord}_v(s_1) \leq \mathrm{ord}_v(c_v)$), there exists a solution s_2 to the congruence if and only if $\mathrm{ord}_v(s_1) \leq \mathrm{ord}_v(n_v)$. If this condition holds, then the number of solutions s_2 is

$$|\frac{c_v}{s_1}\mathcal{O}_v/c_v\mathcal{O}_v| = |\mathcal{O}_v/s_1\mathcal{O}_v| = N(s_1).$$

Thus the cardinality of (6.1) is

$$\sum_{\substack{s \in \mathcal{O}_v/c_v\mathcal{O}_v \\ \mathrm{ord}_v(s) \leq \mathrm{ord}_v(n_v)}} N(s) = \sum_{\ell=0}^{\mathrm{ord}_v n_v} N(\varpi_v^\ell) \cdot |(\mathcal{O}_v/\frac{c_v}{\varpi_v^\ell}\mathcal{O}_v)^*|$$

$$\leq \sum_{\ell=0}^{\mathrm{ord}_v n_v} N(\varpi_v^\ell)N(c_v/\varpi_v^\ell) = N(c_v)(\mathrm{ord}_v(n_v) + 1) \leq N(c_v)N(n_v) \tag{6.2}$$

as desired. $\square$

Proposition 6.2. *As a function of* $\mathfrak{N}$, *the contribution* (5.12) *of the second Bruhat cell is* $\ll \frac{\psi(\mathfrak{N})}{N(\mathfrak{N})^{2-\varepsilon}}$ *for any* $0 < \varepsilon < 1$ *as* $N(\mathfrak{N}) \to \infty$. *Here the implied constant depends on* $\mathfrak{n}$, k *and* ε.

Remark: The middle expression of (6.2) is $\ll_{\varepsilon'} N(c_v)N(n_v)^{\varepsilon'}$ for any $\varepsilon' > 0$. Using this, the dependence of (5.12) on $\mathfrak{n}$ can easily be shown to be $\ll N(\mathfrak{n})^{3/2+\varepsilon'-\varepsilon/2}$ for $0 < \varepsilon < 1$.

Proof. By Lemma 6.1, (5.12) is

$$\ll \psi(\mathfrak{N}) \sum_{i=1}^{t} \sum_{\substack{u \in U \\ \eta_i u \in F^+}} \sum_{\substack{s \in \mathfrak{b}_i \mathfrak{N}/\pm \\ s \neq 0}} \frac{N(\eta_i u)N(s)}{N(\mathfrak{b}_i)^3} \frac{N(\eta_i u)^{1/2}}{N(s)}$$

$$\cdot \prod_{j=1}^{r} J_{k_j-1}\left(\frac{4\pi\sqrt{\sigma_j(\eta_i u m_1 m_2)}}{|\sigma_j(s)|}\right)$$

$$\ll \psi(\mathfrak{N}) \sum_{i=1}^{t} \sum_{\substack{u \in U \\ \eta_i u \in F^+}} \sum_{\substack{s \in \mathfrak{b}_i \mathfrak{N}/\mathcal{O}^* \\ s \neq 0}} \sum_{a \in \mathcal{O}^*/\{\pm 1\}} \prod_{j=1}^{r} J_{k_j-1}\left(\frac{4\pi\sqrt{\sigma_j(\eta_i u m_1 m_2)}}{|\sigma_j(s)||\sigma_j(a)|}\right).$$

We remark that $|\sigma_j(s)|$ is not well-defined for $s \in \mathfrak{b}_i \mathfrak{N}/\mathcal{O}^*$, so the summands depend on a choice of representatives s, which we regard as fixed. By the units theorem, the set

$$\{(\log|\sigma_1(a)|, \ldots, \log|\sigma_r(a)|) : a \in \mathcal{O}^*\}$$

is a full lattice Λ in the $(r-1)$-dimensional hyperplane $x_1 + \cdots + x_r = 0$ in $\mathbf{R}^r$. Thus the summation over $a \in \mathcal{O}^*/\{\pm 1\}$ can be replaced by a sum over $\lambda = (\lambda_1, \ldots, \lambda_r) \in \Lambda$. Recall that

$$J_{k_j-1}(x) \ll \min(|x|^{-1/2}, |x|^{k_j-1}) \leq \frac{1}{\max(|x|^{1/2}, |x|^{-2})} \leq \frac{2}{|x|^{1/2} + |x|^{-2}}.$$

As a result, if we let $\tau_j = \log\left|\frac{\sigma_j(s)}{4\pi\sqrt{\sigma_j(\eta_i u m_1 m_2)}}\right|$ and $\lambda_j = \log|\sigma_j(a)|$, we can bound $J_{k_j-1}(e^{-(\lambda_j+\tau_j)})$ to give

$$\sum_{a \in \mathcal{O}^*/\{\pm 1\}} \prod_{j=1}^{r} J_{k_j-1}\left(\frac{4\pi\sqrt{\sigma_j(\eta_i u m_1 m_2)}}{|\sigma_j(s)||\sigma_j(a)|}\right)$$

$$\ll \sum_{\lambda \in \Lambda} \frac{1}{\prod_{j=1}^{r}(e^{-(\lambda_j+\tau_j)/2} + e^{2(\lambda_j+\tau_j)})}.$$

Let $\beta = 2 - \varepsilon$ for $0 < \varepsilon < 1$. Using $\sum_j \lambda_j = 0$, the right-hand sum is

$$= \sum_{\lambda \in \Lambda} \frac{1}{\prod_{j=1}^{r} e^{\beta(\lambda_j+\tau_j)} \prod_{j=1}^{r}(e^{-(1/2+\beta)(\lambda_j+\tau_j)} + e^{\varepsilon(\lambda_j+\tau_j)})}$$

$$= \sum_{\lambda \in \Lambda} \frac{1}{e^{(\sum_{j=1}^r \tau_j)\beta} \prod_{j=1}^r (e^{-(1/2+\beta)(\lambda_j+\tau_j)} + e^{\varepsilon(\lambda_j+\tau_j)})}$$

$$= \left| \frac{(4\pi)^r \sqrt{\mathbb{N}(\eta_i u m_1 m_2)}}{\mathbb{N}(s)} \right|^\beta \sum_{\lambda \in \Lambda} \frac{1}{\prod_{j=1}^r (e^{-(1/2+\beta)(\lambda_j+\tau_j)} + e^{\varepsilon(\lambda_j+\tau_j)})}.$$

We claim that this last sum over λ is bounded, independently of s. Let L_0 be the hyperplane $x_1 + \cdots + x_r = 0$ containing Λ. Then

$$\sum_{\lambda \in \Lambda} \frac{1}{\prod_{j=1}^r (e^{-(1/2+\beta)(\lambda_j+\tau_j)} + e^{\varepsilon(\lambda_j+\tau_j)})}$$

$$\ll \int_{L_0} \frac{dx_1 \cdots dx_{r-1}}{\prod_{j=1}^r (e^{-(1/2+\beta)(x_j+\tau_j)} + e^{\varepsilon(x_j+\tau_j)})}.$$

The rth factor of the integrand is bounded, so the above is

$$\ll \int_{\mathbf{R}^{r-1}} \frac{dx_1 \cdots dx_{r-1}}{\prod_{j=1}^{r-1} (e^{-(1/2+\beta)(x_j+\tau_j)} + e^{\varepsilon(x_j+\tau_j)})}$$

$$= \prod_{j=1}^{r-1} \int_{\mathbf{R}} \frac{dx}{e^{-(1/2+\beta)x} + e^{\varepsilon x}} < \infty$$

as claimed. Therefore (5.12) is

$$\ll \psi(\mathfrak{N}) \sum_{i=1}^t \sum_{\substack{u \in U \\ \eta_i u \in F^+}} \sum_{\substack{s \in \mathfrak{b}_i \mathfrak{N}/\mathcal{O}^* \\ s \neq 0}} \left| \frac{(4\pi)^r \sqrt{\mathbb{N}(\eta_i u m_1 m_2)}}{\mathbb{N}(s)} \right|^{2-\varepsilon}.$$

The first two sums are taken over finite sets. The set of $s \in \mathfrak{b}_i \mathfrak{N}/\mathcal{O}^*$ is in 1-1 correspondence with the set of principal ideals (s) divisible by $\mathfrak{b}_i \mathfrak{N}$. Hence we can replace the sum over s by a larger set: all integral ideals $\mathfrak{a}$ divisible by $\mathfrak{N}$. Therefore (5.12) is

$$\ll \psi(\mathfrak{N}) \sum_{\mathfrak{N} | \mathfrak{a}} \frac{1}{\mathbb{N}(\mathfrak{a})^{2-\varepsilon}} \ll \frac{\psi(\mathfrak{N})}{\mathbb{N}(\mathfrak{N})^{2-\varepsilon}} \sum_{\mathfrak{a}'} \frac{1}{\mathbb{N}(\mathfrak{a}')^{2-\varepsilon}} \ll \frac{\psi(\mathfrak{N})}{\mathbb{N}(\mathfrak{N})^{2-\varepsilon}}.$$

(We use the absolute convergence of the zeta function $\zeta_F(s) = \sum_{\mathfrak{a}'} \frac{1}{\mathbb{N}(\mathfrak{a}')^s}$ for $\operatorname{Re} s > 1$.) $\qquad \square$

6.2 Weighted equidistribution

Let (X, μ) be a Borel measure space. For each $i = 1, 2, \ldots$ let $\mathcal{F}_i$ be a finite nonempty index set, and let $S_i = \{x_{ij}\}_{j \in \mathcal{F}_i}$ be a finite sequence of points of X.

Suppose each $j \in \mathcal{F}_i$ is assigned a weight $w_{ij} \in \mathbf{R}^+$. Define

$$d\mu_i = \frac{\sum_{j \in \mathcal{F}_i} w_{ij} \delta_{x_{ij}}}{\sum_{j \in \mathcal{F}_i} w_{ij}},$$

where $\delta_{x_{ij}}$ is the Dirac measure at x_{ij}. We say that the sequence $\{S_i\}$ is **w-equidistributed** with respect to the measure $d\mu$ if

$$\lim_{i \to \infty} d\mu_i = \lim_{i \to \infty} \frac{\sum_{j \in \mathcal{F}_i} w_{ij} \delta_{x_{ij}}}{\sum_{j \in \mathcal{F}_i} w_{ij}} = d\mu.$$

This means that for any continuous function $f : X \to \mathbf{C}$, we have

$$\lim_{i \to \infty} \int_X f(x) d\mu_i(x) = \lim_{i \to \infty} \frac{\sum_{j \in \mathcal{F}_i} w_{ij} f(x_{ij})}{\sum_{j \in \mathcal{F}_i} w_{ij}} = \int_X f(x) d\mu(x).$$

If $w_{ij} = 1$ for all i, j, then this definition reduces to that of equidistribution given in §1 of [Se].

6.3 The distribution theorem

Let $\mathfrak{p}$ be a prime ideal, not dividing the level $\mathfrak{N}$. In this section we will apply the main theorem in the case where $\mathfrak{n} = \mathfrak{p}^\ell$ ($\ell \geq 0$), and $m = m_1 = m_2 \in \mathfrak{d}_+^{-1}$.

For each irreducible summand π of $H_k(\mathfrak{N}, \omega)$ (cf. (3.4)), let $\mathcal{F}_\pi$ be an orthogonal basis for the nonzero finite-dimensional subspace $\pi \cap A_k(\mathfrak{N}, \omega)$. Let

$$\mathcal{F} = \bigcup_\pi \mathcal{F}_\pi \tag{6.3}$$

be the resulting orthogonal basis for $A_k(\mathfrak{N}, \omega)$.

Lemma 6.3. *For any $\varphi \in \mathcal{F}$ and any prime ideal $\mathfrak{p} \nmid \mathfrak{N}$, the cuspidal representation (π, V) containing φ is unramified at $\mathfrak{p}$. Furthermore, φ is an eigenfunction of the global Hecke operator $T_{\mathfrak{p}^\ell}$, and the associated eigenvalue $\lambda_{\mathfrak{p}^\ell}^\varphi$ coincides with the local eigenvalue $\lambda_{\mathfrak{p}_v^\ell}$ attached to $\pi_{\mathfrak{p}}$ in Prop. 4.4.*

Proof. Write $\pi_{\mathrm{fin}} = \pi_{\mathfrak{p}} \otimes \pi'$, where $\pi' = \otimes_{\substack{v < \infty \\ v \neq v_{\mathfrak{p}}}} \pi_v$ is a representation of the restricted direct product $G' = \prod'_{\substack{v < \infty \\ v \neq v_{\mathfrak{p}}}} G_v$. Let $K' = G' \cap K_1(\mathfrak{N})$. Let $f = f_{\mathfrak{n}}$ with $\mathfrak{n} = (1)$. Then $\pi_{\mathrm{fin}}(f)$ is the projection operator of V_{fin} onto $\pi_{\mathrm{fin}}^{K_1(\mathfrak{N})}$, and

$$\pi_{\mathrm{fin}}^{K_1(\mathfrak{N})} = \pi_{\mathrm{fin}}(f) V_{\mathrm{fin}} = \pi_{\mathfrak{p}}(f_{v_{\mathfrak{p}}}) V_{\mathfrak{p}} \otimes \pi'(f') V' = \pi_{\mathfrak{p}}^{K_{\mathfrak{p}}} \otimes \pi'^{K'}. \tag{6.4}$$

The middle equality holds e.g. by Prop. 13.17 of [KL2].

Now take $\mathfrak{n} = \mathfrak{p}^\ell$ and $f = f_\infty \times f_\mathfrak{n}$. By the definition of $A_k(\mathfrak{N}, \omega)$ (eq. (3.5)) we can write $\varphi = w_\infty \otimes w_{\text{fin}}$, where $w_\infty = \otimes v_{\pi_{\infty_j}}$ and $w_{\text{fin}} \in \pi_{\text{fin}}^{K_1(\mathfrak{N})}$. Because $0 \neq w_{\text{fin}} \in \pi_{\text{fin}}^{K_1(\mathfrak{N})}$, it follows immediately from (6.4) that $\pi_\mathfrak{p}$ is unramified. Because $\pi_\mathfrak{p}$ is also unitary, it follows that $\pi_\mathfrak{p} = \pi_\chi$ is induced from some unramified character $\chi(\left(\begin{smallmatrix} a & b \\ 0 & d \end{smallmatrix}\right)) = \chi_1(a)\chi_2(d)$ of $B(F_v)$. Therefore $\pi_\mathfrak{p}^{K_\mathfrak{p}} = \mathbf{C}\phi_0$ is one-dimensional (for notation see (4.4)), and we can write $w_{\text{fin}} = \phi_0 \otimes w'$ as in (6.4). Thus $\varphi = w_\infty \otimes \phi_0 \otimes w'$, and

$$T_{\mathfrak{p}^\ell}\varphi = R(f)\varphi = \pi_\infty(f_\infty)w_\infty \otimes \pi_\mathfrak{p}(f_{v_\mathfrak{p}})\phi_0 \otimes \pi'(f')w'$$
$$= w_\infty \otimes \lambda_{\mathfrak{p}_v^\ell}\phi_0 \otimes w' = \lambda_{\mathfrak{p}_v^\ell}\varphi.$$

$\square$

The local Langlands class of $\pi_\mathfrak{p}$ is the $GL_2(\mathbf{C})$-conjugacy class of the matrix $g(\pi_\mathfrak{p}) = \left(\begin{smallmatrix} \chi_1(\varpi_\mathfrak{p}) & \\ & \chi_2(\varpi_\mathfrak{p}) \end{smallmatrix}\right)$. Let $q = \mathbf{N}(\mathfrak{p})$. By the above lemma and Prop. 4.4, the trace of $g(\pi_\mathfrak{p})$ is given by

$$\chi_1(\varpi_\mathfrak{p}) + \chi_2(\varpi_\mathfrak{p}) = q^{-1/2}\lambda_\mathfrak{p}^\varphi. \tag{6.5}$$

If χ is unitary (i.e. $\pi_\mathfrak{p}$ is not complementary series), then it is clear that $|\lambda_\mathfrak{p}^\varphi| \leq 2q^{1/2}$. In fact by the Ramanujan conjecture this is always the case.[1]

When $\mathfrak{n} = \mathfrak{p}^\ell$, the operator $\omega_\mathfrak{p}(\varpi_\mathfrak{p})^{-\ell/2}R(f)$ is self-adjoint (cf. Prop. 4.3), so its eigenvalues are real numbers. We let

$$\nu_{\mathfrak{p}^\ell}^\varphi \stackrel{\text{def}}{=} \omega_\mathfrak{p}(\varpi_\mathfrak{p})^{-\ell/2}q^{-\ell/2}\lambda_{\mathfrak{p}^\ell}^\varphi \in \mathbf{R}$$

denote this normalized Hecke eigenvalue. We sometimes write $\nu_{\mathfrak{p}^\ell}^\pi$ to emphasize that it depends only on the cuspidal representation π. Note that $\nu_\mathfrak{p}^\varphi \in [-2, 2]$ by the Ramanujan conjecture. By Prop. 4.5,

$$\nu_{\mathfrak{p}^\ell}^\varphi = X_\ell(\nu_\mathfrak{p}^\varphi). \tag{6.6}$$

We will adapt the argument of [Li1] for finding the asymptotic weighted distribution of the set of $\nu_\mathfrak{p}^\varphi$.

Lemma 6.4. *In the notation of Theorem 5.11, for any $\ell \geq 0$ and any $m \in \mathfrak{d}_+^{-1}$, $\widehat{T}(m, m, \mathfrak{p}^\ell) \neq 0$ if and only if both of the following hold:*

(i) $\ell = 2\ell'$ *is even*

(ii) $0 \leq \ell' \leq \text{ord}_\mathfrak{p}(m\mathfrak{d})$.

---

[1] The Ramanujan conjecture was proven at all but a finite number of unspecified places for holomorphic cuspidal representations π of $GL_2(\mathbf{A})$ with all weights ≥ 2 by Brylinski and Labesse ([BL], Theorem 3.4.6). Recently the full conjecture (at all places, when all weights are ≥ 2) was proven by Blasius, with a parity condition on the weights [Bl]. The parity requirement was removed in the thesis of his student L. Nguyen [Ng]. However, as we remark after Theorem 6.6 below, our results do not actually depend on this deep theorem.

Proof. The two conditions of Lemma 5.2 specialize respectively to the two conditions above in the special case where $m_1 = m_2 = m$ and $\mathfrak{n} = \mathfrak{p}^\ell$. $\qquad\square$

Proposition 6.5. *Fix $m \in \mathfrak{d}_+^{-1}$ and a prime ideal $\mathfrak{p} \nmid \mathfrak{N}$. For each $\varphi \in \mathcal{F}$, define a weight $w_\varphi = \frac{|W_m^\varphi(1)|^2}{\|\varphi\|^2}$. Then for any $\ell \geq 0$ and $0 < \varepsilon < 1$,*

$$\sum_{\varphi \in \mathcal{F}} X_\ell(\nu_\mathfrak{p}^\varphi) w_\varphi =$$

$$\begin{cases} J\psi(\mathfrak{N}) + O(\frac{\psi(\mathfrak{N})}{\mathrm{N}(\mathfrak{N})^{2-\varepsilon}}) & \text{if } \ell = 2\ell' \text{ with } 0 \leq \ell' \leq \mathrm{ord}_v\, m\mathfrak{d} \\ O(\frac{\psi(\mathfrak{N})}{\mathrm{N}(\mathfrak{N})^{2-\varepsilon}}) & \text{otherwise,} \end{cases}$$

where X_ℓ is the Chebyshev polynomial defined in Prop. 4.5, and

$$J = \frac{d_F^{1/2}}{e^{4\pi\, \mathrm{tr}_\mathbb{Q}^F(m)}} \prod_{j=1}^r \frac{(4\pi\sigma_j(m))^{k_j-1}}{(k_j-2)!}.$$

The implied constant depends only on m, $\mathfrak{p}$, ℓ, k and ε.

Remark: This shows in particular that when $\mathrm{N}(\mathfrak{N})$ is sufficiently large,

  (i) $A_k(\mathfrak{N}, \omega)$ is nontrivial
  (ii) $W_m^\varphi(1)$ is nonzero for some $\varphi \in \mathcal{F}$.

Proof. This follows from the generalized Petersson trace formula. We use the form developed in the proof, rather than the final statement in Theorem 5.11. The spectral side (5.2) of the trace formula with $\mathfrak{n} = \mathfrak{p}^\ell$ and $m_1 = m_2 = m \in \mathfrak{d}_+^{-1}$ gives

$$\sum_{\varphi \in \mathcal{F}} \frac{\lambda_{\mathfrak{p}^\ell}^\varphi |W_m^\varphi(1)|^2}{\|\varphi\|^2} = \sum_{\varphi \in \mathcal{F}} \omega_\mathfrak{p}(\varpi_\mathfrak{p})^{\ell/2} q^{\ell/2} X_\ell(\nu_\mathfrak{p}^\varphi) w_\varphi.$$

The above is equal to the geometric side, which by Prop. 5.4 and Prop. 6.2 is

$$= \frac{\widehat{T}(m, m, \mathfrak{p}^\ell)}{\omega_\mathfrak{N}(m/s)\omega_{\mathrm{fin}}(s)} q^{\ell/2} J\psi(\mathfrak{N}) + O(\frac{\psi(\mathfrak{N})}{\mathrm{N}(\mathfrak{N})^{2-\varepsilon}})$$

for J as above. By the above lemma, the first term is nonzero if and only if $\ell = 2\ell'$ with $0 \leq \ell' \leq \mathrm{ord}_\mathfrak{p}(m\mathfrak{d})$. In this case we can take

$$s = (\ldots, m, m, \overset{\mathfrak{p}^{\mathrm{th}}}{m\varpi_\mathfrak{p}^{-\ell'}}, m, m, \ldots).$$

Note that $\omega_\mathfrak{N}(m/s) = \prod_{\mathfrak{p}|\mathfrak{N}} \omega_\mathfrak{p}(1) = 1$, so

$$\omega_\mathfrak{N}(m/s)\omega_{\mathrm{fin}}(s) = \omega_{\mathrm{fin}}(s) = \omega_{\mathrm{fin}}(m)\omega_\mathfrak{p}(\varpi_\mathfrak{p})^{-\ell'}.$$

Furthermore $\omega_{\text{fin}}(m) = \omega_\infty(m)^{-1} = 1$ since $m \in F^+$. Thus in this case the geometric side is

$$= \omega_{\mathfrak{p}}(\varpi_{\mathfrak{p}})^{\ell/2} q^{\ell/2} J\psi(\mathfrak{N}) + O\Big(\frac{\psi(\mathfrak{N})}{\mathbb{N}(\mathfrak{N})^{2-\varepsilon}}\Big).$$

The proposition now follows by equating the spectral side with the geometric side and dividing by $\omega_{\mathfrak{p}}(\varpi_{\mathfrak{p}})^{\ell/2} q^{\ell/2}$. □

Theorem 6.6. *Let $\mathfrak{p}$ be a prime ideal of F, and let $\mathrm{k} = (\mathrm{k}_1, \dots, \mathrm{k}_r)$ be a weight vector with all $\mathrm{k}_j > 2$. For each $i = 1, 2, \dots$*

- *let $\mathfrak{N}_i$ be an ideal coprime to $\mathfrak{p}$, with $\lim\limits_{i \to \infty} \mathbb{N}(\mathfrak{N}_i) = \infty$,*
- *let ω_i be a unitary character as in §3.3 relative to $\mathfrak{N}_i$ and k,*
- *let $\mathcal{F}_i$ be an orthogonal basis for $A_{\mathrm{k}}(\mathfrak{N}_i, \omega_i)$ as in (6.3).*

Fix any $m \in \mathfrak{d}_+^{-1}$, and define weights $w_\varphi = |W_m^\varphi(1)|^2 / \|\varphi\|^2$ for $\varphi \in \mathcal{F}_i$. For each i, define a sequence

$$S_i = \{v_{\mathfrak{p}}^\varphi\}_{\varphi \in \mathcal{F}_i}$$

in the interval $[-2, 2]$. Then the sequence S_i is w_φ-equidistributed relative to the measure

$$d\mu(x) = \sum_{\ell'=0}^{\mathrm{ord}_{\mathfrak{p}}(\partial m)} X_{2\ell'}(x) d\mu_\infty(x),$$

where $d\mu_\infty(x)$ is the Sato-Tate measure defined in the Introduction. In other words, for any continuous function h on $\mathbf{R}$,

$$\lim_{i \to \infty} \frac{\sum_{\varphi \in \mathcal{F}_i} h(v_{\mathfrak{p}}^\varphi) w_\varphi}{\sum_{\varphi \in \mathcal{F}_i} w_\varphi} = \int_{\mathbf{R}} h(x) d\mu(x).$$

Remarks: (1) When $\mathfrak{p} \nmid m\mathfrak{d}$, the measure μ coincides with μ_∞ and is independent of $\mathfrak{p}$. Further taking all k_j even and ω_i trivial, we immediately obtain Theorem 1.1.

(2) The above result (and its proof) is actually independent of the Ramanujan conjecture. All we need is the existence of a finite interval $I_{\mathfrak{p}}$ which contains all of the eigenvalues $v_{\mathfrak{p}}^\varphi$. This is elementary ([Ro], Prop. 2.9).

(3) The theorem shows in particular that the Satake traces $v_{\mathfrak{p}}^\pi$ are dense in the interval $[-2, 2]$. The referee has pointed out that this can also be seen directly by considering CM cusp forms.

Proof. Setting $\ell = 0$ in the previous proposition, we have

$$\sum_{\varphi \in \mathcal{F}_i} w_\varphi = J\psi(\mathfrak{N}_i) + O\Big(\frac{\psi(\mathfrak{N}_i)}{\mathbb{N}(\mathfrak{N}_i)^{2-\varepsilon}}\Big).$$

Therefore for any $\ell \geq 0$

$$\lim_{i \to \infty} \frac{\sum_{\varphi \in \mathcal{F}_i} X_\ell(v_{\mathfrak{p}}^\varphi) w_\varphi}{\sum_{\varphi \in \mathcal{F}_i} w_\varphi} = \begin{cases} 1 & \text{when } \ell = 2\ell', \text{ with } 0 \leq \ell' \leq \mathrm{ord}_v \, m\mathfrak{d}, \\ 0 & \text{otherwise} \end{cases}$$

$$= \int_{\mathbf{R}} X_\ell(x) d\mu(x). \tag{6.7}$$

This last equality holds by the orthonormality of the polynomials $X_n(x)$

$$\int_{\mathbf{R}} X_i(x) X_j(x) d\mu_\infty(x) = \delta_{ij}$$

(cf. [Se]). Because $\deg X_\ell = \ell$, the set $\{X_\ell\}$ spans the space of all polynomials. Because the space of polynomials is dense in $L^\infty([-2, 2])$, we can replace X_ℓ by any continuous function h in (6.7) and thus obtain the result (see §29.3 of [KL2] for details). $\qquad \square$

7 Variants and special cases

We have four corresponding types of parameters (under various hypotheses):

$$\underset{v_{\mathfrak{p}}^\pi}{\overset{\text{Satake}}{}} \quad \longleftrightarrow \quad \underset{\lambda_{\mathfrak{n}}^\varphi}{\overset{\text{Hecke}}{}} \quad \longleftrightarrow \quad \underset{W_m^\varphi(y)}{\overset{\text{Whittaker}}{}} \quad \longleftrightarrow \quad \underset{a_m(h)}{\overset{\text{Classical}}{}}.$$

For convenience, we give the correspondences explicitly here, so that when possible anyone can rewrite the spectral terms $\dfrac{\lambda_{\mathfrak{n}}^\varphi W_{m_1}^\varphi(1) \overline{W_{m_2}^\varphi(1)}}{\|\varphi\|^2}$ in the main formula purely in terms of their parameter of choice. Let $\mathfrak{n}\widehat{\mathcal{O}} = \widehat{\mathfrak{n}}$ and $\mathfrak{d}\widehat{\mathcal{O}} = \widehat{\mathfrak{d}}$. If $W_m^\varphi(1/d) = 1$, then by Corollaries 4.7 and 4.8,

$$\lambda_{\mathfrak{n}}^\varphi = \mathbb{N}(\mathfrak{n}) W_1^\varphi(\mathfrak{n}/d) \qquad \text{and} \qquad W_m^\varphi(1) = \frac{e^{2\pi r} \prod_{j=1}^r \sigma_j(m)^{k_j/2-1}}{d_F e^{2\pi \, \mathrm{tr}(m)}} \lambda_{m\mathfrak{d}},$$

either of which can be used if an orthogonal basis of such φ is given. Also, using (6.6) we have

$$\lambda_{\mathfrak{n}}^\varphi = \omega_{\mathrm{fin}}(\mathfrak{n})^{1/2} \mathbb{N}(\mathfrak{n})^{1/2} \prod_{\mathfrak{p}|\mathfrak{n}} X_{\mathrm{ord}_{\mathfrak{p}}(\mathfrak{n})}(v_{\mathfrak{p}}^\pi). \tag{7.1}$$

The classical picture is given explicitly at the end of this section.

The case of narrow class number 1

From now on we assume that F has narrow class number 1, i.e. that every fractional ideal in F has a totally positive generator. This implies that every

totally positive unit is the square of a unit ([CH], Lemma 11.6). In this case
Theorem 5.11 simplifies substantially, and a classical interpretation follows
from (3.11).

Fix $\eta, N, d \in F^+$ such that

$$(\eta) = \mathfrak{n}, \quad (N) = \mathfrak{N}, \quad (d) = \mathfrak{d}. \tag{7.2}$$

Proposition 7.1. *Suppose F has narrow class number* 1. *For* $\delta = \begin{pmatrix} m_2/m_1 \\ & 1 \end{pmatrix}$,

$$I_\delta(f) =$$

$$\frac{T(dm_1, dm_2, \eta) d_F^{1/2} \mathbb{N}(\eta)^{1/2} \psi(\mathfrak{N})}{e^{2\pi \operatorname{tr}_\mathbb{Q}^F (m_1 + m_2)} \omega_\mathfrak{N}(\sqrt{m_1 \eta/m_2})} \left[\prod_{j=1}^{r} \frac{\left(4\pi \sqrt{\sigma_j(m_1 m_2)}\right)^{k_j-1}}{(k_j - 2)! \, \operatorname{sgn}(\sigma_j(\sqrt{\frac{m_1 m_2}{\eta}}))^{k_j}} \right],$$

where

$$T(a_1, a_2, a_3) = \begin{cases} 1 & \text{if } a_i a_j / a_k \text{ is a square in } \mathcal{O} \text{ for all} \\ & \text{distinct } i, j, k \in \{1, 2, 3\} \\ 0 & \text{otherwise,} \end{cases}$$

and the square roots are chosen compatibly so that

$$\sqrt{m_1 m_2 / \eta} \sqrt{m_1 \eta / m_2} = m_1.$$

Proof. By Prop. 5.4, we know that I_δ is nonzero only if

- $\frac{m_1}{\mathsf{s}}, \frac{m_2}{\mathsf{s}} \in \widehat{\mathcal{O}}$ for some $\mathsf{s} \in \widehat{\mathfrak{d}}^{-1}$
- $\frac{m_1 m_2}{\mathsf{s}^2} \widehat{\mathcal{O}} = \widehat{\mathfrak{n}}$.

If these hold, then because the class number is 1, we can actually take $\mathsf{s} \in \mathfrak{d}^{-1}$, so we change fonts to s as a reminder that $s \in F$. Furthermore the second condition is equivalent to $m_1 m_2 = s^2 u \eta$ for some unit u. Because $m_1 m_2, s^2, \eta \in F^+$, it follows that $u \in F^+$, and hence u is the square of a unit. We can absorb this unit into s, so the condition becomes

$$m_1 m_2 = s^2 \eta. \tag{7.3}$$

Write $s = s'/d$ for d as in (7.2). Then the above conditions are equivalent to

(i) $\frac{dm_1}{s'}, \frac{dm_2}{s'} \in \mathcal{O}$ for some $s' \in \mathcal{O}$

(ii) $\frac{(dm_1)(dm_2)}{s'^2} = \eta$.

It is easy to check that these conditions hold if and only if $T(dm_1, dm_2, \eta) = 1$.

Now suppose the above holds, so that by Prop. 5.4,

$$I_\delta = \left[\prod_{j=1}^{r} \frac{\left(4\pi \sqrt{\sigma_j(m_1 m_2)}\right)^{k_j-1}}{(k_j - 2)!} \right] \frac{d_F^{1/2} \mathbb{N}(\mathfrak{n})^{1/2} \psi(\mathfrak{N})}{e^{2\pi \, \mathrm{tr}_Q^F (m_1+m_2)} \omega_{\mathfrak{N}}(m_1/s) \omega_{\mathrm{fin}}(s)}.$$

By (7.3), we can write $s = \sqrt{m_1 m_2/\eta}$. This element is defined up to ± 1, and there is no canonical choice since we cannot guarantee that $s \in F^+$. (E.g. $(1 + \sqrt{2})^2$ has no totally positive square root.) In any case, the final result is independent of this choice by the remark on page 165. Using this expression for s, the proposition follows by the fact that

$$\omega_{\mathrm{fin}}(s) = \omega_\infty(s)^{-1} = \prod_{j=1}^{r} \mathrm{sgn}(\sigma_j(s))^{k_j}.$$

$$\square$$

For the second type of I_δ (with $\delta = \left(\begin{smallmatrix} 1 & \mu \\ & \end{smallmatrix} \right)$), the condition on μ becomes $\mu = -\eta u/c^2$ for some totally positive unit $u \in U$ and nonzero $c \in N\mathcal{O}$. As in the previous case, u is a square, so it can be absorbed into c. Thus

$$\mu = -\eta/c^2$$

for some $c \in \mathfrak{N}$. Under this condition, by Prop. 5.7 the contribution is

$$I_\delta = \frac{\psi(\mathfrak{N})}{2^r e^{2\pi \, \mathrm{tr}_Q^F (m_1+m_2)}} \omega_{\mathrm{fin}}(c) S_{\omega_{\mathfrak{N}}}(m_1, m_2; \eta; c) \frac{\mathbb{N}(\eta)^{1/2}}{\mathbb{N}(c)}$$

$$\times \prod_{j=1}^{r} \frac{(-4\pi i)^{k_j} \sqrt{\sigma_j(m_1 m_2)}^{k_j-1}}{(k_j - 2)!} J_{k_j-1}\left(4\pi \frac{\sqrt{\sigma_j(\eta m_1 m_2)}}{|\sigma_j(c)|}\right).$$

Once again, we cannot assume $c \in F^+$ so $\omega_{\mathrm{fin}}(c)$ may be nontrivial. In fact

$$\omega_{\mathrm{fin}}(c) = \prod_{j=1}^{r} \mathrm{sgn}(\sigma_j(c))^{k_j}.$$

Summing over all μ amounts to summing over nonzero $c \in \mathfrak{N}/\pm$, and we obtain the following.

Theorem 7.2. *With notation as above, suppose F has narrow class number 1, $k = (k_1, \ldots, k_r)$ with all $k_j > 2$, and $\mathcal{F}$ is an orthogonal basis for $A_k(\mathfrak{N}, \omega)$ consisting of eigenfunctions for the Hecke operator $T_{\mathfrak{n}}$. Then for any $m_1, m_2 \in \mathfrak{d}_+^{-1}$,*

$$\frac{e^{2\pi \, \mathrm{tr}_Q^F (m_1+m_2)}}{d_F^{1/2} \mathbb{N}(\eta)^{1/2} \psi(\mathfrak{N})} \left[\prod_{j=1}^{r} \frac{(k_j - 2)!}{(4\pi \sqrt{\sigma_j(m_1 m_2)})^{k_j-1}} \right] \sum_{\varphi \in \mathcal{F}} \frac{\lambda_{\mathfrak{n}}^\varphi \, W_{m_1}^\varphi(1) \overline{W_{m_2}^\varphi(1)}}{\|\varphi\|^2} =$$

$$T(dm_1, dm_2, \eta) \left[\prod_{j=1}^{r} (\operatorname{sgn}\sigma_j(\sqrt{m_1 m_2/\eta}))^{k_j} \right] \omega_{\mathfrak{N}}(\sqrt{m_1 \eta/m_2})^{-1}$$

$$+ \frac{1}{d_F^{1/2}} \sum_{\substack{c \in N\mathcal{O}/\pm \\ c \neq 0}} S_{\omega\mathfrak{N}}(m_1, m_2; \eta; c) \frac{(2\pi)^r}{\mathbb{N}(c)} \prod_{j=1}^{r} \frac{J_{k_j-1}(4\pi \frac{\sqrt{\sigma_j(\eta m_1 m_2)}}{|\sigma_j(c)|})}{(i \operatorname{sgn}(\sigma_j(c)))^{k_j}}.$$

Remarks: (1) Because $\widehat{\mathcal{O}}/c\widehat{\mathcal{O}} \cong \mathcal{O}/c\mathcal{O}$, we see easily that for $c \in \mathfrak{N}$,

$$S_{\omega\mathfrak{N}}(m_1, m_2; \eta; c) = \sum_{\substack{t_1, t_2 \in \mathcal{O}/c\mathcal{O} \\ t_1 t_2 \equiv \eta \bmod c\mathcal{O}}} \omega_{\mathfrak{N}}(t_2)^{-1} \theta_{\text{fin}}(\frac{m_1 t_1 + m_2 t_2}{c}).$$

The formula can be further simplified if we replace $c \in \mathfrak{N}$ by $c = Nu\tau$ for $u \in \mathcal{O}^*/\{\pm 1\}$ and $\tau \in \mathcal{O}/\mathcal{O}^*$. Then substituting $t_1' = u^{-1} t_1$, $t_2' = u^{-1} t_2$, we have

$$S_{\omega\mathfrak{N}}(m_1, m_2; \eta; c) = \omega_{\mathfrak{N}}(u)^{-1} S_{\omega\mathfrak{N}}(m_1, m_2; u^{-2}\eta; N\tau),$$

so we can break the sum over c into sums over u and τ and group together the terms with the same τ.

(2) If $F = \mathbf{Q}$, then $r = 1$. The sum over $c \in N\mathbf{Z}/\pm$ is simply a sum over $c > 0$, $N|c$, and we recover the generalized Petersson trace formula of [KL1].

Now take $\eta = 1$ so $\mathfrak{n} = \mathcal{O}$. Then $\lambda_{\mathfrak{n}}^{\varphi} = 1$ for all φ. Because the narrow class number is 1, $A_k(\mathfrak{N}, \omega)$ corresponds (isometrically) to a classical space $S_k(\mathfrak{N}, \omega')$ of cusp forms for $\Gamma_0(\mathfrak{N})$ as in (3.10). Here $\omega'((\begin{smallmatrix} a & b \\ c & d \end{smallmatrix})) = \omega'(d) = \omega_{\mathfrak{N}}(d)^{-1}$ for any $(\begin{smallmatrix} a & b \\ c & d \end{smallmatrix}) \in \Gamma_0(\mathfrak{N}) = \mathrm{SL}_2(F) \cap K_0(\mathfrak{N})$. Conversely, given a character $\omega' : \Gamma_0(\mathfrak{N})/\Gamma_1(\mathfrak{N}) \to \mathbf{C}^*$, the space $S_k(\mathfrak{N}, \omega')$ is isometric to $A_k(\mathfrak{N}, \omega)$, where ω is the Hecke character determined by ω' using strong approximation $\mathbf{A}^* = F^*(F_\infty^+ \times \widehat{\mathcal{O}}^*)$ resulting from narrow class number 1. By (3.11), if $\varphi \leftrightarrow h$, then $W_m^{\varphi}(1) = d_F^{1/2} e^{-2\pi \operatorname{tr}(m)} a_m(h)$, and we obtain the following direct generalization of Petersson's original formula and (8) of [Lu].

Corollary 7.3. *For any orthogonal basis $\mathcal{F}$ for $S_k(\mathfrak{N}, \omega')$ and any $m_1, m_2 \in \mathfrak{d}_+^{-1}$ we have*

$$\frac{d_F^{1/2}}{\psi(\mathfrak{N})} \left[\prod_{j=1}^{r} \frac{(k_j - 2)!}{(4\pi \sqrt{\sigma_j(m_1 m_2)})^{k_j-1}} \right] \sum_{h \in \mathcal{F}} \frac{a_{m_1}(h)\overline{a_{m_2}(h)}}{\|h\|^2} =$$

$$\chi(m_1, m_2)$$

$$+ \frac{1}{d_F^{1/2}} \sum_{\substack{c \in N\mathcal{O}/\pm \\ c \neq 0}} S_{\omega'}(m_1, m_2; c) \frac{(2\pi)^r}{\mathbb{N}(c)} \prod_{j-1}^r \frac{J_{k_j-1}(4\pi \frac{\sqrt{\sigma_j(m_1 m_2)}}{|\sigma_j(c)|})}{(i \operatorname{sgn}(\sigma_j(c))^{k_j}},$$

where $\chi(m_1, m_2) \in \{0, 1\}$ *is nonzero if and only if* $m_2 = m_1 u^2$ *for some* $u \in \mathcal{O}^*$ *and*

$$S_{\omega'}(m_1, m_2; c) = \sum_{s \in (\mathcal{O}/c\mathcal{O})^*, s\bar{s}=1} \exp(2\pi i \operatorname{tr}_{\mathbb{Q}}^F(\frac{sm_1 + \bar{s}m_2}{c}))\omega'(s).$$

Bibliography

[Bi] B. J. Birch, *How the number of points of an elliptic curve over a fixed prime field varies,* J. London Math. Soc., 43 (1968), 57–60.

[Bl] D. Blasius, *Hilbert modular forms and the Ramanujan conjecture,* in "Noncommutative geometry and number theory," Vieweg Verlag, (2006), 35-56.

[BJ] A. Borel and H. Jacquet, *Automorphic forms and automorphic representations,* Proc. Symp. Pure Math., 33, Part 1, Amer. Math. Soc., Providence, RI, (1979), 189–207.

[BL] J.-L. Brylinski and J.-P. Labesse, *Cohomologie d'intersection et fonctions L de certaines variétés de Shimura,* Ann. Sci. École Norm. Sup. (4) 17, no. 3, (1984), 361–412.

[BMP] R. Bruggeman, R. Miatello, and I. Pacharoni, *Estimates for Kloosterman sums for totally real number fields,* J. Reine Angew. Math. 535 (2001), 103–164.

[Br] R. Bruggeman, *Fourier coefficients of cusp forms,* Invent. Math. 45 (1978), no. 1, 1–18.

[Bu] D. Bump, *Automorphic forms and representations,* Cambridge Studies in Advanced Mathematics 55, Cambridge University Press, 1998.

[CDF] B. Conrey, W. Duke, and D. Farmer, *The distribution of the eigenvalues of Hecke operators,* Acta Arith. 78 (1997), no. 4, 405–409.

[CH] P. E. Conner and J. Hurrelbrink, *Class number parity,* Series in Pure Mathematics, 8. World Scientific Publishing Co., Singapore, 1988.

[CPS] J. Cogdell and I. Piatetski-Shapiro, *The arithmetic and spectral analysis of Poincaré series,* Perspectives in Mathematics, 13. Academic Press, Inc., Boston, MA, 1990.

[Gu] K.-B. Gundlach, *Über die Darstellung der ganzen Spitzenformen zu den Idealstufen der Hilbertschen Modulgruppe und die Abschätzung ihrer Fourierkoeffizienten,* Acta Math. 92 (1954), 309–345.

[HC] Harish-Chandra, *Automorphic forms on semisimple Lie groups,* Notes by J. G. M. Mars. Lecture Notes in Mathematics, No. 62, Springer-Verlag, Berlin-New York, 1968.

[Hi] H. Hida, *Elementary theory of L-functions and Eisenstein series,* London Math. Soc. Student Texts 26, Cambridge Univ. Press, Cambridge, 1993.

[Ja]   H. Jacquet, *A guide to the relative trace formula*, in: Automorphic representations, *L*-functions and applications: progress and prospects, 257–272, Ohio State Univ. Math. Res. Inst. Publ., 11, de Gruyter, Berlin, 2005.

[KL1]   A. Knightly and C. Li, *A relative trace formula proof of the Petersson trace formula*, Acta Arith. 122 (2006), no. 3, 297–313.

[KL2]   ——, *Traces of Hecke operators*, Mathematical Surveys and Monographs, 133. Amer. Math. Soc., 2006.

[Ku]   N. V. Kuznetsov, *The Petersson conjecture for cusp forms of weight zero and the Linnik conjecture. Sums of Kloosterman sums*, (Russian) Mat. Sb. (N.S.) 111 (153) (1980), no. 3, 334–383, 479. English translation: Math. USSR Sbornik, 39 (1981), 299–342.

[Li1]   C. Li, *Kuznietsov trace formula and weighted distribution of Hecke eigenvalues*, J. Number Theory 104 (2004), no. 1, 177–192.

[Li2]   ——, *On the distribution of Satake parameters of* GL$_2$ *holomorphic cuspidal representations*, Israel J. Math., to appear.

[Lu]   W. Luo, *Poincaré series and Hilbert modular forms*, Rankin memorial issue, Ramanujan J. 7 (2003), no. 1–3, 129–140.

[MW]   R. Miatello and N. Wallach, *Kuznetsov formulas for products of groups of R-rank one*, Festschrift in honor of I. Piatetski-Shapiro on the occasion of his sixtieth birthday, Part II (Ramat Aviv, 1989), 305–320, Israel Math. Conf. Proc., 3, Weizmann, Jerusalem, 1990.

[Ng]   L. Nguyen, *The Ramanujan conjecture for Hilbert modular forms*, Ph.D. thesis, UCLA, 2005.

[Ro]   J. Rogawski, *Modular forms, the Ramanujan conjecture and the Jacquet-Langlands correspondence*, appendix in "Discrete Groups, Expanding Graphs and Invariant Measures," by A. Lubotzky, Birkhäuser, Basel, 1994, pp. 135–176.

[Sa]   Sarnak, P., *Statistical properties of eigenvalues of the Hecke operators*, in: *Analytic number theory and Diophantine problems (Stillwater, OK, 1984)*, Progr. Math., 70, Birkhäuser, Boston, MA, 1987, 321–331.

[Se]   J.-P. Serre, *Répartition asymptotique des valeurs propres de l'opérateur de Hecke T$_p$*, J. Amer. Math. Soc. 10 (1997), no. 1, 75–102.

[Ta]   R. Taylor, *Automorphy for some l-adic lifts of automorphic mod l Galois representations II*, preprint, May 24, 2006.

Modular shadows and the Lévy–Mellin ∞–adic transform

Yuri I. Manin and Matilde Marcolli

Abstract

This paper continues the study of the structures induced on the "invisible boundary" of the modular tower and extends some results of [MaMar1]. We start with a systematic formalism of pseudo–measures generalizing the well–known theory of modular symbols for SL(2). These pseudo–measures, and the related integral formula which we call the Lévy–Mellin transform, can be considered as an "∞–adic" version of Mazur's p–adic measures that have been introduced in the seventies in the theory of p–adic interpolation of the Mellin transforms of cusp forms, cf. [Ma2]. A formalism of iterated Lévy–Mellin transform in the style of [Ma3] is sketched. Finally, we discuss the invisible boundary from the perspective of non–commutative geometry.

Introduction

When the theory of modular symbols for the SL(2)–case had been conceived in the 70's (cf. [Ma1], [Ma2], [Sh1], [Sh2]), it was clear from the outset that it dealt with the Betti homology of some basic moduli spaces (modular curves, Kuga varieties, $\overline{M}_{1,n}$, and alike), whereas the theory of modular forms involved the de Rham and Hodge cohomology of the same spaces.

However, the combinatorial skeleton of the formalism of modular symbols is so robust, depending essentially only on the properties of continued fractions, that other interpretations and connections naturally suggest themselves.

In this paper, we develop the approach to the modular symbols which treats them as a special case of some structures supported by the "invisible boundary" of the tower of classical modular curves along the lines of [MaMar1], [MaMar2].

Naively speaking, this boundary is (the tower of) quotient space(s) of $\mathbb{P}^1(\mathbb{R})$ modulo (finite index subgroups of) $PSL(2, \mathbb{Z})$. The part of it consisting of orbits of "cusps", $\mathbb{P}^1(\mathbb{Q})$, has a nice algebraic geometric description, but irrational points are not considered in algebraic geometry, in particular, since the action of $PSL(2, \mathbb{Z})$ is highly non–discrete. This is why we call this part "invisible".

"Bad quotients" of this type can be efficiently studied using tools of Connes' non–commutative geometry. Accordingly, the Theorem 4.4.1 of [MaMar1] identified the modular homology complex with (a part of) Pimsner's exact sequence in the K_*–theory of the reduced crossed product algebra $C(\mathbb{P}^1(\mathbb{R})) \rtimes PSL(2, \mathbb{Z})$. Moreover, the Theorem 0.2.2 of the same paper demonstrated the existence of a version of Mellin transform (from modular forms to Dirichlet series) where the integrand was supported by the real axis rather than upper complex half–plane.

In [MaMar2], these and similar results were put in connection with the so called "holography" principle in modern theoretical physics. According to this principle, quantum field theory on a space may be faithfully reflected by an appropriate theory on the boundary of this space. When this boundary, rather than the interior, is interpreted as our observable space–time, one can proclaim that the ancient Plato's cave metaphor is resuscitated in this sophisticated guise. This metaphor motivated the title of the present paper.

Here is a review of its contents.

The sections 1–4 address the first of the two basic themes:

(i) *How does the holomorphic geometry of the upper complex half plane project itself onto the invisible boundary?*

In the first and the second sections, we introduce and develop a formalism of general pseudo–measures. They can be defined as finitely additive functions with values in an abelian group W supported by the Boolean algebra generated by segments with rational ends in $\mathbb{P}^1(\mathbb{R})$. Although the definitions (and proofs) are very elementary, they capture some essential properties of the modular boundary.

In particular, in subsections 1.9–1.11 we show that the *generalized Dedekind symbols* studied by Sh. Fukuhara in [Fu1], [Fu2] are essentially certain sequences of pseudo–measures. Example 1.12 demonstrates that modular symbols are pseudo–measures. (In fact, much more general integrals along geodesics connecting cusps produce pseudo–measures; this is why we use the imagery of "projecting the holomorphic geometry of the upper complex half plane onto the invisible boundary".) Finally, subsection 2.3.1 establishes that *rational period functions* in the sense of [Kn1], [Kn2], [A], [ChZ], are pseudo–measures as well.

In the third section, we define the Lévy–Mellin transform of a pseudo–measure and prove the Theorem 3.4 generalizing the Theorem 0.2.2 of [MaMar1]. This shows that some specific Lévy functions involving pseudo–measures can serve as an efficient replacement of (non–existent) restrictions of cusp forms to the real boundary of H.

In the fourth section, a non–commutative version of pseudo–measures is developed. The new formalism was suggested by the theory of iterated integrals of modular forms introduced in [Ma3], [Ma4]. The iterated Lévy–Mellin transform appears naturally in this context producing some interesting multiple Dirichlet series related to but different from those discussed in [Ma3]. We hope to return to this subject later.

The second theme developed in the fifth section is:

(ii) *With what natural structure(s) is the invisible boundary endowed?*

We start with a reformulation of pseudo–measures in terms of currents on the tree $\mathcal{T}$ of $PSL(2, \mathbb{Z})$. The boundary action of $PSL(2, \mathbb{Z})$ can be best visualized in terms of the space of ends of this tree. This set of endpoints is a compact Hausdorff space which maps continuously to $\mathbb{P}^1(\mathbb{R})$ through a natural map that is 1 : 1 on the irrational points and 2 : 1 on the rationals.

In terms of noncommutative geometry, this boundary action is described by a crossed product C^*-algebra, and W-valued pseudo-measures can be interpreted as homomorphisms from the K_0 of the crossed product C^*-algebra to W.

We also introduce another C^*-algebra naturally associated to the boundary action, described in terms of the generalized Gauss shift of the continued fraction expansion. We show that this can be realized as a subalgebra of the crossed product algebra of the action of $PSL(2, \mathbb{Z})$ on $\partial\mathcal{T}$.

We then consider an extension of pseudo–measures to limiting pseudo–measures that parallels the notion of limiting modular symbols introduced in [MaMar1]. The limiting modular symbols can then be realized as limits of pseudo–measures associated to the ordinary modular symbols or as averages of currents on the tree of $PSL(2, \mathbb{Z})$.

1 Pseudo–measures: commutative case

In the following we collect the basic facts from the theory of Farey series in the form that we will use throughout the paper.

1.1 Conventions.

We will consider $\mathbb{Q} \subset \mathbb{R} \subset \mathbb{C}$ as points of an affine line with a fixed coordinate, say, z. Completing this line by one point ∞, we get points of the projective line $\mathbb{P}^1(\mathbb{Q}) \subset \mathbb{P}^1(\mathbb{R}) \subset \mathbb{P}^1(\mathbb{C})$. *Segments* of $\mathbb{P}^1(\mathbb{R})$ are defined as non–empty connected subsets of $\mathbb{P}^1(\mathbb{R})$. The boundary of each segment generally consists of an (unordered) pair of points in $\mathbb{P}^1(\mathbb{R})$. In marginal cases, the boundary might be empty or consist of one point in which case we may speak of an *improper segment*. A proper segment is called *rational* if its ends are in $\mathbb{P}^1(\mathbb{Q})$. It is called infinite if ∞ is in its closure, otherwise it is called finite.

Definition 1.2. A pseudo–measure on $\mathbb{P}^1(\mathbb{R})$ with values in a commutative group (written additively) W is a function $\mu : \mathbb{P}^1(\mathbb{Q}) \times \mathbb{P}^1(\mathbb{Q}) \to W$ satisfying the following conditions: for any $\alpha, \beta, \gamma \in \mathbb{P}^1(\mathbb{Q})$,

$$\mu(\alpha, \alpha) = 0, \quad \mu(\alpha, \beta) + \mu(\beta, \alpha) = 0, \tag{1.1}$$

$$\mu(\alpha, \beta) + \mu(\beta, \gamma) + \mu(\gamma, \alpha) = 0. \tag{1.2}$$

There are two somewhat different ways to look at μ as a version of measure.

(i) We can uniquely extend the map $(\alpha, \beta) \mapsto \mu(\alpha, \beta)$ to a finitely additive function on the Boolean algebra consisting of finite unions of *positively oriented from α to β* rational segments. On improper segments, in particular *points* and the whole $\mathbb{P}^1(\mathbb{R})$, this function vanishes.

Here positive orientation is defined by the increasing z. Thus, the ordered pair $(1, 2)$ corresponds to the segment $1 \leq z \leq 2$ (one or both ends can be excised, the pseudo–measure remains the same). However the pair $(2, 1)$ corresponds to the union $\{2 \leq z \leq \infty\} \cup \{-\infty \leq z \leq 1\}$ in the traditional notation (which can be somewhat misleading since $-\infty = \infty$ in $\mathbb{P}^1(\mathbb{R})$.) Thus, $(2, 1)$ is an infinite segment.

(ii) Another option is to restrict oneself to finite segments and assign to (α, β) the segment with these ends *oriented from α to β*.

We will freely use both viewpoints, and allow ourselves some laxity about which end belongs to our segment and which not whenever it is not essential.

Moreover, working with pseudo–measures as purely combinatorial objects, we will simply identify *segments* and *ordered pairs* $(\alpha, \beta) \in \mathbb{P}^1(\mathbb{Q})$. We will call α (resp. β) the *ingoing* (resp. *outgoing*) end.

1.3 The group of pseudo–measures.

W–valued pseudo–measures form a commutative group M_W with the composition law

$$(\mu_1 + \mu_2)(\alpha, \beta) := \mu_1(\alpha, \beta) + \mu_2(\alpha, \beta). \tag{1.3}$$

If W is an A–module over a ring A, the pseudo–measures form an A–module as well.

1.4 The universal pseudo–measure.

Let $\mathbb{Z}[\mathbb{P}^1(\mathbb{Q})]$ be a free abelian group freely generated by $\mathbb{P}^1(\mathbb{Q})$, and $\nu : \mathbb{P}^1(\mathbb{Q}) \to \mathbb{Z}[\mathbb{P}^1(\mathbb{Q})]$ the tautological map. Put $\mu^U(\alpha, \beta) := \nu(\beta) - \nu(\alpha)$. Clearly, this is a pseudo–measure taking its values in the subgroup $\mathbb{Z}[\mathbb{P}^1(\mathbb{Q})]_0$, kernel of the augmentation map $\sum_i m_i \nu(\alpha_i) \mapsto \sum_i m_i$. This pseudo–measure is universal in the following sense: for any pseudo–measure μ with values in W, there is a homomorphism $w : \mathbb{Z}[\mathbb{P}^1(\mathbb{Q})]_0 \to W$ such that $\mu = w \circ \mu^U$. In fact, we have to put $w(\nu(\beta) - \nu(\alpha)) = \mu(\alpha, \beta)$.

1.5 The action of $GL(2, \mathbb{Q})$.

The group $GL(2, \mathbb{Q})$ acts from the left upon $\mathbb{P}^1(\mathbb{R})$ by fractional linear automorphisms $z \mapsto gz$, mapping $\mathbb{P}^1(\mathbb{Q})$ to itself. It acts on pseudo–measures with values in W from the right by the formula

$$(\mu g)(\alpha, \beta) := \mu(g(\alpha), g(\beta)). \tag{1.4}$$

This action is compatible with the structures described in 1.3.

The map $z \mapsto -z$ defines an involution on M_W, whose invariant, resp., antiinvariant points can be called even, resp. odd measures.

1.6 Primitive segments and primitive chains.

A segment I is called *primitive*, if it is rational, and if its ends are of the form $\left(\dfrac{a}{c}, \dfrac{b}{d}\right)$, $a, b, c, d \in \mathbb{Z}$ such that $ad - bc = \pm 1$. In other words, $I = (g(\infty), g(0))$ where $g \in GL(2, \mathbb{Z})$ and the group $GL(2)$ acts upon $\mathbb{P}^1$ by fractional linear transformations.

If $\det g = -1$, we can simultaneously change signs of the entries of the second column. The segment $I = (g(\infty), g(0))$ will remain the same. The intersection of the stationary subgroups of ∞ and 0 in $SL(2, \mathbb{Z})$ is $\pm \mathrm{id}$. Hence

the set of oriented primitive segments is a principal homogeneous space over
PSL(2, $\mathbb{Z}$).

The rational ends of I above are written in lowest terms, whereas ∞ must
be written as $\dfrac{\pm 1}{0}$. Signs of the numerators/denominators are generally not normalized (can be inverted simultaneously), but it is natural to use $\dfrac{1}{0}$ for $+\infty$ and
$\dfrac{-1}{0}$ for $-\infty$ whenever we imagine our pairs as ends of oriented segments.

Primitive segments with one infinite end are thus $(-\infty, m)$ and (n, ∞),
$m, n \in \mathbb{Z}$. If a segment with finite ends (α, β) is primitive, then $|\alpha - \beta| = n^{-1}$
for some $n \in \mathbb{Z}, n \geq 1$.

Shifting a primitive segment $I = (\alpha, \beta)$ by any integer, or changing signs
to $-I = (-\alpha, -\beta)$ we get a primitive segment. Hence any finite primitive
segment can be shifted into $[0, 1]$, and any primitive segment I of length
(Lebesgue measure) $|I| \leq \dfrac{1}{2}$ after an appropriate shift lands entirely either
in $\left[0, \dfrac{1}{2}\right]$, or in $\left[\dfrac{1}{2}, 1\right]$. Moreover, if $I = (\alpha, \beta) \subset \left[0, \dfrac{1}{2}\right]$ is primitive, then
$1 - I =: (1 - \alpha, 1 - \beta) \subset \left[\dfrac{1}{2}, 1\right]$ is primitive, and vice versa.

Generally, let $\alpha, \beta \in \mathbb{P}^1(\mathbb{Q})$. Let us call *a primitive chain of length n
connecting α to β* any non–empty fully ordered family of proper primitive
segments $I_1, \ldots, I_n$ such that the ingoing end of I_1 is α, outgoing end of I_n is
β, and for each $1 \leq k \leq n - 1$, the outgoing end of I_k coincides with the ingoing end of I_{k+1}. The numeration of the segments is not a part of the structure,
only their order is. We will call chains with $\alpha = \beta$ *primitive loops*, and allow
one improper segment (α, α) to be considered as a primitive loop of length 0.

This notion is covariant with respect to the SL(2, $\mathbb{Z}$)-action: if $I_1, \ldots, I_n$ is a
primitive chain connecting α to β, then for any $g \in$ SL(2, $\mathbb{Z}$), $g(I_1), \ldots, g(I_n)$
is a primitive chain connecting $g(\alpha)$ to $g(\beta)$.

Lemma 1.6.1. *(a) If I_1, I_2 are two open primitive segments and at least one
of them is finite, then either $I_1 \cap I_2 = \emptyset$, or one of them is contained in
another.*

(b) Any two points $\alpha, \beta \in \mathbb{P}^1(\mathbb{Q})$ can be connected by a primitive chain.

Proof. This is well known. We reproduce an old argument showing (b) from
[Ma1] in order to fix some notation. Consider the following sequence of
normalized convergents to α

$$\frac{p_{-1}(\alpha)}{q_{-1}(\alpha)} := \frac{1}{0} = \infty, \quad \frac{p_0(\alpha)}{q_0(\alpha)} := \frac{p_0}{1}, \quad \ldots, \quad \frac{p_n(\alpha)}{q_n(\alpha)} = \alpha \qquad (1.5)$$

Here $\alpha \in \mathbb{Q}$, $p_0 := [\alpha]$ the integer part of α, and convergents are calculated from

$$\alpha = k_0(\alpha) + [k_1(\alpha), \ldots, k_n(\alpha)] := p_0 + \cfrac{1}{k_1(\alpha) + \cfrac{1}{k_2(\alpha) + \ldots \frac{1}{k_n(\alpha)}}}$$

with $1 \leq k_i(\alpha) \in \mathbb{Z}$ for $i \geq 1$ and $k_n(\alpha) \geq 2$ whenever $\alpha \notin \mathbb{Z}$ so that $n \geq 1$.

The sequence

$$I_k(\alpha) := \left(\frac{p_k(\alpha)}{q_k(\alpha)}, \frac{p_{k+1}(\alpha)}{q_{k+1}(\alpha)} \right) \tag{1.6}$$

is a primitive chain connecting ∞ to α.

Applying this construction to β and reversing the sequence of ends (1.5), we get a primitive chain connecting β to ∞. Joining chains from α to ∞ and from ∞ to β, we get a chain from α to β.

We put also

$$g_k(\alpha) := \begin{pmatrix} p_k(\alpha) & (-1)^{k+1} p_{k+1}(\alpha) \\ q_k(\alpha) & (-1)^{k+1} q_{k+1}(\alpha) \end{pmatrix} \in \mathrm{SL}(2, \mathbb{Z}) \tag{1.7}$$

so that

$$I_k(\alpha) = (g_k(\alpha)(\infty), g_k(\alpha)(0)) . \tag{1.8}$$

$\square$

Corollary 1.7. *Each pseudo–measure is completely determined by its values on primitive segments.*

In fact, for any α, β and any primitive chain $I_1, \ldots, I_n$ connecting α to β, we must have

$$\mu(\alpha, \beta) = \sum_{k=1}^{n} \mu(I_k). \tag{1.9}$$

Definition 1.7.1. A pre–measure $\widetilde{\mu}$ is any W–valued function defined on primitive segments and satisfying relations (1.1) and (1.2) for all primitive chains of length ≤ 3.

Clearly, restricting a pseudo–measure to primitive segments, we get a pre–measure. We will prove the converse statement.

Theorem 1.8. *Every pre–measure $\widetilde{\mu}$ can be uniquely extended to a pseudo-measure μ.*

Proof. If such a pseudo–measure exists, it is defined by the (family of) formula(s) (1.9):

$$\mu(\alpha, \beta) := \sum_{k=1}^{n} \widetilde{\mu}(I_k). \tag{1.10}$$

We must only check that this prescription is well defined, that is, does not depend on the choice of $\{I_k\}$. Our argument below is somewhat more elaborate than what is strictly needed here. Its advantage is that it can be applied without changes to the proof of the Theorem 4.3 below, which is a non–commutative version of the Theorem 1.8.

We will first of all define four types of *elementary moves* which transform a primitive chain $\{I_k\}$ connecting α to β to another such primitive chain without changing the r.h.s. of (1.10).

(i) Choose k (if it exists) such that $I_k = (\gamma, \delta)$, $I_{k+1} = (\delta, \gamma)$, and delete I_k, I_{k+1} from the chain. This does not change (1.10) because $\widetilde{\mu}(\gamma, \delta) + \widetilde{\mu}(\delta, \gamma) = 0$.

(ii) A reverse move: choose a point γ which is the outgoing end of some I_k (resp. $\gamma = \alpha$), choose a primitive segment (γ, δ), and insert the pair (γ, δ), (δ, γ) right after I_k (resp. before I_1.) This move can also be applied to the empty loop connecting γ to γ, then it produces a chain of length two. Again, application of such a move is compatible with (1.10).

(iii) Choose I_k, I_{k+1}, I_{k+2} (if they exist) such that these three segments have the form (γ_1, γ_2), (γ_2, γ_3), (γ_3, γ_1), and delete them from the chain.

This move is compatible with (1.10) as well because we have postulated (1.2) on such chains.

(iv) A reverse move: choose a point γ which is the outgoing end of some I_k (resp. $\gamma = \alpha$), and insert any triple of segments as above right after I_k (resp. before I_1.)

Now we will show that

(*) *any primitive loop, that is, a primitive chain with $\alpha = \beta$, can be transformed into an empty loop by a sequence of elementary moves.*

Suppose that we know this. If $I_1, \ldots, I_n, J_1, \ldots, J_m$ are two chains connecting α to β, we can produce from them a loop connecting α to α by putting after I_n the segments $J_m, \ldots, J_1$ with reversed orientations. The r.h.s. part of (1.10) calculated for this loop must be zero because it is zero for the empty loop. Hence $I_1, \ldots, I_n$ and $J_1, \ldots, J_m$ furnish the same r.h.s. of (1.10).

We will establish (*) by induction on the length of a loop. We will consider several cases separately.

Case 1: loops of small length. The discussion of the elementary moves above shows that loops of length 2 or 3 can be reduced to an empty loop by one elementary move. Loops of length 1 do not exist.

Case 2: existence of a subloop. Assume that the ingoing end of some $I_k, k \geq 1$, coincides with the outgoing end of some $I_l, k + 1 \leq l \leq n - 1$. Then the segments $I_k, \ldots, I_l$ form a subloop of lesser length. By induction, we may assume that it can be reduced to an empty loop by a sequence of elementary moves. The same sequence of moves diminishes the length of the initial loop.

From now on, we may and will assume that our loop $I_1, \ldots, I_n$ is of length $n \geq 4$ and does not contain proper subloops.

Applying an appropriate $g \in \mathrm{PSL}(2, \mathbb{Z})$, we may and will assume that $\alpha = \beta = \infty$. Thus our loop starts with $I_1 = (\infty, a)$ and ends with $I_n = (b, \infty)$, $a, b \in \mathbb{Z}$, $a \neq b$ because of absence of subloops.

Case 3: $|a - b| = 1$. We will consider the case $b = a + 1$; the other one reduces to this one by the change of orientation. We may also assume that $I_2, \ldots, I_{n-1} \subset [a, a + 1]$, because otherwise ∞ would appear once again as one of the vertices.

Since $n \geq 4$, the primitive chain $I_2, \ldots, I_{n-1}$ connecting a to $a + 1$ has length at least 2. We will apply to the initial loop the following elementary moves: (i) insert $((a, a + 1), (a + 1, a)$ after $I_1 = (\infty, a)$; (ii) delete $(\infty, a), ((a, a+1), (a+1, \infty)$. The resulting loop connecting a to a has length $n - 1$ and can be reduced to the empty loop by the inductive assumption.

Case 4: $|a - b| \geq 2$. Again, we may assume that $a < b$ and that $I_2, \ldots, I_{n-1} \subset [a, b]$. Consider two subcases.

It can be that the Lebesgue measure of I_2 is 1, so that $I_2 = (a, a + 1)$. Then we apply two elementary moves: (i) insert $(a + 1, \infty), (\infty, a + 1)$ after I_2; (ii) delete $(\infty, a) = I_1, (a, a + 1) = I_2, (a + 1, \infty)$. We will get a loop of length $n - 1$.

If the Lebesgue measure of I_2 is < 1, then some initial subchain $I_2, \ldots, I_k$ of length ≥ 2 will connect a to $a + 1$. In this subcase, we will apply the following sequence of elementary moves: (i) insert $(a, a + 1), (a + 1, a)$ after $I_1 = (\infty, a)$; (ii) insert $(a + 1, \infty), (\infty, a + 1)$ after I_k; (iii) delete $(\infty, a), (a, a + 1), (a + 1, \infty)$.

The resulting loop will have length $n+1$, however, it will also have a subloop $I_2, \ldots, I_k, (a + 1, a)$ of length $\leq n - 1$ and ≥ 3. The latter can be deleted by elementary moves in view of the inductive assumption leaving the loop of length $\leq n - 2$.

This completes the proof of (*) and of the Theorem 1.8. $\qquad\square$

1.9 Pseudo–measures and generalized Dedekind symbols.

Let $V := \{(p, q) \in \mathbb{Z}^2 \mid p \geq 1, \gcd(p, q) = 1\}$.

Slightly extending the definition given in [Fu1], [Fu2], we will call *a W–valued generalized Dedekind symbol* any function $D : V \to W$ satisfying the functional equation

$$D(p, q) = D(p, q + p). \qquad (1.11)$$

The symbol D can be reconstructed (at least, in the absence of 2–torsion in W) from its *reciprocity function* R defined on the subset $V_0 := \{(p, q) \in V \mid q \geq 1\}$ by the equation

$$R(p, q) := D(p, q) - D(q, -p). \qquad (1.12)$$

From 1.11 we get a functional equation for R:

$$R(p + q, q) + R(p, p + q) = R(p, q). \qquad (1.13)$$

Fukuhara in [Fu1] considers moreover the involution $(p, q) \mapsto (p, -q)$. If D is even with respect to such an involution, its reciprocity function satisfies the additional condition $R(1, 1) = 0$, which together with (1.13) suffices for existence of D with reciprocity function R.

In the following, we will work with reciprocity functions only.

1.9.1 From pseudo–measures to reciprocity functions.

Consider the set Π consisting of all primitive segments contained in $[0, 1]$.

If $a/p < b/q$ are ends of $I \in \Pi$ written in lowest terms with $p, q > 0$, we have $(p, q) \in V_0$. We have thus defined a map $\Pi \to V_0$. One easily sees that it is a bijection.

The involution on Π which sends (p, q) to (q, p) corresponds to $I \mapsto 1 - I$.

Let μ be a W–valued pseudo–measure. Define a function $R_{\mu,0}$ which in the notation of the previous paragraph is given by

$$R_{\mu,0}(p, q) := \mu\left(\frac{a}{p}, \frac{b}{q}\right). \qquad (1.14)$$

Furthermore, for each $n \in \mathbb{Z}$ put

$$R_{\mu,n}(p, q) := \mu\left(n + \frac{a}{p}, n + \frac{b}{q}\right). \qquad (1.15)$$

Proposition 1.9.2. *(a) Equations 1.14 and 1.15 determine reciprocity functions.*

(b) We have $R_{\mu,n}(1, 1) = 0$ iff $\mu(n, n + 1) = 0$.

Proof. In order to prove (a) it suffices to establish the equation (1.13) for $R_{\mu,n}$. When $n = 0$, this follows from (1.2) applied to the Farey triple

$$\alpha = \frac{a}{p}, \ \beta = \frac{a+b}{p+q}, \ \gamma = \frac{b}{q}.$$

To get the general case, one simply shifts this triple by n.

Since $R_{\mu,0}(1, 1) = \mu(0, 1)$, we get (b).   □

1.10 From reciprocity functions to pseudo–measures.

Consider now any sequence of reciprocity functions R_n, $n \in \mathbb{Z}$, and an element $\omega \in W$. Construct from this data a W–valued function $\widetilde{\mu}$ on the set of all primitive segments by the following prescriptions. For positively oriented infinite segments we put:

$$\widetilde{\mu}(-\infty, 0) := \omega, \tag{1.16}$$

and moreover, when $n \in \mathbb{Z}$, , $n \geq 1$,

$$\widetilde{\mu}(-\infty, n) := \omega + R_0(1, 1) + R_1(1, 1) + \cdots + R_{n-1}(1, 1), \tag{1.17}$$

$$\widetilde{\mu}(-\infty, -n) := \omega - R_{-1}(1, 1) - \cdots - R_{-n}(1, 1), \tag{1.18}$$

$$\widetilde{\mu}(-n, \infty) := R_{-n}(1, 1) + R_{-n+1}(1, 1) + \cdots + R_{-1}(1, 1) - \omega, \tag{1.19}$$

$$\widetilde{\mu}(n, \infty) := -R_{n-1}(1, 1) - R_{n-2}(1, 1) - \cdots - R_0(1, 1) - \omega. \tag{1.20}$$

For positively oriented finite segments we put:

$$\widetilde{\mu}(n + \alpha, n + \beta) := R_n(p, q) \tag{1.21}$$

if $0 \leq \alpha = a/p < \beta = p/q \leq 1$.

Finally, for negatively oriented primitive segments we prescribe the sign change, in concordance with (1.2):

$$\widetilde{\mu}(\beta, \alpha) = -\widetilde{\mu}(\alpha, \beta). \tag{1.22}$$

Theorem 1.11. *The function $\widetilde{\mu}$ is a pre–measure. Therefore it can be uniquely extended to a pseudo–measure μ.*

The initial family $\{R_n, \omega\}$ can be reconstructed from μ with the help of (1.15) and (1.16).

Proof. In view of the Theorem 1.8, it remains only to check the relations (1.2) for primitive loops of length 3.

If all ends in such a loop are finite, they form a Farey triple in $[0, 1]$ or a Farey triple shifted by some $n \in \mathbb{Z}$. In this case (1.2 follows from the functional equations for R_n.

If one end in such a loop is ∞, then up to an overall change of orientation it has the form

$$(-\infty, n), (n, n+1), (n+1, \infty).$$

Straightforward calculations using (1.16)–(1.21) complete the proof. □

1.12 Example: pseudo–measures associated with holomorphic functions vanishing at cusps.

The spaces $\mathbb{P}^1(\mathbb{Q})$ and $\mathbb{P}^1(\mathbb{R})$ are embedded into the Riemann sphere $\mathbb{P}^1(\mathbb{C})$ using the complex values of the same affine coordinate z. The infinity point acquires one more traditional notation $i\infty$. The upper half plane $H = \{z \mid \mathrm{Im}\, z > 0\}$ is embedded into $\mathbb{P}^1(\mathbb{C})$ as an open subset with the boundary $\mathbb{P}^1(\mathbb{R})$. The metric $ds^2 := \dfrac{|dz|^2}{(\mathrm{Im}\, z)^2}$ has the constant negative curvature -1, and the fractional linear transformations $z \mapsto gz$, $g \in \mathrm{GL}^+(2, \mathbb{Z})$, act upon H by holomorphic isometries.

Let $\mathcal{O}(H)_{\mathrm{cusp}}$ be the space of holomorphic functions f on H having the following property: *when we approach a cusp* $\alpha \in \mathbb{P}^1(\mathbb{Q})$ *along a geodesic leading to this cusp*, $|f(z)|$ *vanishes faster than* $l(z_0, z)^{-N}$ *for all integers N, where z_0 is any fixed reference point in H, and $l(z_0, z)$ is the exponential of the geodesic distance from z_0 to z.*

If $f \in \mathcal{O}(H)_{\mathrm{cusp}}$, the map $\mathbb{P}^1(\mathbb{Q})^2 \to \mathbb{C}$:

$$(\alpha, \beta) \mapsto \int_\alpha^\beta f(z)dz \qquad (1.23)$$

is well defined and satisfies (1.1), (1.2). (Here and henceforth we always tacitly assume that the integration path, at least in some vicinity of α and β, follows a geodesic.)

Considering the r.h.s. of (1.23) as a linear functional of f i.e., an element of the linear dual space $\mathbf{W} := (\mathcal{O}(H)_{\mathrm{cusp}})^*$, we get our basic $\mathbf{W}$–valued pseudo–measure μ. The classical constructions with modular forms introduce some additional structures and involve passage to a quotient of $\mathbf{W}$, cf. 2.6 below.

The real and imaginary parts of (1.23) are pseudo–measures as well.

More generally, one can replace the integrand in (1.23) by a closed real analytic (or even smooth) 1–form in H satisfying the same exponential vanishing condition along cusp geodesics as above.

1.13 *p*–adic analogies.

The segment $[-1, 1]$ is determined by the condition $|\alpha|_\infty \leq 1$ so that it traditionally is considered as an analog of the ring of *p*–adic integers $\mathbb{Z}_p$ determined by $|\alpha|_p \leq 1$. Less traditionally, we suggest to consider the open primitive (Farey) segments of length $\leq 1/2$ in $[-1, 1]$ as analogs of residue classes a mod p^m. Both systems of subsets share the following property: any two subsets either do not intersect, or one of them is contained in another one. The number m then corresponds to the Farey depth of the respective segment that is, to the length of the continued fraction decomposition of one end.

The notion of a pseudo–measure is similar to that of *p*–adic measure: see [Ma2], sec. 8 and 9. This list of analogies will be continued in the section 3.5 below.

2 Modular pseudo–measures.

2.1 Modular pseudo–measures.

Let $\Gamma \subset \mathrm{SL}(2, \mathbb{Z})$ be a subgroup of finite index. In this whole section we assume that the group of values W of our pseudo–measures is a left Γ–module. The action is denoted $\omega \mapsto g\omega$ for $g \in \Gamma$, $\omega \in W$.

Definition 2.1.1. A pseudo–measure μ with values in W is called modular with respect to Γ if $\mu g(\alpha, \beta) = g[\mu(\alpha, \beta)]$ for any (α, β) and $g \in \Gamma$. Here $\mu g(\alpha, \beta)$ is defined by (1.4).

We denote by $M_W(\Gamma)$ the group of all such pseudo–measures. More generally, if χ is a character of Γ and if multiplication by $\chi(g)$, $g \in \Gamma$, makes sense in W (e. g., W is a module over a ring where the values of χ lie), we denote by $M_W(\Gamma, \chi)$ the group of pseudo–measures μ such that $\mu g(\alpha, \beta) = \chi(g) \cdot g\mu(\alpha, \beta)$.

Example 2.1.2. Let $\Gamma = \mathrm{SL}(2, \mathbb{Z})$. Then from the Corollary 1.7 one infers that any Γ–modular pseudo–measure μ is uniquely determined by its single value $\mu(\infty, 0)$, because if $I = g(\infty, 0)$ is a primitive interval, then we must have $\mu(I) = g\mu(\infty, 0)$. In particular,

$$\mu(\infty, \alpha) = \sum_{k=-1}^{n-1} g_k(\alpha)\mu(\infty, 0) \qquad (2.1)$$

where the matrices $g_k(\alpha)$ are defined in (1.7).

For a general Γ, denote by $\{h_k\}$ a system of representatives of the coset space $\Gamma \setminus \mathrm{SL}(2, \mathbb{Z})$. Then the similar reasoning shows that any Γ–modular pseudo–measure μ is uniquely determined by its values $\mu(h_k(\infty), h_k(0))$.

2.1.3 *Modularity and generalized Dedekind symbols.*

Assume that W is fixed by the total group of shifts in $\mathrm{SL}(2, \mathbb{Z})$ fixing ∞: $z \mapsto z + n$, $n \in \mathbb{Z}$. Assume moreover that μ is modular with respect to some subgroup (non–necessarily of finite index) containing the total group of shifts. Then from (1.15) one sees that that all the reciprocity functions $R_{\mu,n}$ coincide with $R_\mu := R_{\mu,0}$, and from (1.17)–(1.20) it follows that $R(1, 1) = 0$. Hence μ is associated with a generalized Dedekind symbol. Conversely, starting with such a symbol, we can uniquely reconstruct a shift–invariant pseudo–measure.

2.2 A description of $M_W(\mathrm{SL}(2, \mathbb{Z}))$.

Consider now the map

$$M_W(\mathrm{SL}(2, \mathbb{Z})) \to W : \mu \mapsto \mu(\infty, 0). \qquad (2.2)$$

From 2.1.1 it follows that this is an injective group homomorphism. Its image is constrained by several conditions.

Firstly, because of modularity, we have

$$(-\mathrm{id})\mu(\infty, 0) = \mu(\infty, 0). \qquad (2.3)$$

Therefore, the image of (2.2) is contained in W_+, the subgroup of fixed points of $-\mathrm{id}$. Hence we may and will consider it as a module over $\mathrm{PSL}(2, \mathbb{Z})$. The latter group is the free product of its two subgroups $\mathbf{Z}_2$ and $\mathbf{Z}_3$ generated respectively by the fractional linear transformations with matrices

$$\sigma = \begin{pmatrix} 0 & -1 \\ 1 & 0 \end{pmatrix}, \quad \tau = \begin{pmatrix} 0 & -1 \\ 1 & -1 \end{pmatrix}. \qquad (2.4)$$

Now, $\sigma(\infty) = 0$, $\sigma(0) = \infty$. Hence secondly,

$$(1 + \sigma)\mu(\infty, 0) = 0. \qquad (2.5)$$

And finally,

$$\tau(0) = 1, \ \tau(1) = \infty, \ \tau(\infty) = 0.$$

so that the modularity implies

$$(1 + \tau + \tau^2)\mu(\infty, 0) = 0. \qquad (2.6)$$

To summarize, we get the following complex

$$0 \to M_W(\mathrm{SL}(2, \mathbb{Z})) \to W_+ \to W_+ \times W_+ \qquad (2.7)$$

where the first arrow is injective, and the last arrow is $(1 + \sigma, 1 + \tau + \tau^2)$.

Theorem 2.3. *The sequence (2.7) is exact. In other words, the map $\mu \mapsto \mu(\infty, 0)$ induces an isomorphism*

$$M_W(\mathrm{SL}(2, \mathbb{Z})) \cong \mathrm{Ker}\,(1 + \sigma) \cap \mathrm{Ker}\,(1 + \tau + \tau^2)|_{W_+}. \qquad (2.8)$$

Proof. Choose an element ω in the r.h.s. of (2.8). Define a W–valued function $\widetilde{\mu}$ on primitive segments by the formula

$$\widetilde{\mu}(g(\infty), g(0)) := g\omega. \qquad (2.9)$$

We will check that it is a pre–measure in the sense of 1.7.1. This will show that any such ω comes from a (unique) measure μ. It is then easy to check that μ is in fact $\mathrm{SL}(2, \mathbb{Z})$–modular: from the formula (1.9) and the independence of its r.h.s. from the choice of a primitive chain $I_k = g_k(\infty, 0)$ we get using (2.9):

$$\mu(g(\alpha), g(\beta)) = \sum_{k=1}^{n} \widetilde{\mu}(g(I_k)) = \sum_{k=1}^{n} \widetilde{\mu}(gg_k(\infty, 0))$$

$$= \sum_{k=1}^{n} gg_k(\omega) = \sum_{k=1}^{n} g\widetilde{\mu}(I_k) = g\mu(\alpha, \beta).$$

The property (1.1) holds for $\widetilde{\mu}$ in view of (2.5). In fact, if $(\alpha, \beta) = g(\infty, 0)$, then

$$(\beta, \alpha) = g(0, \infty) = g\sigma(\infty, 0)$$

so that

$$\widetilde{\mu}(\alpha, \beta) + \widetilde{\mu}(\beta, \alpha) = g\omega + g\sigma\omega = 0.$$

Similarly, the property (1.2) follows from (2.6). In order to deduce this, we must check that both types of primitive loops of length 3 considered in the proof of the Theorem 1.8 can be represented in the form

$$g(\infty, 0), \ g\tau(\infty, 0), \ g\tau^2(\infty, 0)$$

for an appropriate g. We leave this as an easy exercise. □

2.3.1 Rational period functions as pseudo–measures.

Fix an integer $k \geq 0$ and denote by W_k the space of higher differentials $\omega(z) := q(z)(dz)^k$ where $q(z)$ is a rational function. Define the left action of $\mathrm{PSL}(2, \mathbb{Z})$ on W_k by $(g\omega)(z) := \omega(g^{-1}(z))$. Then the right hand side of (2.8) consists of such $q(z)(dz)^k$ that $q(z)$ satisfies the equations

$$q(z) + z^{-2k} q\left(\frac{-1}{z}\right) = 0, \quad q(z) + z^{-2k} q\left(1 - \frac{1}{z}\right) + (z - 1)^{-2k} q\left(\frac{1}{1 - z}\right) = 0.$$

These equations define the space of *rational period functions of weight 2k*. M. Knopp introduced them in [Kn1] and showed that such functions can have poles only at 0 and real quadratic irrationalities (the latter are "shadows" of closed geodesics on modular curves.) Y. J. Choie and D. Zagier in [ChZ] provided a very explicit description of them. Finally, A. Ash studied their generalizations for arbitrary Γ, which can be treated as pseudo–measures using the construction described in the next subsection.

2.4 Induced pseudo–measures.

By changing the group of values W, we can reduce the description of $M_W(\Gamma)$ for general Γ to the case $\Gamma = \mathrm{SL}(2, \mathbb{Z})$.

Concretely, given Γ and a Γ–module W, put

$$\widehat{W} := \mathrm{Hom}_\Gamma(\mathrm{PSL}(2, \mathbb{Z}), W).\tag{2.10}$$

An element $\varphi \in \widehat{W}$ is thus a map $g \mapsto \varphi(g) \in W$ such that $\varphi(\gamma g) = \gamma\varphi(g)$ for all $\gamma \in \Gamma$ and $g \in \mathrm{SL}(2, \mathbb{Z})$. Such functions form a group with pointwise addition, and $\mathrm{PSL}(2, \mathbb{Z})$ acts on it from the left via $(g\varphi)(\gamma) := \varphi(\gamma g)$.

Any $\widehat{W}$–valued $\mathrm{SL}(2, \mathbb{Z})$–modular measure $\hat{\mu}$ induces a W–valued Γ–modular measure:

$$\mu(\alpha, \beta) := (\hat{\mu}(\alpha, \beta))(1_{\mathrm{SL}(2,\mathbb{Z})}).\tag{2.11}$$

Conversely, any W–valued Γ–modular measure μ induces a $\widehat{W}$–valued $\mathrm{SL}(2, \mathbb{Z})$–modular measure $\hat{\mu}$:

$$(\tilde{\mu}(\alpha, \beta))(g) := \mu(g(\alpha), g(\beta)).\tag{2.12}$$

Proposition 2.4.1. *The maps (2.11), (2.12) are well defined and mutually inverse. Thus, they produce a canonical isomorphism*

$$M_W(\Gamma) \cong M_{\widehat{W}}(\mathrm{SL}(2, \mathbb{Z})).\tag{2.13}$$

The proof is straightforward.

2.4.2 Modular pseudo–measures and cohomology.

Given $\mu \in M_W(\Gamma)$ and $\alpha \in \mathbb{P}^1(\mathbb{Q})$, consider the function

$$c_\alpha^\mu = c_\alpha : \Gamma \to W, \quad c_\alpha(g) := \mu(g\alpha, \alpha).$$

From (1.1), (1.2) and the modularity of μ it follows that this is an 1–cocycle in $Z^1(\Gamma, W)$:

$$c_\alpha(gh) = c_\alpha(g) + gc_\alpha(h).$$

Changing α, we get a cohomologous cocycle:

$$c_\alpha(g) - c_\beta(g) = g\mu(\alpha, \beta) - \mu(\alpha, \beta).$$

If we restrict c_α upon the subgroup Γ_α fixing α, we get the trivial cocycle, so the respective cohomology class vanishes. Call a cohomology class in $H^1(\Gamma, W)$ *cuspidal* if it vanishes after restriction on each Γ_α.

Thus, we have a canonical map of Γ–modular pseudo–measures to the cuspidal cohomology

$$M_W(\Gamma) \to H^1(\Gamma, W)_{\text{cusp}}.$$

If $\Gamma = \text{PSL}(2, \mathbb{Z})$, both groups have more compact descriptions. In particular, the map

$$Z^1(\text{PSL}(2, \mathbb{Z}), W_+) \mapsto$$
$$\text{Ker}\,(1 + \sigma) \times \text{Ker}\,(1 + \tau + \tau^2) \subset W_+ \times W_+ : c \mapsto (c(\sigma), c(\tau))$$

is a bijection.

The value $\mu(\infty, 0) = -c_\infty(\sigma)$ furnishes the connection between the two descriptions which takes the following form:

Proposition 2.4.3. *(i) For any* $\text{PSL}(2, \mathbb{Z})$*–modular pseudo–measure* μ*, we have*

$$c_\infty^\mu(\sigma) = c_\infty^\mu(\tau) = -\mu(\infty, 0) \in W_+.$$

(ii) This correspondence defines a bijection

$$M_W(\text{PSL}(2, \mathbb{Z})) \cong Z^1(\text{PSL}(2, \mathbb{Z}), W_+)_{\text{cusp}}$$

where the latter group by definition consists of cocycles with equal components $c(\sigma) = c(\tau)$.

In the sec. 4.8.2, this picture will be generalized to the noncommutative case.

2.5 Action of the Hecke operators on pseudo–measures.

To define the action of the Hecke operators upon $M_W(\Gamma)$, we have to assume that the group of values W is a left module not only over Γ (as up to now) but in fact over $\text{GL}^+(2, \mathbb{Q})$. We adopt this assumption till the end of this section.

Then $M_W(\Gamma)$ becomes a $\text{GL}^+(2, \mathbb{Q})$–bimodule, with the right action (1.4) and the left one

$$(g\mu)(\alpha, \beta) := g[\mu(\alpha, \beta)]. \tag{2.14}$$

Let Δ be a double coset in $\Gamma \backslash GL^+(2, \mathbb{Q}) / \Gamma$, or a finite union of such cosets. Denote by $\{\delta_i\}$ a complete finite family of representatives of $\Gamma \backslash \Delta$.

Proposition 2.5.1. *The map*

$$T_\Delta : \mu \mapsto \mu_\Delta := \sum_i \delta_i^{-1} \mu \delta_i \tag{2.15}$$

restricted to $M_W(\Gamma)$ depends only on Δ and sends $M_W(\Gamma)$ to itself.

Proof. If $\{\delta_i\}$ are replaced by $\{g_i \delta_i\}$, $g_i \in \Gamma$, the r.h.s. of (2.15) gets replaced by

$$\sum_i \delta_i^{-1} g_i^{-1} \mu g_i \delta_i .$$

But $g_i^{-1} \mu g_i = \mu$ on $M_W(\Gamma)$.

To check that $\mu_\Delta g = g \mu_\Delta$ for $g \in \Gamma$, notice that the right action of Γ induces a permutation of the set $\Gamma \backslash \Delta$ so that for each $g \in \Gamma$,

$$\delta_i g = g'(i, g) \cdot \delta_{j(i,g)}, \qquad g'(i, g) \in \Gamma,$$

and $i \mapsto j(i, g)$ is a permutation of indices.

Hence

$$\delta_i^{-1} \mu \delta_i g = \delta_i^{-1} \mu g'(i, g) \cdot \delta_{j(i,g)} = \delta_i^{-1} g'(i, g) \mu \cdot \delta_{j(i,g)}.$$

But $\delta_i^{-1} g'(i, g) = g \delta_{j(i,g)}^{-1}$. Therefore,

$$\delta_i^{-1} \mu \delta_i g = g \delta_{j(i,g)}^{-1} \mu \delta_{j(i,g)},$$

and while g is kept fixed, the summation over i produces the same result as summation over $j(i, g)$. This completes the proof. $\qquad\square$

Notice in conclusion that all elements of one double coset Δ have the same determinant, say, D. Usually one normalizes Hecke operators by choosing $\Delta \subset M(2, \mathbb{Z})$ and replacing T_Δ by $D T_\Delta$.

2.5.2 Classical Hecke operators T_n.

They correspond to the case $\Gamma = SL(2, \mathbb{Z})$. Let Δ_n be the finite union of the double classes contained in $M_n(2, \mathbb{Z})$: matrices with integer entries and determinant n. It is well known that the following matrices form a complete system of representatives of $\Gamma \backslash \Delta_n$:

$$\begin{pmatrix} a & b \\ 0 & d \end{pmatrix}, \quad ad = n, \ 1 \le b \le d . \tag{2.16}$$

We put $T_n := T_{\Delta_n}$.

2.6 Pseudo–measures associated with cusp forms.

The classical theory of modular symbols associated with cusp forms of arbitrary weight with respect to a group Γ furnishes the basic examples of modular pseudo–measures. They are specializations of the general construction of sec. 1.12.

We will recall the main formulas and notation.

First of all, one easily sees that the space $\mathcal{O}(H)_{\text{cusp}}$ is a $\mathbb{C}[z]$–module, and simultaneously a right $GL^+(2, \mathbb{Z})$–module with respect to the "variable change" action. The change from the right to the left comes as a result of dualization, cf. the last paragraph of sec. 1.

In the classical theory, one restricts the pseudo–measure (1.23) onto subspaces of $\mathcal{O}(H)_{\text{cusp}}$ consisting of the products $f(z)P(z)$ where f is a classical cusp form of a weight $w + 2$ and P a polynomial of degree w. The action of $GL^+(2, \mathbb{Z})$ is rather arbitrarily redistributed between f and P.

To present this part of the structure more systematically, we have to consider four classes of linear representations (here understood as right actions) of $GL^+(2, \mathbb{Q})$:

(i) the one–dimensional determinantal representation $g \mapsto \det g \cdot \text{id}$;

(ii) the symmetric power of the basic two–dimensional representation;

(iii) the "variable change" action upon $\mathcal{O}(H)_{\text{cusp}}$, that is, the inverse image g^* with respect to the fractional linear action of $PGL^+(2, \mathbb{Q})$ $g : H \to H, g \mapsto g(z)$;

(iv) the similar inverse image map on polydifferentials $\Omega_1(H)_{\text{cusp}}^{\otimes r}$, that is, holomorphic tensors $f(z)(dz)^r, r \in \mathbb{Z}$;

The latter action is traditionally translated into a "higher weight" action on functions $f(z)$ from $\mathcal{O}(H)_{\text{cusp}}$ via dividing by $(dz)^r$, and further tensoring by a power of the determinantal representation. As a result, the picture becomes somewhat messy, and moreover, in different expositions different normalizations and distributions of determinants between f and P are adopted. Anyway, we will fix our choices by the following conventions (the same as in [He1], [He2].)

For an integer $w \geq 0$, define the right action of weight $w + 2$ upon holomorphic (or meromorphic) functions on H by

$$f|[g]_{w+2}(z) := (\det g)^{w+1} f(gz) j(g, z)^{-(w+2)} \tag{2.17}$$

where we routinely denote $j(g, z) := cz + d$ for

$$g = \begin{pmatrix} a & b \\ c & d \end{pmatrix}.$$

Moreover, define the right action of $GL(2, \mathbb{R})$ on polynomials in two variables by

$$(Pg)(X, Y) := P((\det g)^{-1}(aX + bY), (\det g)^{-1}(cX + dY)). \quad (2.18)$$

Let now α, β be two points in $H \cup \mathbb{P}^1(\mathbb{Q})$. Then we have, for any homogeneous polynomial $P(X, Y)$ of degree w and $g \in GL^+(2, \mathbb{Q})$, in view of (2.17) and (2.18),

$$\int_{g\alpha}^{g\beta} f(z) P(z, 1) dz = \int_\alpha^\beta f(gz) P(gz, 1) d(gz)$$

$$= \int_\alpha^\beta f|[g]_{w+2}(z) \cdot (\det g)^{-w-1} \cdot j(g, z)^{w+2}$$

$$\cdot P\left(\frac{az+b}{cz+d}, 1\right) \cdot j(g, z)^{-2} \cdot \det g \cdot dz$$

$$= \int_\alpha^\beta f|[g]_{w+2}(z) P((\det g)^{-1}(az + b), (\det g)^{-1}(cz + d)) dz$$

$$= \int_\alpha^\beta f|[g]_{w+2}(z)(Pg)(z, 1) dz. \quad (2.19)$$

In particular, if for given f, g, and a constant ε we have

$$f|[g]_{w+2}(z) = \varepsilon f(z), \quad (2.20)$$

then

$$\int_{g\alpha}^{g\beta} f(z) P(z, 1) dz = \int_\alpha^\beta f(z) \varepsilon (Pg)(z, 1) dz. \quad (2.21)$$

More generally, if for a finite family of $g_k \in GL^+(2, \mathbb{Q})$, $c_k \in \mathbb{C}$, we have

$$\sum_k c_k f|[g_k]_{w+2}(z) = \varepsilon f(z), \quad (2.22)$$

then

$$\sum_k c_k \int_{g_k\alpha}^{g_k\beta} f(z) P(z, 1) dz = \int_\alpha^\beta f(z) \varepsilon \sum_k c_k (Pg_k)(z, 1) dz. \quad (2.23)$$

These equations are especially useful in the standard context of modular and cusp forms and Hecke operators.

Let $\Gamma \subset SL(2, \mathbb{Z})$ be a subgroup of finite index, $w \geq 0$ an integer. Recall that a cusp form $f(z)$ of weight $w + 2$ with respect to Γ is a holomorphic function on the upper half–plane H, vanishing at cusps and such that for all

$g \in \Gamma$, (2.20) holds with constant $\varepsilon = 1$. More generally, we can consider constants $\varepsilon = \chi(g)$, where χ is a character of Γ.

When w is odd, nonvanishing cusp form can exist only if $-I \notin \Gamma$ where I is the identity matrix. In particular, for $\Gamma = \mathrm{SL}(2, \mathbb{Z})$ there can exist only cusp forms of even weight.

Denote by $S_{w+2}(\Gamma)$ (resp. $S_{w+2}(\Gamma, \chi)$) the complex space of cusp forms of weight $w + 2$ for Γ (resp. cusp forms with character χ). Let F_w be the space of polynomial forms of degree w in two variables. Let W be the space of linear functionals on $S_{w+2}(\Gamma) \otimes F_w$. Then the function on $(\alpha, \beta) \in \mathbb{P}^1(\mathbb{Q})^2$ with values in the space W

$$\mu(\alpha, \beta) : f \otimes P \mapsto \int_\alpha^\beta f(z) P(z, 1) dz \qquad (2.24)$$

is a pseudo–measure.

We can call this pseudo–measure "the shadow" of the respective modular symbol.

Formulas (2.17) and (2.18) define the structure of a right Γ–module upon $S_{w+2}(\Gamma) \otimes F_w$ and hence the dual structure of a left Γ–module upon W. Formula (2.21) then shows that that the pseudo–measure (2.24) is modular with respect to Γ.

3 The Lévy functions and the Lévy–Mellin transform

3.1 The Lévy functions.

A classical Lévy function (see [L]) $L(f)(\alpha)$ of a real argument α is given by the formula

$$L(f)(\alpha) := \sum_{n=0}^\infty f(q_n(\alpha), q_{n+1}(\alpha)) \qquad (3.1)$$

where $q_n(\alpha)$, $n \geq 0$, is the sequence of denominators of normalized convergents to α (see (1.5)), and f is a function defined on pairs $(q', q) \in \mathbb{Z}^2$, $1 \leq q' \leq q$, $\gcd(q', q) = 1$, taking values in a topological group and sufficiently quickly decreasing so that (3.1) absolutely converges. Then $L(f)(\alpha)$ is continuous on irrational numbers. Moreover, it has period 1.

It was remarked in [MaMar1] that for certain simple f related to modular symbols, the integral $\int_0^1 L(f)(\alpha) d\alpha$ is a Dirichlet series directly related to the Mellin transform of an appropriate cusp form.

In this section, we will develop this remark and considerably generalize it in the context of modular pseudo–measures. We will call the involved integral representations *the Lévy–Mellin transform*. They are "shadows" of the classical Mellin transform.

As P. Lévy remarked in [L], for any $(q', q) \neq (1, 1)$ as above all $\alpha \in [0, 1/2]$ such that $(q', q) = (q_n(\alpha), q_{n+1}(\alpha))$ fill a primitive semi–interval I of length $(q(q + q'))^{-1}$. In $[0, 1]$, such α fill in addition the semi–interval $1 - I$ (in [MaMar1], sec. 2.1, we have inadvertently overlooked this symmetry).

Therefore the class of functions $L(f)$ (restricted to irrational numbers) is contained in a more general class of (formal) infinite linear combinations of characteristic functions of primitive segments I:

$$L(f) := \sum_I f(I) \chi_I, \tag{3.2}$$

where $\chi_I(\alpha) = 1$ for $\alpha \in I$, 0 for $\alpha \notin I$.

A family of coefficients, i.e. a map $I \mapsto f(I)$ from the set of all primitive segments (positively oriented, or without a fixed orientation) in $[0,1]$ to an abelian group M, will also be referred to as a Lévy function.

To treat this generalization systematically, we will display in the next subsection the relevant combinatorics of primitive segments.

3.2 Various enumerations of primitive segments.

A matrix in $GL(2, \mathbb{Z})$ is called *reduced* if its entries are non–negative, and non–decreasing to the right in the rows and downwards in the columns. For a more thorough discussion, see [LewZa].

Clearly, each row and column of a matrix in $GL(2, \mathbb{Z})$, in particular, reduced ones, consists of co–prime entries. There is exactly one reduced matrix with lower row $(1, 1)$. This is one marginal case which does not quite fit in the pattern of the following series of bijections between several sets described below.

The set L. It consists of pairs (c, d), $1 \leq c < d \in \mathbb{Z}$, $\gcd(c, d) = 1$.

The set R. It consists of pairs of reduced matrices (g^-, g^+) with one and the same lower row. The choice of superscript $\pm$ is described in the description of the set S below.

It is easy to see that for any pair $(c, d) \in L$, there exists exactly one pair $(g_{c,d}^-, g_{c,d}^+) \in R$ with lower row (c, d) so that we have a natural bijection $L \to R$.

In the marginal case, only one reduced matrix $g_{1,1}^-$ exists.

The set S. It consists of pairs of primitive segments (I^-, I^+) of length $< 1/2$

such that $I^- \subset [0, \frac{1}{2}]$, $I^+ = 1 - I^- \subset [\frac{1}{2}, 1]$. There is a well defined map $R \to S$ which produces from each pair $(g_{c,d}^-, g_{c,d}^+) \in R$ the pair

$$I_{c,d}^- := [g_{c,d}^-(0), g_{c,d}^-(1)], \quad I_{c,d}^+ := [g_{c,d}^+(0), g_{c,d}^+(1)].$$

Again, it is a well defined bijection.

In the marginal case, we get one primitive segment $I_{1,1}^- = [0, \frac{1}{2}]$; it is natural to complete it by $I_{1,1}^+ := [\frac{1}{2}, 1]$.

The set C. One element of this set is defined as a pair $(q', q) \in L$ such that there exists an α, $0 < \alpha \leq \frac{1}{2}$, for which (q', q) is a pair of denominators $(q_n(\alpha), q_{n+1}(\alpha))$ of two consecutive convergents to α.

The sequence of consecutive convergents to a rational number α stabilizes at some p_{n+1}/q_{n+1}, so only a finite number of pairs (q', q) is associated with α. Integers, in particular 0 and 1, correspond to the marginal pair $(1,1)$.

According to a lemma, used by P. Lévy in [L], all such α fill precisely the semi–interval $I_{q',q}^-$, with the end $g_{q',q}^-(1)$ excluded.

Moreover, those α which belong to the other half of $[0, 1]$ and as well admit (q', q) as a pair of denominators of two consecutive convergents, fill the semi–interval $[g_{q',q}^+(0), g_{q',q}^+(1))$. This produces one more bijection $C = L \to S$, or rather an interpretation of a formerly constructed one.

The pairs of convergents involved are also encoded in this picture: they are $(g_{q',q}^-(0), g_{q',q}^-(\infty))$ and $(g_{q',q}^+(0), g_{q',q}^+(\infty))$ respectively.

Returning now to Lévy functions, we can rewrite (3.2) as

$$L(f)(\alpha) := f(I_{1,1}) + \sum_{I_{c,d}^- \ni \alpha} f(I_{c,d}^-) + \sum_{I_{c,d}^+ \ni \alpha} f(I_{c,d}^+). \qquad (3.3)$$

or else as a classical Lévy function

$$L(f)(\alpha) := f(1, 1) + \sum_{n \geq 0} f^-(q_n(\alpha), q_{n+1}(\alpha)) \chi_{(0,1/2]}(\alpha)$$

$$+ \sum_{n \geq 0} f^+(q_n(\alpha), q_{n+1}(\alpha)) \chi_{[1/2,1)}(\alpha) \qquad (3.4)$$

if the usual proviso about convergence is satisfied.

Notice that for any $(c, d) \in L$, the number n such that $(c, d) = (q_n(\alpha), q_{n+1}(\alpha))$ is uniquely defined by (c, d) and a choice of sign $\pm$ determining the position of α in the right/left half of $[0, 1]$. The sequence of convergents preceding the $(n + 1)$–th one is determined as well. This means that choosing $f^{\pm}$, we may explicitly refer to all these data, including the sequence of incomplete quotients up to the $(n + 1)$–th one. Cf. examples in sec. 2.1 and 2.2.1 of [MaMar1].

3.3 The Lévy–Mellin transform.

Below we will use formal Dirichlet series.

Let $\mathcal{A}$ be an abelian group. With any sequence $A = \{a_1, \ldots, a_n, \ldots\}$ of elements of $\mathcal{A}$ we associate a formal expression

$$L_A(s) := \sum_{n=1}^{\infty} \frac{a_n}{n^s}.$$

If $\mathcal{B}, \mathcal{C}$ are abelian groups, and we are given a composition $\mathcal{A} \times \mathcal{B} \to \mathcal{C}$: $(a, b) \mapsto a \cdot b$ which is a group homomorphism, we have the induced composition

$$(a_1, \ldots, a_n, \ldots) \cdot (b_1, \ldots, b_n, \ldots) = (c_1, \ldots, c_n, \ldots)$$

$$\text{with} \qquad c_n := \sum_{d_1, d_2:\, d_1 d_2 = n} a_{d_1} \cdot b_{d_2},$$

which is interpreted as the multiplication of the respective formal Dirichlet series:

$$L_C(s) = L_A(s) \cdot L_B(s).$$

In particular, we can multiply any $L_A(s)$ by a Dirichlet series with integer coefficients.

Finally, the operator $\{a_1, \ldots, a_n, \ldots\} \mapsto \{1^w a_1, 2^w a_2, \ldots, n^w a_n, \ldots\}$ corresponds to the argument shift $L_A(s - w)$. Here generally $w \in \mathbb{Z}$, $w \geq 0$, but one can use also negative w if $\mathcal{A}$ is a linear $\mathbb{Q}$–space.

Now let W be a left $\mathrm{GL}^+(2, \mathbb{Q})$–module: $w \mapsto g[w]$. We will apply this formalism to the Dirichlet series with coefficients in $\mathbb{Z}[\mathrm{GL}^+(2, \mathbb{Q})]$, in W, and the composition induced by the above action.

Let $\mu \in M_W(\mathrm{SL}(2, \mathbb{Z}))$ (for notation, see sec. 2.1.1.) Consider the following Lévy function f_μ non–vanishing only on the primitive segments in $[0, 1/2]$:

$$f_\mu(I_{c,d}^-)(s) := \frac{1}{|I_{c,d}^-| d^s} \begin{pmatrix} 1 & -cd^{-1} \\ 0 & d^{-1} \end{pmatrix} \left[\mu(\infty, \frac{c}{d}) \right], \qquad (3.5)$$

where the matrix preceding the square brackets acts upon $\mu(\infty, \frac{c}{d})$. This Lévy function takes values in the group of formal Dirichlet series with coefficients in W, with only one non–vanishing term of the series.

Definition 3.3.1. The Mellin–Lévy transform of the pseudo–measure μ is the formal Dirichlet series with coefficients in W:

$$LM_\mu(s) := \int_0^{1/2} L(f_\mu)(\alpha, s)d\alpha . \tag{3.6}$$

(The formal argument s is included in the notation for future use.)

Moreover, introduce the following two formal Dirichlet series with coefficients in $\mathbb{Z}[GL^+(2, \mathbb{Z})]$:

$$Z_-(s) := \sum_{d_1=1}^\infty \begin{pmatrix} 1 & 0 \\ 0 & d_1^{-1} \end{pmatrix} \frac{1}{d_1^s}, \quad Z_+(s) := \sum_{d_2=1}^\infty \begin{pmatrix} d_2^{-1} & 0 \\ 0 & 1 \end{pmatrix} \frac{1}{d_2^s}.$$

Theorem 3.4. *We have the following identity between formal Dirichlet series:*

$$Z_+(s) \cdot Z_-(s) \cdot LM_\mu(s) = \sum_{n=1}^\infty \frac{(T_n\mu)(\infty, 0)}{n^s} \tag{3.7}$$

where T_n is the Hecke operator described in 2.5.2 and 2.5.1.

Proof. Put $\overline{L} := L \cup \{(1, 1)\}$, where the set L was defined in sec. 3.2. Each matrix from (2.16) representing a coset in $SL(2, \mathbb{Z}) \setminus M_n(2, \mathbb{Z})$ can be uniquely written as

$$\begin{pmatrix} d_2 & cd_1 \\ 0 & dd_1 \end{pmatrix} \tag{3.8}$$

where d, d_1, d_2 are natural numbers with $dd_1d_2 = n$ and $(c, d) \in \overline{L}$. Acting from the left upon $(\infty, 0)$ this matrix produces (∞, cd^{-1}). In the respective summand of $(T_n\mu)(\infty, 0)$ (cf. (2.15)), the inverse matrix to (3.8) acts on the left. We have

$$\begin{pmatrix} d_2 & cd_1 \\ 0 & dd_1 \end{pmatrix}^{-1} = \begin{pmatrix} d_2^{-1} & 0 \\ 0 & 1 \end{pmatrix} \cdot \begin{pmatrix} 1 & 0 \\ 0 & d_1^{-1} \end{pmatrix} \cdot \begin{pmatrix} 1 & -cd^{-1} \\ 0 & d^{-1} \end{pmatrix} . \tag{3.9}$$

Summing over all d, d_1, d_2, c we get the identity (3.7) between the formal Dirichlet series. $\square$

This Theorem justifies the name of the Mellin–Lévy transform. In fact, applying (3.7) to a pseudo–measure associated with the space of $SL(2, \mathbb{Z})$–cusp forms of a fixed weight, and taking the value of the r.h.s. of (3.7) on an eigen–form for all Hecke operators, we get essentially the usual Mellin transform of this form. The l.h.s then furnishes its representation as an integral over a real segment (up to extra $Z_\pm$–factors) replacing the more common integral along the upper imaginary half–line.

Of course, one can establish versions of this theorem for the standard congruence subgroups $\Gamma_1(N)$, but (3.7) is universal in the same sense as the Fourier expansion presented in [Me] is universal.

3.5 *p*–adic analogies continued.

Returning to the discussion in sec. 1.13, we now suggest the reader to compare the formula (3.5) with the definition of the *p*–adic measure given by the formula (39) of [Ma2].

We hope that this list makes convincing our suggestion that the theory of pseudo–measures can be considered as an ∞–adic phenomenon.

4 Non–commutative pseudo–measures and iterated Lévy–Mellin transform

Definition 4.1. A pseudo–measure on $\mathbb{P}^1(\mathbb{R})$ with values in a non (necessarily) commutative group (written multiplicatively) U is a function $J : \mathbb{P}^1(\mathbb{Q}) \times \mathbb{P}^1(\mathbb{Q}) \to U$, $(\alpha, \beta) \mapsto J_\alpha^\beta \in U$ satisfying the following conditions: for any $\alpha, \beta, \gamma \in \mathbb{P}^1(\mathbb{Q})$,

$$J_\alpha^\alpha = 1, \quad J_\beta^\alpha J_\alpha^\beta = 1, \quad J_\gamma^\alpha J_\beta^\gamma J_\alpha^\beta = 1. \tag{4.1}$$

The formalism of the sections 1 and 2 can be partially generalized to the non–commutative case.

4.1.1 *Identical and inverse pseudo–measures.*

The set M_U of U–valued pseudo–measures generally does not form a group, but only a set with involution and a marked point: $(J^{-1})_\alpha^\beta := (J_\beta^\alpha)^{-1}$ and $J_\alpha^\beta \equiv 1_U$ are pseudo–measures.

4.1.2 *Universal pseudo–measure.*

Let $\mathbf{U}$ be a free group freely generated by the set $\mathbb{Q}$. Let $\langle \alpha \rangle \in \mathbf{U}$ be the generator corresponding to α. The map

$$(\alpha, \beta) \mapsto \langle \beta \rangle \langle \alpha \rangle^{-1}, \quad (\infty, \beta) \mapsto \langle \beta \rangle ,$$

is a pseudo–measure. It is universal in an evident sense (cf. sec. 1.4.)

4.1.3 Primitive chains.

Let $I_j = (\alpha_j, \alpha_{j+1})$, $j = 1, \ldots, n$, be a primitive chain connecting $\alpha := \alpha_1$ to $\beta := \alpha_{n+1}$ as at the end of sec. 1.6. Then

$$J_\alpha^\beta = J_{\alpha_n}^{\alpha_{n+1}} J_{\alpha_{n-1}}^{\alpha_n} \cdots J_{\alpha_1}^{\alpha_2} . \tag{4.2}$$

In particular, each pseudo–measure is determined by its values on primitive segments.

Definition 4.2. An U–valued pre–measure $\widetilde{J}$ is a function on primitive segments satisfying the relations (4.1) written for primitive chains only.

Theorem 4.3. *Every pre–measure $\widetilde{J}$ can be uniquely extended to a pseudo–measure J.*

The proof proceeds exactly as that of the Theorem 1.8, with only minor local modifications. We define J by any of the formulas

$$J_\alpha^\beta = \widetilde{J}_{\alpha_n}^{\alpha_{n+1}} \widetilde{J}_{\alpha_{n-1}}^{\alpha_n} \cdots \widetilde{J}_{\alpha_1}^{\alpha_2} . \tag{4.3}$$

using a primitive chain as in 4.1.3, and then prove that this prescription does not depend on the choice of this chain using the argument with elementary moves. One should check only that elementary moves are compatible with non–commutative relations (4.1) which is evident.

4.4 Non–commutative reciprocity functions.

By analogy with 1.9, we can introduce the notion of a non–commutative U–valued reciprocity function R_p^q. The extension of the functional equation (1.13) should read

$$R_{p+q}^q R_p^{p+q} = R_p^q . \tag{4.4}$$

The analysis and results of 1.9–1.11 (not involving Dirichlet symbols) can be easily transported to this context.

4.5 Right action of $\mathrm{GL}^+(2, \mathbb{Q})$.

This group acts upon the set of pseudo–measures M_U from the right as in the commutative case:

$$(Jg)_\alpha^\beta := J_{g\alpha}^{g\beta} .$$

4.6 Modular pseudo–measures.

Assume now that a subgroup of finite index $\Gamma \subset SL(2, \mathbb{Z})$ acts upon U from the left by group automorphisms, $u \mapsto gu$. As in the commutative case, we call an U–valued pseudo–measure J *modular with respect to* Γ if it satisfies the condition

$$(Jg)_\alpha^\beta = g[J_\alpha^\beta]$$

for all $g \in \Gamma$, $\alpha, \beta \in \mathbb{P}^1(\mathbb{Q})$. Denote by $M_U(\Gamma)$ the pointed set of such measures. We can repeat the argument and the construction of 2.1.2 showing that any element $J \in M_U(\Gamma)$ is uniquely determined by the values $J_{h_k(\infty)}^{h_k(0)} \in U$ where $\{h_k\}$ runs over a system of representatives of $\Gamma \setminus SL(2, \mathbb{Z})$.

An analog of 2.2 and of the Theorem 2.3 holds as well.

Namely, consider the map

$$M_U(SL(2, \mathbb{Z})) \to U : J \mapsto J_\infty^0 .$$

From the argument above it follows that this is an injective map of pointed sets with involution. Its image is constrained by the similar conditions as in the abelian case.

Firstly, because of modularity, we have

$$(-\mathrm{id})[J_\infty^0] = J_\infty^0 .$$

Therefore, the image of (3.1) is contained in U_+, $(-\mathrm{id})$–invariant subgroup of U. Hence we may and will consider this subgroup as a (non–commutative) module over $PSL(2, \mathbb{Z})$. Secondly,

$$\sigma[J_\infty^0] \cdot J_\infty^0 = 1_U . \tag{4.5}$$

And finally,

$$\tau^2[J_\infty^0] \cdot \tau[J_\infty^0] \cdot J_\infty^0 = 1_U . \tag{4.6}$$

To summarize, we get the following morphisms of pointed sets

$$M_U(SL(2, \mathbb{Z})) \to U_+ \to U_+ \times U_+ \tag{4.7}$$

where the first arrow is the embedding described above and the second one, say φ, sends u to $(\sigma u \cdot u, \tau^2 u \cdot \tau u \cdot u)$.

Theorem 4.7. *The sequence of pointed sets (4.7) is exact. In other words, the map $J \mapsto J_\infty^0$ induces a bijection*

$$M_U(SL(2, \mathbb{Z})) \cong \varphi^{-1}(1_U, 1_U) \tag{4.8}$$

Sketch of proof. The proof follows exactly the same plan as that of Theorem 2.3. The only difference is that now the relations (1.1), (1.2) are replaced by their non–commutative versions (4.1) and the prescription (2.1) must be replaced by its non–commutative version based upon (4.2)

4.8 Induced pseudo–measures.

As in the commutative case, by changing the group of values U, we can reduce the description of $M_U(\Gamma)$ for general Γ to the case $\Gamma = \mathrm{SL}(2, \mathbb{Z})$.

Namely, put

$$\widehat{U} := \mathrm{Map}_\Gamma(\mathrm{PSL}(2, \mathbb{Z}), U). \tag{4.9}$$

An element $\varphi \in \widehat{U}$ is thus a map $g \mapsto \varphi(g) \in U$ such that $\varphi(\gamma g) = \gamma\varphi(g)$ for all $\gamma \in \Gamma$ and $g \in \mathrm{SL}(2, \mathbb{Z})$. Such functions form a group with pointwise multiplication, and $\mathrm{PSL}(2, \mathbb{Z})$ acts on it via $(g\varphi)(\gamma) := \varphi(\gamma g)$.

Any $\widehat{U}$–valued $\mathrm{SL}(2, \mathbb{Z})$–modular measure $\widehat{J}$ induces an U–valued Γ–modular measure J:

$$J_\alpha^\beta := \widehat{J}_\alpha^\beta(1_{\mathrm{SL}(2,\mathbb{Z})}). \tag{4.10}$$

Conversely, any U–valued Γ–modular measure J produces a $\widehat{U}$–valued $\mathrm{SL}(2, \mathbb{Z})$–modular measure $\widehat{J}$:

$$\widehat{J}_\alpha^\beta(g) := J_{g(\alpha)}^{g(\beta)}. \tag{4.11}$$

Proposition 4.8.1. *The maps (4.10), (4.11) are well defined and mutually inverse. Thus, they produce a canonical isomorphism*

$$M_U(\Gamma) \cong M_{\widehat{U}}(\mathrm{SL}(2, \mathbb{Z})). \tag{4.12}$$

The proof is straightforward.

4.8.2 Pseudo–measures and cohomology.

As in the commutative case, given $J \in M_U(\Gamma)$ and $\alpha \in \mathbb{P}^1(\mathbb{Q})$, consider the function

$$c_\alpha^J = c_\alpha : \Gamma \to U, \quad c_\alpha(g) := J_{g\alpha}^\alpha.$$

From (4.1) and the modularity of J it follows that this is a noncommutative 1–cocycle (we adopt the normalization as in [Ma3]):

$$c_\alpha(gh) = c_\alpha(g) \cdot gc_\alpha(h).$$

Changing α, we get a cohomologous cocycle:

$$c_\beta(g) = J_\alpha^\beta c_\alpha(g) \left(g J_\alpha^\beta\right)^{-1}.$$

As in the commutative case, if we restrict c_α upon the subgroup Γ_α fixing α, we get the trivial cocycle, so the respective cohomology class vanishes. This furnishes a canonical map of modular measures to the cuspidal cohomology

$$M_U(\Gamma) \to H^1(\Gamma, U)_{\text{cusp}}. \tag{4.13}$$

Again, for $\Gamma = \text{PSL}(2, \mathbb{Z})$ we have two independent descriptions of these sets, connected by the value $\mu(\infty, 0) = c_\infty(\sigma)$.

Proposition 4.8.3. *(i) For any* $\text{PSL}(2, \mathbb{Z})$*–modular pseudo–measure* J*, we have*

$$c_\infty^J(\sigma) = c_\infty^J(\tau) = J_0^\infty \in U_+.$$

(ii) This correspondence defines a bijection

$$M_W(\text{PSL}(2, \mathbb{Z})) \cong Z^1(\text{PSL}(2, \mathbb{Z}), U_+)_{\text{cusp}}$$

where the latter group by definition consists of cocycles with equal components $c(\sigma) = c(\tau)$*.*

This easily follows along the lines of [Ma4], Proposition 1.2.1.

4.9 Iterated integrals I.

In this and the next subsection we describe a systematic way to construct non–commutative pseudo–measures J based upon iterated integration. We start with the classical case, when J is obtained by integration along geodesics connecting cusps. This construction generalizes that of sec. 1.12 above, which furnishes the "linear approximation" to it.

Consider a pseudo–measure $\mu : \mathbb{P}^1(\mathbb{Q})^2 \to W$ given by the formulas (1.23) restricted to a finite–dimensional subspace $W^* \subset \mathcal{O}(H)_{\text{cusp}}$, that is, induced by the respective finite–dimensional quotient of $\mathbf{W}$ from the last paragraph of sec. 1.

Consider the ring of formal non–commutative series $\mathbb{C}\langle\langle W \rangle\rangle$ which is the completion of the tensor algebra of W modulo powers of the augmentation ideal (W). Put $U = 1 + (W)$. This is a multiplicative subgroup in this ring which will be the group of values of the following pseudo–measure J:

$$J_\alpha^\beta := 1 + \sum_{n=1}^\infty J_\alpha^\beta(n). \tag{4.14}$$

Here $J_\alpha^\beta(n) \in W^{\otimes n}$ is defined as the linear functional upon $W^{*\otimes n}$ whose value at $f_1 \otimes \cdots \otimes f_n$, $f_i \in W^*$, is given by the iterated integral

$$J_\alpha^\beta(n)(f_1, \ldots, f_n) := \int_\alpha^\beta f_1(z_1)\, dz_1 \int_\alpha^{z_1} f_2(z_2)\, dz_2 \ldots \int_\alpha^{z_{n-1}} f(z_n)\, dz_n \,.$$
(4.15)

The integration here is done over the simplex $\sigma_n(\alpha, \beta)$ consisting of the points $\beta > z_1 > z_2 > \ldots z_n > \alpha$, the sign $<$ referring to the ordering along the geodesic oriented from α to β.

The basic properties of (4.14), including the pseudo–measure identities (4.1) are well known, cf. a review in sec. 1 of [Ma3]. In particular, all J_α^β belong to the subgroup of the so called *group–like* elements of U. This property compactly encodes all the *shuffle relations* between the iterated integrals (4.15).

Moreover, (4.13) is functorial with respect to the variable–change action of $GL^+(2, \mathbb{Q})$ so that if the linear term of (4.13), that is, the pseudo–measure (1.23) is Γ–modular, then J is Γ–modular as well.

4.10 Iterated integrals II.

We can now imitate this construction using iterated integrals of, say, piecewise continuous functions along segments in $\mathbb{P}^1(\mathbb{R})$ in place of geodesics. The formalism remains exactly the same, and the general results as well; of course, in this generality it has nothing to do with specifics of the situation we have been considering so far.

To re–introduce these specifics, we will iterate integrals of Lévy functions (3.1)–(3.4) and particularly integrands at the r.h.s. of (3.6) We will get in this way the "shadow" analogs of the more classical multiple Dirichlet series considered in [Ma3], and shuffle relations between them. It would be interesting to see whether one can in this way get new relations between the classical series. Here we only say a few words about the structure of the resulting series.

Consider an iterated integral of the form (4.15) in which $f_k(z)$ are now characteristic functions of finite segments I_k in $\mathbb{P}^1(\mathbb{R})$. The following lemma is easy.

Lemma 4.10.1. *The integral (4.15) as a function of α, β and $2n$ ends of $I_1, \ldots, I_n$ is a piecewise polynomial form of degree n. This form depends only on the relative order of all its arguments in $\mathbb{R}$.*

Consider now a Γ–modular W–valued pseudo–measure μ and a family of Lévy functions with coefficients in formal Dirichlet series as in (3.5)

Then we can interpret an iterated integral of the form (4.15). with integrands $f_\mu(I^-_{c_k,d_k})(s_k) \cdot \chi_{I^-_{c_k,d_k}}(z)$, $k = 1, \ldots, n$, as taking values directly in $W^{\otimes n}$. Accordingly, the iterated version of the Mellin–Lévy transform (3.6) will be represented by a formal Dirichlet–like series involving coefficients which are polynomial forms of the ends of primitive segments involved.

5 Pseudo–measures and limiting modular symbols

5.1 Pseudo–measures and the disconnection space.

We recall the following definition of *analytic* pseudo–measures on totally disconnected spaces and their relation to currents on trees, cf. [vdP].

Definition 5.1.1. Let Ω be a totally disconnected compact Hausdorff space and let W be an abelian group. Let $C(\Omega, W) = C(\Omega, \mathbb{Z}) \otimes_{\mathbb{Z}} W$ denote the group of locally constant W–valued continuous functions on Ω. An analytic W–valued pseudo–measure on Ω is a map $\mu : C(\Omega, W) \to W$ satisfying the properties:

(i) $\mu(V \cup V') = \mu(V) + \mu(V')$ for $V, V' \subset \Omega$ clopen subsets with $V \cap V' = \emptyset$, where we identify a set with its characteristic function.

(ii) $\mu(\Omega) = 0$.

Equivalently, one can define analytic pseudo–measures as finitely additive functions on the Boolean algebra generated by a basis of clopen sets for the topology of Ω satisfying the conditions of Definition 5.1.1.

In particular, we will be interested in the case where the space $\Omega = \partial \mathcal{T}$ is the boundary of a tree. One defines currents on trees in the following way.

Definition 5.1.2. Let W be an abelian group and let $\mathcal{T}$ be a locally finite tree. We denote by $\mathcal{C}(\mathcal{T}, W)$ the group of W–valued *currents* on $\mathcal{T}$. These are W–valued maps $\mathbf{c}$ from the set of oriented edges of $\mathcal{T}$ to W that satisfy the following properties:

(i) Orientation reversal: $\mathbf{c}(\bar{e}) = -\mathbf{c}(e)$, where $\bar{e}$ denotes the edge e with the reverse orientation.

(ii) Momentum conservation:

$$\sum_{s(e)=v} \mathbf{c}(e) = 0, \tag{5.1}$$

where $s(e)$ (resp. $t(e)$) denote the source (resp. target) vertex of the oriented edge e.

One can then identify currents on a tree with analytic pseudo-measures on its boundary, as in [vdP].

Lemma 5.1.3. *The group $C(T, W)$ of currents on T is canonically isomorphic to the group of W–valued finitely additive analytic pseudo–measures on ∂T.*

Proof. The identification is obtained by setting

$$\mu(V(e)) = \mathbf{c}(e), \qquad (5.2)$$

where $V(e) \subset \partial T$ is the clopen subset of the boundary of T defined by all infinite admissible (i.e. without backtracking) paths in T starting with the oriented edge e. $\qquad \square$

The group of W–valued currents on T can also be characterized in the following way. We let $\mathcal{A}(T, W)$ denote the W–valued functions on the oriented edges of T satisfying $\mu(\bar{e}) = -\mu(e)$ and let $C(T^{(0)}, W)$ denote the W–valued functions on the set of vertices of T. The following result is also proved in [vdP].

Lemma 5.1.4. *Let $d : \mathcal{A}(T, W) \to C(T^{(0)}, W)$ be given by*

$$d(f)(v) = \sum_{s(e)=v} f(e).$$

Then the group of W–valued currents is given by $C(T, W) = Ker(d)$, so that one has an exact sequence

$$0 \to C(T, W) \to \mathcal{A}(T, W) \to C(T^{(0)}, W) \to 0.$$

5.2 The tree of PSL(2, ℤ) and its boundary.

We will apply these results to the tree T of $PSL(2, \mathbb{Z})$ embedded in the hyperbolic plane $\mathbb{H}$. Its vertices are the elliptic points $\tilde{I} \cup \tilde{R}$, where $\tilde{I}$ is the $PSL(2, \mathbb{Z})$–orbit of i and $\tilde{R}$ is the orbit of $\rho = e^{2\pi i/3}$. The set of edges is given by the geodesic arcs $\{\gamma(i), \gamma(\rho)\}$, for $\gamma \in SL(2, \mathbb{Z})$.

The relation between ∂T and $\mathbb{P}^1(\mathbb{R})$ can be summarized as follows.

Lemma 5.2.1. *The boundary ∂T is a compact Hausdorff space. There is a natural continuous $PSL(2, \mathbb{Z})$–equivariant surjection $\Upsilon : \partial T \to \partial \mathbb{H} = \mathbb{P}^1(\mathbb{R})$, which is one–to–one on the irrational points $\mathbb{P}^1(\mathbb{R}) \cap (\mathbb{R} \setminus \mathbb{Q})$ and two–to–one on rational points.*

Proof. Consider the Farey tessellation of the hyperbolic plane $\mathbb{H}$ by $PSL(2, \mathbb{Z})$ translates of the ideal triangle with vertices $\{0, 1, \infty\}$. The tree $T \subset \mathbb{H}$ has a vertex of valence three in each triangle, a vertex of valence two bisecting each edge of the triangulation and three edges in each triangle joining the valence three vertex to each of the valence two vertices on the sides of the triangle.

If we fix a base vertex in the tree, for instance the vertex $v = \rho = e^{2\pi i/3}$, then we can identify the boundary $\partial \mathcal{T}$ with the set of infinite admissible paths (i.e. paths without backtracking) in the tree $\mathcal{T}$ starting at v.

Any such path traverses an infinite number of triangles and can be encoded by the sequence of elements $\gamma_n \in \mathrm{PSL}(2, \mathbb{Z})$ that determine the successive three–valent vertices crossed by the path, $v_n = \gamma_n v$.

This sequence of points $v_n \in \mathbb{H}$ accumulates at some point $\theta \in \mathbb{P}^1(\mathbb{R}) = \partial \mathbb{H}$. If the point is irrational, $\theta \in \mathbb{P}^1(\mathbb{R}) \smallsetminus \mathbb{P}^1(\mathbb{Q})$, then the point θ is not a vertex of any of the Farey triangles and there is a unique admissible sequence of vertices $v_n \in \tilde{R} \subset \mathbb{H}$ with the property that $\lim_{n \to \infty} v_n = \theta$. To see this, consider the family of segments $I_n \subset \mathbb{P}^1(\mathbb{R})$ given by all points that are ends of an admissible path in $\mathcal{T}$ starting at v_n and not containing v_{n-1}.

The intersection $\cap_n I_n$ consists of a single point which is identified with the point in the limit set $\mathbb{P}^1(\mathbb{R})$ of $\mathrm{PSL}(2, \mathbb{Z})$ given by the infinite sequence $\gamma_1 \gamma_2 \cdots \gamma_n \cdots$.

Consider now a rational point $\theta \in \mathbb{P}^1(\mathbb{Q})$. Then θ is a vertex of some (in fact infinitely many) of the Farey triangles. In this case, one can see that there are two distinct sequences of 3-valent vertices v_n of $\mathcal{T}$ with the property that $\lim_{n \to \infty} v_n = \theta$, due to the fact that, beginning with the first v_k such that θ is a vertex of the triangle containing v_k, there are two adjacent triangles that also have θ as a vertex, see Figure 1.

This defines a map $\Upsilon : \partial \mathcal{T} \to \partial \mathbb{H}$, given by $\Upsilon(\{v_n\}) = \lim_{n \to \infty} v_n = \theta$. By construction it is continuous, $1 : 1$ on the irrationals and $2 : 1$ on the rationals. $\qquad \square$

5.3 The disconnection space.

An equivalent description of the space $\partial \mathcal{T}$ of ends of the tree $\mathcal{T}$ of $\mathrm{PSL}(2, \mathbb{Z})$ can be given in terms of the *disconnection* spaces considered in [Spi]. We discuss it here briefly, as it will be useful later in describing the noncommutative geometry of the boundary action of $\mathrm{PSL}(2, \mathbb{Z})$.

Given a subset $U \subset \mathbb{P}^1(\mathbb{R})$ one considers the abelian C^*-algebra $\mathcal{A}_U$ generated by the algebra $C(\mathbb{P}^1(\mathbb{R}))$ and the characteristic functions of the positively oriented intervals (in the sense of § 1.1 above) with endpoints in U. If the set U is dense in $\mathbb{P}^1(\mathbb{R})$ then this is the same as the closure in the supremum norm of the $*$-algebra generated by these characteristic functions.

The Gelfand-Naimark correspondence $X \leftrightarrow C_0(X)$ furnishes an equivalence of categories of locally compact Hausdorff topological spaces X and

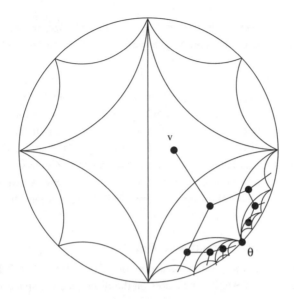

Figure 1.

commutative C^*-algebras respectively. Thus, we have $\mathcal{A}_U = C(D_U)$, where the topological space D_U is called *the disconnection* of $\mathbb{P}^1(\mathbb{R})$ along U. It is a compact Hausdorff space and it is totally disconnected if and only if U is dense in $\mathbb{P}^1(\mathbb{R})$.

In particular, one can consider the set $U = \mathbb{P}^1(\mathbb{Q})$ and the resulting disconnection $D_{\mathbb{Q}}$ of $\mathbb{P}^1(\mathbb{R})$ along $\mathbb{P}^1(\mathbb{Q})$.

Lemma 5.3.1. *The map* $\Upsilon : \partial \mathcal{T} \to \mathbb{P}^1(\mathbb{R})$ *of Lemma 5.2.1 factors through a homeomorphism* $\tilde{\Upsilon} : \partial \mathcal{T} \to D_{\mathbb{Q}}$, *followed by the surjective map* $D_{\mathbb{Q}} \to \mathbb{P}^1(\mathbb{R})$ *determined by the inclusion of the algebra* $C(\mathbb{P}^1(\mathbb{R}))$ *inside* $C(D_{\mathbb{Q}})$.

Proof. The compact Hausdorff space $\partial \mathcal{T}$ is dual to the abelian C^*-algebra $C(\partial \mathcal{T})$. After the choice of a base vertex $v \in \mathcal{T}$, the topology of $\partial \mathcal{T}$ is generated by the clopen sets $V(v')$, of ends of admissible paths starting at a vertex v' and not passing through v.

In fact, it suffices to consider only 3–valent vertices, because for a vertex v' of valence two we have $V(v') = V(v'')$ with v'' being the next 3–valent vertex in the direction away from v. The C^*–algebra $C(\partial \mathcal{T})$ is thus generated by the characteristic functions of the $V(v')$, since $\partial \mathcal{T}$ is a totally disconnected space. Since $v' = \gamma v$ for some $\gamma \in \mathrm{PSL}(2, \mathbb{Z})$, the set $V(v') = \Upsilon^{-1}(I(v')) \subset \partial \mathcal{T}$, where $I(v') \subset \mathbb{P}^1(\mathbb{R})$ is a segment of the form $I(v') = \gamma I$, where I is one of the segments $[\infty, 0]$ or $[0, 1]$ or $[1, \infty]$. These are the segments in $\mathbb{P}^1(\mathbb{R})$

of the form $[p/q, r/s]$ for $p, q, r, s \in \mathbb{Z}$ with $ps - qr = \pm 1$, that is, the primitive segments on the boundary, as defined in § 1.6, corresponding to sides of triangles of the Farey tessellation.

On the other hand the disconnection $D_{\mathbb{Q}}$ of $\mathbb{P}^1(\mathbb{R})$ along $\mathbb{P}^1(\mathbb{Q})$ is dual to the abelian C^*-algebra generated by all the characteristic functions of the oriented intervals with endpoints in $\alpha, \beta \in \mathbb{P}^1(\mathbb{Q})$. Since the sets $V(e)$ that generate the topology of $\partial \mathcal{T}$ are of this form, this shows that there is an injection $C(\partial \mathcal{T}) \to C(D_{\mathbb{Q}})$. The primitive intervals of § 1.6 in fact give a basis for the topology of $D_{\mathbb{Q}}$, since the characteristic functions that generate $C(D_{\mathbb{Q}})$ can be written as combinations of characteristic functions of such intervals, using primitive chains as in § 1.6. This shows that the map is in fact an isomorphism. $\square$

5.4 Analytic pseudo–measures on $D_{\mathbb{Q}}$.

We show here that the pseudo–measures on $\mathbb{P}^1(\mathbb{R})$ (Definition 2.1.1) can be regarded as analytic pseudo–measures on the disconnection space $D_{\mathbb{Q}}$.

Lemma 5.4.1. *There is a natural bijection between* W*–valued pseudo–measures on* $\mathbb{P}^1(\mathbb{R})$ *and* W*–valued analytic pseudo–measures on* $D_{\mathbb{Q}}$ *(or equivalently,* W*–valued currents on* $\mathcal{T}$*.)*

Proof. Recall that a basis for the topology of $D_{\mathbb{Q}}$ is given by the preimages under the map Υ of the primitive (Farey) segments in $\mathbb{P}^1(\mathbb{R})$. These are segments of the form $(g(\infty), g(0))$ for some $g \in G = \mathrm{PSL}(2, \mathbb{Z})$. Let e be an oriented edge in the tree $\mathcal{T}$ of $\mathrm{PSL}(2, \mathbb{Z})$. A basis of the topology of $\partial \mathcal{T}$ is given by sets of the form $V(e)$. These are in fact Farey intervals and all such intervals arise in this way. Thus, an analytic pseudo–measure on $D_{\mathbb{Q}}$ is a finitely additive function on the Boolean algebra generated by the Farey intervals with the properties of Definition 5.1.1.

A pseudo–measure on $\mathbb{P}^1(\mathbb{R})$, on the other hand, is a map $\mu : \mathbb{P}^1(\mathbb{Q}) \times \mathbb{P}^1(\mathbb{Q}) \to W$ with the properties (1.1) and (1.2). Such a function in fact extends, as we have seen, to a finitely additive function on the Boolean algebra consisting of finite unions of (positively oriented) intervals (α, β) with $\alpha, \beta \in \mathbb{P}^1(\mathbb{Q})$.

We then just set $\mu_{an}(V(e)) := \mu(\alpha, \beta)$, where $\mathbb{P}^1(\mathbb{R}) \supset [\alpha, \beta] = \Upsilon(V(e))$. This is then a finitely additive function on the Boolean algebra of the $V(e)$. The condition $\mu(\beta, \alpha) = -\mu(\alpha, \beta)$ implies that $\mu_{an}(\partial \mathcal{T}) = 0$ and similarly this combined with the property $\mu(\alpha, \beta) + \mu(\beta, \gamma) + \mu(\gamma, \alpha) = 0$ implies that $\mu_{an}(V \cup V') = \mu_{an}(V) + \mu_{an}(V')$ for $V \cap V' = \emptyset$.

Conversely, start with an analytic pseudo–measure μ_{an} on $D_{\mathbb{Q}}$ with values in W. For $\Upsilon V(e) = [\alpha, \beta]$ we write $\mu(\alpha, \beta) := \mu_{an}(V(e))$. Then the properties

$\mu_{an}(V \cup V') = \mu_{an}(V) + \mu_{an}(V')$ for $V \cap V' = \emptyset$ and $\mu_{an}(\partial T) = 0$ imply the corresponding properties $\mu(\beta, \alpha) = -\mu(\alpha, \beta)$ and $\mu(\alpha, \beta) + \mu(\beta, \gamma) + \mu(\gamma, \alpha) = 0$ for the pseudo–measure.

Under this correspondence, Γ–modular W–valued pseudo–measures correspond to Γ–modular analytic W–valued pseudo–measures. Thus, in the following we often use the term pseudo–measure equivalently for the analytic ones on $D_{\mathbb{Q}}$ or for those defined in § 2 on $\mathbb{P}^1(\mathbb{R})$. □

5.5 K–theoretic interpretation.

In [MaMar1] we gave an interpretation of the modular complex of [Ma1] in terms of K-theory of C^*-algebras, by considering the crossed product C^*-algebra of the action of $G = \mathrm{PSL}(2, \mathbb{Z})$ on its limit set $\mathbb{P}^1(\mathbb{R})$. Here we consider instead the action of $\mathrm{PSL}(2, \mathbb{Z})$ on the ends $\partial T = D_{\mathbb{Q}}$ of the tree and we obtain a similar K-theoretic interpretation of pseudo–measures.

More generally, let $\Gamma \subset \mathrm{PSL}(2, \mathbb{Z})$ be a finite index subgroup and we consider the action of $G = \mathrm{PSL}(2, \mathbb{Z})$ on $\partial T \times \mathbb{P}$, where $\mathbb{P} = \Gamma \backslash \mathrm{PSL}(2, \mathbb{Z})$. We let $\mathcal{A} = C(\partial T \times \mathbb{P})$ with the action of $G = \mathrm{PSL}(2, \mathbb{Z}) = \mathbb{Z}/2\mathbb{Z} * \mathbb{Z}/3\mathbb{Z}$ by automorphisms.

Lemma 5.5.1. *The K–groups of the crossed product C^*–algebra $\mathcal{A} \rtimes G$ have the following structure:*

$$K_0(\mathcal{A} \rtimes G) = Coker(\alpha), \quad K_1(\mathcal{A} \rtimes G) = Ker(\alpha),$$

where

$$\alpha : C(\partial T \times \mathbb{P}, \mathbb{Z}) \to C(\partial T \times \mathbb{P}, \mathbb{Z})^{G_2} \oplus C(\partial T \times \mathbb{P}, \mathbb{Z})^{G_3}$$

is given by $\alpha : f \mapsto (f + f \circ \sigma, f + f \circ \tau + f \circ \tau^2)$ for σ and τ, respectively, the generators of G_2 and G_3 defined by (2.4)).

Proof. First recall that, for a totally disconnected space Ω, one can identify the locally constant integer valued functions $C(\Omega, \mathbb{Z})$ with $K_0(C(\Omega))$, whereas $K_1(C(\Omega)) = 0$.

The six–terms exact sequence of [Pim] for groups acting on trees gives

$$0 \to K_1(\mathcal{A} \rtimes G) \to K_0(\mathcal{A}) \to K_0(\mathcal{A} \rtimes G_2) \oplus K_0(\mathcal{A} \rtimes G_3) \to K_0(\mathcal{A} \rtimes G) \to 0$$

for $G = \mathrm{PSL}(2, \mathbb{Z})$ and G_i equal to $\mathbb{Z}/2\mathbb{Z}$ or $\mathbb{Z}/3\mathbb{Z}$. Let $\alpha : K_0(\mathcal{A} \rtimes G_2) \oplus K_0(\mathcal{A} \rtimes G_3) \to K_0(\mathcal{A} \rtimes G)$ be the map in the sequence above.

Here we use the fact that $K_1(\mathcal{A}) = 0$ and $K_1(\mathcal{A} \rtimes G_i) = K_{G_i}^1(\mathcal{A}) = 0$ so that the remaining terms in the six–terms exact sequence do not contribute.

Moreover, we have $K_0(\mathcal{A}) = C(\partial T \times \mathbb{P}, \mathbb{Z})$. Similarly, we have

$$K_0(\mathcal{A} \rtimes G_i) = K^0_{G_i}(\partial T \times \mathbb{P}) = C(\partial T \times \mathbb{P}, \mathbb{Z})^{G_i}.$$

Following [Pim], we see that the map α can be described as the map

$$\alpha : f \mapsto (f + f \circ \sigma, f + f \circ \tau + f \circ \tau^2).$$

This gives a description of $Ker(\alpha)$ as

$$Ker(\alpha) = \{ f \in C(\partial T \times \mathbb{P}, \mathbb{Z}) \mid f + f \circ \sigma = f + f \circ \tau + f \circ \tau^2 = 0 \}.$$

Similarly, the cokernel of α is the group of coinvariants, that is, the quotient of $C(\partial T \times \mathbb{P}, \mathbb{Z})$ by the submodule generated by the elements of the form $f + f \circ \sigma$ and $f + f \circ \tau + f \circ \tau^2$. By the six–terms exact sequence we know that $K_1(\mathcal{A} \rtimes G) = Ker(\alpha)$ and $K_0(\mathcal{A} \rtimes G) = Coker(\alpha)$.

$\square$

For simplicity let us reduce to the case with $\mathbb{P} = \{1\}$, that is, $\Gamma = G = PSL(2, \mathbb{Z})$.

5.5.2 Integral.

Given a W–valued pseudo–measure μ on $D_{\mathbb{Q}}$, in the sense of Definition 5.1.1, we can define an integral

$$f \mapsto \int f \, d\mu \in W,$$

for $f \in C(D_{\mathbb{Q}}, \mathbb{Z})$ in the following way.

An element $f \in C(D_{\mathbb{Q}}, \mathbb{Z})$ is of the form $f = \sum_{i=1}^n a_i \chi_{I_i}$ with $a_i \in \mathbb{Z}$ and the intervals $I_i \subset D_{\mathbb{Q}}$ with $\Upsilon I_i = g_i(\infty, 0) \subset \mathbb{P}^1(\mathbb{R})$, for some $g_i \in G$. Thus, a natural prescription is

$$\int f \, d\mu = \sum_i a_i \mu(I_i) \in W.$$

To simplify notation, in the following we often do not distinguish between an interval in $\mathbb{P}^1(\mathbb{R})$ and its lift to the disconnection space $D_{\mathbb{Q}}$. So we write equivalently $\mu(I_i)$ or $\mu(g_i(\infty), g_i(0))$.

Lemma 5.5.3. *Let $g \in G$. The following change of variable formula holds:*

$$\int f \circ g \, d\mu = \int f \, d(\mu \circ g^{-1}).$$

Proof. We have $f \circ g = \sum_i a_i \chi_{I_i} \circ g = \sum_i a_i \chi_{g^{-1} I_i}$, so that

$$
\int f \circ g \, d\mu = \sum_i a_i \mu(g^{-1} g_i(\infty), g^{-1} g_i(0))
$$

$$
= \sum_i a_i \mu \circ g^{-1}(g_i(\infty), g_i(0)) = \int f \, d(\mu \circ g^{-1}).
$$

$\square$

The following generalization is also true (and easy):

$$
\int (f \circ g) \, h \, d\mu = \int f \, (h \circ g^{-1}) d(\mu \circ g^{-1}),
$$

for $f, h \in C(D_{\mathbb{Q}}, \mathbb{Z})$.

Proposition 5.5.4. *Let μ be a G-modular W–valued pseudo–measure. For any $h \in K_1(\mathcal{A} \rtimes G)$, there exists a unique G-modular W-valued pseudomeasure μ_h with*

$$
\mu_h(\infty, 0) := \int h \, d\mu
$$

Proof. We will consider the case of $G = \mathrm{PSL}(2, \mathbb{Z})$. The general case can be reduced to this one by proceeding as in 2.4 for the modular pseudomeasures.

In view of the Theorem 2.3, it suffices to check that the element

$$
\int h \, d\mu \in W,
$$

is annihilated by $1 + \sigma$ and $1 + \tau + \tau^2$.

An element $h \in K_1(\mathcal{A} \rtimes G)$ is a function $f \in C(\partial \mathcal{T}, \mathbb{Z})$ satisfying $h + h \circ \sigma = 0$ and $h + h \circ \tau + h \circ \tau^2 = 0$. Therefore

$$
(1 + \sigma) \int h \, d\mu = \int h \, d\mu + \int h \, d\mu \circ \sigma = \int (h + h \circ \sigma) d\mu = 0.
$$

Similarly,

$$
(1 + \tau + \tau^2) \int h \, d\mu = \int (h + h \circ \tau^2 + h \circ \tau) d\mu = 0.
$$

Thus, one obtains a G-modular pseudo-measure μ_h for each $h \in K_1(\mathcal{A} \rtimes G)$.

$\square$

Proposition 5.5.5. *A G-modular W–valued pseudo–measure μ defines a group homomorphism $\mu : K_0(\mathcal{A} \rtimes G) \to W$ induced by integration with respect to μ.*

Proof. Using the identification with analytic pseudomeasures, we can consider the functional from $C(\partial\mathcal{T}, \mathbb{Z})$ to W given by integration

$$\mu(f) = \int f \, d\mu.$$

This descends to the quotient of $C(\partial\mathcal{T}, \mathbb{Z})$ by the relations $f + f \circ \sigma$ and $f + f \circ \tau + f \circ \tau^2$. In fact, it suffices to consider the case where f is the characteristic function of a segment $(g(\infty), g(0))$. We have

$$\int (f + f \circ \sigma) d\mu = \int f \, d\mu + \int f \, d\mu \circ \sigma = (1 + \sigma) \int f \, d\mu$$

and

$$\int \chi_{(g(\infty), g(0))} \, d\mu = \int \chi_{(\infty, 0)} \circ g^{-1} \, d\mu = \int \chi_{(\infty, 0)} \, d\mu \circ g$$

so that

$$\sigma \int \chi_{(\infty, 0)} \, d\mu \circ g = \int \chi_{(\infty, 0)} d\mu \circ g \circ \sigma = \int \chi_{(g\sigma(\infty), g\sigma(0))} \, d\mu,$$

hence

$$(1 + \sigma) \int \chi_{(g(\infty), g(0))} \, d\mu = g(1 + \sigma)\mu(\infty, 0) = 0.$$

The argument for $f + f \circ \tau + f \circ \tau^2$ is similar.   □

5.6 Boundary action and noncommutative spaces.

In [MaMar1], we described the noncommutative boundary of the modular tower in terms of the quotient $\Gamma \backslash \mathbb{P}^1(\mathbb{R})$, which we interpreted as a noncommutative space, described by the crossed product C^*-algebra $C(\mathbb{P}^1(\mathbb{R})) \rtimes \Gamma$ or $C(\mathbb{P}^1(\mathbb{R}) \times \mathbb{P}) \rtimes G$, for $\Gamma \subset G$ a finite index subgroup and $\mathbb{P}$ the coset space.

Here we have seen that, instead of considering $\mathbb{P}^1(\mathbb{R})$ as the boundary, one can also work with the disconnection space $D_\mathbb{Q}$. We have then considered the corresponding crossed product $C(D_\mathbb{Q} \times \mathbb{P}) \rtimes G$. This is similar to the treatment of Fuchsian groups described in [Spi].

There is, however, another way of describing the boundary action of $G = \mathrm{PSL}(2, \mathbb{Z})$ on $\mathbb{P}^1(\mathbb{R})$. As we have also discussed at length in [MaMar1], it uses a dynamical system associated to the Gauss shift of the continued fraction expansion, generalized in the case of $\Gamma \subset G$ to include the action on the coset space $\mathbb{P}$. In [MaMar1] we worked with $\mathrm{PGL}(2, \mathbb{Z})$ instead of $\mathrm{PSL}(2, \mathbb{Z})$. For the $\mathrm{PSL}(2, \mathbb{Z})$ formulation, see [ChMay], [Mayer] and [KeS].

This approach to describing the boundary geometry was at the basis of our extension of modular symbols to limiting modular symbols. We discuss briefly here how this formulation also leads to a noncommutative space, in the form of an Exel–Laca algebra.

5.6.1 The generalized Gauss shift.

The Gauss shift for $\mathrm{PSL}(2, \mathbb{Z})$ is the map $\hat{T} : [-1, 1] \to [-1, 1]$ of the form

$$\hat{T} : x \mapsto -sign(x) T(|x|),$$

where

$$T(x) = \frac{1}{x} - \left[\frac{1}{x}\right].$$

Notice that this differs from the Gauss shift $T : [0, 1] \to [0, 1]$, $T(x) = 1/x - [1/x]$ of $\mathrm{PGL}(2, \mathbb{Z})$ by the presence of the extra sign, as in [ChMay], [KeS].

When one considers a finite index subgroup $\Gamma \subset G = \mathrm{PSL}(2, \mathbb{Z})$ one extends the shift $\hat{T}$ to the generalized Gauss shift

$$\hat{T}_{\mathbb{P}} : \mathcal{I} \times \mathbb{P} \to \mathcal{I} \times \mathbb{P}, \quad (x, s) \mapsto (\hat{T}(x), [gST^k]),$$

where here $\mathcal{I} = [-1, 1] \cap (\mathbb{R} \setminus \mathbb{Q})$ and where $k = sign(x) n_1$. Here $\mathbb{P} = \Gamma \backslash G$ is the coset space and $g \in G$ denotes the representative $\Gamma g = s \in \mathbb{P}$. The n–th iterate of the map $\hat{T}_{\mathbb{P}}$ acts on $\mathbb{P}$ as the $\mathrm{SL}(2, \mathbb{Z})$ matrix

$$\begin{pmatrix} -sign(x_1) p_{k-1}(x) & (-1)^k p_k(x) \\ q_{k-1}(x) & (-1)^{k+1} q_k(x) \end{pmatrix}.$$

5.6.2 Shift happens.

Instead of considering the action of the generalized Gauss shift on the space $\mathcal{I} \times \mathbb{P}$, one can proceed as in [KeS] and introduce a shift space over which $\hat{T}_{\mathbb{P}}$ acts as a shift operator. This is obtained by considering the countable alphabet $\mathbb{Z}^{\times} \times \mathbb{P}$ and the set of *admissible sequences*

$$\Sigma_{\Gamma} = \{((x_1, s_1), (x_2, s_2), \ldots) \mid A_{(x_i, s_i), (x_{i+1}, s_{i+1})} = 1\},$$

where the matrix A giving the admissibility condition is defined as follows. One has $A_{(x,s),(x',s')} = 1$ if $xx' < 0$ and

$$s' = \tau_x(s) := [gST^{x_1}] \in \mathbb{P}, \quad \text{where } s = \Gamma g$$

and $A_{(x,s),(x',s')} = 0$ otherwise. The action of $\hat{T}_{\mathbb{P}}$ described above becomes the action of the one–sided shift $\sigma : \Sigma_{\Gamma} \to \Sigma_{\Gamma}$

$$\sigma : ((x_1, s_1), (x_2, s_2), \ldots) \mapsto ((x_2, s_2), (x_3, s_3), \ldots).$$

The space Σ_Γ can be topologized in the way that is customarily used to treat this type of dynamical systems given by shift spaces. Namely, one considers on Σ_Γ the topology generated by the cylinders (all words in Σ_Γ starting with an assigned finite admissible word in the alphabet). This makes Σ_Γ into a compact Hausdorff space. One can see, in fact, that the topology is metrizable and induced for instance by the metric

$$d((x_k, s_k)_k, (x'_k, s'_k)_k) = \sum_{n=1}^{\infty} 2^{-n} \left(1 - \delta_{(x_n, s_n), (x'_n, s'_n)}\right).$$

As is shown in [KeS], the use of this topology as opposed to the one induced by $\mathbb{P}^1(\mathbb{R}) \times \mathbb{P}$ simplifies the analysis of the associated Perron–Frobenius operator. The latter now falls into the general framework developed in [MaUr] and one obtains the existence of a shift invariant ergodic measure on Σ_Γ from this general formalism. One uses essentially the *finite irreducibility* of the shift space (Σ_Γ, σ), which follows in [KeS] from the ergodicity of the geodesic flow. The σ–invariant measure on Σ_Γ induces via the bijection between these spaces a $\hat{T}_{\mathbb{P}}$–invariant measure on the space $\mathcal{I} \times \mathbb{P}$.

5.6.3 Exel–Laca algebras.

There is a way to associate to a shift space on an alphabet a noncommutative space, in the form of a Cuntz–Krieger algebra in the case of a finite alphabet [CuKrie], or more generally an Exel–Laca algebra for a countable alphabet [ExLa].

Start with a (finite or countable) alphabet $\mathbb{A}$ and a matrix $A : \mathbb{A} \times \mathbb{A} \to \{0, 1\}$ that assigns the admissibility condition for words in the alphabet $\mathbb{A}$. In the case of a finite alphabet $\mathbb{A}$, the corresponding Cuntz–Krieger algebra is the C^*-algebra generated by partial isometries S_a for $a \in \mathbb{A}$ with the relations

$$S_a S_a^* = P_a, \quad \text{with} \sum_a P_a = 1 \tag{5.3}$$

$$S_a^* S_a = \sum_{b \in \mathbb{A}} A_{ab} S_b S_b^*. \tag{5.4}$$

In the case of a countably infinite alphabet $\mathbb{A}$, one has to be more careful, as the summations that appear in the relations (5.3) and (5.4) no longer converge in norm. A version of CK algebras for infinite matrices was developed by Exel and Laca in [ExLa]. One modifies the relations (5.3) and (5.4) in the following way.

Definition 5.6.4. ([ExLa]) For a countably infinite alphabet $\mathbb{A}$ and an admissibility matrix $A : \mathbb{A} \times \mathbb{A} \to \{0, 1\}$, the CK algebra O_A is the universal

C^*-algebra generated by partial isometries S_a, for $a \in \mathbb{A}$, with the following conditions:

(i) $S_a^* S_a$ and $S_b^* S_b$ commute for all $a, b \in \mathbb{A}$.
(ii) $S_a^* S_b = 0$ for $a \neq b$.
(iii) $(S_a^* S_a) S_b = A_{ab} S_b$ for all $a, b \in \mathbb{A}$.
(iv) For any pair of finite subsets $X, Y \subset \mathbb{A}$, such that the product

$$A(X, Y, b) := \prod_{x \in X} A_{xb} \prod_{y \in Y} (1 - A_{yb}) \tag{5.5}$$

vanishes for all but finitely many $b \in \mathbb{A}$, one has the identity

$$\prod_{x \in X} S_x^* S_x \prod_{y \in Y} (1 - S_y^* S_y) = \sum_{b \in \mathbb{A}} A(X, Y, b) S_b S_b^*. \tag{5.6}$$

The conditions listed above are obtained by formal manipulations from the relations (5.3) and (5.4) and are equivalent to them in the finite case.

5.6.5 The algebra of the generalized Gauss shift.

We now consider the shift space (Σ_Γ, σ) of [KeS]. In this case we have the alphabet $\mathbb{A} = \mathbb{Z}^\times \times \mathbb{P}$ and the admissibility matrix given by the condition

$$A_{ab} = 1 \quad \text{iff } nn' < 0 \quad \text{and} \quad s = s' \, ST^{n'},$$

for $a = (n, s)$ and $b = (n', s')$ in $\mathbb{A}$. The corresponding Exel–Laca algebra O_A is generated by isometries $S_{(n,s)}$ satisfying the conditions of Definition 5.6.4 above.

Let $\mathcal{J} = \Upsilon^{-1}[-1, 1] \subset D_\mathbb{Q}$ be the preimage of the interval $[-1, 1]$ under the continuous surjection $\Upsilon : D_\mathbb{Q} \to \mathbb{P}^1(\mathbb{R})$ and let $\mathcal{J}_{+1} = \Upsilon^{-1}[0, 1]$ and $\mathcal{J}_{-1} = \Upsilon^{-1}[-1, 0]$. Also let

$$\mathcal{J}_k = \{ x \in \mathcal{J} \mid \Upsilon(x) = sign(k)[a_1, a_2, \ldots], \ a_1 = |k|, \ a_i \geq 1 \}.$$

Let Γ be a finite index subgroup of $G = \mathrm{PSL}(2, \mathbb{Z})$, $\mathbb{P} = \Gamma \backslash G$. Consider the sets

$$\mathcal{J}_{k,s} := \mathcal{J}_k \times \{s\},$$

for $s \in \mathbb{P}$ and for $k \in \mathbb{Z}^\times$. Let $\chi_{k,s}$ denote the characteristic function of the set $\mathcal{J}_{k,s}$.

Proposition 5.6.6. *The subalgebra of the crossed product $C(D_\mathbb{Q} \times \mathbb{P}) \rtimes G$ generated by elements of the form*

$$S_{k,s} := \chi_{k,s} U_k,$$

where $U_k = U_{\gamma_k}$, for $\gamma_k = T^k S \in \Gamma$, is isomorphic to the Exel–Laca algebra O_A of the shift (Σ_Γ, σ).

Proof. We need to check that the $S_{k,s}$ satisfy the Exel–Laca axioms of Definition 5.6.4. We have $S_{k,s}^* = U_k^* \chi_{k,s} = U_k^*(\chi_{k,s}) U_k^*$ so that $S_{k,s} S_{k,s}^* = \chi_{k,s} U_k U_k^* \chi_{k,s} = \chi_{k,s} =: P_{k,s}$ and $S_{k,s}^* S_{k,s} = U_k^* \chi_{k,s} \chi_{k,s} U_k = U_k^*(\chi_{k,s})$. One sees from this that $S_{k,s}^* S_{k,s}$ and $S_{k',s'}^* S_{k',s'}$ commute, as they both belong to the subalgebra $C(D_{\mathbb{P}^1(\mathbb{Q})} \times \mathbb{P})$, so that the first axiom is satisfied. Similarly, $S_{k,s}^* S_{k',s'} = U_k^* \chi_{k,s} \chi_{k',s'} U_{k'} = 0$ for $(k, s) \neq (k', s')$, which is the second axiom. To check the third axiom, we have

$$(S_{k,s}^* S_{k,s}) S_{k',s'} = U_k^*(\chi_{k,s}) \chi_{k',s'} U_{k'}.$$

We now describe more explicitly the element $U_k^*(\chi_{k,s})$. Since we have $U_k^*(f) = f \circ \gamma_k$, we consider the action of the element $\gamma_k = T^k S \in \mathrm{PSL}_2(\mathbb{Z})$ on the set $\mathcal{J}_k \times \{s\}$. We have $S(\mathcal{J}_k) = \{x \mid \pi(x) = -k - sign(k)[a_2, a_3, \ldots]\}$, and $T^k S(\mathcal{J}_k) = \{x \mid \pi(x) = -sign(k)[a_2, a_3, \ldots]\}$. This is the union of all

$$\{x \mid \pi(x) = sign(k')[|k'|, a_3, \ldots] \; sign(k') = -sign(k)\}.$$

Thus, we have

$$T^k S : \mathcal{J}_k \times \{s\} \to \cup_{k':kk'<0} \mathcal{J}_{k'} \times \{s \, ST^{-k}\}, \tag{5.7}$$

we obtain that

$$U_k^*(\chi_{k,s}) \chi_{k',s'} = A_{(k,s),(k',s')} \chi_{k',s'}. \tag{5.8}$$

Thus, we obtain $(S_{k,s}^* S_{k,s}) S_{k',s'} = A_{(k,s),(k',s')} S_{k',s'}$, which is the third Exel–Laca axiom.

For the last axiom, consider the condition that $A(X, Y, b)$ of (5.5) vanishes for all but finitely many $b \in \mathbb{A}$. Given two finite subsets $X, Y \subset \mathbb{Z}^\times \times \mathbb{P}$, the only way that $A(X, Y, b) = 0$ for all but finitely many $b \in \mathbb{Z}^\times \times \mathbb{P}$ is that $A(X, Y, b) = 0$ for all b.

In fact, for given X and Y, suppose that there exists an element $b = (k, s)$ such that $A(X, Y, b) \neq 0$. This means that, for all $x \in X$ we have $A_{xb} = 1$ and for all $y \in Y$ we have $A_{yb} = 0$. The first condition means that for all $x = (k_x, s_x)$ we have $k_x k < 0$ and $s = s_x ST^{k_x}$, while the second condition means that, for all $y = (k_y, s_y)$ we either have $k_y k > 0$ or $k_y k < 0$ and $s \neq s_y ST^{k_y}$. Consider then elements of the form $b' = (k', s)$ with $k'k > 0$ and the same $s \in \mathbb{P}$ as $b = (k, s)$. All of these still satisfy $A_{xb'} = 1$ for all $x \in X$ and $A_{yb} = 0$ for all $y \in Y$, since the conditions only depend on the sign of k and on $s \in \mathbb{P}$. Thus, there are infinitely many b' such that $A(X, Y, b') \neq 0$. If

$A(X, Y, b) \equiv 0$ for all b, the condition of the fourth axiom reduces to

$$\prod_{x \in X} S_x^* S_x \prod_{y \in Y} (1 - S_y^* S_y) = 0. \tag{5.9}$$

Suppose that $A(X, Y, b) \equiv 0$ for all $b \in \mathbb{A}$ but that the expression in (5.9) is non-zero. This means that both $\prod_{x \in X} S_x^* S_x \neq 0$ and $\prod_{y \in Y} (1 - S_y^* S_y) \neq 0$. We first show by induction that, for $X = \{x_1, \ldots, x_N\}$ we can write first product in the form

$$(\prod_{x \in X} S_x^* S_x) P_b = A_{x_1 b} A_{x_2, b} \cdots A_{x_N b} P_b, \tag{5.10}$$

where $P_b = \chi_b = S_b S_b^*$. We know already that this is true for a single point $X = \{x\}$ by (5.8). Suppose it is true for N points. Then we have

$$S_{x_0}^* S_{x_0} (\prod_{i=1}^N S_{x_i}^* S_{x_i}) P_b = (\prod_{i=1}^N A_{x_i b}) \, S_{x_0}^* S_{x_0} \, P_b$$

$$= A_{x_1 b} A_{x_2, b} \cdots A_{x_N b} \, A_{x_0 b} P_b,$$

which gives the result. Thus, we see that the condition $\prod_{x \in X} S_x^* S_x \neq 0$ implies that, for some $b \in \mathbb{A}$, one has $A_{x,b} \neq 0$ for all $x \in X$, i.e. one has $\prod_{x \in X} A_{x,b} \neq 0$. We analyze similarly the condition $\prod_{y \in Y} (1 - S_y^* S_y) \neq 0$. We show by induction that, for $Y = \{y_1, \ldots, y_M\}$, the product can be written in the form

$$(\prod_{y \in Y} (1 - S_y^* S_y)) P_b = (1 - A_{y_1 b})(1 - A_{y_2 b}) \cdots (1 - A_{y_M b}) P_b. \tag{5.11}$$

First consider the case of a single point $Y = \{y\}$. We have $1 - S_y^* S_y = 1 - U_{k_y}^* (\chi_{k_y, s_y})$ so that using (5.8) we get $(1 - S_y^* S_y) P_b = (1 - A_{yb}) P_b$. We then suppose that the identity (5.11) holds for M points. We obtain, again using (5.8),

$$(1 - S_{y_0}^* S_{y_0})(\prod_{i=1}^M (1 - A_{y_i b})) P_b = (\prod_{i=1}^M (1 - A_{y_i b})) \, (1 - A_{y_0 b}) P_b.$$

Thus, the condition $\prod_{y \in Y} (1 - S_y^* S_y) \neq 0$ implies that there exists $b \in \mathbb{A}$ such that $\prod_{y \in Y} (1 - A_{yb}) \neq 0$, which contradicts the fact that we are assuming $A(X, Y, b) \equiv 0$. This proves that the fourth Exel–Laca axiom is satisfied.

$\square$

To summarize, there are several interesting noncommutative spaces related to the boundary of modular curves X_Γ: the crossed product algebras $C(\mathbb{P}^1(\mathbb{R}) \times \mathbb{P}) \rtimes G$ and $C(D_\mathbb{Q} \times \mathbb{P}) \rtimes G$, and the Exel–Laca algebra O_A. This calls for a more detailed investigation of their relations and of the information about the modular tower that each of these noncommutative spaces captures.

5.7 Limiting modular pseudo–measures.

In [MaMar1] we extended classical modular symbols to include the case of limiting cycles associated to geodesics with irrational endpoints, the limiting modular symbols. The theory of limiting modular symbols was further studied in [Mar] and more recently in [KeS].

We show here that we can similarly define limiting modular pseudo–measures, extending the class of pseudo–measures from finitely additive functions on the Boolean algebra of the $V(e) \subset D_{\mathbb{Q}}$ to a larger class of sets by a limiting procedure.

In the following we assume that the group W where the pseudo–measures take values is also a real (or complex) vector space. If as a vector space W is infinite dimensional, we assume that it is a topological vector space. This is always the case for finite dimension. As before we also assume that W is a left G-module.

We recall the following property of the convergents of the continued fraction expansion, cf. [PoWe]. For $\theta \in \mathbb{R} \smallsetminus \mathbb{Q}$, let q_n denote, as before, the successive denominators of the continued fraction expansion. Then the limit

$$\lambda(\theta) := \lim_{n \to \infty} \frac{2 \log q_n(\theta)}{n} \tag{5.12}$$

is defined away from an exceptional set $\Omega \subset \mathbb{P}^1(\mathbb{R})$ and where it exists it is equal to the Lyapunov exponent of the Gauss shift.

In [MaMar1] we proved that the limiting modular symbol

$$\{\{*, \theta\}\} = \lim_{t \to \infty} \frac{1}{t} \{x_0, y(t)\}_{\Gamma}$$

can be computed by the limit

$$\lim_{n \to \infty} \frac{1}{\lambda(\theta)n} \sum_{k=1}^{n+1} \{g_{k-1}(0), g_{k-1}(\infty)\}_{\Gamma} \in H_1(X_{\Gamma}, \mathbb{R}) \tag{5.13}$$

where $g_k = g_k(\theta)$ is the matrix in G that implements the action of the k–th power of the Gauss shift.

Fix a W–valued pseudo–measure μ on $\mathbb{P}^1(\mathbb{R})$, equivalently thought of as an analytic pseudo–measure on $D_{\mathbb{Q}}$. Consider the class of positively oriented intervals (∞, θ) with $\theta \in \mathbb{P}^1(\mathbb{R}) \smallsetminus \mathbb{P}^1(\mathbb{Q})$ and define the limit

$$\mu^{lim}(\infty, \theta) := \lim_{n \to \infty} \frac{1}{\lambda(\theta)n} \sum_{k=1}^{n+1} \mu(g_{k-1}(\infty), g_{k-1}(0)) \in W. \tag{5.14}$$

This is defined away from the exceptional set Ω in $\mathbb{P}^1(\mathbb{R})$ which contains $\mathbb{P}^1(\mathbb{Q})$ as well as the irrational points where either the limit defining $\lambda(\theta)$ does not exist or the limit of the

$$\frac{1}{n} \sum_{k=1}^{n+1} \mu(g_{k-1}(\infty), g_{k-1}(0))$$

does not exist.

Similarly, for $\theta \in \mathbb{P}^1(\mathbb{R}) \setminus \Omega$ we set

$$\mu^{lim}(\theta, \infty) = \lim_{n \to \infty} \frac{1}{\lambda(\theta)n} \sum_{k=1}^{n+1} \mu(g_{k-1}(0, \infty)). \tag{5.15}$$

We then set

$$\mu^{lim}(\theta, \eta) := \mu^{lim}(\theta, \infty) + \mu^{lim}(\infty, \eta). \tag{5.16}$$

Lemma 5.7.1. *The function μ^{lim} defined as above satisfies*

$$\mu^{lim}(\eta, \theta) = -\mu^{lim}(\theta, \eta) \quad \mu^{lim}(\theta, \eta) + \mu^{lim}(\theta, \zeta) + \mu^{lim}(\zeta, \theta) = 0$$

for all $\theta, \eta, \zeta \in \mathbb{P}^1(\mathbb{R}) \setminus \Omega$.

Proof. Since $(\beta, \alpha) = g(0, \infty) = g\sigma(\infty, 0)$, we have

$$\mu(\beta_k, \alpha_k) = \mu(g_k(0, \infty)) = \mu(g_k\sigma(\infty, 0)) = -\mu(\alpha_k, \beta_k),$$

so that

$$\mu^{lim}(\theta, \infty) = -\mu^{lim}(\infty, \theta).$$

We then have $\mu^{lim}(\eta, \theta) = \mu^{lim}(\infty, \theta) + \mu^{lim}(\eta, \infty) = -\mu^{lim}(\theta, \eta)$. The argument for the second identity is similar. $\qquad \square$

Thus, the limiting pseudo–measure μ^{lim} defines a finitely additive function on the Boolean algebra generated by the intervals (θ, η) with $\theta, \eta \in \mathbb{P}^1(\mathbb{R}) \setminus \Omega$. The limiting pseudo–measure μ^{lim} is G–modular if μ is G–modular.

5.7.2 *Limiting pseudo–measures and currents.*

In terms of currents on the tree $\mathcal{T}$, we can describe the limit computing μ^{lim} as a process of averaging the current **c** over edges along a path.

Lemma 5.7.3. *Let μ be a W-valued pseudo-measure and $\mathbf{c}$ the corresponding current on $\mathcal{T}$. The limiting pseudo-measure μ^{lim} is computed by the limit*

$$\mu^{lim}(\infty, \theta) = \lim_{n \to \infty} \frac{1}{\lambda(\theta)n} \sum_{k=1}^{n} \mathbf{c}(e_k).$$

Proof. An irrational point θ in $\partial\mathcal{T}$ corresponds to a unique admissible infinite path in the tree $\mathcal{T}$ starting from a chosen base vertex. We describe such a path as an infinite admissible sequence of oriented edges $e_1 e_2 \cdots e_n \cdots$. To each such edge there corresponds an open set $V(e_k)$ in $\partial\mathcal{T}$ with the property that $\cap_k V(e_k) = \{\theta\}$. These correspond under the map $\Upsilon : \partial\mathcal{T} \to \mathbb{P}^1(\mathbb{R})$ to intervals $[g_k(\infty), g_k(0)]$. Thus, the expression computing the limiting pseudo-measure can be written equivalently via the limit of the averages

$$\frac{1}{n} \sum_{k=1}^{n} \mathbf{c}(e_k).$$

$\square$

Bibliography

[A]   A. Ash. *Parabolic cohomology of arithmetic subgroups of* SL(2, $\mathbb{Z}$) *with coefficients in the field of rational functions on the Riemann sphere.* Amer. Journ. of Math., 111 (1989), 35–51.

[ChMay]   C.H. Chang, D. Mayer. *Thermodynamic formalism and Selberg's zeta function for modular groups.* Regular and Chaotic Dynamics 15 (2000) N.3, 281–312.

[ChZ]   Y. J. Choie, D. Zagier. *Rational period functions for for* PSL(2, $\mathbb{Z}$). In: A Tribute to Emil Grosswald: Number Theory and relatied Analysis, Cont. Math., 143 (1993), AMS, Providence, 89–108.

[CuKrie] J. Cuntz, W. Krieger. *A class of C^*-algebras and topological Markov chains, I,II.* Invent. Math. 56 (1980) 251–268 and Invent. Math. 63 (1981) 25–40.

[ExLa]   R. Exel, M. Laca. *Cuntz–Krieger algebras for infinite matrices.* J. Reine Angew. Math. 512 (1999), 119–172.

[Fu1]   Sh. Fukuhara. *Modular forms, generalized Dedekind symbols and period polynomials.* Math. Ann., 310 (1998), 83–101.

[Fu2]   Sh. Fukuhara. *Hecke operators on weighted Dedekind symbols.* Preprint math.NT/0412090.

[He1]   A. Herremans. *A combinatorial interpretation of Serre's conjecture on modular Galois representations.* PhD thesis, Catholic University Leuven, 2001.

[He2]   A. Herremans. *A combinatorial interpretation of Serre's conjecture on modular Galois representations.* Ann. Inst. Fourier, Grenoble, 53:5 (2003), 1287–1321.

[KeS]   M. Kessenböhmer, B. O. Stratmann. *Limiting modular symbols and their fractal geometry.* Preprint math.GT/0611048.

[Kn1]   M. Knopp. *Rational period functions of the modular group.* Duke Math. Journal, 45 (1978), 47–62.

[Kn2]   M. Knopp. *Rational period functions of the modular group II.* Glasgow Math. Journal, 22 (1981), 185–197.

[L]   P. Lévy. *Sur les lois de probabilité dont dépendent les quotients complets et incomplets d'une fraction continue.* Bull. Soc. Math. France, 557 (1929), 178–194.

[LewZa]   J. Lewis, D. Zagier. *Period functions and the Selberg zeta function for the modular group.* In: The Mathematical Beauty of Physics, Adv. Series in Math. Physics 24, World Scientific, Singapore, 1997, pp. 83–97.

[Ma1]   Yu. Manin. *Parabolic points and zeta-functions of modular curves.* Russian: Izv. AN SSSR, ser. mat. 36:1 (1972), 19–66. English: Math. USSR Izvestiya, publ. by AMS, vol. 6, No. 1 (1972), 19–64, and Selected Papers, World Scientific, 1996, 202–247.

[Ma2]   Yu. Manin. *Periods of parabolic forms and p–adic Hecke series.* Russian: Mat. Sbornik, 92:3 (1973), 378–401. English: Math. USSR Sbornik, 21:3 (1973), 371–393, and Selected Papers, World Scientific, 1996, 268–290.

[Ma3]   Yu. Manin. *Iterated integrals of modular forms and noncommutative modular symbols.* In: Algebraic Geometry and Number Theory. In honor of V. Drinfeld's 50th birthday. Ed. V. Ginzburg. Progress in Math., vol. 253. Birkhäuser, Boston, pp.565–597. Preprint math.NT/0502576.

[Ma4]   Yu. Manin. *Iterated Shimura integrals.* Moscow Math. Journal, vol. 5, Nr. 4 (2005), 869–881. Preprint math.AG/0507438.

[MaMar1]   Yu. Manin, M. Marcolli. *Continued fractions, modular symbols, and noncommutative geometry.* Selecta math., new ser. 8 (2002), 475–521. Preprint math.NT/0102006.

[MaMar2]   Yu. Manin, M. Marcolli. *Holography principle and arithmetic of algebraic curves.* Adv. Theor. Math. Phys., 5 (2001), 617–650. Preprint hep-th/0201036.

[Mar]   M. Marcolli. *Limiting modular symbols and the Lyapunov spectrum.* Journal of Number Theory, 98 (2003) 348–376.

[MaUr]   R.D. Mauldin, M. Urbański. *Graph directed Markov systems. Geometry and dynamics of limit sets.* Cambridge Tracts in Mathematics, 148. Cambridge University Press, Cambridge, 2003.

[Mayer]   D. Mayer. *Continued fractions and related transformations.* Ergodic theory, symbolic dynamics, and hyperbolic spaces (Trieste, 1989), 175–222, Oxford Sci. Publ., Oxford Univ. Press, New York, 1991.

[Me]   L. Merel. *Universal Fourier expansions of modular forms.* In: Springer LN in Math., vol. 1585 (1994), 59–94.

[Pim]   M.V. Pimsner, KK–*groups of crossed products by groups acting on trees.* Invent. Math. 86 (1986), no. 3, 603–634.

[PoWe]　　M. Pollicott, H. Weiss. *Multifractal analysis of Lyapunov exponent for continued fraction and Manneville-Pomeau transformations and applications to Diophantine approximation.* Comm. Math. Phys. 207 (1999), no. 1, 145–171.

[Sh1]　　　V. Shokurov. *The study of the homology of Kuga varieties.* Math. USSR Izvestiya, 16:2 (1981), 399–418.

[Sh2]　　　V. Shokurov. *Shimura integrals of cusp forms.* Math. USSR Izvestiya, 16:3 (1981), 603–646.

[Spi]　　　J.S. Spielberg. *Cuntz–Krieger algebras associated with Fuchsian groups.* Ergodic Theory Dynam. Systems 13 (1993), no. 3, 581–595.

[vdP]　　　M. van der Put. *Discrete groups, Mumford curves and theta functions.* Ann. Fac. Sci. Toulouse Math. 6 (1992) 399–438.

Jacobi forms of critical weight and Weil representations

Nils-Peter Skoruppa

Abstract

Jacobi forms can be considered as vector valued modular forms, and Jacobi forms of critical weight correspond to vector valued modular forms of weight $\frac{1}{2}$. Since the only modular forms of weight $\frac{1}{2}$ on congruence subgroups of $SL(2, \mathbb{Z})$ are theta series the theory of Jacobi forms of critical weight is intimately related to the theory of Weil representations of finite quadratic modules. This article explains this relation in detail, gives an account of various facts about Weil representations which are useful in this context, and it gives some applications of the theory developed herein by proving various vanishing theorems and by proving a conjecture on Jacobi forms of weight one on $SL(2, \mathbb{Z})$ with character.

Mathematics Subject Classification: 11F03,11F50,11F27

1 Introduction

All Siegel modular forms of degree n and weight $k < \frac{n}{2}$ are singular. Similarly, every orthogonal modular form associated to a quadratic module of signature $(2, n+1)$ and of weight $k < \frac{n}{2}$ is singular. The weights in the given ranges are called singular weights, a terminology, which was introduced by Resnikoff. Spaces of modular forms of singular weight are well-understood (cf. [Fr 91] and the literature cited therein for singular Siegel modular forms, and [Re 75], [Gri 95] for singular orthogonal modular forms). For critical weights much less is known. Here, by *critical weight*, we understand the weight $\frac{n}{2}$ in the theory of Siegel modular forms of degree n and in the theory of orthogonal modular forms of signature $(2, n+1)$ respectively. [We 92] gives some results for Siegel modular forms of degree 2 and critical weight 1. More recently, it was shown

239

in [I-S 07] that there are no cusp forms of degree 2 and weight 1 (with trivial character) on the subgroups $\Gamma_0(l)$ of $\mathrm{Sp}(2, \mathbb{Z})$.

In a joint project Ibukiyama and the author [I-S 08] take up a more systematic study of Siegel and orthogonal modular forms of singular weight, a first result being [I-S 07]. One of the main tools used in this project is the Fourier-Jacobi expansion of the modular forms in question, more precisely, the Fourier-Jacobi expansion with respect to Jacobi forms of degree 1. For modular forms of critical weight the Fourier-Jacobi coefficients will then be Jacobi forms whose index is a symmetric positive definite matrix of size n and whose weight equals $\frac{n+1}{2}$. We shall call this weight *critical* for Jacobi forms of degree 1 whose matrix index is of size n.

Jacobi forms of critical weight play also a role in current work of Gritsenko on the geometry of moduli spaces of polarized abelian surfaces. The explicit construction of Jacobi forms of small weight is an essential tool in his studies. In particular, Gritsenko proposes a construction of special Jacobi forms (called *Jacobi theta quarks*) with scalar index of weight one, hence of critical weight [Gri 05]. The Jacobi cusp forms which are important in the theory of Siegel modular three-folds have *canonical* weight 3 and are products of three weight one forms. The theta quarks have nice product expansions; in fact, they are special instances of so-called *thetablocks* (cf. the discussion before Theorem 11 in section 5 for a more precise statement). It is an interesting question whether all Jacobi forms of small weight (and of scalar index) can be obtained by thetablocks as it is suggested by numerical experiments. For Jacobi forms of weight $1/2$ this question can be answered affirmatively using the description of these forms given in Corollary to Theorem 5 below (cf. [G-S-Z 07] for details). For Jacobi forms of weight 1 (and of scalar index) we are similarly lead to the problem of finding a sufficiently explicit description of these forms. Such a description will drop out as part of the considerations of this article. In fact, we shall show that all Jacobi forms of weight 1 with character ε^8 (see section 2) are linear combinations of theta quarks. A more systematic theory of thetablocks will be developed in a joint article of V. Gritsenko, D. Zagier and the author [G-S-Z 07].

In view of the aforementioned applications it seems to be worthwhile to develop a systematic theory of Jacobi forms of critical weight. These Jacobi forms can be studied as vector valued elliptic modular forms of (critical) weight $\frac{1}{2}$. The first main tool for setting up such a theory is an explicit description of this connection. The second main tool is then the fact that elliptic modular forms of weight $\frac{1}{2}$ are theta series [Se-S 77], and the third main tool is the decomposition of the space of modular forms of weight $\frac{1}{2}$ with respect to the action of the metaplectic double cover of $\mathrm{SL}(2, \mathbb{Z})$ [Sko 85]. The latter makes it possible to describe spaces of vector valued elliptic modular forms of weight $\frac{1}{2}$ in sufficiently explicit form. A deeper study of these explicit descriptions requires a deeper understanding of Weil representations of SL$(2, \mathbb{Z})$ (or rather its metaplectic double cover) associated to finite quadratic modules. In fact, it turns out that spaces of Jacobi forms of critical weight are always naturally isomorphic to spaces of invariants of suitable Weil representations (cf. Theorem 8 and the subsequent remark).

The present article aims to pave the way for a theory of Jacobi forms of critical weight and, in particular, for the aforementioned projects [I-S 08], [G-S-Z 07] and possible other applications. We propose a formal framework for such a theory, we give an account of what one can prove for Jacobi forms of critical weight within this framework, and we describe the necessary tools from the theory of Weil representations and the theory of modular forms of weight $\frac{1}{2}$ which are needed. Some of the following results are new, others are possibly known but are not easily available elsewhere or are unpublished.

The plan of this article is as follows: In section 3 we discuss Weil representations of finite quadratic modules. All the considerations of this section are mainly motivated by the question for the invariants of a given Weil representation. In section 4 the relation between (vector valued) Jacobi forms and vector valued modular forms is made explicit. As an application we obtain (cf. Theorem 6) a dimension formula for spaces of Jacobi forms (of arbitrary matrix index). In section 5 we finally turn to Jacobi forms of critical weight. We shall see that Jacobi forms of critical weight are basically invariants of certain Weil representations associated to finite quadratic modules (cf. Theorem 8 and the subsequent remarks). We apply our theory to prove various vanishing results in Theorems 10, 11, Corollary to Theorem 13, and to prove in Theorem 12 that Jacobi forms of weight 1 and character ε^8 are obtained by theta blocks. In section 6 we append those proofs which have been omitted in the foregoing sections because of their more technical or computational nature. For the convenience of the reader we insert a section 2 which contains a glossary of the main notations.

We are indebted to the referee for several useful hints to the literature.

2 Notation

We use $[a, b; c, d]$ for the 2×2 matrix with first and second row equal to (a, b) and (c, d), respectively. If F is an $n \times n$ matrix and x a column vector of size n, we write $F[x]$ for $x^t F x$. We use $e(X)$ for $\exp(2\pi i X)$, and $e_m(X)$ for $\exp(2\pi i X/m)$. For integers a and b, the notation $a|b^\infty$ indicates that every prime divisor of a is also a divisor of b. For relatively prime a and b we use $\left(\frac{a}{b}\right)$

for the usual generalized Jacobi-Legendre Symbol with the (additional) convention $\left(\frac{a}{2^\beta}\right) = -1$ if $a \equiv \pm 3 \bmod 8$ and β is odd and $\left(\frac{a}{2^\beta}\right) = +1$ if $a \equiv \pm 1 \bmod 8$ or β is even. We summarize (in roughly alphabetical order) the most important notations of this article:

$\mathbb{C}(\chi)$:   For a character χ of the metaplectic cover $\mathbb{M} = \mathrm{Mp}(2, \mathbb{Z})$, the $\mathbb{M}$-module with underlying vector space $\mathbb{C}$ and with $\mathbb{M}$-action $(g, z) \mapsto \chi(z)g$.

D_F: For a symmetric half integral matrix F or a non-zero integer F, the quadratic module $\left(\mathbb{Z}^n/2F\mathbb{Z}^n, x \mapsto F^{-1}[x]/4\right)$, where n is the size of F.

$\vartheta(\tau, z)$:   The Jacobi form
$$q^{1/8}\left(\zeta^{1/2} - \zeta^{-1/2}\right) \prod_{n \geq 1} \left(1 - q^n\right)\left(1 - q^n\zeta\right)\left(1 - q^n\zeta^{-1}\right).$$

ε: The one dimensional character of $\mathbb{M} = \mathrm{Mp}(2, \mathbb{Z})$ given by $\varepsilon(A, w) = \eta(A\tau)/\left(w(\tau)\eta(\tau)\right)$, where η is the Dedekind eta function.

$\widetilde{\Gamma}, \Gamma'$:   For a subgroup Γ of $\mathrm{SL}(2, \mathbb{Z})$ or $\mathrm{Mp}(2, \mathbb{Z})$, the inverse image of Γ in $\mathrm{Mp}(2, \mathbb{Z})$, resp. the image of Γ in $\mathrm{SL}(2, \mathbb{Z})$ under the natural projection onto the first factor.

$\Gamma(4m)^*$:   The (normal) subgroup of $\mathrm{Mp}(2, \mathbb{Z})$ of all pairs $\left(A, j(A, \tau)\right)$, where A is in the principal congruence subgroup $\Gamma(4m)$ of matrices which are 1 modulo $4m$, and where $j(A, \tau)$ stands for the standard multiplier system from the theory of modular forms of half-integral weight, i.e. $j(A, \tau) = \theta(A\tau)/\theta(\tau)$ where $\theta(\tau) = \sum_{r \in \mathbb{Z}} \exp(2\pi i \tau r^2)$.

$\mathbb{H}$:   The upper half plane of complex numbers.

$\mathrm{Inv}(V)$:   The space of $\mathrm{Mp}(2, \mathbb{Z})$-invariant elements in an $\mathrm{Mp}(2, \mathbb{Z})$-module V.

$M_k(\Gamma)$:   For a subgroup Γ of $\mathrm{SL}(2, \mathbb{Z})$, the space of modular forms of weight k on Γ; if $k \in \frac{1}{2} + \mathbb{Z}$ then it is assumed that Γ is contained in $\Gamma_0(4)$, and every f in $M_k(\Gamma)$ satisfies $f(A\tau)j(A, \tau)^{-2k} = f(\tau)$ for $A \in \Gamma$ (cf. above for $j(A, \tau)$).

$\mathbb{M}$:   Abbreviation for $\mathrm{Mp}(2, \mathbb{Z})$, see below.

$\mathrm{Mp}(2, \mathbb{Z})$:   The metaplectic double cover of $\mathrm{SL}(2, \mathbb{Z})$, i.e. the group of all pairs (A, w), where $A \in \mathrm{SL}(2, \mathbb{Z})$ and w is a holomorphic function on $\mathbb{H}$ such that $w(\tau)^2 = c\tau + d$, equipped with the composition law $(A, w) \cdot (B, v) = \left(AB, w(B\tau)v(\tau)\right)$.

S: The matrix $[0, -1; 1, 0]$.

T: The matrix $[1, 1; 0, 1]$.

$\vartheta_{F,x}$:   For a symmetric, half-integral positive definite matrix F of size n and a column vector $x \in \mathbb{Z}^n$,

$$\vartheta_{F,x}(\tau, z) = \sum_{\substack{r \in \mathbb{Z}^n \\ r \equiv x \bmod 2F\mathbb{Z}^n}} e\left(\tau \frac{1}{4} F^{-1}[r] + r^t z\right)$$

(with $\tau \in \mathbb{H}$ and $z \in \mathbb{C}^n$).

V^*:   For a left G-module V, the right G-module with the dual of the complex vector space V as underlying space equipped with the G-action $(\lambda, g) \mapsto \big(v \mapsto \lambda(gv)\big)$.

V^c: For a left G-module V, the left G-module with the dual of the complex vector space V as underlying space equipped with the G-action $(g, \lambda) \mapsto \big(v \mapsto \lambda(g^{-1}v)\big)$.

w_A: For a matrix $A \in \mathrm{SL}(2, \mathbb{Z})$, the function $w_A(\tau) = \sqrt{a\tau + b}$, where the square root is chosen in the right half plane or on the nonnegative imaginary axes.

3 Weil Representations of $\mathrm{Mp}(2, \mathbb{Z})$

In this section we recall those basic facts about Weil representations of $\mathbb{M} = \mathrm{Mp}(2, \mathbb{Z})$ associated to finite quadratic modules which we shall need in the sequel. These representations have been studied by [Kl 46], [N-W 76], [Ta 67] et al.. A more complete account of this theory as well as some deeper facts which can not be found in the literature will be given in [Sko 08].

By a finite quadratic module M we understand a finite abelian group M endowed with a quadratic form $Q_M : M \to \mathbb{Q}/\mathbb{Z}$. Thus, by definition, we have $Q_M(ax) = a^2 Q_M(x)$ for all x in M and all integers a, and the application $B_M(x, y) := Q_M(x + y) - Q_M(x) - Q_M(y)$ defines a $\mathbb{Z}$-bilinear map $B_M : M \times M \to \mathbb{Q}/\mathbb{Z}$. All quadratic modules occurring in the sequel will be assumed to be non-degenerate if not otherwise stated. Recall that M is called non-degenerate if $B_M(x, y) = 0$ for all y is only possible for $x = 0$.

Denote by $\mathbb{C}[M]$ the complex vector space of all formal linear combinations $\sum_x \lambda(x) \, e_x$, where e_x, for $x \in M$, is a symbol, where $\lambda(x)$ is a complex number and where the sum is over all x in M. We define an action of (T, w_T) and (S, w_S) on $\mathbb{C}[M]$ by

$$(T, w_T) \, e_x = e(Q_M(x)) \, e_x$$

$$(S, w_S) \, e_x = \sigma \, |M|^{-\frac{1}{2}} \sum_{y \in M} e_y \, e(-B_M(y, x)),$$

where

$$\sigma = \sigma(M) = |M|^{-\frac{1}{2}} \sum_{x \in M} e(-Q_M(x)).$$

This can be extended to an action of the metaplectic group $\mathbb{M}$ [Sko 08], and we shall use $W(M)$ to denote the $\mathbb{M}$-module with underlying vector space $\mathbb{C}[M]$. This action factors through $SL(2, \mathbb{Z})$ if and only if $\sigma^4 = 1$ [Sko 08]; in general, σ is an eighth root of unity. That these formulas define an action of $SL(2, \mathbb{Z})$ if $\sigma^4 = 1$ is well-known (cf. e.g. [N 76]).

It follows immediately from the defining formulas for the Weil representations that the pairing $\{-, -\} : W(M) \otimes W(-M) \to \mathbb{C}$ given by $\{e_x, e_y\} = 1$ if $x = y$ and $\{e_x, e_y\} = 0$ otherwise is invariant under $\mathbb{M}$. Here $-M$ denotes the quadratic module with the same underlying group as M but with the quadratic form $x \mapsto -Q_M(x)$. The $\mathbb{M}$-invariance of this pairing is just another way to state that the matrix representation of $\mathbb{M}$ afforded by $W(M)$ with respect to the basis e_x ($x \in M$) is unitary. The perfect pairing induces a natural isomorphism of the $\mathbb{M}$-modules $W(M)^c$ and $W(-M)$.

A standard example for a quadratic module is the determinant group D_F of a symmetric non-degenerate half-integral matrix F. By half-integral we mean that $2F$ has integer entries and even integers on the diagonal. The quadratic module has $D_F = \mathbb{Z}^n / 2F\mathbb{Z}^n$ as underlying abelian group. The quadratic form on D_F is the one induced by the quadratic form $x \mapsto \frac{1}{4}F^{-1}[x]$ on $\mathbb{Z}$. We shall henceforth write $W(F)$ for $W(D_F)$. Special instances are the quadratic modules $D_m = (\mathbb{Z}/2m, x \mapsto \frac{x^2}{4m})$ for integers $m \neq 0$ and their associated Weil representations $W(m)$. The decomposition of $W(m)$ into irreducible $\mathbb{M}$-modules was given in [Sko 85, Theorem 1.8, p.22]. We shall recall this in section 6.

The *level l* of a quadratic module is the smallest positive integer l such that $lM = 0$ and $lQ_M = 0$. The level of D_F coincides with the level of $2F$ as defined in the theory of quadratic forms, i.e. it coincides with the smallest integer $l > 0$ such that $lF^{-1}/4$ is half-integral. If l denotes the level of M and if $\sigma(M)^4 = 1$, i.e. if $W(F)$ can be viewed as $SL(2, \mathbb{Z})$-module, then the group $\Gamma(l)$ acts trivially on $W(M)$; if $\sigma(M)^4 \neq 1$ then l is divisible by 4 and $\Gamma(l)^*$ acts trivially on $W(M)$ [Sko 08].

By $O(M)$ we denote the orthogonal group of a quadratic module M, i.e. the group of all automorphisms of the underlying abelian group of M such that $Q_M \circ \alpha = Q_M$. The group $O(D_m)$, for an integer $m > 0$, is the group of left multiplications of $\mathbb{Z}/2m$ by elements a in $(\mathbb{Z}/2m)^*$ satisfying $a^2 = 1$. Its order equals the number of prime factors of m. The group $O(M)$ acts on $\mathbb{C}[M]$ in the obvious way. It is easily verified from the defining equations for the action of (S, w_S) and (T, w_T) on $\mathbb{C}[M]$ that the action of $O(M)$ intertwines with the action of $\mathbb{M}$ on $W(M)$. In particular, if H is a subgroup of $O(M)$ then the subspace $W(M)^H$ of elements in $W(F)$ which are invariant by H is a $\mathbb{M}$-submodule of $W(M)$.

It will turn out that spaces of Jacobi forms of critical weight are intimately related to the spaces of invariants of Weil representations. For a Weil representation $W(M)$ we use $\mathrm{Inv}(M)$ for the subspace of elements in $W(M)$ which are invariant under the action of $\mathbb{M}$. For a matrix F, we also write $\mathrm{Inv}(F)$ for $\mathrm{Inv}(D_F)$. If $\sigma(M)^4 \neq 1$ then the action of $\mathbb{M}$ on $W(M)$ does not factor through $\mathrm{SL}(2, \mathbb{Z})$, hence $(1, -1)$ does not act trivially on $W(M)$. It can be checked (or cf. [Sko 08]) that $(1, -1)$ acts as nontrivial homothety, whence $\mathrm{Inv}(M) = 0$.

If $\sigma(M)$ is a fourth root of unity then the question for $\mathrm{Inv}(M)$ is much more subtle. Roughly speaking there will be invariants if M is big enough, and the spaces $\mathrm{Inv}(M)$ fall into several natural categories according to certain local invariants of quadratic modules [Sko 08].

There is one obvious way to construct invariants. Namely, suppose the quadratic module M contains an isotropic self-dual subgroup U, i.e a subgroup U such that $Q_M(x) = 0$ for all x in U and such that the *dual* U^* of U equals U (where, for a submodule U, the dual U^* is, by definition, the submodule of all y in M satisfying $B_M(x, y) = 0$ for all x in U). Then the element $I_U := \sum_{x \in U} e_x$ is invariant under $\mathrm{SL}(2, \mathbb{Z})$ (as follows immediately from the defining equations for the action of S and T and the fact that these matrices generate $\mathrm{SL}(2, \mathbb{Z})$). Note that here $|M|$ must be a perfect square (since, for a subgroup U of M one always has an isomorphism of abelian groups $M/U \cong \mathrm{Hom}(U^*, \mathbb{Q}/\mathbb{Z})$). Also, it is not hard to check that here $\sigma(M) = 1$.

There is one important case where this construction exhausts all invariants. We cite some of the results of [Sko 08] which will clarify this a bit and which will supplement the considerations in section 5. For a prime p, let $M(p)$ be the quadratic module with the p-part of the abelian group M as underlying space, equipped with the quadratic form inherited from M.

We set $\sigma_p(M) := \sigma(M(p))$. If F is half-integral and non-degenerate then

$$\sigma_p(D_F) = \begin{cases} e_8\left(p\text{-excess}(2F)\right) & \text{for } p \geq 3, \\ e_8\left(-\text{oddity}(2F)\right) & \text{for } p = 2 \end{cases}.$$

(For p-excess and oddity cf. [Co-S 88, p. 370].) The proof will be given in [Sko 08]; we use these formulas in this article only for the case where F is a scalar matrix, say $F = (n)$. Here these formulas read

$$\sigma_p(D_n) = \begin{cases} \sqrt{\left(\frac{-4}{q}\right)}\left(\frac{-n/q}{q}\right) & \text{if } p \neq 2 \\ e_8(-n/q)\left(\frac{-n/q}{2q}\right) & \text{if } p = 2 \end{cases},$$

where q denotes the exact power of p dividing n. These identities follow directly from the well-known theory of Gauss sums.

Theorem 1. *Let M be a quadratic module whose order is a perfect square and such that $\sigma_p(M) = 1$ for all primes p. Then $\mathrm{Inv}(M)$ is different from zero. Moreover, $\mathrm{Inv}(M)$ is generated by all $I_U = \sum_{x \in U} e_x$, where U runs through the isotropic self-dual subgroups of M.*

The proof of this theorem is quite tedious. The theorem can be reduced to a special case of a more general theory concerning invariants of the Clifford-Weil groups of certain form rings [N-S-R 07, Theorem 5.5.7]. A more direct proof tailored to the Weil representations considered here will be given in [Sko 08]. It is not hard to show that the assumptions of Theorem 1 are necessary for $\mathrm{Inv}(M)$ being generated by invariants of the form I_U (in this article we do not make use of this). In general, if M does not satisfy the hypothesis of Theorem 1, there might still be invariants. However, there is one important case, where this is not the case.

Theorem 2. *Let M be a quadratic module whose order is a power of the prime p. Suppose $\dim_{\mathbb{F}_p} M \otimes \mathbb{F}_p \leq 2$. Then $\mathrm{Inv}(M) \neq 0$ if and only if $|M|$ is a perfect square and $\sigma(M) = 1$*

The proof of this theorem will be given in 6. In general, as soon as $\dim_{\mathbb{F}_p} M \otimes \mathbb{F}_p > 2$ the space of invariants of D_F is nontrivial [Sko 08].

If M and N are quadratic modules, we denote by $M \perp N$ the orthogonal sum of M and N, i.e. the quadratic module whose underlying abelian group is the direct sum of the abelian groups M and N and whose quadratic form is given by $x \oplus y \mapsto Q_M(x) + Q_N(y)$. It is obvious that every M is the orthogonal sum of its p-parts. From the product formula for quadratic forms [Co-S 88, p. 371] the sum of the numbers $p\text{-excess}(2F)$, taken over all odd p, minus the oddity of $2F$ plus the signature of $2F$ add up to 0 modulo 8. Hence we obtain [1] $\sigma(D_F) = e_8(-\mathrm{signature}(2F))$.

A nice functorial (and almost obvious) property is that the $\mathbb{M}$-modules $W(M \perp N)$ and $W(M) \otimes W(N)$ are isomorphic. In particular, $W(M)$ is isomorphic to $\bigotimes W(M(p))$, taken over, say, all p dividing the exponent of M. If l denotes the level of M then, for each prime p, the level of $M(p)$ equals the p-part l_p of l, and then $W(M)$, $W(M(2))$ and $W(M(p))$, for odd p,

---

[1] In the literature, this formula is sometimes cited as Milgram's formula.

factor through $\Gamma(l)^*$, $\Gamma(l_2)^*$ and $\Gamma(l_p)$, respectively (here we view $M(p)$ as $SL(2, \mathbb{Z})$-module which is possible since $\sigma\left(M(p)\right)^4 = 1$ as is obvious from the definition of σ in terms of Gauss sums). Since $\mathbb{M}/\Gamma(l)^*$ is isomorphic to the product of the groups $\mathbb{M}/\Gamma(l_2)^*$ and $SL(2, \mathbb{Z})/\Gamma(l_p)$ we deduce that in fact $W(M)$, viewed as $\mathbb{M}/\Gamma(l)^*$-module, is naturally isomorphic to the outer tensor product of the $\mathbb{M}/\Gamma(l_2)^*$-module $W\left(M(2)\right)$ and the $SL(2, \mathbb{Z})/\Gamma(l_p)$-modules $W\left(M(p)\right)$. In particular, we have a natural isomorphism

$$\text{Inv}(M) \cong \bigotimes_{p \mid l} \text{Inv}\left(M(p)\right).$$

The preceding Theorem thus implies

Theorem 3. *Let F be half-integral. Suppose that* $\dim_{\mathbb{F}_p} D_F \otimes F_p \leq 2$ *for all primes p. Then* $\text{Inv}(F) \neq 0$ *if and only if $\det(2F)$ is a perfect square and $\sigma_p(D_F) = 1$ for all primes p.*

Remark. If F is positive-definite, say, of size n, then $\sigma(D_F) = e_8(-n)$. Thus, if $\dim_{\mathbb{F}_p} D_F \otimes F_p \leq 2$ for all primes p then, by the theorem, $\text{Inv}(F) = 0$ unless $\det(D_F)$ is a perfect square and n is divisible by 8.

The meaning of the numbers $\sigma_p(M)$ becomes clearer if one introduces the Witt group of finite quadratic modules (see [Sch 84, Ch. 5, §1], or [Sko 08] for a discussion more adapted to the current situation). This group generalizes the well-known Witt group of quadratic spaces, say, over the field $\mathbb{F}_p$, which can be viewed as special quadratic modules. We call two quadratic modules M and N *Witt equivalent* if they contain isotropic subgroups U and V, respectively, such that U^*/U and V^*/V are isomorphic as quadratic modules. Here, for an isotropic subgroup U of a quadratic module M, we use U^*/U for the quadratic module with underlying group U^*/U (as quotient of abelian groups) and quadratic form $x + U \mapsto Q_M(x)$ (note that $Q_M(x)$, for $x \in U^*$ does depend on x only modulo U). Note that a quadratic module M is Witt equivalent to the trivial module 0 if and only if it contains an isotropic self-dual subgroup. Is is not hard to see that Witt equivalence defines indeed an equivalence relation, and that the orthogonal sum $\perp$ induces the structure of an abelian group on the set of Witt equivalence classes. One can prove then [Sko 08] (but this can also be read off from [Sch 84, Ch. 5 §1, §2]):

Theorem 4. *Two quadratic modules M and N are Witt equivalent if and only if their orders are equal up to a rational square and $\sigma_p(M) = \sigma_p(N)$ for all primes p.*

That Witt equivalent modules have the same order in $\mathbb{Q}^*/\mathbb{Q}^{*2}$ and the same σ_p-invariants is obvious (for proving equality of the sigma invariants, say, for modules of prime power order, split, for an isotropic submodule of M, the sum in the definition of $\sigma(M)$ into a double sum over a complete set of representatives y for M/U and a sum over x in U). The converse statement is not needed in this article and for its proof we refer the reader to [Sko 08] or [Sch 84].

The connection between Witt equivalence and Weil representations is given by the following functorial property of quadratic modules. If U is an isotropic subgroup of M then U^*/U is again non-degenerate, and hence we can consider its associated Weil representation. The map $e_{x+U} \mapsto \sum_{y \in x+U} e_y$ defines an $\mathbb{M}$-equivariant embedding of $W(U^*/U)$ into $W(M)$. Again, this is an immediate consequence of the defining equations for the action of $(T, 1)$ and (S, w_S).

4 Jacobi Forms and Vector Valued Modular Forms

If Γ denotes a subgroup of $SL(2, \mathbb{Z})$ we use $J_n(\Gamma)$ for the *Jacobi group* $J_n(\Gamma) = \Gamma \ltimes (\mathbb{Z}^n \times \mathbb{Z}^n)$. Thus, $J_n(\Gamma)$ consists of all pairs $(A, (\lambda, \mu))$ with $A \in \Gamma$, and $\lambda, \mu \in \mathbb{Z}^n$, equipped with the composition law

$$(A, (\lambda, \mu)) \cdot (A', (\lambda', \mu')) = (AA', (\lambda, \mu)A' + (\lambda', \mu')).$$

Here and in the following elements of $\mathbb{Z}^n$ will be considered as column vectors. Moreover, we use $(\lambda, \mu)A$ for $(\lambda a + \mu c, \lambda b + \mu d)$ if $A = [a, b; c, d]$.

We identify Γ with the subgroup $\Gamma \ltimes (0 \times 0)$, and for $\lambda, \mu \in \mathbb{Z}^n$ we use $[\lambda, \mu]$ for the element $(1, (\lambda, \mu))$ of $J_n(\Gamma)$. Then any element $g \in J_n(\Gamma)$ can be written uniquely as $g = A[\lambda, \mu]$ with suitable $A \in \Gamma$ and λ, μ in $\mathbb{Z}^n$.

Let F be a symmetric, half-integral $n \times n$ matrix. For every integer k, we have an action of the Jacobi group $J_n(\Gamma)$ on functions ϕ defined on $\mathbb{H} \times \mathbb{C}^n$ which is given by the formulas:

$$\phi|_{k,F} A(\tau, z) = \phi\left(A\tau, \frac{z}{c\tau + d}\right) (c\tau + d)^{-k} \, e\left(\frac{-cF[z]}{c\tau + d}\right),$$

$$\phi|_{k,F}[\lambda, \mu](\tau, z) = \phi(\tau, z + \lambda\tau + \mu) \, e\left(\tau F[\lambda] + 2z^t F\lambda\right)$$

where $A = [a, b; c, d] \in \Gamma$ and $\lambda, \mu \in \mathbb{Z}^n$.

For the following its is convenient to admit also half-integral k. To this end we consider the group $J_n(\Gamma) = \Gamma \ltimes (\mathbb{Z}^n \times \mathbb{Z}^n)$ for subgroups Γ of $\mathbb{M}$, which is defined as in the case of a subgroup of $SL(2, \mathbb{Z})$ with respect to the action $((A, w), x) \mapsto xA$ of $\mathbb{M}$ on $\mathbb{Z}^2$. For half-integral k, we then define $\phi|_{k,F}(A, w)$ as in the formulas above but with the factor $(c\tau + d)^{-k}$ replaced by $w(\tau)^{-2k}$. In this way the symbol $|_{k,F}$ defines a right action of $J_n(\Gamma)$ on functions defined on $\mathbb{H} \times \mathbb{C}$. If k is integral this action factors through the action of $J_n(\Gamma')$ defined above, where Γ' denotes the projection of Γ onto its first coordinate.

Definition. Let F be a symmetric, half-integral positive definite $n \times n$ matrix, let k be a half-integral integer and, for a subgroup Γ of finite index in $\mathbb{M}$, let V be a complex finite dimensional Γ-module. A *Jacobi form of weight k and index F with type (Γ, V)* is a holomorphic function $\phi : \mathbb{H} \times \mathbb{C}^n \to V$ such that the following two conditions hold true:

(i) For all $\mathrm{J}_n(\Gamma)$ and all $\tau \in \mathbb{H}$, $z \in \mathbb{C}^n$, one has $\left(\phi|_{k,F} g \right)(\tau, z) = g\left(\phi(\tau, z) \right)$, where we view V as a $\mathrm{J}_n(\Gamma)$-module by letting act $\mathbb{Z}^n \times \mathbb{Z}^n$ trivially on V.

(ii) For all $\alpha \in \mathbb{M}$ the function $\phi|_{k,F} \alpha$ possesses a Fourier expansion of the form

$$\phi|_{k,F}\alpha = \sum_{\substack{l \in \mathbb{Q},\, r \in \mathbb{Z}^n \\ 4l - F^{-1}[r] \geq 0}} c(l, r)\, q^l\, e(z^t r).$$

We shall use $J_{k,F}(\Gamma, V)$ for the complex vector space of Jacobi forms of weight k and index F of type (Γ, V)[2]. If, for a Jacobi form ϕ, in condition (i) of the definition, for all α, the stronger inequality $4l - F^{-1}[r] > 0$ holds true then we call ϕ a cusp form. The subspace of cusp forms in $J_{k,F}(\Gamma, V)$ will be denoted by $J_{k,F}^{\mathrm{cusp}}(\Gamma, V)$.

If V is a Γ-module for a subgroup Γ of $SL(2, \mathbb{Z})$, then we may view V as a $\widetilde{\Gamma}$-module $\widetilde{V}$ by setting $(A, w)v := Av$, and we simply write $J_{k,F}(\Gamma, V)$ for the space $J_{k,F}(\widetilde{\Gamma}, \widetilde{V})$. Note that in this case $J_{k,F}(\Gamma, V) = 0$ unless k is integral (since then the element $(1, -1)$ of $\widetilde{\Gamma}$ acts trivially on $\widetilde{V}$ and it acts as multiplication by $(-1)^{2k}$ on $J_{k,F}(\Gamma, V)$ by the very definition of the operator $|_{k,F}$). Thus, if V is a Γ-module for a subgroup Γ of $SL(2, \mathbb{Z})$, we may confine our considerations to integral k and then the first condition in the definition of $J_{k,F}(\Gamma, V)$ is equivalent to the statement that $\phi|_{k,F} A(\tau, z) = A\left(\phi(\tau, z) \right)$ for all $A \in \Gamma$. Similarly, if V is an irreducible Γ-module for a subgroup Γ of $\mathbb{M}$ then, for integral k, we have $J_{k,F}(\Gamma, V) = 0$ unless the action of Γ on V factors through an action of Γ' (since, if the action of Γ does not factor then Γ contains $(1, -1)$, which must act non trivially on V, and, since V is irreducible, the central element $(1, -1)$ acts then as multiplication by -1). Finally, we may in principle always confine to irreducible V. Indeed, if $V = \bigoplus V_j$ is the decomposition of the Γ-module V into irreducible parts, then $J_{k,F}(\Gamma, V) \cong \bigoplus J_{k,F}(\Gamma, V_j)$.

[2] It is sometimes useful to consider more general types of Jacobi forms. In particular, the definition does not include the case of Jacobi forms of half-integral scalar index, the basic example for such type being $\vartheta(\tau, z)$ (cf. section 2). However, $\vartheta(\tau, 2z)$ defines an element of $J_{\frac{1}{2},2}(\mathbb{M}, \mathbb{C}(\varepsilon^3))$ and thus our omission merely amounts to ignoring a certain additional invariance with respect to the bigger lattice $\frac{1}{2}\mathbb{Z} \times \frac{1}{2}\mathbb{Z}$.

If $\mathbb{C}(1)$ denotes the trivial Γ-module for a subgroup Γ of $SL(2, \mathbb{Z})$ or $\mathbb{M}$ we simply write $J_{k,F}(\Gamma)$ for $J_{k,F}\big(\Gamma, \mathbb{C}(1)\big)$, and we write $J_{k,F}$ for $J_{k,F}(\Gamma)$ if Γ is the full modular group. Note that for a positive integer m and integral k, the space $J_{k,m}(\Gamma)$ coincides with the usual space of Jacobi forms of weight k and index m on Γ as defined in [E-Z 85].

It is an almost trivial but useful observation that, in the theory of vector valued Jacobi or modular forms, one can always restrict to forms on the full group $\mathbb{M}$. In fact, one has

Lemma 1. *Let Γ be a subgroup of $\mathbb{M}$ and let V be a Γ-module. Then there is a natural isomorphism*

$$J_{k,F}(\Gamma, V) \xrightarrow{\sim} J_{k,F}(\mathbb{M}, \mathrm{Ind}_\Gamma^\mathbb{M} V).$$

(Here $\mathrm{Ind}_\Gamma^\mathbb{M} V = \mathbb{C}[\mathbb{M}] \otimes_{\mathbb{C}[\Gamma]} V$ denotes the $\mathbb{M}$-module induced by V.)

Proof. Consider the natural map $\phi \mapsto \sum_{g\Gamma\in\mathbb{M}/\Gamma} g \otimes \phi|_{k,F} g^{-1}$. It is easily verified that the summands do not depend on the choice of the representatives g and that this map defines an isomorphism as claimed in the theorem. $\square$

Jacobi forms may be regarded as vector valued elliptic modular forms. For stating this more precisely we denote by M_k the space of holomorphic functions h on $\mathbb{H}$ such that h is a modular form of weight k on some subgroup Γ of $\mathbb{M}$ (where we assume that Γ is contained in $\Gamma_0(4)$ if k is not integral). Note that M_k is a $\mathbb{M}$-module with respect to the action $|_k$. By M_k^{cusp} we denote the submodule of cusp forms. Using the $\mathbb{M}$-module $W(F)$ introduced in section 3 we then have

Theorem 5. *For a subgroup Γ of $\mathbb{M}$, let V be a finite dimensional Γ-module. Assume that the image of Γ under the associated representation is finite. Then, for any half-integral k and any half-integral F there is a natural isomorphism*

$$J_{k,F}(\Gamma, V) \xrightarrow{\sim} \left(M_{k-\frac{n}{2}} \otimes W(F)^*\right) \otimes_{\mathbb{C}[\mathbb{M}]} \mathrm{Ind}_\Gamma^\mathbb{M} V.$$

Proof. Using the isomorphism of the preceding lemma we can assume that V is a $\mathbb{M}$-module.

Denote by $\mathcal{J}_{k,F}$ the space of all functions ϕ, where ϕ is in $J_{k,F}(\Gamma)$ for some subgroup Γ. The space $\mathcal{J}_{k,F}$ is clearly a $\mathbb{M}$-module under the action $|_{k,F}$. The bilinear map $(\phi, v) \mapsto \phi \cdot v$ induces an isomorphism

$$\mathcal{J}_{k,F} \otimes_{\mathbb{C}[\mathbb{M}]} V \xrightarrow{\sim} J_{k,F}(\mathbb{M}, V).$$

For verifying the surjectivity of this map we need that some subgroup Γ of finite index in $\mathbb{M}$ acts trivially on V (or equivalently that the image of $\mathbb{M}$ under the representation afforded by the $\mathbb{M}$-module V is finite). Namely, if we write a $\phi \in J_{k,F}(\Gamma, V)$ with respect to some basis e_j of V as $\phi = \sum_j \phi_j \cdot e_j$, then, by our assumption, the ϕ_j are scalar valued Jacobi forms on $J_{k,F}(\Gamma)$.

Assume now that ψ is an element of $J_{k,F}(\Gamma)$. Then it is well-know (and follows easily from the invariance of ψ under $\mathbb{Z}^n \times \mathbb{Z}^n$) that we can expand ψ in the form

$$\psi(\tau, z) = \sum_{x \in \mathbb{Z}^n / 2F\mathbb{Z}^n} h_x(\tau)\, \vartheta_{F,x}(\tau, z)$$

with functions h_x which are holomorphic in $\mathbb{H}$, and where the $\vartheta_{F,x}$ denote the theta series defined in section 2. The space of functions $\Theta(F)$ spanned by the $\vartheta_{F,x}$ is a $\mathbb{M}$-(right)module with respect to the action $|_{\frac{n}{2},F}$ and is acted on trivially by $\Gamma(4N)^*$ for some integer N (see e.g. [Kl 46]). Accordingly, the h_x are then invariant under $\Gamma(4N)^* \cap \Gamma$ under the action $|_{k-\frac{n}{2}}$ (one needs here also the linear independence of the $\vartheta_{F,x}(\tau, z)$, where x runs through a set of representatives for $\mathbb{Z}^n / 2F\mathbb{Z}^n$). In fact, one easily deduces from the regularity condition for Jacobi forms at the cusps (i.e. from condition (ii) of the definition) that the h_x are elements of $M_{k-\frac{n}{2}}$. Thus, the bilinear map $(h, \theta) \mapsto h \cdot \theta$ induces an isomorphism of $\mathbb{M}$-right modules

$$M_{k-\frac{n}{2}} \otimes \Theta(F) \xrightarrow{\sim} J_{k,F}.$$

Finally we note that $\Theta(F)$ is isomorphic as $\mathbb{M}$-module to $W(F)^*$ via the map $W(F)^* \ni \lambda \mapsto \sum_{e \in D_F} \lambda(e)\, \vartheta_{F,e}$ (cf. [Kl 46]). From this the theorem is now immediate. $\qquad\square$

Remark. A review of the foregoing proof shows that in the statement of the theorem we may replace $M_{k-\frac{1}{2}}$ by $M_{k-\frac{n}{2}}^{cs}$, the subspace of all h in $M_{k-\frac{n}{2}}$ which are modular forms on some congruence subgroup of $\mathbb{M}$, provided $\Gamma(4N)^*$, for some N, acts trivially on V. This remark will become important in the next section.

Using the isomorphism of Theorem 5 we can define the subspace $J_{k,F}^{\text{Eis}}(\Gamma, V)$ of *Jacobi-Eisenstein series* as the preimage of the subspace which is obtained by replacing $M_{k-\frac{1}{2}}$ on the right hand side of this isomorphism by the space of Eisenstein series $M_{k-\frac{1}{2}}^{\text{Eis}}$ in $M_{k-\frac{1}{2}}$ (i.e. by the orthogonal complement of the cusp forms in $M_{k-\frac{1}{2}}$).

Theorem 5 implies that there are no Jacobi forms of index F and weight $k < \frac{n}{2}$ (since then $M_{k-\frac{n}{2}} = 0$). For $k = \frac{n}{2}$ we have $M_k = \mathbb{C}$, and hence, identifying $W(F)^* \otimes_{\mathbb{C}[\mathbb{M}]} \text{Ind}\, V$ with $\text{Hom}_{\mathbb{M}}(W(F), \text{Ind}\, V)$ via the map $\lambda \otimes w \mapsto \lambda(\cdot)w$, the theorem implies

Corollary. *There exists a natural isomorphism*

$$J_{\frac{n}{2},F}(\Gamma, V) \cong \operatorname{Hom}_{\mathbb{M}}(W(F), \operatorname{Ind}_{\Gamma}^{\mathbb{M}} V).$$

The corollary reduces the study of $J_{\frac{n}{2},F}(\Gamma, V)$ to the problem of describing the decomposition of $W(F)$ and $\operatorname{Ind}_{\Gamma}^{\mathbb{M}} V$ into irreducible parts.

For $k \geq \frac{n}{2} + 2$, the last theorem makes it possible to derive an explicit formula for the dimension of the space of Jacobi forms of weight k. In fact, the isomorphism of the theorem can be rewritten as

$$J_{k,F}(\Gamma, V) \cong M_{k-\frac{n}{2}} \otimes_{\mathbb{C}[\mathbb{M}]} (W(F)^c \otimes \operatorname{Ind}_{\Gamma}^{\mathbb{M}} V).$$

But here (using the obvious map $h \otimes v \mapsto h \cdot v$) the right hand side can be identified with the space $M_{k-\frac{n}{2}}(\rho)$ of holomorphic maps $h : \mathbb{H} \to W(F)^c \otimes$ Ind V which satisfy the transformation law $h(A\tau)w(t)^{-2k} = \rho(\alpha)(h(\tau))$ for all (A, w) in $\mathbb{M}$, where ρ denotes the representation afforded by the $\mathbb{M}$-module $W(F)^c \otimes$ Ind V, and which satisfy the usual regularity conditions at the cusps. The dimensions of the spaces $M_k(\rho)$ have been computed in [Sko 85, p. 100] (see also [E-S 95, Sec. 4.3]) using the Eichler-Selberg trace formula. In our context these formulas read as follows:

Theorem 6. *Let F be a half-integral positive definite $n \times n$ matrix, let $k \in \frac{1}{2}\mathbb{Z}$, and, for a subgroup Γ of $\mathbb{M}$, let V be a Γ-module such that the associated representation of Γ has finite image in $\operatorname{GL}(V)$. Then one has*

$$\dim J_{k,F}(\Gamma, V) - \dim M_{2+\frac{n}{2}-k}^{\text{cusp}} \otimes_{\mathbb{C}[\mathbb{M}]} X(i^{n-2k})^c$$

$$= \frac{k - \frac{n}{2} - 1}{2} \dim X(i^{n-2k}) + \frac{1}{4} \operatorname{Re}\left(e^{\pi i(k-\frac{n}{2})/2} \operatorname{tr}((S, w_S), X(i^{n-2k}))\right)$$

$$+ \frac{2}{3\sqrt{3}} \operatorname{Re}\left(e^{\pi i(2k-n+1)/6} \operatorname{tr}((ST, w_{ST}), X(i^{n-2k}))\right) - \sum_{j=1}^{r}(\lambda_j - \frac{1}{2}).$$

Here $X(i^{n-2k})$ denotes the $\mathbb{M}$-submodule of all v in $W(F)^c \otimes \operatorname{Ind}_{\Gamma}^{\mathbb{M}} V$ such that $(-1, i)v = i^{n-2k}v$; moreover, $0 \leq \lambda_j < 1$ are rational numbers such that $\prod_{j=1}^{r}(t - e^{2\pi i\lambda_j}) \in \mathbb{C}[t]$ equals the characteristic polynomial of the automorphism of $X(i^{n-2k})$ given by $v \mapsto (T, 1)v$. (Recall that $T = [1, 1; 0, 1]$ and $S = [0, -1; 1, 0]$).

Note that the traces on the space $X(i^{n-2k})$ occurring on the right hand side of the dimension formula can be easily computed from the explicit formulas for the action of (S, w_S) and (T, w_T) on $W(F)^c \otimes$ Ind V. Note also that the second term occurring on the left hand side vanishes for $k \geq \frac{n}{2} + 2$. In general, this second term may be interpreted as the dimension of the space $J_{2+n-k,F}^{\text{skew}}(\Gamma, V)$

of *skew-holomorphic* Jacobi forms, which can be defined by requiring that Theorem 5 holds true for skew-holomorphic Jacobi forms if one replaces the $\mathbb{M}$-module M_k on the right hand side by M_k^c.

It remains to discuss the case of weights $k = \frac{n}{2} + \frac{1}{2}, \frac{n}{2} + 1, \frac{n}{2} + \frac{3}{2}$. For weight $k = \frac{n}{2} + 1$ the second term on the left hand side refers to elliptic modular forms of weight 1. Accordingly, this term is in general unknown.

If $k = \frac{n}{2} + \frac{3}{2}$ then the second term on the left hand side equals the dimension of the space of cusp forms in $J^{\text{skew}}_{\frac{n+1}{2}, F}(\Gamma, V)$, which can be investigated by the methods of the next section. However, we shall not pursue this any further in this article. Finally, the case of critical weight $k = \frac{n+1}{2}$ will be considered by different methods in the next section.

Concluding remarks. We end this section by some remarks on the literature related to the results discussed here and on possible directions for generalization.

Theorem 5 remains true if we replace on both sides of the isomorphism the space of Jacobi forms and modular forms by the respective subspaces of cusp forms (as can be easily read off from the proof of the theorem). Moreover, Theorem 6, which gives an explicit formula for the difference $\dim J_{k,F}(\Gamma, V) - \dim J^{\text{skew,cusp}}_{2+n-k,m}(\Gamma, V)$, remains true if the left hand side is replaced by $\dim J^{\text{cusp}}_{k,F}(\Gamma, V) - \dim J^{\text{skew}}_{2+n-k,F}(\Gamma, V)$ and if we subtract from the right hand side the dimension of the maximal subspace of $X(i^{n-2k})$ which is invariant under $(T, 1)$. For deducing this from Theorem 5 we refer the reader again to [Sko 85, p. 100]).

Explicit dimension formulas for $J_{k,F}(\Gamma) = J_{k,F}(\Gamma, \mathbb{C}(1))$ were already computed for arbitrary integral scalar index F, integral k and $\Gamma = \text{SL}(2, \mathbb{Z})$ in [Sko 85, Satz 6.1, 6.2, 6.3], for scalar F, arbitrary integral k and arbitrary Γ in [Sko 06], for arbitrary half integral F, integral $k > \frac{n}{2} + 2$ and Γ containing -1 in [Ar 92, Theorem 5.2] (to be precise, Arakawa, using Selberg zeta functions to derive dimension formulas, states these for the corresponding subspaces of cusp forms only). Moreover, in [Ar 92, Theorem 5.2] Arakawa states formulas relating the dimensions of $J_{k,F}(\Gamma)$ (or of the subspaces of cusp forms) for integral weights $\frac{n}{2} < k \leq \frac{n}{2} + 2$ and special values of Selberg zeta functions.

As already pointed out (see the footnote after the definition of *vector valued Jacobi forms*) we did not treat here Jacobi forms whose index is a not necessarily half integral matrix. In fact, there should be a slightly more general theory of Jacobi forms whose index is a symmetric positive definite matrix F with entries from $\frac{1}{2}\mathbb{Z}$. Namely, the set of these matrices is in one to one correspondence with the set of positive Chern classes in $H^2(\Lambda_\tau, \mathbb{Z})$ (where $\Lambda_\tau = \mathbb{C}^n / \mathbb{Z}^n \tau + \mathbb{Z}^n$), which in turn describe the positive line bundles on X_τ, and any positive line bundle gives rise to non trivial Jacobi forms by studying its holomorphic sections. For details of this see [Sko 07].

A theory of Jacobi forms with (scalar) index in $\frac{1}{2}\mathbb{Z}$ was systematically developed in [G-N 98]. In this article the authors construct in particular some remarkable Siegel (and Jacobi cusp forms) of critical weight one using essentially Jacobi forms whose (scalar) index is not integral.

The significance of the theory Jacobi forms with general matrix index for the theory of orthogonal modular forms (and in particular for the theory of moduli spaces of abelian and K3 surfaces) is shown in [Gri 95]. The theta decomposition of Jacobi forms, which we used in the proof of Theorem 5, is explained in this paper in more detail (cf. Lemma 2.3). It is used to deduce amongst others as Corollary 4.2 that $\frac{n-1}{2}$ is the only singular weight in the theory of orthogonal modular forms (with Fourier-Jacobi expansion) of quadratic modules of signature $(2, n + 1)$ which admits non-zero forms.

5 Jacobi Forms of Critical Weight

In this section we study the spaces $J_{\frac{n+1}{2},F}(\Gamma, V)$, where n denotes the size of F. We are mainly interested in $J_{\frac{n+1}{2},F} = J_{\frac{n+1}{2},F}(\mathbb{M})$ and, for positive integers N, in $J_{1,N}(\varepsilon^a) = J_{1,N}(\mathbb{M}, \mathbb{C}(\varepsilon^a))$. Here ε denotes the character of $\mathbb{M} = \mathrm{Mp}(2, \mathbb{Z})$ given by $\varepsilon(A, w) = \eta(A\tau)/\bigl(w(\tau)\eta(\tau)\bigr)$, where $\eta(\tau)$ is the Dedekind eta-function. It is a well-known fact that ε generates the group of one dimensional characters of $\mathbb{M}$. At the end of this section we add a remark concerning the case where $\Gamma = \Gamma_0(l)$ and where V is the trivial Γ-module $\mathbb{C}(1)$; a more thorough discussion of this case with complete proofs will be given in [I-S 08].

We assume that Γ is a congruence subgroup. By Theorem 5 and the subsequent remark we need first of all a description of the Γ-module $M^{cs}_{\frac{1}{2}}$ of all modular forms of weight $\frac{1}{2}$ on congruence subgroups of Γ. Starting with the observation that $M^{cs}_{\frac{1}{2}}$ is generated by theta series (cf. [Se-S 77]) the decomposition of the $\mathbb{M}$-module $M^{cs}_{\frac{1}{2}}$ into irreducible parts was calculated in [Sko 85, p. 101]. As an immediate consequence of the result loc. cit. we obtain

$$M_{\frac{1}{2}}(\Gamma(4m)) \cong \bigoplus_{\substack{l \mid m \\ m/l \text{ squarefree}}} \bigl(W(l)^\epsilon\bigr)^*,$$

$$M_{\frac{1}{2}}^{\mathrm{Eis}}(\Gamma(4m)) \cong \bigoplus_{\substack{l \mid m \\ m/l \text{ squarefree}}} \bigl(W(l)^{O(l)}\bigr)^*$$

Here ϵ denotes the element $\epsilon : x \mapsto -x$ of $O(l) := O(D_l)$, and $W(l)^\epsilon$ and $W(l)^{O(l)}$ denote the subspaces of elements of $W(l)$ which are invariant under ϵ and $O(l)$, respectively. (For deducing this from [Sko 85, p. 101, Theorem 5.2] one also needs [Sko 85, Theorem 1.8, p. 22]).

If we insert this into the isomorphism of Theorem 5, we obtain

Theorem 7. *For a congruence subgroup Γ of $\mathbb{M}$, let V be a finite dimensional Γ-module, and let F be half-integral of size n and level f. Assume that, for some m, the group $\Gamma(4m)^*$ acts trivial on V and that f divides $4m$. Then there are natural isomorphisms*

$$J_{\frac{n+1}{2},F}(\Gamma, V) \longrightarrow \bigoplus_{\substack{l \mid m \\ m/l \, squarefree}} \left(W\big((l) \oplus F\big)^{\epsilon \times 1} \right)^* \otimes_{\mathbb{C}[\mathbb{M}]} \mathrm{Ind}_\Gamma^\mathbb{M} V,$$

$$J_{\frac{n+1}{2},F}^{\mathrm{Eis}}(\Gamma, V) \longrightarrow \bigoplus_{\substack{l \mid m \\ m/l \, squarefree}} \left(W\big((l) \oplus F\big)^{O(l) \times 1} \right)^* \otimes_{\mathbb{C}[\mathbb{M}]} \mathrm{Ind}_\Gamma^\mathbb{M} V.$$

Here $O(l) \times 1$ denotes the subgroup of $O\big(D_{(l) \oplus F}\big)$ of all elements of the form $(x, y) \mapsto (\alpha(x), y)$ $(x \in D_l, y \in D_F, \alpha \in O(l))$, and $\epsilon \times 1$ denotes the special element of $O(l) \times 1$ given by $(x, y) \mapsto (-x, y)$.

Thus the description of Jacobi forms of critical weight reduces to a problem in the theory of finite dimensional representations of $\mathbb{M}$. Actually, the description of Jacobi forms of critical weight reduces to an even more specific problem, namely to the problem of determining the invariants of Weil representations associated to finite quadratic modules. To make this more precise we rewrite the first isomorphism of Theorem 7 as

Theorem 8. *Under the same assumptions as in Theorem 7 the applications*

$$\sum_j e_{x_j} \otimes e_{y_j} \otimes w_j \mapsto \sum_j \vartheta_{l,x_j}(\tau, 0)\, \vartheta_{F,y_j}(\tau, z)\, w_j$$

define an isomorphism

$$J_{\frac{n+1}{2},F}(\Gamma, V) \longleftarrow \bigoplus_{\substack{l \mid m \\ m/l \, squarefree}} \mathrm{Inv}\left(W(-l)^\varepsilon \otimes W(-F) \otimes \mathrm{Ind}_\Gamma^\mathbb{M} V \right).$$

Here we used that, for any group G, any G-right module A and G-left module B, the spaces $A \otimes_{\mathbb{C}[G]} B$ and $(A' \otimes B)^G$ are naturally isomorphic (where A' denotes the G-left module with underlying space A and action $(g, a) \mapsto a \cdot g^{-1}$), we used the isomorphism (of vector spaces) of $W(G)^c$ with $W(-G)$ (cf. section 3), and we wrote out explicitly the isomorphism constructed in the proof of Theorem 5.

Remark. Next, consider a decomposition $\mathrm{Ind}_{\Gamma}^{\mathbb{M}} V = \bigoplus_j W_j$ into irreducible $\mathbb{M}$-submodules W_j. By the results in [N-W 76] every irreducible representation of $\mathbb{M}$ is equivalent to a subrepresentation of a Weil representation $W(M)$ for a suitable finite quadratic module M[3]. Hence we may replace the W_j by submodules of Weil representations $W(M_j)$, and at the end we find a natural injection

$$J_{\frac{n+1}{2},F}(\Gamma, V) \hookrightarrow \bigoplus_j \bigoplus_{\substack{l \mid m \\ m/l \text{ squarefree}}} \mathrm{Inv}\left(D_{-l} \perp D_{-F} \perp M_j\right).$$

Here the precise image can also be characterized by the action of the groups $O(-l)$ on D_{-l} and $O(M_j)$ on $W(M_j)$ (and certain additional intertwiners of the $\mathbb{M}$-action), so that the last isomorphism could be written in an even more explicit form. We shall not pursue this any further in this article.

We note a special case of Theorem 7. Namely, if V is the trivial $\mathbb{M}$-module then the right hand side of, say, the first formula of Theorem 7 becomes the space of elements in $W\big((l) \oplus F\big)^*$ which are invariant under $\mathbb{M}$ and $\epsilon \times 1$. Using again the natural isomorphism between the spaces $W\big((l) \oplus F\big)^c$ and $W\big((-l) \oplus (-F)\big)$ (cf. sect. 3) we thus obtain

Theorem 9. *For any half-integral F of size n and level dividing $4m$ there are natural isomorphisms*

$$J_{\frac{n+1}{2},F} \cong \bigoplus_{\substack{l \mid m \\ m/l \text{ squarefree}}} \mathrm{Inv}\left((-l) \oplus (-F)\right)^{\epsilon \times 1},$$

$$J_{\frac{n+1}{2},F}^{\mathrm{Eis}} \cong \bigoplus_{\substack{l \mid m \\ m/l \text{ squarefree}}} \mathrm{Inv}\left((-l) \oplus (-F)\right)^{O(-l) \times 1}.$$

(Here $O(-l) \times 1$ and $\epsilon \times 1$ are as explained in Theorem 7.)

---

[3] Actually it is proven in [N-W 76] that, for prime powers m, every finite dimensional irreducible $\mathrm{SL}(2, \mathbb{Z}/m)$-module is contained in a Weil representation associated to some finite quadratic module. Since $\mathrm{SL}(2, \mathbb{Z}/m)$, for arbitrary m, is the direct product of the groups $\mathrm{SL}(2, \mathbb{Z}/p^n)$, where p^n runs through the exact prime powers dividing m, and since the category of Weil representations $W(M)$ is closed under tensor products we can dispense with the assumption of m being a prime power. Hence every irreducible $\mathrm{SL}(2, \mathbb{Z})$-module which is acted on trivially by some congruence subgroup is isomorphic to a submodule of some $W(M)$. Finally, if ρ is an irreducible representation of $\mathbb{M}$ which does not factor through a representation of $\mathrm{SL}(2, \mathbb{Z})$, i.e. which satisfies $\rho(1, -1) \neq 1$, but which factors through some congruence subgroup $\Gamma(4N)^*$ then ρ/ε factors through a representation of $\mathrm{SL}(2, \mathbb{Z}/4N')$ for some N'. Hence by the preceding argument, ρ/ε is afforded by a submodule of a Weil representation $W(M)$. But ε is afforded by a submodule in D_6 (cf. section 6), and hence ρ is afforded by a submodule of the Weil representation $D_6 \otimes W(M)$.

Note that $O(-l) \times 1$ acts on $\mathrm{Inv}\left((-l) \oplus (-F)\right)$ since the action of $O(-l) \times 1$ intertwines with the action of $\mathbb{M}$ on the space $W\left((-l) \oplus (-F)\right)$. Note also that this Theorem is a trivial statement if n is even since then both sides of the claimed isomorphism are 0 for trivial reasons (consider the action of $(-1, 1)$). If the size of F is odd then the level of F is divisible by 4, and hence we may choose $4m$ equal to the level of F.

If we take F equal to a number m, then $W\left((-l) \oplus (-m)\right)$ does not contain nontrivial invariants (see the remark following Theorem 3.). The theorem thus implies $J_{1,N} = 0$, a result which was proved in [Sko 85, Satz 6.1]. More generally, Theorem 9 and the remark following Theorem 3 imply

Theorem 10. *Let F be half-integral of size n. If $n \not\equiv 7 \bmod 8$ and F has at most one nontrivial elementary divisor then $J_{\frac{n+1}{2},F} = 0$.*

Note that we cannot dispense with the assumption $n \not\equiv 7 \bmod 8$. A counterexample can be constructed as follows. Let $2G$ denote a Gram matrix of the E_8-lattice then

$$\theta(\tau, z) := \sum_{x \in \mathbb{Z}^8} e(\tau G[x] + 2z^t Gx) \quad (z \in \mathbb{C}^8)$$

defines an element of $J_{4,G}$. If M is an integral 8×7-matrix then $F := M^t GM$ is half-integral positive definite of size 7 and it is easily checked that

$$\theta | U_M(\tau, w) := \theta(\tau, Mw) \quad (w \in \mathbb{C}^7)$$

defines a non-zero element of $J_{4,F}$, hence a Jacobi form of critical weight. Suitable choices of G and M yield an F with exactly one elementary divisor[4], i.e. an F satisfying the second assumption of the theorem.

For doing explicit calculations it is worthwhile to write out explicitly the isomorphisms of Theorem 9. If $v = \sum_{x,y} \lambda(x, y) e_{(x,y)}$ is an element of $\mathrm{Inv}\left((-l) \oplus (-F)\right)$ then

---

[4] One can take

$$2G = \begin{pmatrix} 4 & -2 & 0 & 0 & 0 & 0 & 0 & 1 \\ -2 & 2 & -1 & 0 & 0 & 0 & 0 & 0 \\ 0 & -1 & 2 & -1 & 0 & 0 & 0 & 0 \\ 0 & 0 & -1 & 2 & -1 & 0 & 0 & 0 \\ 0 & 0 & 0 & -1 & 2 & -1 & 0 & 0 \\ 0 & 0 & 0 & 0 & -1 & 2 & -1 & 0 \\ 0 & 0 & 0 & 0 & 0 & -1 & 2 & 0 \\ 1 & 0 & 0 & 0 & 0 & 0 & 0 & 2 \end{pmatrix}$$

and $2F$ equal to the matrix which is obtained by deleting the last row and column of $2G$, which has 4 as sole nontrivial elementary divisor.

$$\phi_v(\tau, z) := \sum_{\substack{x \in \mathbb{Z}/2l \\ y \in \mathbb{Z}^n/2F\mathbb{Z}^n}} \lambda(x, y)\, \vartheta_{l,x}(\tau, 0)\, \vartheta_{F,y}(\tau, z)$$

defines a Jacobi form in $J_{\frac{n+1}{2}, F}$. It vanishes unless $\lambda(x, y)$ is even in x, i.e. unless v is invariant under $\epsilon \times 1$, and it defines an Eisenstein series if $\lambda(x, y)$ is invariant under $O(-l) \times 1$.

We now turn to the case $J_{1,N}(\epsilon^a)$. Using the Jacobi forms (of weight $\frac{1}{2}$, index $\frac{a^2}{2}$ on $\mathbb{M}$ and with character ϵ^3)

$$\vartheta(\tau, az) = q^{1/8}\left(\zeta^{a/2} - \zeta^{-a/2}\right) \prod_{n \geq 1} \left(1 - q^n\right)\left(1 - q^n\zeta^a\right)\left(1 - q^n\zeta^{-a}\right)$$

for positive natural numbers a and trying to build *thetablocks*, i.e. Jacobi forms which are products or quotients of these forms and powers of the Dedekind eta-function, it turns out [G-S-Z 07] that there are indices N and nontrivial Jacobi forms in $J_{1,N}(\epsilon^a)$ if $a \equiv 2, 4, 6, 8, 10, 14 \bmod 24$. Moreover, extensive computer aided search suggests that, for all other a modulo 24, there are no Jacobi forms, and that all Jacobi forms of weight 1 can be obtained by the indicated procedure built on the $\vartheta(\tau, az)$. Note that in fact $J_{1,N} = 0$ for all N [Sko 85, Satz 6.1]. As an application of the theory developed so far we shall prove in section 6

Theorem 11. *For all positive integers N, one has $J_{1,N}(\epsilon^{16}) = 0$.*

Actually, we shall prove more. Namely,

Theorem 12. *For every positive integers N, the space $J_{1,N}(\epsilon^8)$ is spanned by the series*

$$\vartheta_\rho := \sum_{\alpha \in \mathcal{O}} \left(\frac{x(\alpha)}{3}\right) q^{|\alpha|^2/3} \zeta^{y(\alpha\rho)}$$

Here $\mathcal{O} = \mathbb{Z}\left[\frac{-1+\sqrt{-3}}{2}\right]$, and we use $x(\alpha) = \alpha + \overline{\alpha}$ and $y(\alpha) = (\alpha - \overline{\alpha})/\sqrt{-3}$. Moreover, ρ runs through all numbers in $\mathcal{O}$ with $|\rho|^2 = N$.

Remark. Let $\rho = \frac{p+\sqrt{-3}q}{2}$ be a number in $\mathcal{O}$ with $|\rho|^2 = N$. By multiplying ρ by a suitable 6th root of unity (which will not change ϑ_ρ) we may assume that $q \geq |p| > 0$. For $q = |p|$ one has $\vartheta_\rho = 0$. For $q > |p|$, one can show by elementary transformations of the involved series that ϑ_ρ has the factorization [G-S-Z 07]

$$\vartheta_\rho(\tau, z) = -\vartheta\left(\tau, \frac{q+p}{2}z\right)\vartheta\left(\tau, \frac{q-p}{2}z\right)\vartheta\left(\tau, qz\right)/\eta(\tau).$$

Similar theorems can be produced for the spaces $J_{1,N}(\varepsilon^a)$ for arbitrary even integers a modulo 24. Since the analysis of the invariants in the corresponding Weil representations becomes quite involved this will eventually be treated elsewhere.

Finally, we mention the case $\Gamma = \Gamma_0(l)$ acting trivially on $V = \mathbb{C}(1)$. On investigating the representation Ind V occurring on the right hand side of Theorem 7 it is possible to deduce the following [I-S 08]

Theorem 13. *Let F be a symmetric, half-integral and positive definite matrix of size $n \times n$ with odd n. Then, for every positive integer l, we have*

$$J_{\frac{n+1}{2},F}(\Gamma_0(l)) = J_{\frac{n+1}{2},F}(\Gamma_0(l_1)).$$

Here we write $l = l_1 l_2$ with $l_1 | \det(2F)^\infty$ and with l_2 relatively prime to $\det(2F)$.

For $n = 1$, i.e. for the case of ordinary Jacobi forms in one variable, the theorem was already stated and proved in [I-S 07].

As immediate consequence we obtain

Corollary. *Suppose $J_{\frac{n+1}{2},F} = 0$. Then we have $J_{\frac{n+1}{2},F}(\Gamma_0(l)) = 0$ for all l which are relatively prime to $\det(2F)$.*

This corollary might be meaningful for the study of Siegel modular forms of critical weight on subgroups $\Gamma(l)$. In [I-S 07] we used it to prove that there are no Siegel cusp forms of weight one on $\Gamma_0(l)$ for any l.

6 Proofs

In this section we append the proofs of Theorems 2, 11 and 12.

We start with the description of the decomposition of Weil representations of rank 1 modules into irreducible parts. For an odd prime power $q = p^\alpha \geq 1$ let L_q be the quadratic module $(\mathbb{Z}/q, \frac{x^2}{q})$. The submodule $U := p^{\lceil \alpha/2 \rceil} L_q$ is isotropic, for its dual one finds $U^* = p^{\lfloor \alpha/2 \rfloor} L_q$ and, for $\alpha \geq 2$, the quotient module U^*/U is isomorphic (as quadratic module) to L_{q/p^2}. Hence $W(L_{q/p^2})$ embeds naturally as $\mathbb{M}$-submodule into $W(L_q)$ (see the discussion at the end of section 3). Let $W_1(L_q)$ be the orthogonal complement of $W(L_{q/p^2})$ (with respect to the $\mathbb{M}$-invariant scalar product $(-, -)$ given by $(e_x, e_y) = 1$ if $x = y$ and $(e_x, e_y) = 0$ otherwise). Then $W_1(L_q)$ is invariant under $\mathbb{M}$ and under $O(L_q)$. The latter group is generated by the involution $\alpha : x \mapsto -x$. For $\epsilon = \pm 1$, let $W_1^\epsilon(L_q)$ be the ϵ-eigenspace of α viewed as involution on $W_1(L_q)$.

Since α intertwines with $\mathbb{M}$ these eigenspaces are $\mathbb{M}$-submodules of $W(L_q)$. By induction we thus obtain the decomposition of $W(L_q)$ into $\mathbb{M}$-submodules

$$W(L_q) = \bigoplus_{d^2|q} \left(W_1^+(L_{q/d^2}) \oplus W_1^-(L_{q/d^2}) \right) \oplus \begin{cases} W(L_1) & \text{if } q \text{ is a square} \\ 0 & \text{otherwise} \end{cases}$$

Note that $W(L_1) \cong \mathbb{C}(1)$.

Finally, for an integer a which is relatively prime to q we let $L_q(a)$ be the quadratic module which has the same underlying abelian group as L_q but has quadratic form $Q_{L_q(a)}(x) = a Q_{L_q}(x)$. With $W_1^\epsilon(L_q(a))$ being defined similarly as before we obtain a corresponding decomposition of $W(L_q(a))$ as for $W(L_q)$.

Note that $\sigma(L_q(a)) = 1$ if q is a square, and otherwise $\sigma(L_q(a)) = \left(\frac{-a}{p}\right)$ if $p \equiv 1 \bmod 4$ and $\sigma(L_q(a)) = \left(\frac{-a}{p}\right) i$ if $p \equiv 3 \bmod 4$. We may restate this as $\sigma(L_q(a)) = \left(\frac{-a}{p}\right)^\alpha \sqrt{\left(\frac{-4}{p}\right)^\alpha}$ for $q = p^\alpha$. In particular, $W(L_q(a))$ can be viewed as $\mathrm{SL}(2,\mathbb{Z})$-module. The level of $L_q(a)$ equals q, whence $\Gamma(q)$ acts trivially on $W(L_q(a))$.

Lemma 2. *The* $\mathrm{SL}(2,\mathbb{Z})$-*modules* $W_1^\epsilon(L_q(a))$ *are irreducible. The exact level of* $W_1^\epsilon(L_q(a))$ *equals* q *(i.e.* q *is the smallest positive integer such that* $\Gamma(q)$ *acts trivially on* $W_1^\epsilon(L_q(a))$*). Two* $\mathbb{M}$-*modules* $W_1^\epsilon(L_q(a))$ *and* $W_1^{\epsilon'}(L_{q'}(a'))$ *are isomorphic if and only if* $q = q'$, $\epsilon = \epsilon'$ *and* aa' *is a square module* p.

Proof. It is easy to see that the modules $W_1^\epsilon(L_q(a))$ are non-zero. Thus the given decomposition of $W(L_q(a))$ contains $\sigma_0(q)$ non-zero parts. Hence, for proving the irreducibility of these parts is suffices to show that the number of irreducible components of $W(L_q(a))$ is exactly $\sigma_0(q)$. But the number of irreducible components equals the number of invariant elements of $W(L_q(a)) \otimes W(L_q(-a))$. i.e. the dimension of $\mathrm{Inv}(L_q(a) \perp L_q(-a))$. The latter space is spanned by the elements $I_U := \sum_{x \in U} e_x$, where U runs through the isotropic self-dual modules of $L_q(a) \perp L_q(-a)$ (cf. Theorem 1).

Alternatively, for avoiding the use of Theorem 1, which we did not prove in this article, (but see [N-S-R 07, Theorem 5.5.7]), one can show that $W(L_q(-a)) = \mathrm{Inv}(L_q(a) \perp L_q(-a))$ is isomorphic to the permutation representation of $\mathrm{SL}(2,\mathbb{Z})$ given by the right action of $\mathrm{SL}(2,\mathbb{Z})$ on the row vectors of length 2 with entries in $\mathbb{Z}/q$ (see e.g. [N-W 76, § 3]). The number of invariants equals thus the number of orbits of this action, which in turn are naturally parameterized by the divisors of q (all vectors (x, y) with $\gcd(x, y, q) = d$, for fixed $d|q$ constitute an orbit).

We already saw that $\Gamma(q)$ acts trivially on $W_1^\epsilon(L_q(a))$. That q is the smallest integer with this property follows from the fact that $e(a/q)$ is an eigenvalue of T acting on $W_1^\epsilon(L_q(a))$. The latter can directly be deduced from the definition of $W_1^\epsilon(L_q(a))$ (see also [Sko 85, p. 33] for details of this argument).

Finally, if $W_1^\epsilon(L_q(a))$ and $W_1^{\epsilon'}(L_{q'}(a'))$ are isomorphic then, by comparing the levels of the $SL(2, \mathbb{Z})$-modules in question, we conclude $q = q'$. The eigenvalues of T on each of these spaces are of the form $e_q(ax^2)$ and $e_q(a'x^2)$, respectively, for suitable integers x, with at least one x relatively prime to q; whence aa' must be a square modulo q. But then $W_1^{\epsilon'}(L_{q'}(a'))$, (and hence, by assumption, $W_1^\epsilon(L_q(a))$), is isomorphic to $W_1^{\epsilon'}(L_q(a))$. This finally implies $\epsilon = \epsilon'$, since $-1 = S^2$ acts by $e_x \mapsto \sigma^2 e_{-x} = \sigma^2(\alpha \cdot e_x)$ on $L_q(a)$ (where $\sigma = \sigma(L_q(a))$) as follows from the explicit formula for the action of S. $\qquad\square$

Similarly, we can decompose the representations associated to the modules D_{2^α} or, more generally, to the modules

$$D_{2^\alpha}(a) := \left(\mathbb{Z}/2q, ax^2/4q\right),$$

where a is an odd number and $q = 2^\alpha \geq 1$. As before, $W(D_{q/4}(a))$ embeds naturally into $W(D_q(a))$ and we define $W_1^\epsilon(D_q(a))$ analogously to $W_1^\epsilon(L_q(a))$. Again, $W(D_{2^\alpha}(a))$ decomposes as direct sum of all the modules $W_1^\epsilon(D_{2^\alpha/d^2}(a))$ (plus $W(D_1)$ if 2^α is square) with d^2 running through the square divisors of 2^α and $\epsilon = \pm 1$. Moreover, the preceding lemma still holds true with suitable modifications when replacing $L_q(a)$ by $D_{2^\alpha}(a)$. Note, however, that $D_{2^\alpha}(a)$ does not factor through a representation of $SL(2, \mathbb{Z})$ since $\sigma(D_{2^\alpha}(a)) = e_8(-a).^5$

Lemma 3. *Let q be a power of 2. The $SL(2, \mathbb{Z})$-modules $W_1^\epsilon(D_q(a))$ are irreducible. The exact level of $W_1^\epsilon(D_q(a))$ equals $4q$ (i.e. $4q$ is the smallest positive integer such that $\Gamma(q)^*$ acts trivially). Two $\mathbb{M}$-modules $W_1^\epsilon(D_q(a))$ and $W_1^{\epsilon'}(D_{q'}(a'))$ are isomorphic if and only if $q = q'$, $\epsilon = \epsilon'$ and aa' is a square modulo $4q$.*

The proof is essentially the same as for the preceding lemma and we leave it to the reader.

Note that the modules D_m, for arbitrary non-zero m, and, more generally, the modules $D_m(a) := (\mathbb{Z}/2m, ax^2/4m)$, for a relatively prime to m, have as p-parts modules of the form $L_q(a')$ and $D_{2^\alpha}(a'')$, whence $W(D_m(a))$ is isomorphic to the outer tensor product of $SL(2, \mathbb{Z})/\Gamma(q)$-modules $W(L_q(a'))$

---

[5] For $\alpha = 0$, this formula holds only true for $a = \pm 1$, which, however, can always be assumed without loss of generality.

and a suitable $\mathbb{M}/\Gamma(4 \cdot 2^{\alpha})^*$-module $W(D_{2^{\alpha}}(a''))$. We conclude that the $\mathbb{M}$-module $W(D_m(a))$ decomposes as

$$W(D_m(a)) \cong \bigoplus_{\substack{fd^2 \mid m \\ f \text{ squarefree}}} W_1^f(D_{m/d^2}(a)).$$

Here $W_1^f(D_m(a))$ is the subspace of all $v = \sum_{x \in D_m} \lambda(x) \, e_x$ in $W(D_m(a))$ such that, for all isotropic submodules U of $D_m(a)$ and all $y \in U^*$, one has $\sum_{x \in y+U} \lambda(x) = 0$, and such that $gv = \chi_f(g)v$ for all g in $O(m)$. Moreover, χ_f denotes that character of $O(m)$ which maps g_p to -1 if $p \mid f$ and to $+1$ otherwise, where g_p is the orthogonal map which corresponds to the residue class x in $\mathbb{Z}/2m$ (under the correspondence described in section 3) which satisfies $x \equiv -1 \bmod 2p^{\alpha}$ and $x \equiv +1 \bmod 2m/p^{\alpha}$ with p^{α} being the exact power of p dividing m. Note that, for a fixed U, the vanishing condition can be restated as v being orthogonal to the image of the quadratic module U^*/U under the embedding of $W(U^*/U)$ into $W(D_m(a))$ and with respect to the scalar product as described above. Note also that $U_d := \frac{m}{d} D_m$ runs through all isotropic subgroups of D_m if d runs through all positive integers whose square divides m, and that, for such d, the quadratic module U_d^*/U_d is isomorphic to D_{m/d^2}.

The decomposition of $W(m)$ was already given (and essentially deduced by the same methods as explained here) in [Sko 85, Theorem 1.8, p. 22].

After these preparations we can now give the proofs of Theorems 2, 11 and 12.

Proof of Theorem 2. Suppose first of all that p is odd. The assumptions on M imply a decomposition $M \cong L_{p^{\alpha}}(a) \perp L_{p^{\beta}}(b)$ with, say, $0 \le \alpha \le \beta$ [N 76, Satz 3]. If $|M|$ is a square and $\sigma(M) = 1$ then α, β are both even or they are both odd and $\left(\frac{-ab}{p}\right) (= \sigma(M)) = 1$. In the first case M contains the trivial submodule, as follows immediately from Lemma 2. In the second case we may assume $b = -a$; but then, as follows again from Lemma 2, $A := W(L_{p^{\alpha}}(a))$ is a SL$(2, \mathbb{Z})$-submodule of $B := W(L_{p^{\beta}}(-b))$, whence $W(M) \cong A \otimes B^c$ contains non-zero invariants.

Conversely, if α and β do not have the same parity, or if $-ab$ is not a square modulo p, then by Lemma 2, A and B have no irreducible representation in common, whence $W(M) \cong A \otimes B^c$ contains no invariants.

If $p = 2$ and $M \cong D_{2^{\alpha}}(a) \perp D_{2^{\beta}}(b)$ with, say, $\alpha \le \beta$, then $\sigma(M) = e_8(-a - b)$ (using the convention that a or b is chosen to be ± 1 if α or β are equal to zero, respectively). Hence $\sigma(M) = 1$ if and only if $b \equiv -ax^2 \bmod 4 \cdot 2^{\beta}$ for some x. We can therefore follow the same line of reasoning as before to deduce the claim.

Finally, if $p = 2$ and M is not of the form as in the preceding argument then [N 76, Satz 4]

$$M \cong \left((\mathbb{Z}/2^\alpha)^2, \frac{xy}{2^\alpha} \right) \quad \text{or} \quad M \cong \left((\mathbb{Z}/2^\alpha)^2, \frac{x^2 + xy + y^2}{2^\alpha} \right).$$

In particular, $|M|$ is a square. In the first case $\sigma(M) = 1$ and $W(M)$ possesses an invariant, namely $I_U = \sum_{x \in U} e_x$, where, e.g. $U = \mathbb{Z}/2^\alpha \times 0$. In the second case $\sigma(M) = (-1)^\alpha$. If α is even then $I_V = \sum_{x \in V} e_x$, for $V = 2^{\alpha/2} M$, is an invariant. If α is odd then $W(M)$ possesses no invariants (see the complete decomposition of $W(M)$ in [N-W 76, pp. 519], which is called N_α loc. cit.). This completes the proof of the theorem. $\qquad\square$

Proof of Theorems 11 and 12. The $\mathbb{M}$-module $L_3^-(s)$ ($s = \pm 1$) is one-dimensional, and T acts on it by multiplication with $e_3(s)$. Hence the character afforded by $L_3^-(s)$ equals ϵ^{8s}. By Theorem 8 the space $J_{1,N}\left(\epsilon^{8s} \right)$ embeds injectively into the direct sum of the spaces $\text{Inv}(M_l)$, where

$$M_l := D_{-l} \perp D_{-N} \perp L_3(s)$$

and where l runs through the divisors of $N' := \text{lcm}(3, N)$ such that N'/l is squarefree.

Suppose that $\text{Inv}(M_l) \neq 0$. Since $\text{Inv}(M_l) = \bigotimes_k \text{Inv}\left(M_l(p) \right)$, where, $M_l(p)$, for a prime p, denotes the p-part of M_l, we conclude $\text{Inv}(M_l(p)) \neq 0$. For $p \neq 3$ this implies, by Theorem 3, that the p-parts of $|M_l| = 3Nl$ are perfect squares and that $\sigma_p(M_l) = 1$. From the first condition and since N'/l is squarefree we conclude $l = N'$ or $l = N'/3$. But

$$\sigma_2(M_l) = e_8(-l/q - N/q) \left(\frac{lN/q^2}{2q} \right),$$

where q is the exact power of 2 dividing N, and this is real if only if lN/q^2 is not a square mod 4. Thus $l = N/3$ or $l = 3N$ accordingly as 3 divides N or not. For this l and $p \neq 3$ we have $\sigma_p(M_l) = \left(\frac{-3}{q_p} \right)$, where q_p, for any p, denotes the exact power of p dividing N. Since $\sigma_p(M) = 1$ we find that $\left(\frac{-3}{q_p} \right) = 1$ for all $p \neq 3$. In particular, $N/q_3 \equiv +1$ mod 3. Finally, if, say, $3|N$ and $l = 3N$, we find

$$M_l(3) = \left(\mathbb{Z}/3q_3 \times \mathbb{Z}/3q_3 \times \mathbb{Z}/q_3, \frac{-N/q_3 x^2 - 3N/q_3 y^2 + q_3 s z^2}{3q_3} \right).$$

If $N/q_3 \equiv +1$ mod 3 then, for $s = -1$, this quadratic module contains no nontrivial isotropic element, hence T does not afford the eigenvalue 1 on $W\left(M_l(3) \right)$. Hence $\text{Inv}(M_l(3)) \neq 0$ implies $s = +1$. The same holds true, by a similar argument, if N is not divisible by 3. Note that, for $s = +1$ and $N/q_3 \equiv +1$ mod 3 we have $\sigma_3(M_l) = 1$.

Summing up we thus have proved that $J_{1,N}(\varepsilon^{8s}) = 0$ unless $s = +1$ and $\left(\frac{-3}{q_p}\right) = 1$ for all $p \neq 3$. Moreover, if the latter conditions hold true then

$$J_{1,N}(\varepsilon^8) \cong \begin{cases} \mathrm{Inv}(M_{3N})^{+,-} & \text{if } 3 \nmid N \\ \mathrm{Inv}(M_{N/3})^{+,-} & \text{if } 3|N \end{cases}.$$

Here the superscripts indicate the subspaces of all elements in $W(M_{3N})$ resp. $W(M_{N/3})$ which are even in the first and odd in the third argument.

By Theorem 1 the space $\mathrm{Inv}(M_{3N})$ is spanned by the special elements $e_U = \sum_{m \in U} e_m$, where U runs through the isotropic self-dual subgroups of M_{3N}. Accordingly, $J_{1,N}(\varepsilon^8)$ is the spanned by the Jacobi forms

$$\vartheta_U(\tau, z) := \sum_{(x,y,z) \in U} \vartheta_{3N,x}(\tau, 0)\, \vartheta_{N,y}(\tau, z) \left(\frac{z}{3}\right).$$

Here we used the application of Theorem 8 (and the identification $L_3^- \xrightarrow{\sim} \mathbb{C}(\varepsilon^8)$ given by $e_z - e_{-z} \mapsto \left(\frac{z}{3}\right)$).

The statements of the last paragraph still hold true if N is divisible by 3. In fact, a literal application of Theorem 8 and the preceding considerations imply that $J_{1,N}(\varepsilon^8)$ is spanned by the the Jacobi forms which are given by the same formula as the ϑ_U but with $\vartheta_{3N,x}$ replaced by $\vartheta_{N/3,x}$ and where U runs through the isotropic self-dual subgroups of $M_{N/3}$. But the quadratic module $D_{-N/3}$ is isomorphic to X^*/X (via $x \mapsto 3x + X$), where $X^* = 3\mathbb{Z}/6N$. This map induces an embedding of $\mathbb{M}$-modules $W(-N/3) \to W(-3N)$ (via $e_x \mapsto \sum_{y \equiv 3x \bmod 2N} e_y$; see the end of section 3), and it induces a map $U \mapsto U'$ from the set of isotropic self-dual subgroups of $M_{n/3}$ into the set of isotropic self-dual subgroups of M_{3N} (via pullback). Note that this map is one to one (since $-x^2 - 3y^2 + 4Nz^2 \equiv 0 \bmod 12N$ implies $3|x$). Finally, the diagram formed by the maps $\mathrm{Inv}(M_{N/3}) \ni U \mapsto e_U \mapsto \vartheta_U$ and $\mathrm{Inv}(M_{N/3}) \ni U \mapsto U' \mapsto \vartheta_{U'}$ commute as follows on using the formulas

$$\vartheta_{N/3,x} = \sum_{y \equiv 3x \bmod 2N} \vartheta_{3N,y}.$$

It remains to determine, for arbitrary N, the isotropic self-dual subgroups U of M_{3N}. The map $U \ni (x, y, z) \mapsto (x, y)$ is injective (since $-\frac{x^2+3y^2}{4N} + z^2 \in 3\mathbb{Z}$) and thus maps U onto an isotropic subgroup U' of the (degenerate) quadratic module $M := \left(\mathbb{Z}/6N \times \mathbb{Z}/2N, -\frac{x^2+3y^2}{4N}\right)$ of order $|U'| = 6N$. For determining the set S of isotropic subgroups U' of M whose order is $6N$ let $\mathcal{O}$ be the maximal order of $\mathbb{Q}(\sqrt{-3})$. The image of the map $\mathcal{O} \ni \alpha = \frac{x+\sqrt{-3}y}{2} \mapsto \left(x + 6N\mathbb{Z}, y + 2N\mathbb{Z}\right)$ contains the isotropic vectors of M

(since $\frac{x^2+3y^2}{4N} \in \mathbb{Z}$ implies that x and y have the same parity) and its kernel equals $2\sqrt{-3}N\mathbb{Z}[\sqrt{-3}]$. The U' in S thus correspond to the subgroups $2\sqrt{-3}N\mathbb{Z}[\sqrt{-3}] \subset I \subset \mathcal{O}$ of index N in $\mathcal{O}$ and such that $N||\alpha|^2$ for all $\alpha \in I$. Let I be such a subgroup and let I' be the $\mathcal{O}$-ideal generated by I. Clearly $N||\alpha|^2$ for all α in I'. Since $I' = \mathcal{O}\rho$ for some ρ we find $N||\rho|^2$ and $|\rho|^2 \leq N$. But then $|\rho|^2 = N$ and $I' = I$. Thus all isotropic self-dual subgroups U of M_{3N} are of the form

$$U = \left\{ \left(x(\alpha) + 6N\mathbb{Z}, y(\alpha) + 2N\mathbb{Z}, \psi(\alpha)\right) : \alpha \in \mathcal{O}\rho \right\},$$

where $\mathcal{O}\rho$ is an ideal of norm N, where ψ is a group homomorphism $\mathcal{O}\rho \rightarrow \mathbb{Z}/3$ such that $-|\alpha|^2/N + \psi(\alpha)^2 \in 3\mathbb{Z}$, and where we use $x(\alpha) = \alpha + \overline{\alpha}$, $y(\alpha) = (\alpha - \overline{\alpha})/\sqrt{-3}$. It is easily checked that the only possible ψ are $\psi(\alpha) = \epsilon x(\alpha/\rho)$ ($\epsilon = \pm 1$). In view of the formula for ϑ_U it suffices to consider only those $U = U_\rho$ where $\psi(\alpha) = x(\alpha/\rho)$. If we write ϑ_ρ for ϑ_U then by a simple calculation

$$\vartheta_\rho = \sum_{\alpha \in \mathcal{O}} \left(\frac{x(\alpha)}{3}\right) q^{|\alpha|^2/3} \zeta^{y(\alpha\rho)},$$

We have thus have proved that $J_{1,N}(\varepsilon^8)$ is indeed spanned by the series ϑ_ρ as stated in Theorem 12, where $\mathcal{O}\rho$ runs through the ideals of norm N in $\mathcal{O}$. $\quad \square$

Bibliography

[Ar 92]   Arakawa T., Selberg Zeta Functions and Jacobi Forms. in Adv. Stud. Pure Math. 21 (1992) Zeta Functions in Geometry, 181–218.

[Co-S 88]   Conway,J. H. and Sloane, N. J. A., Sphere Packings, Lattices and Groups. Grundlehren 290, Springer, 1988.

[E-S 95]   Eholzer, W. and Skoruppa, N-P., Modular invariance and uniqueness of conformal characters, Comm. Math. Phys. 174 (1995), 117–136.

[E-Z 85]   Eichler, M. and Zagier, D., The Theory of Jacobi Forms. Birkhäuser, 1985.

[Fr 91]   Freitag, E., Singular Modular Forms and Theta Relations, LNM 1487, Springer, 1991.

[Gri 95]   Gritsenko, V., Modular Forms and Moduli Spaces of Abelian and K3 Surfaces, St. Petersburg Math. J. 6 (1995), 1179–1208.

[G-N 98]   Gritsenko, V. and Nikulin, V., Automorphic Forms and Lorentzian Kac-Moody Algebras. Part II, Internat. J. Math. 9 (1998), 201–275.

[Gri 05]   Gritsenko, V., Unpublished notes for a talk given 2005 in the Max-Planck Institut for Mathematics in Bonn.

[G-S-Z 07]   Gritsenko, V., Zagier, D., and Skoruppa, N-P., The theory of thetablocks, in preparation.

[I-S 07]   Ibukiyama, T. and Skoruppa, N-P., A Vanishing Theorem for Siegel
             Modular Forms of Weight One, Abh. Math. Sem. Univ. Hamburg 77
             (2007), 229–235.

[I-S 08]   Ibukiyama, T. and Skoruppa, N-P., Siegel and Orthogonal Modular
             Forms of Critical Weight, in preparation.

[Kl 46]   Kloosterman, H. D., The behaviour of general theta functions under
             the modular group and the characters of binary modular congruence
             groups I, II, Ann. of Math. 47 (1946), 317–447.

[N 76]   Nobs, A., Die irreduziblen Darstellungen der Gruppe SL(2, $\mathbb{Z}_p$), ins-
             besondere SL(2, $\mathbb{Z}_2$) I, Comment. Math. Helv. 39 (1976), 465–490.

[N-W 76]   Nobs, A., Wolfart, J., Die irreduziblen Darstellungen der Gruppe
             SL(2, $\mathbb{Z}_p$), insbesondere SL(2, $\mathbb{Z}_2$) I, II, Comment. Math. Helv. 39
             (1976), 491–526.

[N-S-R 07]   Nebe G., Sloane, N. J. A. and Rains E. M., Self-dual codes and
             invariant theory. Algorithms and Computation in Mathematics, 17.
             Springer-Verlag, Berlin, 2006.

[Re 75]   Resnikoff, H. L., Automorphic forms of singular weight are singular
             forms, Math. Ann. 215 (1975), 173–193.

[Sch 84]   Scharlau, W., Quadratic and Hermitian Forms. Grundlehren 270,
             Springer, Berlin 1985.

[Se-S 77]   Serre, J-P. and Stark H., Modular forms of weight 1/2. in Modular
             Functions of one variable VI, Lecture Notes 627, Springer, 1977.

[Sko 85]   Skoruppa N-P., Über den Zusammenhang zwischen Jacobiformen und
             Modulformen halbganzen Gewichts, Bonner Mathematische Schriften
             Nr. 159, Bonn 1985.

[Sko 06]   Skoruppa N-P., Memorandum on Dimension Formulas for Spaces of
             Jacobi Forms. in Proceedings of the Conference on Automorphic Rep-
             resentations, L-Functions and Periods, Research Institute for Mathe-
             matical Sciences, Kyoto 2006, 172–183, arXiv:math.NT/0711.0632.

[Sko 07]   Skoruppa N-P., Jacobi Forms of Degree One and Weil Representations.
             in Proceedings of the Conference *Siegel Modular Forms and Abelian
             Varieties* near Lake Hamana 2007, arXiv:math.NT/0711.0525.

[Sko 08]   Skoruppa N-P., Finite Quadratic Modules and Weil Representations, in
             preparation.

[Ta 67]   Tanaka, S., Irreducible representations of the binary congruence groups
             mod p^λ, J. Math. Kyoto Univ. 7-2 (1967), 123–132.

[We 92]   Weissauer, R., Modular forms of genus 2 and weight 1, Math. Z. 210
             (1992), 91–96.

Tannakian Categories attached to
abelian varieties

Rainer Weissauer

Let k be either the algebraic closure of a finite field k or an algebraically closed field of characteristic zero. Let l be a prime different from the characteristic of k.

Notation. For a variety X over k let $D^b_c(X, \overline{\mathbb{Q}}_l)$ denote the triangulated category of complexes of étale $\overline{\mathbb{Q}}_l$-sheaves on X in the sense of [5]. For a complex $K \in D^b_c(X, \overline{\mathbb{Q}}_l)$ let $D(K)$ denote its Verdier dual, and let $\mathcal{H}^\nu(K)$ denote its étale cohomology $\overline{\mathbb{Q}}_l$-sheaves with respect to the standard t-structure. If $\dim(S_\nu) \leq \nu$ holds for all integers $\nu \in \mathbb{Z}$, where S_ν denotes the support of the cohomology sheaf $\mathcal{H}^{-\nu}(K)$, then the complex $K \in D^b_c(X, \overline{\mathbb{Q}}_l)$ is called semi-perverse. The abelian subcategory $\mathrm{Perv}(X)$ of middle perverse sheaves is the full subcategory of all $K \in D^b_c(X, \overline{\mathbb{Q}}_l)$, for which K and its Verdier dual $D(K)$ are contained in the full subcategory ${}^pD^{\leq 0}(X)$ of semi-perverse sheaves.

If k is the algebraic closure of a finite field κ a complex K of étale $\overline{\mathbb{Q}}_l$-Weil sheaves is mixed of weight $\leq w$ if all its cohomology sheaves $\mathcal{H}^\nu(K)$ are mixed étale $\overline{\mathbb{Q}}_l$-sheaves with upper weights $w(\mathcal{H}^\nu(K)) - \nu \leq w$ for all integers ν. It is called pure of weight w if K and its Verdier dual $D(K)$ are mixed of weight $\leq w$. Concerning base fields of characteristic zero, we assume mixed sheaves to be sheaves of geometric origin in the sense of the last chapter of [1], so we still dispose over the notion of the weight filtration and purity and Gabber's decomposition theorem in this case. In this sense let $\mathrm{Perv}_m(X)$ denote the abelian category of mixed perverse sheaves on X. The full subcategory $P(X)$ of $\mathrm{Perv}_m(X)$ of pure perverse sheaves is a semisimple abelian category.

Abelian varieties. Let X be an abelian variety of dimension g over an algebraically closed field k. Let $a : X \times X \to X$ be the addition on the abelian variety. Given objects K and L of $D_c^b(X, \overline{\mathbb{Q}}_l)$, define their convolution product $K * L \in D_c^b(X, \overline{\mathbb{Q}}_l)$ to be

$$K * L = Ra_*(K \boxtimes L).$$

This product is commutative in the sense, that there exists functorial isomorphisms $K * L \cong L * K$. For the skyscraper sheaf δ_0 concentrated at the zero element 0 notice $K * \delta_0 = K$.

Translation-invariant sheaf complexes. More generally $K * \delta_x = T_{-x}^*(K)$, where δ_x is a skyscraper sheaf with support in a k-valued point $x \in X(k)$, and where $T_x(y) = y + x$ denotes the translation $T_x : X \to X$ by x. In fact $T_x^*(K * L) \cong T_x^*(K) * L \cong K * T_x^*(L)$ holds for all $x \in X(k)$. A complex K is called translation-invariant if $T_x^*(K) \cong K$ holds for all $x \in X(k)$. If $f : X \to Y$ is a surjective homomorphism between abelian varieties, then the direct image $Rf_*(K)$ of a translation-invariant complex is translation-invariant. As a consequence of the formulas above, the convolution of an arbitrary $K \in D_c^b(X, \overline{\mathbb{Q}}_l)$ with a translation-invariant complex on X is a translation-invariant complex. A translation-invariant perverse sheaf K on X is of the form $K = E[g]$, for an ordinary étale translation-invariant $\overline{\mathbb{Q}}_l$-sheaf E. For a translation-invariant complex $K \in D_c^b(X, \overline{\mathbb{Q}}_l)$ the irreducible constituents of the perverse cohomology sheaves ${}^pH^\nu(K)$ are translation-invariant.

Multipliers. The subcategory $T(X)$ of $\mathrm{Perv}(X)$ of all perverse sheaves whose irreducible perverse constituents are translation-invariant is a Serre subcategory of the abelian category $\mathrm{Perv}(X)$. Let denote $\overline{\mathrm{Perv}}(X)$ its abelian quotient category and $\overline{P}(X)$ the image of $P(X)$, which is a full subcategory with semisimple objects. The full subcategory of $D_c^b(X, \overline{\mathbb{Q}}_l)$ of all K, for which ${}^pH^\nu(K) \in T(X)$, is a thick subcategory of the triangulated category $D_c^b(X, \overline{\mathbb{Q}}_l)$. Let

$$\overline{D}_c^b(X, \overline{\mathbb{Q}}_l)$$

be the corresponding triangulated quotient category, which contains $\overline{\mathrm{Perv}}(X)$. Since ${}^pH^\nu(L) \in T(X)$ for all $\nu \in \mathbb{Z}$ implies ${}^pH^\mu(K * L) \in T(X)$ for all $\mu \in \mathbb{Z}$, the convolution product

$$* : \overline{D}_c^b(X, \overline{\mathbb{Q}}_l) \times \overline{D}_c^b(X, \overline{\mathbb{Q}}_l) \to \overline{D}_c^b(X, \overline{\mathbb{Q}}_l)$$

is well defined.

Definition. A perverse sheaf K on X is called a multiplier if the convolution induced by K

$$*K : \overline{D_c^b}(X, \overline{\mathbb{Q}_l}) \to \overline{D_c^b}(X, \overline{\mathbb{Q}_l})$$

preserves the abelian subcategory $\overline{\mathrm{Perv}}(X)$.

Obvious from this definition are the following properties of multipliers: If K and L are multipliers, so are the product $K * L$ and the direct sum $K \oplus L$. Direct summands of multipliers are multipliers. If for a distinguished triangle (K, L, M) in $D_c^b(X, \overline{\mathbb{Q}_l})$ the complexes K, M are multipliers, then so is L. If K is a multiplier, then the Verdier dual $D(K)$ is a multiplier and also the dual

$$K^\vee = (-\mathrm{id}_X)^*(D(K)) .$$

Proposition. *The following perverse sheaves are multipliers*

(i) *Skyscraper sheaves.*

(ii) *If $i : C \hookrightarrow X$ is a projective curve, which generates the abelian variety X, and E is an étale $\overline{\mathbb{Q}_l}$-sheaf on C with finite monodromy, then the intersection cohomology sheaf attached to (C, E) is a multiplier.*

(iii) *If $i : Y \hookrightarrow X$ is a smooth ample divisor, then the intersection cohomology sheaf δ_Y of Y is a multiplier.*

Proof. (i) is obvious. For (ii) a proof using reduction mod p is given in [7]. This is based on the Čebotarev density theorem and an argument involving counting of points.

Concerning (iii). Let g denote the dimension of X. Then $\delta_X = \overline{\mathbb{Q}}_{l,X}[g]$ is a translation-invariant perverse sheaf on X. The morphism $j : U = X \setminus Y \hookrightarrow X$ is an affine open embedding for an arbitrary irreducible divisor Y on X. Notice Y is a Cartier divisor, since X is smooth. Hence by [KW], corollary III.6.2 and remark III.6.4 the sheaf complexes $\lambda_U = Rj_!\overline{\mathbb{Q}}_l[g]$ and $\lambda_Y = i_*\overline{\mathbb{Q}}_{l,Y}[g-1]$ are perverse sheaves on X. These perverse sheaves are related by an exact sequence of perverse sheaves

$$0 \to \lambda_Y \to \lambda_U \to \delta_X \to 0 .$$

Hence λ_Y and λ_U become isomorphic in $\overline{\mathrm{Perv}}(X)$.

Now consider the morphism $\pi = a \circ (j \times \mathrm{id}_X)$. We claim that $\pi : U \times X \to X$ is affine, i.e. for affine open subsets V of X the inverse image $W = \pi^{-1}(V)$ is affine. Indeed under the isomorphism $(u, v) \mapsto (u, u + v)$ of X^2 the open subset W becomes isomorphic to the product $U \times V$. By assumption Y is ample. Hence its complement U is affine, and therefore also $W \cong U \times V$.

By the affine vanishing theorem of Artin $R\pi_!(\mathrm{Perv}(U \times X))$ is therefore contained in $^p D^{\geq 0}(X, \overline{\mathbb{Q}_l})$. For an arbitrary perverse sheaf $L \in \mathrm{Perv}(X)$ then

$R\pi_!(\overline{\mathbb{Q}}_{l,U}[g] \boxtimes L) = Ra_*(\lambda_U \boxtimes L) = \lambda_U * L$ implies ${}^pH^\nu(\lambda_U * L) = 0$ for all integers $\nu < 0$. The distinguished triangle $(\lambda_Y * L, \lambda_U * L, \delta_X * L)$ and the induced long exact perverse cohomology sequence gives isomorphisms ${}^pH^{\nu-1}(\delta_X * L) \cong {}^pH^\nu(\lambda_Y * L)$ for all integers $\nu < 0$. Since the perverse cohomology sheaves of $\delta_X * L$ are contained in $T(X)$, we conclude that ${}^pH^\nu(\lambda_Y * L)$ becomes zero in $\overline{\mathrm{Perv}}(X)$ for $\nu < 0$. In particular $\lambda_Y * \lambda_Y^\vee$ has image in $\overline{\mathrm{Perv}}(X)$.

If Y is smooth, the intersection cohomology sheaf δ_Y of Y coincides with the perverse sheaf $\lambda_Y = i_*\overline{\mathbb{Q}}_{l,Y}[g - 1]$. Since the intersection cohomology sheaf δ_Y is selfdual, the perverse sheaf λ_Y is selfdual. By Verdier duality this implies $D({}^pH^\nu(\lambda_Y * L)) \cong {}^pH^{-\nu}(D(\lambda_Y * L)) \cong {}^pH^{-\nu}(D(\lambda_Y) * D(L)) \cong {}^pH^{-\nu}(\lambda_Y * D(L)) \in T(X)$ again for all $\nu > 0$. Since $T(X)$ is stable under Verdier duality, the image of $\lambda_Y * L$ therefore is in $\overline{\mathrm{Perv}}(X)$. Thus for a smooth ample divisor Y the perverse sheaf $\delta_Y = \lambda_Y$ is a multiplier.   □

Let $M(X) \subseteq P(X)$ denote the full category of semisimple multipliers. Let $\overline{M}(X)$ denote its image in the quotient category $\overline{P}(X)$ of $P(X)$. Then, by the definition of multipliers, the convolution product preserves $\overline{M}(X)$

$$* : \overline{M}(X) \times \overline{M}(X) \to \overline{M}(X) .$$

Theorem. *With respect to the convolution product* $*$ *the category* $\overline{M}(X)$ *is a semisimple super-Tannakian* $\overline{\mathbb{Q}}_l$*-linear tensor category, hence as a tensor category* $\overline{M}(X)$ *is equivalent to the category of representations* $\mathrm{Rep}(G, \varepsilon)$ *of a projective limit*

$$G = G(X)$$

of supergroups.

Outline of proof. The convolution product obviously satisfies the usual commutativity and associativity constraints compatible with unit objects. See [7] 2.1. By [7], Corollary 3 furthermore one has functorial isomorphisms

$$\mathrm{Hom}_{\overline{M}(X)}(K, L) \cong \Gamma_{\{0\}}(X, \mathcal{H}^0(K * L^\vee)^*) ,$$

where $\mathcal{H}^0$ denotes the degree zero cohomology sheaf and $\Gamma_{\{0\}}(X, -)$ is the space of sections with support in the neutral element. Let $L = K$ be simple and nonzero. Then the left side becomes $\mathrm{End}_{\overline{M}(X)}(K) \cong \overline{\mathbb{Q}}_l$. On the other hand $K * L^\vee$ is a direct sum of a perverse sheaf P and translates of translation-invariant perverse sheaves. Therefore $\mathcal{H}^0(K * L^\vee)^\vee$ is the direct sum of a skyscraper sheaf S and translation-invariant étale sheaves. Hence $\Gamma_{\{0\}}(X, \mathcal{H}^0(K * L^\vee)^\vee) = \Gamma_{\{0\}}(X, S)$. By a comparison of both sides therefore

$S = \delta_0$. Notice δ_0 is the unit element 1 of the convolution product. Using this we not only get

$$\mathrm{Hom}_{\overline{M}(X)}(K, L) \cong \mathrm{Hom}_{\overline{M}(X)}(K * L^\vee, 1) \,,$$

but also find a nontrivial morphism

$$\mathrm{ev}_K : K * K^\vee \to 1 \,.$$

By semisimplicity, δ_0 is a direct summand of the complex $K * K^\vee$. In particular the Künneth formula implies, that the étale cohomology groups do not all vanish identically, i.e.

$$H^\bullet(X, K) \neq 0 \,.$$

Therefore the arguments of [7] 2.6 show, that the simple perverse sheaf K is dualizable. Hence $\overline{M}(X)$ is a rigid $\overline{\mathbb{Q}_l}$-linear tensor category. Let $\mathcal{T}$ be a finitely $\otimes$-generated tensor subcategory with generator, say A. To show $\mathcal{T}$ is super-Tannakian, by [4] it is enough to show for all n

$$\mathrm{length}_{\mathcal{T}}(A^{*n}) \leq N^n \,,$$

where N is a suitable constant. For any $B \in \overline{M}(X)$ let B, by abuse of notation, also denote a perverse semisimple representative in $\mathrm{Perv}(X)$ without translation invariant summand. Put

$$h(B, t) = \sum_\nu \dim_{\overline{\mathbb{Q}_l}}(H^\nu(X, B))t^\nu.$$

Then $\mathrm{length}_{\mathcal{T}}(B) \leq h(B, 1)$, since every summand of B is a multiplier and therefore has nonvanishing cohomology. For $B = A^{*n}$ the Künneth formula gives $h(B, 1) = h(A, 1)^n$. Therefore the estimate above holds for $N = h(A, 1)$. This completes the outline for the proof of the theorem. $\quad\square$

Remark. More generally, using [7], Theorem 2, one can define a super-Tannakian group for the category of all mixed perverse sheaves whose irreducible constituents are multipliers.

Principally polarized abelian varieties. Suppose Y is a divisor in X defining a principal polarization. Suppose the intersection cohomology sheaf δ_Y of Y is a multiplier. Then a suitable translate of Y is symmetric, and again a multiplier. So we may assume $Y = -Y$ is symmetric. Let $\overline{M}(X, Y)$ denote the super-Tannakian subcategory of $\overline{M}(X)$ generated by δ_Y. The corresponding super-group $G(X, Y)$ attached to $\overline{M}(X, Y)$ acts on the superspace $W = \omega(\delta_Y)$ defined by the underlying super-fiber functor ω of $\overline{M}(X)$. By assumption δ_Y

is self-dual in the sense, that there exists an isomorphism $\varphi : \delta_Y^\vee \cong \delta_Y$. Obviously $\varphi^\vee = \pm\varphi$. This defines a non-degenerate pairing on W, and the action of $G(X, Y)$ on W respects this pairing.

Curves. The Jacobian X of a smooth projective curve C of genus $g \geq 2$ over k carries a natural principal polarization $Y = W_{g-1}$. The divisor Y is a multiplier. If we replace this divisor by a symmetric translate, the corresponding group $G(X, Y)$ is a semisimple algebraic group G. And this group is either $G = \mathrm{Sp}(2g - 2, \overline{\mathbb{Q}}_l)/\mu_{g-1}[2]$ or $G = \mathrm{Sl}(2g - 2, \overline{\mathbb{Q}}_l)/\mu_{g-1}$ depending on whether the curve C is hyperelliptic or not. Here $\mu_{g-1}[2]$ denotes the two-torsion subgroup of μ_{g-1}. The faithful representation W of G defined by δ_Y is the unique irreducible $\overline{\mathbb{Q}}_l$-representation ρ_G of highest weight in the $(g-1)$-th exterior power of the $(2g - 2)$-dimensional standard representation of G. See [7], section 7.6.

Some recursion formulas. Let $\mathcal{L}^{-1} = O_X(-Y)$ be the ideal sheaf of a smooth irreducible divisor Y on an abelian variety X of dimension $g \geq 2$. The tangent sheaf of X is trivial, hence

$$0 \to (\mathcal{L}|_Y)^{-1} \otimes_{O_Y} \Omega_Y^{\nu-1} \to \binom{g}{\nu} \cdot O_Y \to \Omega_Y^\nu \to 0 \qquad (*)$$

are exact sequences obtained from the adjunction formula. The dualizing sheaf of Y is $\omega_Y = \Omega_Y^{g-1} \cong \mathcal{L}|_Y$. Define $\chi_M(m) = \sum_i(-1)^i h^i(Y, M \otimes_{O_Y} (\mathcal{L}|_Y)^m)$ for coherent sheaves M on Y. Then $\chi(m) := \chi_{O_Y}(m) = cm^g - c(m-1)^g$ by the Riemann-Roch formula $\chi(X, \mathcal{L}^m) = cm^g$ for $c = \deg(\mathcal{L}^g/g!)$, using the exact sequence $0 \to \mathcal{L}^{m-1} \to \mathcal{L}^m \to (\mathcal{L}|_Y)^m \to 0$. Notice $c = 1$ for a theta divisor. The sequences (*) give the recursion relations $\chi_{\Omega_Y^\nu}(m) = \binom{g}{\nu}\chi(m) - \chi_{\Omega_Y^{\nu-1}}(m-1)$. Hence $\chi_{\Omega_Y^\nu}(m) = \sum_{k=0}^\nu (-1)^{k-\nu}\binom{g}{k}\chi(m+k-\nu)$ by induction.

The analytic case. If X is defined over $\mathbb{C}$ of complex numbers the topological Euler characteristic $\chi(Y) = \sum_{i=0}^{g-1}(-1)^i \chi_{\Omega_Y^i}(0)$ of a smooth divisor Y can be computed via Hodge decomposition. This gives

$$c \sum_{0 \leq k \leq g-1} (-1)^k \binom{g}{k} \sum_{k \leq i \leq g-1} [(k-i)^g - (k-i-1)^g],$$

hence

$$\chi(Y) = -c \sum_{0 \leq k \leq g-1} (-1)^k \binom{g}{k}(k-g)^g = -c \sum_{1 \leq \nu \leq g} (-1)^\nu \binom{g}{\nu}\nu^g .$$

Since this is $-c(\partial_t)^g(1-t)^g|_{t=1} = (-1)^{g-1}cg! = (-1)^{g-1}\deg(\mathcal{L}^g)$, we get

Lemma. *The equality* $\chi(\delta_Y) = c \cdot g! = \deg(\mathcal{L}^g)$ *holds for smooth Y in the analytic case.*

A consequence. For an ample smooth divisor of an abelian variety X over $\mathbb{C}$ of dimension $g \geq 2$ the faithful irreducible representation $W = \omega(\delta_Y)$ of the Tannakian supergroup $G(X, Y)$ has super-dimension $\chi(\delta_Y) = c \cdot g!$. In particular

$$\chi(\delta_Y) = 2, \ 6, \ 24, \ \ldots$$

for $g = 2, 3, 4, \ldots$ in the case of a principal polarization.

For the first two values $g = 2$ and $g = 3$ this coincides with $\chi(\delta_{W_{g-1}}) = \binom{2g-2}{g-1}$ for the theta divisor of the Jacobian of a generic curve C of genus g (see [7]) in accordance with the fact, that principally polarized abelian varieties of dimension $g \leq 3$ generically are Jacobian varieties. For $g = 4$ on the other hand

$$\chi(\delta_{W_{g-1}}) = 20 \ < \ \chi(\delta_Y) = 24 \, .$$

Since the groups $\mathrm{Sp}(6)$ and $\mathrm{Sl}(6)/\mu_3$ do not admit an irreducible representation of dimension 24, the supergroup $G(X, Y)$ has to be different from these.

It is therefore tempting to ask the question whether a principally polarized abelian variety (X, Y) of dimension g is isomorphic to a Jacobian variety $(\mathrm{Jac}(C), W_{g-1})$ of a smooth projective curve C (up to translates of the polarization Y in X as explained above) if and only if Y is a multiplier, such that the corresponding super-Tannakian group $G(X, Y)$ is one of the two groups

$$\mathrm{Sp}(2g - 2, \overline{\mathbb{Q}_l})/\mu_{g-1}[2] \ \text{ or } \ \mathrm{Sl}(2g - 2, \overline{\mathbb{Q}_l})/\mu_{g-1}$$

and the representation $W = \omega(\delta_Y)$ is the representation ρ_G.

The bound N. In the situation of the last lemma the bound $N = h(\delta_Y, 1)$ defined above is $N = c \cdot g! + 2^{2g} - 2\binom{2g}{g}$, since $\dim_{\mathbb{C}}(H^\nu(Y, \mathbb{C})) = \dim_{\mathbb{C}}(H^\nu(X, \mathbb{C})) = \binom{2g}{\nu}$ holds for $\nu \leq \dim(Y) - 1$. The proof of the last claim uses the following.

Kodaira vanishing. For ample line bundles L on a complex manifold M one has $H^p(M, \Omega_M^q \otimes_{O_M} L) = 0$ for all $p + q > \dim(M)$. Alternative versions are $H^p(M, \omega_M \otimes_{O_M} L) = 0$ for all $p > 0$, or $H^{\dim(M)-p}(M, L^{-1}) = 0$ for $p > 0$ by Serre duality. Similarly, $H^{\dim(M)-p}(M, L^{-\mu}) = 0$ for all $\mu > 0$ and $p > 0$. Kodaira vanishing implies $H^a(Y, \mathcal{L}|_Y^{-\mu}) = 0$ for $a < g - 1 = \dim(Y)$

and $\mu > 0$ for an ample smooth divisor Y on X. For $\mu > 0$ the long exact cohomology sequence attached to

$$0 \to (\mathcal{L}|_Y)^{-1-\mu} \otimes_{O_Y} \Omega_Y^{\nu-1} \to (\mathcal{L}^{-\mu}|_Y)^{\binom{g}{\nu}} \to (\mathcal{L}|_Y)^{-\mu} \otimes_{O_Y} \Omega_Y^{\nu} \to 0$$

gives $H^i(Y, \mathcal{L}|_Y^{-\mu} \otimes_{O_Y} \Omega_Y^{\nu}) \cong H^{i+1}(Y, \mathcal{L}|_Y^{-1-\mu} \otimes_{O_Y} \Omega_Y^{\nu-1})$ for $0 \le i \le g-3$, and $H^0(Y, \mathcal{L}|_Y^{-1-\mu} \otimes_{O_Y} \Omega_Y^{\nu-1}) = 0$. Hence

$$H^a(Y, \mathcal{L}|_Y^{-\mu} \otimes_{O_Y} \Omega_Y^b) \cong H^{a+1}(Y, \mathcal{L}|_Y^{-\mu-1} \otimes_{O_Y} \Omega_Y^{b-1})$$

$$\cong \cdots \cong H^{a+b}(Y, \mathcal{L}|_Y^{-\mu-b})$$

for $\mu > 0$, $a \le g-2$ and $a+b \le g-2 < \dim(Y)$. Since $H^{a+b}(Y, \mathcal{L}|_Y^{-\mu-b})$ is zero again by Kodaira vanishing for $a+b \le g-2$, this gives

$$H^a(Y, \mathcal{L}|_Y^{-\mu} \otimes_{O_Y} \Omega_Y^b) = 0 \quad \text{for} \quad \mu > 0, \ a+b \le g-2.$$

For $\mu = 1$ this vanishing and the sequence

$$0 \to (\mathcal{L}|_Y)^{-1} \otimes_{O_Y} \Omega_Y^{b-1} \to O_Y^{\binom{g}{b}} \to \Omega_Y^b \to 0$$

implies the existence of isomorphisms

$$H^a(Y, O_Y)^{\binom{g}{b}} \cong H^a(Y, \Omega^b), \quad a+b \le g-2.$$

Indeed $H^a(Y, \mathcal{L}|_Y^{-1} \otimes_{O_Y} \Omega_Y^{b-1}) = 0$ and $H^{a+1}(Y, \mathcal{L}|_Y^{-1} \otimes_{O_Y} \Omega_Y^{b-1}) = 0$. Hence the Hodge numbers are $h^{a,b} = h^{a,0}\binom{g}{b}$ for $a+b \le g-2$. By symmetry $h^{a,b} = h^{b,a}$ then $h^{a,0}\binom{g}{b} = h^{b,0}\binom{g}{a}$. Then $h^{a,0} = \binom{g}{a}$ for $b = 0$ and $a \le g-2$ implies $h^{a,b} = \dim_{\mathbb{C}}(H^a(Y, \Omega_Y^b)) = \binom{g}{a}\binom{g}{b}$ for $a+b < g-1 = \dim(Y)$. In particular $\dim_{\mathbb{C}}(H^\nu(Y, \mathbb{C})) = \binom{2g}{\nu}$ for $\nu < g-1 = \dim(Y)$. $\qquad\square$

We remark, that the Hodge numbers $h^{a,g-1-a}$ in the middle dimension can be computed from the formulas for $\chi_{\Omega_Y^\nu}(0)$ stated earlier.

Bibliography

[1] Beilinson A., Bernstein J., Deligne P., Faisceaux pervers, Astérisque 100 (1982).

[2] Deligne P., Milne J.S., Tannakian categories, in Lecture Notes in Math 900, pp. 101–228

[3] Deligne P., Catégories tannakiennes, The Grothendieck Festschrift, vol. II, Progr. Math., vol. 87, Birkhäuser (1990), 111–195.

[4] Deligne P., Catégories tensorielles, Moscow Math. Journal 2 (2002) no.2, 227–248

[5] Kiehl R., Weissauer R., Weil conjectures, perverse sheaves and *l*-adic Fourier transform, Ergebnisse der Mathematik und ihrer Grenzgebiete 42, Springer (2001).

[6] Weissauer R., Torelli's theorem from the topological point of view, this volume, pp. 275–284.

[7] Weissauer R., Brill-Noether Sheaves, arXiv math.AG/0610923.

Torelli's theorem from the topological point of view

Rainer Weissauer

Torelli's theorem states, that the isomorphism class of a smooth projective curve C of genus $g \geq 2$ over an algebraically closed field k is uniquely determined by the isomorphism class of the associated pair (X, Θ), where X is the Jacobian variety of C and Θ is the canonical theta divisor. The aim of this note is to give a 'topological' proof of this theorem. Although Torelli's theorem is not a topological statement, the proof to be presented gives a characterization of C in terms of perverse sheaves on the Jacobian variety X, which are attached to the theta divisor by a 'topological' construction.

For complexes $K, L \in D_c^b(X, \overline{\mathbb{Q}}_l)$ define $K * L \in D_c^b(X, \overline{\mathbb{Q}}_l)$ as the direct image complex $Ra_*(K \boxtimes L)$, where $a : X \times X \to X$ is the addition law of X. Let $K_*^0(X)$ be the tensor product of the Grothendieck group of perverse sheaves of geometric origin on X with the polynomial ring $\mathbb{Z}[t^{1/2}, t^{-1/2}]$. Then $K_*^0(X)$ is a commutative ring with ring structure defined by the convolution product, and therefore the quotient ring $K_*(X)$, obtained by dividing the principal ideal generated by the constant perverse sheaf δ_X on X, is so too. The rings $K_*^0(X)$ and $K_*(X)$ have properties resembling those of the homology ring of X endowed with the $*$-product, but they have a much richer structure. A complex of sheaves $L \in D_c^b(X, \overline{\mathbb{Q}}_l)$ of geometric origin defines a class in $K_*(X)$ by the perverse Euler characteristic

$$\sum_{\nu} (-1)^{\nu} \cdot {}^p H^{\nu}(L) \cdot t^{\nu/2}.$$

Similar to the homology ring every irreducible closed subvariety Y has a class in $K_*(X)$ defined as the class of the perverse intersection cohomology sheaf δ_Y of Y. For details see [W]. This allows to consider the product

$$\delta_{\Theta} * \delta_{\Theta} \in K_*(X) .$$

275

Whereas the corresponding product in the homology ring of X is zero, this product turns out to be nonzero. The product $\delta_\Theta * \delta_\Theta$ is of the form $\sum_{v,\mu} A_{v,\mu} t^{\mu/2}$. Note that the coefficients are irreducible perverse sheaves $A_{v,\mu}$ on X. For a perverse sheaf A on X, which is a complex of sheaves on X, let $\mathcal{H}^i(A)$ denote the associated cohomology sheaves for $i \in \mathbb{Z}$. Let $\kappa \in X(k)$ be the Riemann constant defined by $\Theta = \kappa - \Theta$. It depends on the choice of the Abel-Jacobi map $C \to X$.

Theorem 1. *Let C be a smooth projective curve of genus $g \geq 3$. There exists a unique irreducible perverse sheaf $A = A_{v,0}$, among the coefficients of $\delta_\Theta * \delta_\Theta$, characterized by the following equivalent properties*

1. *$\mathcal{H}^{-1}(A)$ is nonzero, but not a constant sheaf on X.*
2. *$\mathcal{H}^{-1}(A)$ is the skyscraper sheaf $H^1(C) \otimes \delta_{\{\kappa\}}$ with support in the point $\kappa \in X$.*

Furthermore the support of the perverse sheaf A is $\kappa + C - C \subseteq X$.

Taking this for granted, Torelli's theorem is an immediate consequence. In fact, analytically the subvariety $C - C$ uniquely determines the curve C. This is well known. For a brief overview and further references see [Mu] p.78. Since the theorem stated above defines the sheaf complex A in terms of (X, Θ), this determines C via the support $\kappa + C - C$ of A for $g \geq 3$. The cases $g = 1, 2$ are trivial. Another argument to recover C from A, relying on minimal models, will be given below.

Remark. The sheaf A is a direct summand of the complex $\delta_C * \delta_{\kappa-C}$. If C is not hyperelliptic it splits into two irreducible perverse summands $\delta_{\{\kappa\}} \oplus A$. If C is hyperelliptic, then it splits into the three irreducible perverse summands $\delta_{\{\kappa\}} \oplus \delta_{\kappa+C-C} \oplus A$.

We first give a sketch of the theorem in the non-hyperelliptic case. For all integers $r \geq 0$ let $\delta_r \in K_*(X)$ be the class of the direct image complex $Rp_{r,*}\delta_{C^{(r)}}$, where $p_r : C^{(r)} = C^r/\Sigma_r \to X$ are the Abel-Jacobi maps from the symmetric quotient of C^r to X. For $r \leq g - 1$ the image of p_r is the Brill-Noether subvariety $W_r = C + \cdots + C$ (r copies) of X. If C is not hyperelliptic, then

$$\delta_{W_r} = \delta_r \ ,$$

since p_r is a small morphism by the theorem of Martens [M] for $r \leq g - 1$. In particular $\delta_\Theta = \delta_{g-1}$, which will be used in the proof. (In the hyperelliptic case for $r \leq g - 1$ the morphism p_r is only semi-small. The theorem of Martens then gives $\delta_{W_r} = \delta_r - \delta_{r-2}$. In particular $\delta_\Theta = \delta_{g-1} - \delta_{g-3}$. For this and further details we refer to [W], lemma 21).

Proof of Theorem 1. Suppose C is not hyperelliptic.

1) Suppose $i = \min(i, j)$. Since the canonical morphism

$$\tau : C^{(i)} \times C^{(j)} \to C^{(i+j)}$$

is a finite ramified covering map, the direct image $R\tau_* \delta_{C^{(i)} \times C^{(j)}}$ decomposes into a direct sum of etale sheaves $\bigoplus_{\nu=0}^{i} m(i, j, \nu) \cdot \mathcal{F}_{i+j-\nu,\nu}$ by keeping track of the underlying action of the symmetric group Σ_{i+j} for the map $C^{i+j} \to C^{(i+j)}$ (see [W].4.1). The intertwining algebra of the induced representation $\mathrm{Ind}_{\Sigma_i \times \Sigma_j}^{\Sigma_{i+j}}(\overline{\mathbb{Q}}_l)$ is commutative, since inversion in the group induces an anti-involution, which is the identity on the representatives

$$\tau_\nu = \prod_{\mu=1}^{\nu} (i - \mu, i + \mu)$$

of the $\Sigma_i \times \Sigma_j$ double-cosets of Σ_{i+j}. Hence the multiplicities are $m(i, j, \nu) = 1$.

If we apply $Rp_{i+j,*}$, this gives a formula for $\delta_i * \delta_j$. From the identity

$$p_{i+j} \circ \tau = a \circ (p_i \times p_j),$$

where $a : X \times X \to X$ is the addition law of X, one obtains for $i \geq j$ that the convolution $\delta_i * \delta_j$ is $\delta_{i+j} \oplus \delta_{i+j-1,1} \oplus \cdots \oplus \delta_{i-j,j}$, where $\delta_{r,s} = Rp_{i+j,*}(\mathcal{F}_{r,s})$. A special case is

$$\delta_\Theta * \delta_\Theta = \delta_{g-1} * \delta_{g-1} = \delta_{2g-2} \oplus \delta_{2g-3,1} \oplus \cdots \oplus \delta_{g-1,g-1} .$$

Another case is $\delta_1 * \delta_{2g-3} = \delta_{2g-2} \oplus \delta_{2g-3,1}$, and together this implies

$$\boxed{\delta_{2g-3} * \delta_1 \hookrightarrow \delta_\Theta * \delta_\Theta} .$$

2) The morphism $f : C \times C \to \kappa + C - C \subseteq X$, defined by $(x, y) \mapsto \kappa + x - y$, is semi-small. If C is not hyperelliptic, then f is a birational map, which blows down the diagonal to the point κ, and is an isomorphism otherwise. Hence the direct image $Rf_*(\delta_C \boxtimes \delta_C)$ is perverse on X, and necessarily decomposes $Rf_*(\delta_C \boxtimes \delta_C) = \delta_C * \delta_{\kappa - C} = \delta_{\{\kappa\}} \oplus \delta_{\kappa + C - C}$ such that

$$\boxed{\mathcal{H}^{-1}(\delta_{\kappa+C-C}) \cong H^1(C) \otimes \delta_{\{\kappa\}}} .$$

For this we refer to the section on cohomology of blow-ups further below.

3) We claim $\delta_{2g-3} \equiv \delta_{\kappa-C}$ and $\delta_{2g-2} \equiv \delta_{\{\kappa\}}$ in $K_*(X)$ (ignoring Tate twists). These are the simplest cases of the duality theorem [W] 5.3. The first implies

$$\boxed{\delta_{2g-3} * \delta_1 \equiv \delta_{\{\kappa\}} + \delta_{\kappa+C-C}} ,$$

in $K_*(X)$ using step 2.

Proof of the claim. By the theorem of Riemann-Roch $C^{(2g-3)} \xrightarrow{p} X$ is a $\mathbb{P}^{g-2}$-bundle over $\kappa - C$ and a $\mathbb{P}^{g-3}$-bundle over the open complement $X \setminus (\kappa - C)$. Hence $Rp_* \delta_{C^{(2g-3)}}$ is a direct sum of $\delta_{\kappa-C}$ and a sum of translates of constant sheaves on X. Similarly $p_{2g-2}^{-1}(\{\kappa\}) = \mathbb{P}^{g-1}$, and p_{2g-2} is a $\mathbb{P}^{g-2}$-bundle over the open complement $X \setminus \{\kappa\}$. Hence $\delta_{2g-2} \equiv \delta_{\{\kappa\}}$ in $K_*(X)$. $\qquad\qquad\square$

4) Since $\Theta = \kappa - \Theta$ is a principal polarization, the definition of convolution product implies, that $\mathcal{H}^{-1}(\delta_\Theta * \delta_\Theta)$ is a skyscraper sheaf with support in κ. Its stalk at κ can be computed by a spectral sequence, whose relevant E_2-term $\bigoplus_{a,b \geq 0} H^{a+b-1}(X, \mathcal{H}^{-a}(\delta_\Theta) \otimes \mathcal{H}^{-b}(\delta_\Theta))$ is isomorphic to $H^{2g-3}(\Theta)$. The last isomorphism follows from the IC-conditions and the fact, that Θ is non-singular in codimension one ([W], 2.8). The latter also implies, that $H^{2g-3}(\Theta)$ is isomorphic to the intersection cohomology group $IH^{2g-3}(\Theta)$, hence it is pure of weight $2g - 3$. Thus the spectral sequence degenerates in this degree

$$\mathcal{H}^{-1}(\delta_\Theta * \delta_\Theta) \cong IH^{2g-3}(\Theta) \otimes \delta_{\{\kappa\}} \ .$$

Ignoring Tate twists this implies

$$\boxed{\mathcal{H}^{-1}(\delta_\Theta * \delta_\Theta) \cong H^1(X) \otimes \delta_{\{\kappa\}}} \ .$$

In fact, this follows from

$$IH^\bullet(W_d) = H^\bullet(X, Rp_{d,*}\delta_{C^{(d)}}[-d]) = H^\bullet(C^{(d)}) = \left(\bigotimes^d H^\bullet(C) \right)^{\Sigma_d}$$

for non-hyperelliptic curves, where the first equality holds since p_d is a small morphism. Thus

$$IH^{d+\bullet}(W_d) \cong \bigoplus_{a+b=d} \mathrm{Sym}^a \Big(H^0(C)[1] \oplus H^2(C)[-1] \Big) \otimes \Lambda^b(H^1(C)).$$

For $IH^{2d-1}(W_d)$ only $a = d - 1$ contributes, hence $IH^{2d-1}(W_d) \cong H^1(C) \cong H^1(X)$. The case $d = g - 1$ implies the result above. For the hyperelliptic case we refer to [W] 4.2. For a generalization to the case of principally polarized abelian varieties see [W] 2.9.

Conclusion. For curves C, which are not hyperelliptic, the perverse sheaf A defined by $\delta_{\kappa+C-C}$ satisfies all the assertions of the theorem. $\delta_{\kappa+C-C}$ is a direct summand of $\delta_\Theta * \delta_\Theta$ by step 1 and 3. By step 2 and 4 we obtain modulo constant sheaves on X

$$\mathcal{H}^{-1}(\delta_\Theta * \delta_\Theta) \equiv H^1(X) \otimes \delta_{\{\kappa\}} \equiv H^1(C) \otimes \delta_{\{\kappa\}} \equiv \mathcal{H}^{-1}(\delta_{\kappa+C-C}) \ .$$

Since $K_*(X)$ is a quotient of $K_*^0(X)$, the last identity only holds modulo constant sheaves on X. But this suffices to imply the theorem. $\qquad\qquad\square$

The corresponding proof for hyperelliptic curves is less elementary, and will not be given here in full detail. The argument for uniqueness (step 2 and 4) is the same, however the embedding of A to be explained later is more delicate.

Remark. In [W] we constructed a $\overline{\mathbb{Q}}_l$-linear Tannakian category $\mathcal{BN}$ attached to C that is equivalent to the category of finite dimensional $\overline{\mathbb{Q}}_l$-representations $\mathrm{Rep}(G)$, where G is $\mathrm{Sp}(2g-2, \overline{\mathbb{Q}}_l)$ or $\mathrm{Sl}(2g-2, \overline{\mathbb{Q}}_l)$ depending on whether C is hyperelliptic or not. In this category δ_Θ corresponds to the alternating power $\Lambda^{g-1}(st)$ of the standard representation, and A corresponds to the adjoint representation. This intrinsically characterizes A.

The cohomology of a blow-up. Let Y be a variety of dimension two with isolated singularities with a desingularization $\pi : Z \to Y$. Then $\pi : Z \to Y$ is a semi-small morphism, hence for a smooth etale $\overline{\mathbb{Q}}_l$-sheaf E on Z of geometric origin the direct image complex $R\pi_*(E)$ is a perverse sheaf on Y, and by the Gabber decomposition theorem necessarily of the form

$$R\pi_*(E) = \delta_E \oplus \bigoplus_{y \in Y_{\mathrm{sing}}} m(y, E) \cdot \delta_{\{y\}} \,,$$

where δ_E is the intersection cohomology sheaf on Y defined by setting $E|_{\pi^{-1}(Y_{\mathrm{reg}})}$, and where $m(y, E) = \sum_{i \in I} \dim(H^2(F_i, E|_F))$ for the irreducible curves F_i of the reduced fiber $F = \pi^{-1}(y)_{\mathrm{red}}$ over y over the singular point y. Hence

$$\mathcal{H}^{-1}(\delta_E)_y = H^1(F, E|_F) \,.$$

The case $E = \overline{\mathbb{Q}}_l$. Then $H^\bullet(Z)$ is the direct sum of $IH^\bullet(Y)$ and a space of dimension $\sum_{y \in Y_{\mathrm{sing}}} m(y)$ in degree two. The weight filtration of $H^1(F) = \mathcal{H}^{-1}(\delta_Y)_y$ induces an exact sequence

$$0 \to \mathrm{Gr}_0^W(H^1(F)) \to \mathcal{H}^{-1}(\delta_Y)_y \to H^1(\tilde{F}) \to 0 \,,$$

where $\tilde{F} = \coprod_{i \in I} F_i$ is the normalization of F ([D], (10.3.6)). Let Σ be the set of intersection points of the irreducible components F_i. For $y \in \Sigma$ let $e(y)$ be the number of components through y. Since F is connected (Zariski's main theorem) the dimension of $\mathrm{Gr}_0^W(H^1(F))$ is $1 - \#I - \#\Sigma + \sum_{y \in \Sigma} e(y)$. This number vanishes (purity) if and only if the components of F_i intersect in the form of a tree.

Minimal models. The support of the irreducible perverse sheaf A on X is an irreducible closed subvariety Y of X of dimension two. Suppose Y is an arbitrary irreducible closed subvariety of dimension two of an abelian variety X over k, and $Z \to Y$ a desingularization. Then Z is neither rational nor ruled, since any morphism from $\mathbb{P}^1$ to X is trivial. Therefore every relatively minimal model of Y is minimal by a result of Zariski, hence Y has a unique minimal model (see [Z], or [Sh], p.132ff). Let $\pi : Z \to Y$ be the minimal model. If $Y = -Y$, then the automorphism $\sigma(x) = -x$ of X induces an isomorphism $\sigma : Y \cong Y$ and uniquely extends to an isomorphism $\sigma : Z \to Z$ of the minimal model.

Example 1. We let $Y = C - C$ in $X = \text{Jac}(C)$ for C not hyperelliptic. Put $Z = C \times C$ and $\pi(x, y) = x - y$. Then Y is not regular, and $\pi : Z \to Y$ is the minimal model. If Y were regular, then $A = \delta_Y$ would be $\overline{\mathbb{Q}}_{l,Y}[2]$ contradicting $\mathcal{H}^{-1}(A) \cong H^1(C) \otimes \delta_{\{\kappa\}}$, as shown in the proof of the theorem. The morphism π contracts the diagonal of $C \times C$ to a point, and is an isomorphism outside the diagonal by Clifford's theorem. Hence Y has an isolated singularity at the origin. The singular fiber $F = \pi^{-1}(0)$ is irreducible and isomorphic to C. Thus Z is a relatively minimal model, hence minimal by Zariski's results.

Example 2. We let $Y = C - C$ in $X = \text{Jac}(C)$ for hyperelliptic C. Then $-C = C - e$ for some point e defined by the hyperelliptic involution $\theta : C \to C$. Hence $Y = C + C - e \cong C + C$. Then the map $a : C \times C \to X$ defined by $a(x, y) = x + y$ is branched double covering outside the exceptional divisor $a^{-1}(e) \cong C$, which is the image of $\text{id} \times \theta : C \to C \times C$. The corresponding covering automorphism on $C \times C$ for a is defined by $\sigma(x, y) = (y, x)$

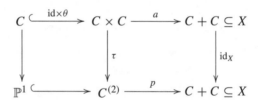

and induces $\theta : C \to C$ on the left side of the diagram. The morphism p is a birational map, and blows down the fiber $p^{-1}(e) = \mathbb{P}^1$. Hence

$$Rp_*(\delta_{C^{(2)}}) = \delta_{e+C-C} \oplus \delta_{\{e\}} = \delta_{C+C} \oplus \delta_{\{e\}} = \delta_{W_2} \oplus \delta_{\{e\}} \ .$$

On the other hand $Ra_*(\delta_{C^2}) = \delta_2 \oplus \delta_{1,1}$ by the decomposition into σ-equivariant perverse sheaves with $\delta_2 = Rp_*(\delta_{C^{(2)}})$. A comparison implies that $\delta_C * \delta_C$ decomposes into the direct sum $\delta_{W_2} \oplus \delta_{\{e\}} \oplus A$ of three irreducible perverse sheaves, where $A = \delta_{1,1}$ is the intersection cohomology sheaf on $C^{(2)}$ attached to the rank one etale sheaf $E = \mathcal{F}_{1,1}$ defined by the double covering τ. Then E ramifies at the τ-image of the diagonal in $C \times C$. This diagonal intersects $C \cong \tau^{-1}(\mathbb{P}^1)$ in the fixed points of $\sigma = \theta$. Hence the restriction of $E|_{\mathbb{P}^1}$ of E to the exceptional divisor of p corresponds to the etale hyperelliptic covering of $\mathbb{P}^1$ minus the $2g + 2$ image points of the fixed points on $\tau^{-1}(\mathbb{P}^1)$. We claim, that $C + C$ is not regular for $g \geq 3$. Hence $p : C^{(2)} \to C + C$

must be the minimal model. The criterion of Castelnuovo gives a necessary and sufficient criterion for the image $C + C = p(C^{(2)})$ to be a regular variety, namely the exceptional divisor $F = \mathbb{P}^1$ of p has to be of first kind, i.e. F must have self intersection $(F, F) = -1$. To compute the self-intersection of the covering divisor $D = (i \times \theta)(C)$, observe that $\pi(D) = 1 + \frac{1}{2}(K + D, D)$ is equal to the genus $g(D) = g$ (adjunction formula), since D is smooth. Hence $(D, D) = 2 - 2g$, because $(K, D) = \deg(K_C + \theta^*(K_C)) = 4g - 4$. Then $D = \tau^*(F)$ and $\tau_*(D) = \deg(\tau) \cdot F = 2F$ imply $(D, D) = \tau_*(D, \tau^*(F)) = (\tau_*(D), F) = 2(F, F)$, thus $(F, F) = 1 - g$. So $(F, F) = -1$ only holds for $g = 2$, in which case $C + C = X$. See also [Mu] p. 52ff for the case of low dimensions.

Resume. The discussion above shows how C can be reconstructed from the 'adjoint' perverse sheaf A by considering the minimal resolution of its support Y.

We finally come to the

Proof of the theorem in the hyperelliptic case. We may assume $g \geq 3$. Recall $C = e - C$. For simplicity assume $e = 0$ by a suitable translation of the curve C. Then $W_r = -W_r$. The following result is an immediate consequence of the Theorem of Martens (see [W], lemma 21)

Lemma 1. *The equality* $\delta_{W_r} = \overline{\mathbb{Q}}_{l, W_r}[r]$ *holds for* $1 \leq r \leq g - 1$ *in the hyperelliptic case.*

For $1 \leq r \leq g - 1$ consider now the proper morphisms

$$f_r : W_r \times W_r \to X$$

defined by $f_r(x, y) = x - y$, and

$$g_r : C \times C \times W_{r-1} \longrightarrow W_r \times W_r$$

defined by $(x, y, D) \mapsto (x + D, y + D) \in W_r \times W_r$. Notice $C + W_{r-1} = W_r$. The composition $f_r \circ g_r$ is a surjection onto $C - C$.

By the lemma and $W_r = -W_r$ we have

$$\delta_{W_r} * \delta_{W_r} = \delta_{W_r} * \delta_{-W_r} = Rf_{r*}(\delta_{W_r} \boxtimes \delta_{W_r}) = Rf_{r*}(\overline{\mathbb{Q}}_{l, W_r \times W_r})[2r] \,,$$

hence $\mathcal{H}^{-2}(\delta_{W_r} * \delta_{W_r}) = R^{2r-2} f_{r*}(\overline{\mathbb{Q}}_{l, W_r \times W_r})$.

Proposition 1. *For r as above the product $\delta_{W_r} * \delta_{W_r}$ is a direct sum of a semi-simple perverse sheaf K on X and a sum of translates of constant perverse sheaves, such that*

$$\mathcal{H}^i(\delta_{W_r} * \delta_{W_r}) = \mathcal{H}^i(K) \quad \text{for} \quad i > -2 \,.$$

Taking this for granted the perversity condition implies, that the support of $\mathcal{H}^{-2}(K)$ has dimension ≤ 2. Now K is a finite direct sum of simple intersection cohomology sheaves $K_\nu = IC(Y_\nu, E_\nu)$, where Y_ν are irreducible closed subvarieties Y_ν of X and the E_ν are smooth irreducible coefficient systems on open dense subsets $U_\nu \subseteq Y_\nu$. Let L be the direct sum of all K_ν, for which $Y_\nu = C - C$. Then $K = L \oplus L'$. By the IC-condition for the perverse sheaves K_ν in L' there exists an an open dense subset $U \subseteq C - C$, such that $L'|_U = 0$. In particular $\mathcal{H}^{-2}(K)|_U = \mathcal{H}^{-2}(L)|_U$. The strategy now is to construct A as a summand of L by perverse analytic continuation from $L[-2]|_U = \mathcal{H}^{-2}(\delta_{W_r} * \delta_{W_r})\big|_U$. For this we use

Proposition 2. *Suppose* $1 \leq r \leq g - 1$. *All fibers of* f_r *have dimension* $\leq r$. *There exists an open dense subset* $U \subseteq C - C$, *such that the fibers of* f_r *over* U *have dimension* $r - 1$. *In particular* $\dim(f_r^{-1}(C - C)) = r + 1$, *and* $g_r(C \times C \times W_{r-1})$ *is an irreducible component of* $f_r^{-1}(C - C)$ *of highest dimension.*

From Proposition 2, for U small enough, the usual excision and trace morphism arguments show, that the higher direct image

$$R^{2r-2} f_{r*}(\overline{\mathbb{Q}}_{l, W_r \times W_r})\big|_U$$

contains the following subsheaf as a direct summand

$$R^{2r-2}(f_r \circ g_r)_*(\overline{\mathbb{Q}}_l)\big|_U \subseteq R^{2r-2} f_{r*}(\overline{\mathbb{Q}}_{l, W_r \times W_r})\big|_U = L[-2]|_U .$$

Notice, that the image of g_r is an irreducible component of dimension $r + 1$ in $f_r^{-1}(C - C)$ of relative dimension $r - 1$ over U. The subsheaf so defined is easily computed:

The morphism $(f_r \circ g_r)(x, y, D) = x - y = f_1(x, y)$ is the cartesian product $f_1 \times s : (C \times C) \times W_{r-1} \to (C - C) \times \mathrm{Spec}(k) = C - C$, where $s : W_{r-1} \to \mathrm{Spec}(k)$ denotes the structure morphism. Notice $R^i f_{1*}(\overline{\mathbb{Q}}_l)\big|_U = 0$ for $i \geq 1$ over a suitably small open dense $U \subseteq C - C$, since then f_1 is finite. Hence the Künneth formula for the proper morphisms $f_1 \times s$ implies

$$R^{2r-2}(f_r \circ g_r)_*(\overline{\mathbb{Q}}_l)|_U = f_{1*}(\overline{\mathbb{Q}}_l)\big|_U \otimes R^{2r-2} s_*(\overline{\mathbb{Q}}_l)\big|_U .$$

Since W_{r-1} is irreducible of dimension $r - 1$, ignoring Tate twists the trace map gives an isomorphism $R^{2r-2} s_*(\overline{\mathbb{Q}}_l)\big|_U \cong \overline{\mathbb{Q}}_{l,U}$, hence an inclusion

$$f_{1*}(\overline{\mathbb{Q}}_l)|_U \subseteq L[-2]|_U = \mathcal{H}^{-2}(\delta_{W_r} * \delta_{W_r})\big|_U .$$

Notably for $r = 1$ this inclusion is an isomorphism $f_{1*}(\overline{\mathbb{Q}}_l)|U = \overline{\mathbb{Q}}_{l,U} \oplus E|U$ for $E|_U = \mathcal{F}_{1,1}|_U$ as seen earlier in the second example above. Hence

$\mathbb{Q}_{l,U} \oplus E|_U \cong \mathcal{H}^{-2}(\delta_1 * \delta_1)|_U$ is a summand of $\mathcal{H}^{-2}(\delta_{W_r} * \delta_{W_r})|_U$. By Proposition 1 and perverse continuation from U to $C - C$ the non-constant summand $E[2]|U$ extends to the perverse constituent $A = \delta_E = \delta_{1,1}$ defined in the second example

$$^p H^0(\delta_{W_r} * \delta_{W_r}) = A \oplus \cdots .$$

For $r = g - 1$ then obviously A has the property $\mathcal{H}^{-1}(A) = H^1(C) \otimes \delta_{\{\kappa\}}$ using $\kappa = (g - 1)e$, which proves the theorem in the hyperelliptic case. $\quad\square$

Proof of Proposition 2. Since $f_r : W_r \times W_r \to X$ is proper, for closed points x of X the proper base change theorem implies that $\dim(f_r^{-1}(x)) > m$ if and only if there exists an $n > 2m$ such that $R^n f_{r*}(\overline{\mathbb{Q}_l})_x \neq 0$. Recall that $Rf_{r*}(\overline{\mathbb{Q}_l})[2r] = \delta_{W_r} * \delta_{W_r}$, as we have seen above. Therefore it is enough to show $\mathcal{H}^k(\delta_{W_r} * \delta_{W_r})_x = 0$ for $k > 0$ and all $x \in X$, respectively for $k > -2$ and all $x \in U$ in a suitable open dense subset $U \subseteq C-C$. However this is an immediate consequence of proposition 1 and the perversity condition for K. $\quad\square$

Concerning Proposition 1. Notice $\delta_{W_r} * \delta_{W_r} \hookrightarrow \delta_r * \delta_r$ by Martens' theorem (see [W], lemma 21). Recall

$$\delta_r * \delta_r = \bigoplus_\alpha \delta_\alpha$$

for certain $\alpha = (\alpha_1, \alpha_2) \in \mathbb{N}^2$ with $\deg(\alpha) = 2r$ and $\alpha_1 \geq \alpha_2$. Now use

Theorem 2. *Each sheaf δ_α is a direct sum of a perverse sheaf $^P\delta_\alpha$ and a sum of translates of constant perverse sheaves $T_\alpha = \bigoplus_{v \in \mathbb{Z}} m(\alpha, v) \cdot \delta_X[v]$, i.e.,*

$$\delta_\alpha = {}^P\delta_\alpha \oplus T_\alpha,$$

such that $m(\alpha, v) = 0$ holds for $|v| > \alpha_1 - g$.

To prove Proposition 1 by this theorem it suffices to observe, that $\deg(\alpha) = 2r$ implies $\alpha_1 \leq 2r$ or $\alpha_1 - g \leq 2r - g \leq r - 1$. This in turn implies $\alpha_1 - g \leq g - 2$ (with equality only for $r = g - 1$ and $\alpha_1 = 2(g - 1)$, i.e. $\delta_\alpha = \delta_{2g-2}$). Therefore only translates $\delta_X[v]$ for $|v| \leq g - 2$ occur. Since $\mathcal{H}^k(\delta_X[v]) = \mathcal{H}^{k+v+g}(\overline{\mathbb{Q}_{l,X}}) = 0$ for $k \neq -v - g$, this proves $\mathcal{H}^k(\bigoplus T_\alpha) = 0$ for $k > -2$. $\quad\square$

The theorem above extends to a larger class of complexes δ_α, which are indexed by arbitrary partitions α and which includes those above as special cases. For a proof of the last theorem we refer to [W] and [W2]. One approach uses the curve lemma [W], section 3.2, which states that curves define multipliers in the sense of [W2]. The proof relies on the Čebotarev density theorem. Another approach (see in [W], cor. 15 and cor. 16 of loc. cit) uses iterated convolutions and a Riemann-Roch type duality theorem ([W], section 5.5). The crucial point here is, that the t-degree of the classes of δ_α in $K_*(X)$ is bounded by a constant independent from α. For the fundamental sheaf complexes $\varepsilon_1, \varepsilon_2, \cdots$ (see [W], section 5.5) this follows from the affine vanishing theorem of Artin-Grothendieck ([W], lemma 27). Notice, in the Tannakian category $\mathcal{BN}$ the ε_i correspond to the symmetric powers $S^i(st)$ of the standard representation. To obtain a bound in the general case one uses elementary invariant theory, which implies that up to translates of constant sheaves any δ_α is a direct summand of a convolution product of fundamental sheaf complexes $\varepsilon_1, \varepsilon_2, \cdots$ (see [W], section 5.5), where the number of factors a priori can be bounded by $2g - 2$.

Bibliography

[D]   Deligne P., *Theorie de Hodge III*, Publ. Math. IHES 44 (1974), 5–77.

[M]   Martens H.H, *On the variety of special divisors on a curve*, Journal reine u. angew. Math. 227 (1967), 111–120.

[Mu]   Mumford D., *Curves and their Jacobians*, The university of Michigan Press (1975).

[Sh]   Shafarevich I.R., *Lectures on minimal models and birational transformations of two dimensional schemes*, Tata Institute (1966).

[W]   Weissauer R., *Brill-Noether sheaves*, arXiv math.AG/0610923.

[W2]   Weissauer R., *Tannakian categories attached to abelian varieties*, this volume, pp. 267–274.

[Z]   Zariski O., *Introduction to the problem of minimal models in the theory of algebraic surfaces*, *The problem of minimal models in the theory of algebraic surfaces*, *On Castelnuovo's criterion of rationality $p_a = P_2 = 0$ of an algebraic surface*, Collected papers, M.I.T. Press, Cambridge.

Existence of Whittaker models related to four dimensional symplectic Galois representations

Rainer Weissauer

Let $\mathbb{A} = \mathbb{A}_{\text{fin}} \otimes \mathbb{R}$ be the ring of rational adeles and GSp(4) be the group of symplectic similitudes in four variables. Suppose that $\Pi \cong \Pi_{\text{fin}} \otimes \Pi_\infty$ is a cuspidal irreducible automorphic representation of the group GSp(4, $\mathbb{A}$), where Π_∞ belongs to a discrete series representation of the group GSp(4, $\mathbb{R}$). The discrete series representations of the group GSp(4, $\mathbb{R}$) are grouped into local L-packets [W1, W2], which have cardinality two and consist of the class of a holomorphic and the class of a nonholomorphic discrete series representation. Two irreducible automorphic representations $\Pi = \otimes_v \Pi_v$ and $\Pi' = \otimes_v \Pi'_v$ are said to be weakly equivalent if $\Pi_v \cong \Pi'_v$ holds for almost all places v. An irreducible cuspidal representation Π is called a CAP representation if it is weakly equivalent to automorphic representation Π', which is induced from an automorphic representation of a proper Levi subgroup. An automorphic representation Π for GSp(4) is called a *weak endoscopic lift* if there exist two irreducible cuspidal automorphic representations τ_1, τ_2 of the group GL(2, $\mathbb{A}$) with central characters $\omega_{\tau_1} = \omega_{\tau_2}$, such that the degree 4 standard L-series of Π coincides with the product of the two L-series of τ_1 and τ_2 for almost all places.

The aim of this article is to prove the following

Theorem 1. *Let Π be a cuspidal irreducible automorphic representation of the group* GSp(4, $\mathbb{A}$). *Suppose Π is not CAP and suppose Π_∞ belongs to the discrete series of the group* GSp(4, $\mathbb{R}$). *Then Π is weakly equivalent to an irreducible globally generic cuspidal automorphic representation Π_{gen} of the group* GSp(4, $\mathbb{A}$), *whose archimedean component $\Pi_{\text{gen},\infty}$ is the nonholomorphic discrete series representation contained in the local archimedean L-packet of Π_∞.*

One can attach to an automorphic representation Π as in Theorem 1 an integer w, which is uniquely determined by Π_∞, and an associated Galois

285

representation $\rho_{\Pi,\lambda}$ by [W2]. The reference [W2] and Theorem 1 imply, that this Galois representation $\rho_{\Pi,\lambda}$ is symplectic in the following sense

Theorem 2. *Suppose* Π *is as in Theorem 1. Then the associated Galois representation* $\rho_{\Pi,\lambda}$ *preserves a nondegenerate symplectic* $\overline{\mathbb{Q}}_l$*-bilinear form* $\langle \cdot, \cdot \rangle$ *such that the Galois group acts with the multiplier* $\omega_\Pi \mu_l^{-w}$

$$\langle \rho_{\Pi,\lambda}(g)v, \rho_{\Pi,\lambda}(g)w \rangle \;=\; \omega_\Pi(g)\mu_l^{-w}(g) \cdot \langle v, w \rangle, \qquad g \in \mathrm{Gal}(\overline{\mathbb{Q}}/\mathbb{Q})$$

where μ_l *is the cyclotomic character.*

A variant of this construction also yields certain orthogonal Galois representations of dimension 4. See the remark at the end of this article.

For an irreducible cuspidal automorphic representation Π of the group $\mathrm{GSp}(4, \mathbb{A})$, which is not CAP and whose archimedean component belongs to the discrete series, we want to show that Π is weakly equivalent to a globally generic representation Π_{gen}, whose archimedean component again belongs to the discrete series. Π not being CAP implies, that Π_{gen} again is cuspidal. Hence [W2], Theorem III now also holds unconditionally, since by [S3] the multiplicity one theorem is known for the generic representation Π_{gen}.

A careful analysis of the proof shows, that the arguments imply more. For this we refer to the forthcoming work of U. Weselmann [Wes].

Proof of Theorem 1. The proof will be based on the hypotheses A,B of [W2] proved in [W1], and Theorem 3 and Theorem 4, which will be formulated further below during the proof of Theorem 1. Theorem 3 is a consequence of results of [GRS]. Theorem 4 is proved in [Wes]. For the following it is important, that under the assumptions made in Theorem 1 Ramanujan's conjecture holds for the representation Π at almost all places, as explained in [W2] Section 1. □

Restriction to $\mathrm{Sp}(4)$. The restriction of Π to $\mathrm{Sp}(4, \mathbb{A})$ contains an irreducible constituent, say $\tilde{\Pi}$. In the notation of [W2], Section 3, consider the degree five standard L-series

$$\zeta^S(\Pi, \chi, s)$$

of $\tilde{\Pi}$ for $\mathrm{Sp}(4)$. For our purposes it suffices to consider this L-series for the primes outside a sufficiently large finite set of places S containing all ramified places. So this partial L-series depends only on Π, and does not depend on the particular chosen $\tilde{\Pi}$. It may also be replaced by any representation weakly equivalent to it.

Euler characteristics. The representation Π is cohomological in the sense of [W2, W1], i.e. Π occurs in the cohomology of the Shimura variety M of principally polarized abelian varieties of genus two for a suitable chosen $\overline{\mathbb{Q}}_l$-adic coefficient system $\mathcal{V}_\mu(\overline{\mathbb{Q}}_l)$. Since Π is of cohomological type and since we excluded CAP-representations Π, the representations Π_{fin} and also Π^S (i.e. $\Pi^S = \otimes_{v \notin S} \Pi_v$ outside a finite set S of bad primes) only contribute to the cohomology $H^i(M, \mathcal{V}_\mu(\overline{\mathbb{Q}}_l))$ for the coefficient system $\mathcal{V}_\mu$ [W2] in the middle degree $i = 3$ and not for the other degrees (see [W2] Hypothesis B(1),(2) and [W1]). This cohomological property is inherited to the subgroup Sp(4, $\mathbb{A}$) by restriction and induction using the following easy observation: Suppose given two irreducible automorphic representations Π_1, Π_2 of GSp(4, $\mathbb{A}$) having a common irreducible constituent after restriction to Sp(4, $\mathbb{A}$). Then, if Π_1 is cuspidal but not CAP, also Π_2 is cuspidal and not CAP.

Therefore, if we consider the generalized Π^S-isotypic subspaces for either of the groups GSp(4, $\mathbb{A}^S$) or Sp(4, $\mathbb{A}^S$)) in the middle cohomology group of degree 3, we may as well replace the middle cohomology group by the virtual representation $H^\bullet(M, \mathcal{V}_\mu(\overline{\mathbb{Q}}_l))$. Up to a minus sign this does not change the traces of Hecke operators, which are later considered in Theorem 4. Here S may be any finite set containing the archimedean place.

First temporary assumption. For the moment suppose that $\tilde{\Pi}$ admits a weak lift $\tilde{\Pi}'$ to an irreducible automorphic representation of the group PGL(5, $\mathbb{A}$) in the sense below (we later show using Theorem 4, that this weak lift always exists). A representation $\tilde{\Pi}'$ of PGL(5, $\mathbb{A}$) can be considered to be a representation of GL(5, $\mathbb{A}$) with trivial central character. In this sense the lifting property just means, that there exists an irreducible automorphic representation $\tilde{\Pi}'$ of PGL(5, $\mathbb{A}$), for which

$$L^S(\tilde{\Pi}' \otimes \chi, s) = \zeta^S(\Pi, \chi, s)$$

holds for the standard L-series of GL(5) and all idele class characters χ and certain sufficiently large finite sets $S = S(\chi, \tilde{\Pi}')$ of exceptional places.

Considered as an irreducible automorphic representation of GL(5, $\mathbb{A}$) the representation $\tilde{\Pi}'$ need not be cuspidal (this non-cuspidal case will be useful in the additional remark on orthogonal representations made after the proof of Theorem 1). There exist irreducible cuspidal automorphic representations σ_i of groups GL(n_i, $\mathbb{A}$) where $\sum_i n_i = 5$, such that $\tilde{\Pi}'$ is a constituent of the representation induced from a parabolic subgroup with Levi subgroup $\prod_i$ GL(n_i, $\mathbb{A}$). See [L], Prop. 2. Each of the cuspidal representations σ_i can be written in the form $\sigma_i = \chi_i \otimes \sigma_i^0$ for some unitary cuspidal representations σ_i^0 and certain one dimensional characters χ_i.

Let ω_{σ_i} denote the central character of σ_i. The identity

$$L^S(\tilde{\Pi}', s) = \zeta^S(\Pi, 1, s)$$

implies that the characters $\omega_{\sigma_i} = \chi_i^{n_i} \omega_{\sigma_i^0}$ are unitary. In fact, since Π^S satisfies the Ramanujan conjecture by [W2], they have absolute value one at all places outside S. Therefore by the approximation theorem all χ_i are unitary. Hence the σ_i itself must have been cuspidal unitary representations.

Now, since the σ_i are cuspidal unitary, the well known theorems of Jaquet-Shalika and Shahidi on L-series for the general linear group [JS] imply the non-vanishing $L^S(\tilde{\Pi}' \otimes \chi, 1) = \prod_i L^S(\sigma_i \otimes \chi, 1) \neq 0$ for arbitrary unitary idele class characters χ. By the temporary assumption, that $\tilde{\Pi}'$ is a lift of $\tilde{\Pi}$, this implies

$$\zeta^S(\Pi, \chi, 1) \neq 0$$

for all unitary idele class characters χ.

Second temporary assumption. Now in addition we suppose, that Π can be weakly lifted to an irreducible automorphic representation (Π', ω) of the group $GL(4, \mathbb{A}) \times \mathbb{A}^*$. By this we mean, that there exists an irreducible automorphic representation Π' of $GL(4, \mathbb{A})$ and an idele class character ω, such that for the central characters of Π and Π' we have

$$\omega_{\Pi'} = \omega^2 \quad \text{and} \quad \omega_\Pi = \omega,$$

and that furthermore

$$L^S(\Pi \otimes \chi, s) = L^S(\Pi' \otimes \chi, s)$$

holds for sufficiently large finite sets of places S containing all ramified places. Here, following the notation of [W2] section three, $L^S(\Pi, s)$ denotes the standard degree four L-series of Π.

These conditions imposed at almost all unramified places of course completely determines the automorphic representation Π' by the strong multiplicity one theorem for $GL(n)$. In particular this implies, that the global lift $\Pi \mapsto (\Pi', \omega)$ commutes with character twists $\Pi \otimes \chi \mapsto (\Pi' \otimes \chi, \omega\chi^2)$. Furthermore it implies $(\Pi')^\vee \cong \Pi' \otimes \omega^{-1}$. In particular this finally holds locally at all places including the archimedean place.

The archimedean place. From the last observation we obtain $\Pi'_\infty \otimes \text{sign} \cong \Pi'_\infty$ from the corresponding $\Pi_\infty \otimes \text{sign} \cong \Pi_\infty$, which is known for discrete series representations Π_∞ of $GSp(4, \mathbb{R})$. Hence it is easy to see, that Π'_∞ respectively Π_∞ are determined by their central characters and their restriction to SL(4, $\mathbb{R}$) respectively Sp(4, $\mathbb{R}$). In fact we will later show, that the lift Π' can be assumed to have a certain explicitly prescribed behavior at the archimedean place. What this means will become clear later.

Properties of Π'. Again the irreducible automorphic representation Π' of GL(4, $\mathbb{A}$) need not be cuspidal. Π' is a constituent of a representation induced from some parabolic subgroup with Levi subgroup $\prod_j$ GL(m_j) for $\sum_j m_j = 4$, with respect to some irreducible cuspidal representations τ_j of GL(m_j, $\mathbb{A}$).

The same argument, as already used for the first temporary assumption, implies the τ_j to be unitary cuspidal. This excludes $m_j = 1$ for some j, since otherwise this would force the existence of a pole of $L^S(\Pi \otimes \chi, s)$ for $\chi = \tau_j^{-1}$ at $s = 1$ again by the results [JS] of Jaquet-Shalika and Shahidi on the analytic behavior of L-series for GL(n) on the line $Re(s) = 1$. Notice, a pole at $s = 1$ would imply Π to be a CAP representation (see [P]). This however contradicts the assumptions of Theorem 1, by which Π is not a CAP representation.

Let us return to Π. Either Π' is cuspidal; or Π' comes by induction from a pair (τ_1, τ_2) of irreducible cuspidal representations τ_i of GL(2, $\mathbb{A}$), for which $\omega_{\tau_1} \omega_{\tau_2} = \omega^2$ holds. In this second case we do have

$$L^S(\Pi', s) = L^S(\tau_1, s) L^S(\tau_2, s) .$$

By $L^S(\Pi, s) = L^S(\Pi', s)$ therefore Π is a weak endoscopic lift, provided the central characters $\omega_{\tau_1} = \omega_{\tau_2}$ coincide. For weak endoscopic lifts Theorem 1 obviously holds. See [W2], Hypothesis A part (2) and (6) combined with [W1]. To complete the discussion of this case, it therefore would be enough to establish the identity $\omega_{\tau_1} = \omega_{\tau_2}$. To achieve this we need some further argument, and therefore we make a digression on theta lifts first.

The theta lift. The group GL(1) acts on GL(4) $\times$ GL(1) such that t maps (h, x) to (ht^{-1}, xt^2). By our temporary assumptions made, the central character of Π' is completely determined so that the representation (Π', ω) of GL(4) $\times$ GL(1) descends to a representation on the quotient group $G(\mathbb{A}) = $ (GL(4, $\mathbb{A}$) $\times$ GL(1, $\mathbb{A}$))/GL(1, $\mathbb{A}$). This quotient group $G(\mathbb{A})$ is isomorphic to the special orthogonal group of similitudes GSO(3, 3)($\mathbb{A}$) attached to the split 6-dimensional Grassmann space $\Lambda^2(\mathbb{Q}^4)$ with the underlying quadratic form given by the cup-product.

The generalized theta correspondence of the pair

$$(\text{GSp}(4), \text{GO}(3, 3))$$

preserves central characters. If we apply the corresponding theta lift to the representation Π of GSp(4, 𝔸), then according to [AG] p.40 the theta lift of Π to GO(3, 3)(𝔸) is nontrivial if and only if Π is a globally generic representation. In this case it is easy to see, that the lift is globally generic. For the converse we need the following result announced by Jaquet, Piateskii-Shapiro and Shalika [JPS]. See also [AG] and [S2], [PSS].

Theorem 3. *An irreducible cuspidal automorphic representation* Π' *of* GL(4, 𝔸) *lifts nontrivially to* GSp(4, 𝔸) *under the generalized theta correspondence of the pair* (GSp(4), GO(3, 3)) *if and only if the alternating square degree six L-series* $L(\Pi', \chi, s, \Lambda^2)$ *or equivalently some partial L-series*

$$L^S(\Pi', \chi, s, \Lambda^2)$$

(for a suitably large finite set of places S containing all bad places) has a pole at $s = 1$ *for some unitary idele class character* χ. *In this case the lift of* Π' *is globally generic, and also* Π *is generic.*

Remark on the degree-6 L-series. To apply this it is enough to observe, that under our second temporary assumptions we have enough control on $L^S(\Pi', \chi, s, \Lambda^2)$ to apply Theorem 3 for the representation Π' of GL(4, 𝔸), attached to the representation Π of GSp(4, 𝔸) subject to our second temporary assumption. Indeed by an elementary computation the second temporary assumption implies the identity

$$L^S(\Pi', \chi, s, \Lambda^2) = L^S(\omega\chi, s)\zeta^S(\Pi, \omega\chi, s),$$

where $\omega = \omega_\Pi$ denotes the central character of Π. Now compare the term $L^S(\Pi', \chi, s, \Lambda^2)$ with the standard L-series of the special orthogonal group SO(3, 3), which is used in the proof of Theorem 3:

We have exact sequences

$$1 \longrightarrow SO(3, 3) \longrightarrow GSO(3, 3) \overset{\lambda}{\longrightarrow} GL(1) \longrightarrow 1 \ ,$$

$$1 \longrightarrow GL(1) \overset{i}{\longrightarrow} GL(4) \times GL(1) \longrightarrow GSO(3, 3) \longrightarrow 1 \ ,$$

where λ is the similitude homomorphism and where $i(t) = (t^{-1} \cdot \text{id}, t^2)$. Hence for the Langlands dual groups, which at the unramified places describe the restriction of spherical representations of GSO(3, 3)(F) to SO(3, 3)(F), we get

$$1 \longrightarrow \mathbb{C}^* \overset{\hat{\lambda}}{\longrightarrow} \widehat{GSO(3, 3)} \longrightarrow \widehat{SO(3, 3)} \longrightarrow 1 \ .$$

Notice $\widehat{GSO(3,3)} \subseteq \hat{G}L(4) \times \hat{G}L(1)$. The 6-dimensional complex representation of $GL(4, \mathbb{C}) \times GL(1, \mathbb{C})$ on $\Lambda^2(\mathbb{C}^4)$ defined by

$$(A, t) \cdot X = t^{-1} \cdot \Lambda^2(A)(X)$$

is trivial on the subgroup $\hat{\lambda}(\mathbb{C}^*)$, hence defines a 6-dimensional representation of the L-group $\widehat{SO(3,3)}$. The L-series of this representation defines the degree 6 standard L-series $L^S(\sigma, s)$ of an irreducible automorphic representation σ of the group $SO(3, 3)(\mathbb{A})$. Apparently, for σ spherical outside S in the restriction of (Π', χ) so that $\omega_{\Pi'} = \chi^2$, we therefore get

$$L^S(\sigma, s) = L^S(\Pi', \chi^{-1}, s, \Lambda^2) .$$

Proof of Theorem 3. Using the remark on the degree six L-series we now can invoke [GRS], Theorem 3.4 to deduce Theorem 3. The condition on genericity made in loc. cit. automatically holds for an cuspidal irreducible automorphic representation σ of $SO(n, n)(\mathbb{A})$ for $n = 3$, since this conditions is true for $GL(4, \mathbb{A})$. Therefore by [GRS] a pole of $L^S(\sigma, s)$ at $s = 1$ implies, that σ has a nontrivial, cuspidal generic theta lift to the group $Sp(2n - 2)(\mathbb{A}) = Sp(4, \mathbb{A})$. This in turn easily implies the same for the extended theta lift from $GSO(3, 3)(\mathbb{A})$ to $GSp(4, \mathbb{A})$. This proves Theorem 3. $\qquad\square$

Additional remark. If in addition the spherical representation Π'^S is a constituent of an induced representation attached to a pair of unitary cuspidal irreducible automorphic representations τ_1, τ_2 of $GL(2, \mathbb{A})$, then furthermore

$$L^S(\Pi', \chi, s, \Lambda^2) = L^S(\tau_1 \times (\tau_2 \otimes \chi), s)L^S(\omega_{\tau_1}\chi, s)L^S(\omega_{\tau_2}\chi, s) .$$

This, as well as $\omega^2 = \omega_{\Pi'} = \omega_{\tau_1}\omega_{\tau_2}$, are rather obvious. But by the wellknown analytic properties of L-series [JS] it implies, that the right side now has poles for $\chi = \omega_{\tau_i}^{-1}$ and $i = 1, 2$. Notice the τ_i are unitary cuspidal.

This being said we now complete the discussion of the case, where Π' is not cuspidal.

Reduction to the case Π' cuspidal. If Π' is not cuspidal, then as already shown Π' is obtained from a pair of irreducible unitary cuspidal representations (τ_1, τ_2) by induction. To cover Theorem 1 in this case we already remarked, that it suffices to show $\omega_{\tau_1} = \omega_{\tau_2}$. If these two characters were different, then $\chi_i = \omega\omega_{\tau_i}^{-1} \neq 1$ for $i = 1$ or $i = 2$. Therefore $L^S(\Pi', \chi, s, \Lambda^2)$ would have a pole at $s = 1$ for $\chi = \omega_{\tau_i}^{-1}$ by the previous 'additional remark'. Hence the identity $L^S(\Pi', \omega_{\tau_i}^{-1}, s, \Lambda^2) = L^S(\chi_i, s)\zeta^S(\Pi, \chi_i, s)$ would imply the existence of poles at $s = 1$ for the L-series $\zeta^S(\Pi, \chi_i, s)$. Theorem 4.2 of [W2], Section 4 then would imply $(\chi_i)^2 = 1$. Since $\chi_1\chi_2 = 1$ holds by definition of Π', therefore $\chi_1 = \chi_2$. Thus $\omega_{\tau_1} = \omega_{\tau_2}$. So we are in the case already considered in [W2]: Π is a weak lift. In this case the statement of Theorem 1 follows from the multiplicity formula for weak endoscopic lifts [W2], hypothesis A (6). Thus we may suppose from now on, that Π' is cuspidal.

Applying Theorem 3. Both our temporary assumptions on the existence of the lifts $\tilde{\Pi}'$ and Π' imply, that from now on we can assume without restriction of generality, that Π' is a unitary cuspidal representation. Since we deduced $\zeta^S(\Pi, \chi, 1) \neq 0$ for all unitary characters χ from our first temporary assumption, the crucial identity $L^S(\Pi', \chi, s, \Lambda^2) = L^S(\omega\chi, s)\zeta^S(\Pi, \omega\chi, s)$ forces the existence of a pole for $L^S(\Pi', \omega^{-1}, s, \Lambda^2)$ at $s = 1$. Therefore we are in a situation where we can apply Theorem 3: Since Π' is cuspidal, the pair (Π', ω) defines a cuspidal irreducible automorphic representation of $GSO(3, 3)(\mathbb{A})$. It nontrivially gives a backward lift from $GSO(3, 3)$ to a globally generic automorphic representation Π_{gen} of $GSp(4, \mathbb{A})$ using Theorem 3. Comparing both lifts at the unramified places gives

$$L^S(\Pi, s) = L^S(\Pi', s) = L^S(\Pi_{\text{gen}}, s) .$$

The first equality holds by assumption. The second equality follows from the behavior of spherical representations under the Howe correspondence [R]. See also [PSS], p.416 for this particular case. Hence Π and the generic representation Π_{gen} are weakly equivalent.

In other words, using two temporary assumptions, we now have almost deduced Theorem 1. In fact the generic representation Π_{gen}, that has been constructed above, is weakly equivalent to Π. Hence it is cuspidal, since Π is not CAP. However, for the full statement of Theorem 1 one also needs control over the archimedean component of Π_{gen}. We postpone this archimedean considerations for the moment and rather explain first, how to establish the two temporary assumptions to hold unconditionally. This will be deduced from the topological trace formula.

Construction of the weak lifts Π' and $\tilde{\Pi}'$ of Π. The existence of these lifts will follow from a comparison of the twisted topological trace formula of a group (G, σ) with the ordinary topological trace formula for a group H for the pairs $H = GSp(4, \mathbb{A})$ and $(G, \sigma) = (GSO(3, 3)(\mathbb{A}), \sigma)$ respectively $\tilde{H} = Sp(4, \mathbb{A})$ and $\tilde{G} = (PGL(5, \mathbb{A}), \tilde{\sigma})$. Here σ respectively $\tilde{\sigma}$ denote automorphisms of order two of G resp. $\tilde{G}$. In both cases the group of fixed points in the center under the automorphism $\sigma, \tilde{\sigma}$ will be a Zariski connected group, a condition imposed in [W3]. Notice G is isomorphic to the quotient of $GL(4) \times GL(1)$

divided by the subgroup S of all central elements of the form $(z \cdot \mathrm{id}, z^{-2})$, hence is isomorphic to $\mathrm{GL}(4)/\{\pm 1\}$. Hence for a local field F

$$\mathrm{GSO}(3, 3)(F) \cong \left(\mathrm{GL}(4, F) \times F^*\right)/F^*,$$

where $t \in F^*$ acts on $(h, x) \in \mathrm{GL}(4, F) \times F^*$ via $(h, x) \mapsto (ht^{-1}, xt^2)$. The group $\mathrm{GSO}(3, 3)(F)$ can be realized to act on the six dimensional Grassmann space $\Lambda^2(F^4)$. See [AG], p. 39ff and [Wa], p. 44f. This identifies the quotient group $G(F)$ with the special orthogonal group of similitudes $\mathrm{GSO}(3, 3)(F)$ attached to the split 6-dimensional quadratic space $\Lambda^2(F^4)$ defined by the cup-product. The similitude character is $\lambda(h, x) = \det(h)x^2$. Let h' denote the transposed matrix of h. The automorphism σ of G is induced by the map $(h, x) \mapsto (\omega(h')^{-1}\omega^{-1}, \det(h)x)$ for a suitable matrix $\omega \in \mathrm{GL}(4, F)$ chosen in such a way, that σ stabilizes a fixed splitting [BWW].1.9. Then σ is the identity on the center $Z(G) \cong \mathrm{GL}(1)$ of G.

Let us start by introducing some notation.

Notations. Let G be a split reductive $\mathbb{Q}$-group with a fixed splitting $(B, T, \{x_\alpha\})$ over $\mathbb{Q}$ and center $Z(G)$. Let σ be a $\mathbb{Q}$-automorphism of G, which stabilizes the splitting, such that the group $Z(G)^\sigma$ of fixed points is connected for the Zariski topology. Let K_∞^+ be the topological connected component of a maximal compact subgroup of $G(\mathbb{R})$, similarly let Z_∞^+ be the connected component of the $\mathbb{R}$-valued points of the maximal $\mathbb{R}$-split subtorus of the center of G. Let $X_G = G(\mathbb{R})/K_\infty^+ Z_\infty^+$ be the associated symmetric domain as in [Wes] (1.18) and (5.22). Let V be an irreducible finite dimensional complex representation of G with highest weight $\chi \in X^*(T)$, which is invariant under σ. It defines a bundle $V_G = G(\mathbb{Q}) \setminus [G(\mathbb{A}_{\mathrm{fin}}) \times X_G \times V]$ over $M_G = G(\mathbb{Q}) \setminus [G(\mathbb{A}_{\mathrm{fin}}) \times X_G]$. See also [Wes] (3.4). Let $\mathcal{V}_\chi$ denote the associated sheaf. For the natural right action of $G(\mathbb{A}_{\mathrm{fin}})$ on M_G and V_G the cohomology groups $H^i(M_G, \mathcal{V}_\chi)$ become admissible $G(\mathbb{A}_{\mathrm{fin}})$-modules, on which σ acts. Let $H^\bullet(M_G, \mathcal{V}_\chi) = \sum_i (-1)^i H^i(M_G, \mathcal{V}_\chi)$ be the corresponding virtual modules. Then for $f \in C_c^\infty(G(\mathbb{A}_{\mathrm{fin}}))$ the traces

$$T(f, \sigma, G, \chi) = \mathrm{Trace}(f \cdot \sigma, H^\bullet(M_G, \mathcal{V}_\chi))$$

are well defined.

Let G_1 be the maximal or stable endoscopic group for $(G, \sigma, 1)$ in the sense of σ-twisted endoscopy (see [KS](2.1) and [Wes], Section 5), where we assume that the character $\omega = \omega_a$ is trivial. The corresponding endoscopic datum $(G_1, \mathcal{H}, s, \xi)$ has the property $\mathcal{H} = {}^L G_1$. The dual group $\hat{G}_1$ is the group of fixed points of $\hat{G}$ under $\hat{\sigma}$ and is again split, defining $\xi : {}^L G_1 \to {}^L G$.

Let T_1 be a maximal $\mathbb{Q}$-torus of G_1, then we can identify the character group $X^*(T_1)$ with the fixed group $X^*(T)^\sigma$. Hence χ defines a coefficient system V_{χ_1} on M_{G_1} and we can similarly define $T(f_1, id, G_1, \chi_1)$. Functions $f = \prod_{v \neq \infty} f_v \in C_c^\infty(G(\mathbb{A}_{\text{fin}}))$ and $f_1 = \prod_{v \neq \infty} f_{1,v} \in C_c^\infty(G_1(\mathbb{A}_{\text{fin}}))$ are called matching functions, if each of the local pairs are matching in the sense of [KS] (5.5.1) – up to z-extensions which can be ignored in our situation – so that in particular f_v and $f_{v,1}$ are characteristic functions of suitable hyperspecial maximal compact subgroups $K_v \subseteq G_v = G(\mathbb{Q}_v)$ resp. $K_{v,1} \subseteq G_{1,v} = G_1(\mathbb{Q}_v)$ for almost all $v \notin S$ (S a suitable finite set of places which may be chosen arbitrarily large). By simplicity here we tacitly neglect, that f_1 and f have to be chosen as in [KS] p. 24 or 70 up to an integration over the central group denoted $Z_1(F)$ in loc. cit. The functions $f_1 = \prod_v f_{1,v}$ and $f = \prod_v f_v$ are said to be *globally matching* functions, if the analog of formula [KS] (5.5.1) holds for the global stable orbital integrals (for the finite adeles $\mathbb{A}_{\text{fin}}$) for all global elements δ, δ_H. This slightly weaker global condition of cause suffices for the comparison of the global trace formulas.

The homomorphism b_ξ. For $v \notin S$ the classes of irreducible $K_{1,v}$-spherical representations $\Pi_{v,1}$ of $G_1(\mathbb{Q}_v)$ are parameterized by their 'Satake parameter' $\alpha_1 \in Y_1(\mathbb{C}) = \hat{T}_1/W(\hat{G}_1)$. Since $Y_1(\mathbb{C})$ can be identified with $(\hat{G}_1)_{ss}/\text{int}(\hat{G}_1)$ (see [Bo], Lemma 6.5) and similar for G, the endoscopic map ξ induces an algebraic morphism $(\hat{G}_1)_{ss}/\text{int}(\hat{G}_1) \to (\hat{G})_{ss}/\text{int}(\hat{G})$, hence a map ξ_Y : $Y_1(\mathbb{C}) \to Y(\mathbb{C})$. Since the spherical Hecke algebra of $(G_{1,v}, K_{1,v})$ resp. (G_v, K_v) can be identified with the ring of regular functions $\mathbb{C}[Y_1]$ resp. $\mathbb{C}[Y]$, we obtain an induced algebra homomorphism $b_\xi = \xi_Y^*$ from the spherical Hecke algebra $C_c^\infty(G_v//K_v)$ to spherical Hecke algebra $C_c^\infty(G_{1,v}//K_{1,v})$. The Satake parameter $\alpha = \xi_Y(\alpha_1)$ defines a K_v-spherical representation of G_v, also denoted $\Pi_v = r_\xi(\Pi_{v,1})$. By construction of the map it satisfies $\Pi_v^\sigma \cong \Pi_v$. Hence the action of G_v on Π_v can uniquely be extended to an action of the semidirect product $G_v \rtimes < \sigma >$ assuming, that σ fixes the spherical vector.

The topological trace formula. We give a review of the σ-twisted topological trace formula of Weselmann [Wes] Section 3, which generalizes the topological trace formula of Goresky and MacPherson [GMP], [GMP2] from the untwisted to the twisted case. Weselmann shows that these trace formulas themselves are 'stable' trace formulas [Wes] Theorem 4.8 and Remark 4.5, i.e. can be written entirely in terms of stable twisted orbital integrals $SO_\gamma^{G,\sigma}(f)$

for the maximal elliptic endoscopic group of (G, σ) in the sense of twisted endoscopy

$$T(f, \sigma, G, \chi) = \sum_{I \subset \Delta, \sigma(I) = I} (-1)^{|(\Delta \setminus I)/\sigma|} \cdot T_I(f, \sigma, G, \chi)$$

$$T_I(f, \sigma, G, \chi) = \sum_{\gamma \in P_I(\mathbb{Q})/\sim}' \alpha_\infty(\gamma, 1) \cdot \mathrm{SO}_\gamma^{G, \sigma}(f) \cdot \mathrm{Trace}(\gamma \cdot \sigma, V)$$

where the summation runs over all stable conjugacy classes γ in $P_I(\mathbb{Q})$ of I-contractive elements, whose norm is in L_∞^I (see loc. cit Theorem 4.8). In principle these trace formulas can be compared without simplifying assumptions (see [Wes], Section 5) except for a different notion of matching functions due to the factors $\alpha_\infty(\gamma, 1)$. See [Wes] (5.15).

Strongly matching functions. Globally matching functions (f, f_1) are said to be *strongly matching*, if there exists a universal constant $c = c(G, \sigma) \neq 0$ such that $T_I(f, \sigma, G, \chi) = c \cdot T_I(f_1, \mathrm{id}, G_1, \chi)$ holds for all I. Then

$$T(f, \sigma, G, \chi) = c \cdot T(f_1, \mathrm{id}, G_1, \chi)$$

holds by definition. In the lemma below we will show for constants $c_I = c(G, \sigma, I) \neq 0$

$$|\alpha_\infty(\gamma_0, 1)| = c_I \cdot |\alpha_\infty(\gamma_1, 1)|$$

for all summands of the sum defining $T_I(f, \sigma, G, \chi)$, resp. $T_I(f_1, \mathrm{id}, G_1, \chi)$ and sufficiently regular γ_0 respectively γ_1. Notice $(\gamma_0, \gamma_1) = (\delta, \delta_H)$ are the global $\mathbb{Q}$-rational elements to be compared in the notions above. For $\gamma = \gamma_0$ (or $\gamma = \gamma_1$) and the respective group G (or G_1) by [Wes] Theorem 4.8

$$|\alpha_\infty(\gamma, 1)| = \frac{O_\sigma^\infty(\gamma, 1) \cdot \#H^1(\mathbb{R}, T)}{d_{\zeta, \gamma}^I \cdot \mathrm{vol}_{db_\infty'}(\widetilde{(G_{\gamma, \sigma}^I)}'(\mathbb{R})/\tilde{\zeta})} .$$

This is the factor for the terms in the sum defining $T_I(f, \sigma, G, \chi)$, i.e. $\gamma \in P_I(\mathbb{Q})$ for the σ-stable $\mathbb{Q}$-rational parabolic subgroup $P_I = M_I U_I$. We now discuss the ingredients of the formula defining $\alpha_\infty(\gamma, 1)$.

Regularity. We will later choose a pair of matching functions (f, f_1), whose stable orbital integrals have strongly σ-regular semisimple support. For the notion of strongly σ-regular see [KS] p.28. Taking this for granted, we therefore analyze the above terms under the assumption, that $\mathrm{Int}(\gamma) \circ \sigma$ is strongly σ-regular semisimple. For simplicity of notation we consider the case (G, θ) in the following. The case of its maximal elliptic σ-endoscopic group (G_1, id) of course is analogous. We will also assume $Z(G)^\theta$ to be Zariski connected, since this suffices for our applications. Although the following discussion holds more generally, at the end we restrict to our two relevant cases for convenience.

The groups $G^I_{\gamma,\sigma}$. Fix a σ-stable rational parabolic subgroup P_I of G (or similar G_1). For a strongly σ-regular semisimple elements $\theta = \mathrm{Int}(\gamma) \circ \sigma$, where $\gamma \in P_I(\mathbb{Q})$, the twisted centralizer $G^I_{\gamma,\sigma} = (P_I)^\theta$ is abelian. In fact, G^θ is abelian by [KS] p.28 and the centralizer of G^θ in G is a maximal torus $\mathrm{T} \subseteq G$. T is θ-stable and $T := G_{\gamma,\sigma}$ is the group of θ-fixed points in T

$$G^\theta = \mathrm{T}^\theta .$$

θ is strongly σ-regular, hence there exists a pair (T, B) (B a Borel containing T defined over the algebraic closure), which is θ-stable. Since T acts transitively on splittings, there exists a t in T (again over the algebraic closure), such that $\theta^* = \mathrm{int}(t)\theta$ respects a fixed splitting $(\mathrm{T}, B, \{x_\alpha\})$. Then

$$\mathrm{T}^\theta = \mathrm{T}^{\mathrm{int}(t)\theta} = \mathrm{T}^{\theta^*} .$$

By [KS] p.14 G^{θ^*} is Zariski connected if and only if T^{θ^*} is Zariski connected. Therefore T^θ is Zariski connected, if G^{θ^*} is Zariski connected. Now $G^{\theta^*} = G^1 \cdot Z(G)^{\theta^*}$ by [KS], p.14, for the Zariski connected component $G^1 = (G^{\theta^*})^0$. By our assumption $Z(G)^{\theta^*} = Z(G)^\sigma$ is Zariski connected. Hence G^{θ^*} and therefore T^θ are both Zariski connected. In other words T^θ is a subtorus of T. Notice $\sigma^2 = 1$ implies $(\theta^*)^2 = 1$, since an inner automorphism fixing a splitting is trivial. Therefore to describe (T, θ^*) over the algebraic closure, we can replace θ^* by our original automorphism σ, which also fixes (some other but conjugate) splitting of G. Over the algebraic closure (T, θ^*) is isomorphic to a direct product $\prod_i (T_i, \theta_i^*)$, where the factors are either $(\mathrm{GL}(1), \mathrm{id})$ or $(\mathrm{GL}(1), \mathrm{inv})$ or $(\mathrm{GL}(1)^2, \theta^*)$, where θ^* is the flip automorphism of the two factors $\theta^*(x, y) = (y, x)$. Which types arise does not depend on I as shown below, and can be directly read of from the way in which σ acts on the σ-stable diagonal reference torus [BWW]. Two cases are relevant: [BWW] Example 1.8 where $G = \mathrm{PGL}_{2n+1}$ with $\sigma(g) = J({}^t g^{-1})J^{-1}$ and J as in loc. cit., and [BWW] Example 1.9 where $G = \mathrm{GL}(2n) \times \mathrm{GL}(1)$ and $\theta(g, a) = (J({}^t g^{-1})J^{-1}, \det(g)a)$. For $n = 2$ these specialize to the cases considered in Theorem 4, except that in Example 1.9 one has to divide by $\mathrm{GL}(1)$ to obtain $\mathrm{GSO}(3, 3)$ as explained already.

Absolute independence from I. T is the unique maximal torus of G, which contains G^θ as subgroup and which is θ-stable. Furthermore $G^I_{\gamma,\sigma} = \mathrm{T}^\theta \cap P_I$.

Since $U_I \cap T^\theta$ is trivial, the projection $P_I \to P_I/U_I \cong M_I$ induces an isomorphism of $G^I_{\gamma,\sigma}$ with its image in M_I. We can find a θ-stable maximal torus in M_I containing the image. By dimension reasons this maximal torus coincides with the image of $G^I_{\gamma,\sigma}$. By the uniqueness of T this determines T and $T = T^\theta$ within the subgroup (M_I, θ) of (G, θ). For strongly regular γ this implies

$$G^I_{\gamma,\sigma} \cong T^\theta = T^{\theta^*} = T ,$$

which is independent from I. In our situation $\sigma = \eta_1$ holds for η_1 as defined in [Wes] (2.1). By the Gauss-Bonnet formula [Wes] (Section 3) the torus T is $\mathbb{R}$-anisotropic modulo the center of M_I. By a global approximation argument therefore the groups $\tilde{\zeta}$ in the formula above are trivial. Similarly, since in our case the centers of G and G_1, hence also the centers of their respective $\mathbb{Q}$-Levi subgroups M_I, are split $\mathbb{Q}$-tori, the groups ζ can be assumed to be trivial. Hence $d^I_{\zeta,\gamma} = 1$ by [Wes] (2.23).

The other factors. By [Wes] Lemma (2.17) and (2.15) and the isomorphism $\pi_0(\tilde{F}(g_\eta, \gamma)^{\tilde{\eta}_\gamma, h_\infty}) \cong \pi_0(\tilde{F}(g_\eta, \gamma))^{\tilde{\eta}_\gamma, h_\infty}$ the factors $O^\infty_\sigma(\gamma, 1) = \#R_{\gamma,\sigma}$ are

$$O^\infty_\sigma(\gamma, 1) = \frac{\#\left(K^{I,m}_\infty / K^I_\infty\right)^{\eta_2}}{\#\pi_0\left(G^I_{\gamma,\sigma}(\mathbb{R})/(G_{\gamma,\sigma}(\mathbb{R}) \cap p_1 K^+_\infty Z^+_\infty A_I p_1^{-1})\right)} .$$

The denominator is trivial, since $G^I_{\gamma,\sigma}(\mathbb{R})$ is a torus and $\mathbb{R}$-anisotropic modulo the center. Hence the factor $O^\infty_\sigma(\gamma, 1)$ becomes $(K^{I,m}_\infty / K^I_\infty)^{\eta_2}$ with notations from [Wes], (2.2). It only depends on I, but does not depend on γ. So we need compare the factors

$$|\alpha_\infty(\gamma, 1)| = \#\left(K^{I,m}_\infty / K^I_\infty\right)^{\eta_2} \cdot \frac{\#H^1(\mathbb{R}, T)}{\mathrm{vol}_{db'_\infty}(T'(\mathbb{R}))} .$$

The σ-stable torus T contains $G^I_{\gamma,\sigma} = T = T^\sigma$ (σ-invariant subtorus). Its $\mathbb{R}$-structure might a priori depend on γ, but in fact does not. $T' = (T^\sigma)'$ is the maximal $\mathbb{R}$-anisotropic subtorus of T, as follows from the description of [Wes] (3.9). Now, since

$$G^I_{\gamma_0,\sigma} = T \cong T^\sigma \longrightarrow T_\sigma \cong T_1 = (G_1)^{I_1}_{\gamma_1,\mathrm{id}}$$

are isogenious tori, the definition of measures db'_∞ ([Wes], (3.9)) and the measures used the definition of matching of Kottwitz-Shelstad ([KS], p. 71) shows, that both factors $\mathrm{vol}_{db'_\infty}((G^I_{\gamma_0,\sigma})'(\mathbb{R}))$ for G and γ_0 and $\mathrm{vol}_{db'_\infty}((G_1)^I_{\gamma_1,\mathrm{id}})'(\mathbb{R}))$ for G_1 and γ_1 differ by a constant independent of (γ_0, γ_1). Recall that $\gamma = \gamma_0 \in G(\mathbb{Q})$ and $\gamma = \gamma_1 \in G_1(\mathbb{Q})$ is a pair of points related by the Kottwitz-Shelstad norm. For such a pair we have the tori $T \subseteq G$ and $T_1 \subseteq G_1$ and $T_1 = T_\sigma$ (σ-coinvariant quotient torus of $T \subseteq G$) by the definition of the Kottwitz-Shelstad norm. See [KS], Chapter 3 and [Wes], Chapter 5. The relative factor $\mathrm{vol}_{db'_\infty}((G^I_{\gamma_0,\sigma})'(\mathbb{R}))/\mathrm{vol}_{db'_\infty}((G_1)^I_{\gamma_1,\mathrm{id}})'(\mathbb{R}))$ turns out to be the degree of the isogeny $T^{\theta^*} \to T_{\theta^*}$ independently from I. Notice, up to conjugacy over $\mathbb{R}$ (hence up to isomorphism over $\mathbb{R}$) the tori T, respectively T_1 only depend on I and not on γ_0, γ_1.

Example. For $G = (\mathrm{PGL}(2n + 1), \sigma)$ and $G_1 = (\mathrm{Sp}(2n), \mathrm{id})$ as in [Wes] (5.18) the quotient $|\alpha_\infty(\gamma_0, 1)|/|\alpha_\infty(\gamma_1, 1)|$ is equal to the relative measure factor 2^n. In fact $K^{I,m}_\infty = K^{I,+}_\infty$ both for (G, σ) and (G_1, id), since $\mathrm{O}(2n + 1, \mathbb{R})/\{\pm\mathrm{id}\} \cong \mathrm{SO}(2n + 1, \mathbb{R})$ and since $\mathrm{U}(n)$ is connected. If it holds for $I = \Delta$, then for all I. In all cases $H^1(\mathbb{R}, T) \cong H^1(\mathbb{R}, T_1)$. For $M_I \cong \mathrm{PGL}(2m + 1) \times \prod_i \mathrm{GL}(r_i) \subseteq G = \mathrm{PGL}(2n + 1)$ and the corresponding $(M_1)_I \cong \mathrm{Sp}(2m) \times \prod_i \mathrm{GL}(r_i) \subseteq G_1 = \mathrm{Sp}(2n)$ only the cases $r_i = 0, 1, 2$ give nonvanishing contributions to the trace formula, using that T and T_1 are anisotropic modulo the center of M_I resp. $(M_1)_I$. The corresponding decomposition $T = T_m \times \prod_i T_{r_i}$ and $T_1 = T_{1,m} \times \prod_i T_{1,r_i}$ easily implies $H^1(\mathbb{R}, T_m) \cong H^1(\mathbb{R}, T_{1,m})$ (being $\mathbb{R}$-anisotropic of the same rank) and $H^1(\mathbb{R}, T_{r_i}) \cong H^1(\mathbb{R}, T_{1,r_i})$ by direct inspection, again using that T and T_1 are anisotropic modulo the center of M_I resp. $(M_1)_I$.

We have shown

Lemma 1. *For matching elements strongly σ-regular elements γ_0 and matching (γ_0, γ_1) the quotient $|\alpha_\infty(\gamma_0)|/|\alpha_\infty(\gamma_1)| = c_I$ only depends on I and (G, σ, G_1), but not on the stable conjugacy class of $\gamma \in P_I(\mathbb{Q})$. For $I = \Delta$ this quotient defines a universal constant $c = c(G, \sigma, G_1) \neq 0$.*

In the other relevant case $(G, \sigma) = (\mathrm{GSO}(3, 3), \sigma)$ and $G_1 = \mathrm{GSp}(4)$ the constants c_I depend on I.

Theorem 4. *For $G = \mathrm{GSO}(3, 3)$ resp. $G = \mathrm{PGL}(5)$ and σ as above, so that G_1 is $\mathrm{GSp}(4)$, respectively G_1 is $\mathrm{Sp}(4)$, the following hold.*

(i) *Suppose Π_1 is a cuspidal automorphic representation of $\mathrm{GSp}(4, \mathbb{A})$ (resp. a suitable irreducible automorphic representation weakly equivalent to a component of its restriction to $\mathrm{Sp}(4, \mathbb{A})$), which is not CAP nor a weak endoscopic lift. Suppose $(\Pi_1)_{\mathrm{fin}}$ contributes to $H^\bullet(M_{G_1}, \mathcal{V}_1)$. Suppose S is a sufficiently large finite set of places for which Π^S_1 defined by $\Pi_1 = \Pi_{1,S}\Pi^S_1$ is nonarchimedean and unramified. Let Ψ be the virtual representation of $G_1(\mathbb{A}_{\mathrm{fin}}) = G_1(\mathbb{A}_S) \times G_1(\mathbb{A}^S)$ on the generalized (cuspidal) eigenspaces contained* in $H^\bullet(M_{G_1}, \mathcal{V}_1)$, on which $G_1(\mathbb{A}^S)$ acts by Π_1^S. Then there exists a pair of globally and strongly matching functions f, f_1 with strongly regular support, so that $\mathrm{Trace}(f_1, (\Pi_1)_{\mathrm{fin}}) \neq 0$ and

$$\mathrm{Trace}(f_1, \Psi) \neq 0 \,.$$

(ii) *The fundamental lemma holds: In the situation of* (i) *there exists a finite set of places* $S = S(\Pi_1, G_1, G)$ *such that for* $v \notin S$ *the representation* $\Pi_{1,v}$ *is unramified, the spherical Hecke algebra* $C_c^\infty(G(\mathbb{Q}_v)/\!/K_v)$ *is defined, so that for all* $f_v \in C_c^\infty(G_v/\!/K_v)$ *the functions* f_v *and* $b_\xi(f_v)$ *are locally matching functions for the endoscopic datum* $(G_1, {}^L G_1, \xi, s)$ *for* $(G, \sigma, 1)$ *in the sense of [KS], (5.5.1). So in particular the following generalized Shintani identities hold for* $v \notin S$

$$\mathrm{Trace}(b_\xi(f_v), \Pi_{1,v}) = \mathrm{Trace}(f_v \cdot \sigma, r_\xi(\Pi_{1,v})) \,.$$

(iii) *For globally and strongly matching functions* (f, f_1) *there exists a universal constant* $c = c(G, G_1) \neq 0$ *such that the trace identity*

$$T(f, \sigma, G, \chi) = c \cdot T(f_1, \mathrm{id}, G_1, \chi_1)$$

holds.

Assertion (4.(iii)) has already been explained.

Assertion (4.(ii)). The comparison (4.(iii)) of trace formulas would be pointless unless there exist matching functions at least of the form (4.(ii)). Assertion (4.(ii)) is a statement, which is enough to be proved for almost all places $v \notin S$ for a suitable large finite set S. For an arbitrary spherical Hecke operator at $v \notin S$ it can be reduced to the case, where $f_v = 1_{K_v}$ is the unit element of the spherical Hecke algebra. In the untwisted case this was shown by Hales [Ha] in full generality. In [W4] this is done for a large class of twisted cases, including those considered above, by extending a method of Clozel and Labesse. We remark, that in the situation of Theorem 4, the arguments of [W4] can be simplified by the use of the topological trace identity (4.(iii)) explained above. Concerning unit elements: The case of the unit element $f_v = 1_{K_v}$ was established by Flicker [Fl] for the first kind of trace comparison involving GSO(3, 3) or more precisely GL(4) × GL(1), and was deduced from that result for the second kind of trace comparison involving PGL(5) in [BWW] Theorem 7.9 and Corollary 7.10. In both cases this is obtained for the unit elements of the spherical Hecke algebras for large enough primes and for sufficiently regular (γ_0, γ_1). That these regularity assumptions do no harm is shown below.

Assertion (4.(i)). We claim that the assertions (4.(ii)) and (4.(iii)) of Theorem 4 imply Assertion (4.(i)). Recall that $(\Pi_1)_{\mathrm{fin}}$ and the coefficient system $\mathcal{V}_\chi$ are fixed. Since $(\Pi_1)_{\mathrm{fin}}$ is cuspidal but not CAP, it only contributes to cohomology in degree 3, if it contributes nontrivially to the Euler characteristic of $\mathcal{V}_\chi$ (in our case this amounts to the assumption that the archimedean component belongs to the discrete series). All constituents of Ψ, being weakly equivalent to Π_1 and isomorphic to Π_1 outside S, are not CAP. Hence the same applies for them.

To construct f_1 such that $T(f_1, \mathrm{id}, G_1, \chi) \neq 0$ and $\mathrm{Trace}(\Psi(f_1)) \neq 0$ the easiest candidate to come into mind is the following: Let N be some principal congruence level, i.e. assume $(\Pi_1)_{\mathrm{fin}}^{K(N)} \neq 0$ where $K(N) \subseteq \mathrm{GSp}(4, \mathbb{Z}_{\mathrm{fin}})$ is defined by the congruence condition $k \equiv id \bmod N$ (and similar for the group $\mathrm{Sp}(4)$). Choose a sufficiently large finite set of places S containing the finite set of divisors of N. Then put $f_1 = \prod_{v \neq \infty} f_{1,v}$, where $f_{1,v}$ is chosen to be the function which is zero for all $g \neq k \cdot z$ for k is in the principal congruence subgroup of level N and chosen equal to $f_{1,v} = \omega(z)^{-1}$ else. Then $f_{1,v}$ is the unit element of the Hecke algebra at almost all places not in S and the required condition $\mathrm{Trace}((\Psi)_{\mathrm{fin}}(f_1)) \geq \mathrm{Trace}((\Pi_1)_{\mathrm{fin}}(f_1)) > 0$ holds by definition. To start to prove (4.(i)) we first have to find a matching function f on $G(\mathbb{A}_{\mathrm{fin}})$. This involves the fundamental lemma, which is known for elements (γ_0, γ_1) sufficiently regular. Therefore we modify our first naive choice f_1 slightly. For this we choose an auxiliary place $w \notin S$, where Π_1 is unramified. At this auxiliary place w our previous function $f_{1,w}$ will be replaced by an elementary function in the sense of [La], [W4]. We will see, that this allows us to restrict ourselves to consider matching conditions at sufficiently regular elements (γ_0, γ_1). Once these data in S and w are fixed, we finally allow for some additional modification at other unramified place $w' \neq w$ in order to construct a good Π_1^S-projector in the sense of [W1].

At the auxiliary place w we choose $f_{1,w}$ to be an elementary function as in [La] §2 attached to some sufficiently regular element t_0 in the split diagonal torus, to be specified later. Then there exists a matching σ-twisted elementary function on $G(\mathbb{Q}_w)$ [W4]. Notice, the stable orbital integrals $\mathrm{SO}_t^{G_1}(f_{1,w})$ vanish, unless t is conjugate in $G_1(F)$ to some element in the torus of diagonal matrices, which differs from t_0 by a unimodular diagonal matrix and a central factor. In other words, the stable orbital integral of $f_{1,w}$ then locally has regular semisimple support for suitably chosen sufficiently regular element t_0. For the global topological trace formulas of G_1 and G this has the effect, that only regular semisimple global elements give a nonzero contribution on the geometric side. Hence for the construction a global matching functions (f_1, f) it suffices to show local matching for $(f_v, f_{1,v})$ at $v \neq w$ by considering stable orbital integrals at regular elements only! Since in our situation the Kottwitz-Shelstad transfer factors are identically one for all regular elements, this means we have to show local matching of stable orbital integrals at all semisimple regular elements γ_1 at the places $v \neq w$. In addition we have to find a function f_w matching with $f_{1,w}$ at the place w supported in strongly σ-regular elements γ_0 of $G(\mathbb{Q}_w)$, and we have to guarantee $\mathrm{Trace}(\Psi_{1,w})(f_{1,w}) \neq 0$ by a suitable choice of t_0. In addition, as already mentioned, one can find a corresponding finite linear combination f_w of σ-twisted elementary functions matching with $f_{1,w}$, and for this choice of $f_{1,w}$ the matching function f_w has support in strongly σ-regular elements (see [W4]).

The modification using t_0 at the fixed auxiliary unramified place w will be discussed in more detail below. At the moment, taking the situation at the place w for granted, let us look at the other places first. For the nonarchimedean places $v \neq w$ not in S we now can use Assertion (4.(ii)) to obtain the globally matching functions f_v by choosing unit elements at all these places. Recall, that by our modification at the auxiliary place w for this it suffices to know the statement of the fundamental lemma for unit elements at sufficiently regular global, hence local elements. Concerning the places $v \in S$. Since matching deals with a purely local question now, of course one has to allow singular elements. However, it is well known that the existence of local matching function can be reduced to the existence of matching germs. This can be achieved by the argument of [LS], §2.2 or the similar argument of [V]. So for the places $v \in S$ one is reduced to the matching of germs of stable orbital integrals. Fortunately for the pairs (G, G_1) under consideration the matching of germs of stable orbital integrals has been proven by Hales [Ha2]. This completes the construction of the functions f_v at the places $v \neq w$, which define the globally matching pair (f_1, f). So let us come back to the auxiliary unramified place w.

Concerning the choice of t_0. The trace $\mathrm{Trace}(f_1, \Pi_1)$ of any irreducible cuspidal automorphic representation Π_1 vanishes unless Π_1 admits a nontrivial fixed vector for the group $K(N, w) = K_w \prod_{v \neq \infty, w} K(N)_v$, where K_w is an Iwahori subgroup. Hence by the finiteness theorems only finitely many irreducible automorphic representations with fixed archimedean component and $\mathrm{Trace}(f_1, \Pi_1) \neq 0$ exist. $\mathrm{Trace}(f_1, \Pi_{\mathrm{fin}})$ considered in Assertion (4.(i)) is a linear combination of the type

$$\sum_i m(\Pi_i)\mathrm{Trace}(f_{1,w}, (\Pi_i)_w)$$

involving only a finite number of local irreducible admissible representa-
tions $(\Pi_i)_w$ of $G_1(\mathbb{Q}_w)$ of Iwahori level with certain constants $m(\Pi_i) = m((f_1)^w, \Pi_i) > 0$. These constants are positive, since they have an interpretation as multiplicities. Since the unramified representation $\Pi_{1,w}$ is one of these,
this sum is nonempty. It can be written as a finite sum $\sum_{\nu=1}^{r} d_\nu \cdot \pi_\nu(f_{1,w})$
of local irreducible admissible inequivalent representations π_ν at the place w,
where $d_\nu > 0$ is the multiplicity of the representation π_ν of $G_1(\mathbb{Q}_w)$ on the
$\Pi_1^{S \cup w}$-isotypic component $V = H^\bullet(M_{G_1}/K(N), \mathcal{V}_1)[\Pi_1^{S \cup w}]$. By the arguments of [La], Propositions 4 and 5, the trace of $f_{1,w}$ on V is the trace of t_0
on the fixed points $(V_N)^{I \cap N}$ of the Iwahori group $I = K_w$ on a corresponding
Jaquet module V_N (for the unipotent radical N of a minimal parabolic, both N
and I suitably adapted to t_0 as explained in loc. cit). By the linear independence
of characters this trace vanishes for all choices of t_0 only if $(V_N)^{I \cap N} = 0$. By
a theorem of Borel and Casselman [B] this would imply $V^I = 0$, contradicting the fact that the spherical representation $\Pi_{1,w}$ occurs nontrivially in V.
Put $S' = S \cup \{w\}$. Since the trace of f_1 on $\Psi' = H^\bullet(M_{G_1}, \mathcal{V}_1)[\Pi_1^{S \cup w}]$ is
$\sum_{\nu=1}^{r} d_\nu \pi_\nu(f_{1,w}) \neq 0$ and therefore does not vanish for a suitable choice of
t_0 as explained above, we have proven the second statement $\text{Trace}(f_1, \Psi) \neq 0$
of (4.(i)), once we write S instead of S'. At least one of the automorphic irreducible representations Π_1, which contribute to Ψ has $\text{Trace}(f_1, \Pi_1) \neq 0$.
Since in the case $G_1 = \text{Sp}(4)$ the choice of Π_1 in its weak equivalence class
is irrelevant, we may assume that our choice of Π_1 was this representation
from beginning. In the case of the group $G_1 = \text{GSp}(4)$, one must argue dif-
ferently. But now G_1 is of adjoint type. Then by the argument of [La] (Section
5) the character $\text{Trace}(\pi_1)$ of the unramified representation π_1 is linear independent from the span of the characters $\text{Trace}(\pi_\nu)$ for $\nu = 2, .., r$ on the space
of test functions $\mathcal{T}$, defined by the span of the elementary functions attached
to strongly regular elements t_0, provided G_1 is of adjoint type. In the adjoint
case we can therefore choose $f_{1,w} \in \mathcal{T}$ so that $\text{Trace}(\pi_1)(f_{1,w})$ is nonzero and
also that the sum is $\sum_\nu \text{Trace}(\pi_\nu)(f_{1,w})$ is nonzero. This remark proves the
existence of f_1 with the required nonvanishing properties $\text{Trace}(f_1, \Pi_1) \neq 0$
and $\text{Trace}((f_1, \Psi) \neq 0$ also in the case $G_1 = \text{GSp}(4)$.

We finally make some further adjustments, but only using spherical functions. This is necessary to obtain strongly matching functions, and this is only
necessary in the case $(G, G_1) = (\text{GSO}(3, 3), \text{GSp}(4))$. For this we consider
the unramified representation $\Pi'^S = r_\xi(\Pi^S)$. Using the method of CAP-
localization [W1] (with levels and coefficient system fixed) it is clear that
we can modify our f at unramified places not in S using a good spherical
Π'^S-projector $\tilde{f}^S$. Similarly modify f_1 by the corresponding good spherical
Π_1^S-projector $\tilde{f}_1^S = b_\xi(\tilde{f}^S)$. Then all contributions $T_I(f_1, id, G_1, \chi)$ in the

trace formula for $P_I \neq G_1$ in the trace formula vanish. This is possible, since we assumed Π^S not to be a CAP-representation. Then, since f is globally matching, we get $T_I(f, G, \sigma, \chi) = c_I \cdot T_I(f_1, \mathrm{id}, G_1, \chi)$ for constants c_I by the last lemma. Hence also $T_I(f, G, \sigma, \chi) = 0$ for $P_I \neq G$. Therefore the modified pair $(f \tilde{f}^S, f_1 \tilde{f}_1^S)$ is not only globally matching, but also a strongly matching pair of functions with the properties required by Assertion (4.(i)). This completes the proof of Theorem 4.

Applying Theorem 4. For an irreducible cuspidal representation Π which is not CAP, suppose Π_∞ belongs to the discrete series. For $G_1(\mathbb{A}) = \mathrm{GSp}(4, \mathbb{A})$ the representation Π_{fin} contributes to $H^3(M_1, \mathcal{V}_{\chi_1})$ for some character χ_1 ([W2], hypothesis A and B) and we can therefore assume that $\mathrm{Trace}(\Pi_{\mathrm{fin}}(f_1); H^\bullet(M_1, \mathcal{V}_{\chi_1}))$ does not vanish for all f_1. Now choose f_1 and a globally and strongly matching function f on the group $G(\mathbb{A}_{\mathrm{fin}})$ for $G = \mathrm{GSO}(3, 3)$ (or its z-extension $\mathrm{GL}(4) \times \mathrm{GL}(1)$) as in Theorem (4.(i)). Then by Theorem (4.(ii)) and (4.(iii)) there must exist a σ-stable irreducible $G(\mathbb{A}_{\mathrm{fin}})$-constituent (Π', ω) of $H^\bullet(M_G, \mathcal{V}_\chi)$, for which $\mathrm{Trace}(\Pi', \omega)(f \cdot \sigma) \neq 0$ holds. By a theorem of Franke [Fr] this representation is automorphic. Any modification of the pair (f_1, f) to $(f_{1,v} \prod_{w \neq v} f_{1,w}, f_v \prod_{w \neq v} f_w)$ at a place $v \notin S$ for $f_{1,v} = b_\xi(f_v)$ and spherical Hecke operators f_v again gives a pair of globally strongly matching functions. Then (4.(ii)) and (4.(iii)) and the linear independence of characters of the group $G(\mathbb{A}_{\mathrm{fin}}) \rtimes < \sigma >$ imply, that the representation identity $(\Pi'_v, \omega_v) = r_\xi(\Pi_v)$ holds for all $v \notin S$. Since the partial L-series outside S can be easily computed from the Satake parameters, we have constructed a weak lift: The L-identity

$$L^S(\Pi' \otimes \chi, s) = L^S(\Pi \otimes \chi, s)$$

holds for all idele class characters χ for a sufficiently large finite set S. This is just a transcription of the fundamental lemma, i.e. follows from part (4.(ii)) and (4.(iii)) of Theorem 4, and defines the weak lift Π'. The weak lift Π', now constructed, is uniquely determined by Π, since by the strong multiplicity 1 theorem for $\mathrm{GL}(n)$ it is determined by the partial L-series $L^S(\Pi' \otimes \chi, s)$ (at the places $v \notin S$), once it exists. This shows, that our second temporary assumption is satisfied. The required identity for the central characters $\omega_\Pi^2 = \omega_{\Pi'}$ holds, since it locally holds at all nonarchimedean places outside S by the L-identity above.

The case $G_1 = \mathrm{Sp}(4)$, $G = \mathrm{PGL}(5)$ is completely analogous and the corresponding considerations now also allows us to get rid of our first temporary assumption.

Concerning the archimedean place. The topological trace formulas compute the traces of the Hecke correspondences on the virtual cohomology of a given coefficient system in terms of orbital integrals. Underlying the trace comparison (4.(iii)) is a corresponding fixed lift of the coefficient system $\mathcal{V}_{\mu_1}$. The coefficient system being fixed, only finitely many archimedean representations contribute to the topological trace formula for this coefficient system. Also notice, the representation Π'_∞ has to be a representation of $GL(4, \mathbb{R})$, which has nontrivial cohomology for the given lift of the coefficient system. In fact this determines Π'_∞ in terms of $\mathcal{V}_\mu$ and the cohomology degree. Furthermore in our relevant case, i.e. for the lift to $GL(4) \times GL(1)$, well known vanishing theorems for Lie algebra cohomology and duality completely determine Π'_∞ in terms of the coefficient system $\mathcal{V}_\mu$. We leave this as an exercise. This uniquely determines Π'_∞ in terms of Π_∞. This yields matching archimedean representations, which have nontrivial cohomology in degree 3 for the group H with σ-equivariant representations, that have nontrivial cohomology for G.

Consider the theta lift from $GO(3, 3)$ to $GSp(4)$, which maps the class of (Π', ω) to the class of Π_{gen}. At the archimedean place we claim, that for this lift $(\Pi'_\infty, \omega_\infty)$ locally (!) uniquely determines $\Pi_{\mathrm{gen}, \infty}$. Since theta lifts behave well with respect to central characters, it is enough to prove this for the dual pair $(O(3, 3), Sp(4))$ by our earlier remarks on Π'_∞. In fact by Mackey's theory the restriction of Π'_∞ to $SL(4, \mathbb{R}) \cdot \mathbb{R}^*$ decomposes into two nonisomorphic irreducible constituents, from which (from each of them) in turn Π'_∞ is obtained by induction. The same holds for the restriction of discrete series representations of $GSp(4, \mathbb{R})$ to $Sp(4, \mathbb{R}) \cdot \mathbb{R}^*$. It is therefore enough to determine the restriction of $\Pi_{\mathrm{gen}, \infty}$ to $Sp(4, \mathbb{R}) \cdot \mathbb{R}^*$. Once we have shown, that this restriction contains an irreducible constituent in the discrete series, we have therefore as a consequence, that the representation $\Pi_{\mathrm{gen}, \infty}$ of $GSp(4, \mathbb{R})$ is uniquely determined by Π'_∞ and that $\Pi_{\mathrm{gen}, \infty}$ belongs to the discrete series. We now still have to understand, why $\Pi_{\mathrm{gen}, \infty}$ should be in the archimedean L-packet of Π_∞.

This being said, we first replace $GL(4, \mathbb{R})$ by $SL(4, \mathbb{R})$, or by the quotient $SO(3, 3)(\mathbb{R}) = SL(4, \mathbb{R})/\mathbb{Z}_2$ (recall the central character was ω^2) and replace Π'_∞ by a suitable irreducible constituent of the restriction. For simplicity of notation still denote it Π'_∞. If the representation on $O(3, 3)(\mathbb{R})$ induced from the irreducible representation Π'_∞ of $SO(3, 3)(\mathbb{R})$ would be irreducible, we can immediately apply [H] to conclude, that its theta lift on $Sp(4, \mathbb{R})$ is uniquely determined by Π'_∞. Otherwise there exist two different extensions of Π'_∞ to $O(3, 3)(\mathbb{R})$, by simplicity denoted Π'_∞ and $\Pi'_\infty \otimes \varepsilon$, where ε is the quadratic character of $O(3, 3)(\mathbb{R})$ defined by the quotient $O(3, 3)(\mathbb{R})/SO(3, 3)(\mathbb{R})$.

However if this happens, we now claim only one of the two possibilities contributes to the theta correspondence, so that again we can apply [H]. The claim made follows from an archimedean version of part c) of the Proposition of [V], p.483. For the convenience of the reader we prove this in the archimedean exercise below.

So passing from Π_∞ to $(\Pi'_\infty, \omega_\infty)$ and then back to $\Pi_{\text{gen},\infty}$ turns out to be a well defined local assignment at the archimedean place; to be accurate, in the first instance only up to twist by the sign-character. That it is well defined then follows a posteriori, once we know the assigned image is contained in the discrete series. For this see the archimedean remarks made following our second temporary assumption. To compute the local assignment - first only up to a possible character twist - it is enough to do this for a single suitably chosen global automorphic cuspidal representation Π. We have to show, that passing forth back with these two lifts locally at the archimedean place, we do not leave the local archimedean L-packet. Since for every archimedean discrete series representation Π_∞ of weight (k_1, k_2) there exists a global weak endoscopic lift Π with this given archimedean component Π_∞ (see [W2]), it is now enough to do this calculation globally for this global weak endoscopic lift Π. For global weak endoscopic lifts there exists a unique globally generic cuspidal representation Π_{gen} in the weak equivalence class of Π by [W2], hypothesis A. So Π_{gen} is uniquely determined by the description of this generic component given in [W2], hypothesis A. Therefore looking at the archimedean place the archimedean component $\Pi_{\text{gen},\infty}$, which is uniquely determined by Π_∞ as we already have seen, has to be this unique generic nonholomorphic member of the discrete series L-packet of $GSp(4, \mathbb{R})$ of weight (k_1, k_2). Since it depends only on the archimedean local representation, and since this holds true in the special case, this holds in general. This proves Theorem 1 modulo the following

Archimedean exercise. We now prove the archimedean analog of proposition [V], p.483, which we used above. See also [W5]. Let V be a nondegenerate real quadratic space of dimension $m = p + q$, let $G = O = O(p, q)$ be its orthogonal group with maximal compact subgroup $K = O(p) \times O(q)$. Let $O^\vee$ be the set of equivalence classes of irreducible $(\text{Lie}(O), K)$-modules. The metaplectic cover $\text{Mp}(2N)$ of $\text{Sp}(2N)$ for $N = m \cdot n$ naturally acts on the Fock space. Its associated Harish-Chandra module $P_V(n) \cong \text{Sym}^\bullet(V \otimes_\mathbb{R} \mathbb{C}^n)$ defines the oscillator representation ω_{Fock}. For the pair $(G, G') = (O, \text{Mp}(2n, \mathbb{R}))$ the restriction of ω_{Fock} to $G \times G' \subseteq \text{Mp}(2N, \mathbb{R})$ induces the theta correspondence. The action of G by ω_{Fock} does not coincide with the natural action on $\text{Sym}^\bullet(V \otimes_\mathbb{R} \mathbb{C}^n)$, except when V is anisotropic. Nevertheless the action of K

does. Let $R(n) \subseteq O^\vee$ denote the classes of irreducible (Lie(O), K)-module quotients of $P_V(n)$. For $\pi \in O^\vee$ let $n(\pi)$ be the smallest integer n, if it exists, such that $\pi \in R(n)$. Since $P_V(n + n') = P_V(n) \otimes_{\mathbb{C}} P_V(n')$ - considered as modules of (Lie(O), K) - for any $\pi \in R(n)$ and $\pi' \in R(n')$ the irreducible quotients of $\pi \otimes_{\mathbb{C}} \pi'$ contribute to $R(n + n')$. For $\pi \in O^\vee$ let $\pi^\vee$ denote the contragredient representation. Since the quadratic character ε is an irreducible quotient of the representation $\pi \otimes (\pi^\vee \otimes \varepsilon)$, we obtain $n(\pi) + n(\pi^\vee \otimes \varepsilon) \geq n(\varepsilon)$. We claim $n(\varepsilon) \geq \dim_{\mathbb{R}}(V)$, which implies as desired

$$n(\pi) + n(\pi^\vee \otimes \varepsilon) \geq \dim_{\mathbb{R}}(V) .$$

For the proof of $n(\varepsilon) \geq m$ put $n = n(\varepsilon)$. The restriction of ε to $K = O(p) \times O(q)$ is $\sigma = \varepsilon \boxtimes \varepsilon$. $M' = \mathrm{Mp}(2n, \mathbb{R}) \times \mathrm{Mp}(2n, \mathbb{R})$ covers the centralizer of K in $\mathrm{Sp}(2N)$ and $M'_0 = \tilde{U}(n, \mathbb{C}) \times \tilde{U}(n, \mathbb{C})$ its maximal compact subgroup. By [KV] and [H], Lemma 3.3 a unique irreducible representation τ' of M'_0 is attached to σ, which is the external tensor product of the irreducible highest weight representations with highest weights $(\frac{p}{2}+1, .., \frac{p}{2}+1, \frac{p}{2}, ..., \frac{p}{2})$ respectively $(\frac{q}{2} + 1, .., \frac{q}{2} + 1, \frac{q}{2}, ..., \frac{q}{2})$ of $\tilde{U}(n, \mathbb{C})$, which is a twofold covering group of $U(n, \mathbb{C})$. Here $\frac{p}{2} + 1$ occurs p times respectively $\frac{q}{2} + 1$ occurs q times. So in particular $n \geq max(p, q)$. The two components are represented by pluriharmonic polynomials in the Fock space of degree p respectively q. As a consequence, τ' is realized in $\mathrm{Sym}^d(V \otimes_{\mathbb{R}} \mathbb{C}^n)$ for $d = p + q = m$. In other words $\deg(\tau') = m$ for the degree deg in the sense of [H]. The restriction of τ' to the maximal compact subgroup $K' = \tilde{U}(N, \mathbb{C}) \cap G'$ is the tensor representation $\det^{\frac{p-q}{2}} \otimes \bigwedge^p(\mathbb{C}^n) \otimes \bigwedge^q(\mathbb{C}^n)^\vee$ of $\tilde{U}(n, \mathbb{C})$, since the negative definite part gives an antiholomorphic embedding ([H], p.541). The highest weights of all irreducible constituents of the representation $\bigwedge^p(\mathbb{C}^n) \otimes \bigwedge^q(\mathbb{C}^n)^\vee$ of $U(n, \mathbb{C})$ are of the form $(1, .., 1, 0, .., 0, -1, ..., -1)$ with $i \leq p$ digits 1, $j \leq q$ digits -1 and $n - i - j$ digits 0. According to [H], Lemma 3.3 there must be a unique irreducible representation σ' of $\tilde{U}(n, \mathbb{C})$ in the restriction of τ' such that $\deg(\sigma') = \deg(\tau') = m$. The representation of $M'_0 \cong \tilde{U}(n, \mathbb{C}) \times \tilde{U}(n, \mathbb{C})$ on the polynomials of degree k in the Fock space $P_V(n)$ is isomorphic to

$$\bigoplus_{a+b=k} \det^{\frac{p}{2}} \cdot \mathrm{Sym}^a(\mathbb{R}^p \otimes_{\mathbb{R}} \mathbb{C}^n) \boxtimes \det^{\frac{q}{2}} \cdot \mathrm{Sym}^b(\mathbb{R}^q \otimes_{\mathbb{R}} \mathbb{C}^n).$$

Its restriction to $K' \cong \tilde{U}(n, \mathbb{C})$ in G' therefore is isomorphic to the representation $\det^{\frac{p-q}{2}} \bigoplus_{a+b=k} \mathrm{Sym}^a(\mathbb{R}^p \otimes_{\mathbb{R}} \mathbb{C}^n) \otimes \mathrm{Sym}^b(\mathbb{R}^q \otimes_{\mathbb{R}} (\mathbb{C}^n)^\vee)$ induced by the natural action of $U(n, \mathbb{C})$ on $\mathbb{C}^n$. Hence for $i \leq p$ and $j \leq q$ (the number of digits 1 resp. -1 of the highest weight) we get $\deg(\det^{\frac{p-q}{2}} \otimes (1, .., 1, 0, 0, .., -1, .., -1)) = i + j$ by considering the representation generated by the product of some $i \times i$-minor in $\mathrm{Sym}^i(\mathbb{R}^i \otimes_{\mathbb{R}} \mathbb{C}^n)$

and some $j \times j$-minor in $\mathrm{Sym}^i(\mathbb{R}^j \otimes_\mathbb{R} \mathbb{C}^n)$. Thus $\deg(\sigma') = m$ implies, that there exists $i \leq p$ and $j \leq q$ such that $\deg(\sigma') = i + j = m$. Therefore $i = p$ and $j = q$, hence $m = i + j \leq n$. So we have shown $n(\varepsilon) \geq m$. In fact it is not hard to see $n(\varepsilon) = m$.

The special case considered. In the situation of Theorem 1 and its proof this gives

$$n(\pi) + n(\pi \otimes \varepsilon) \geq 6 .$$

The underlying representation Π'_∞ of $\mathrm{GO}(3,3)(\mathbb{R})$ decomposes into two non-isomorphic representations π_1, π_2 of $\mathrm{O}(3,3)(\mathbb{R})$ with $n(\pi_1) = n(\pi_2) = 2$. Therefore $n(\pi_1 \otimes \varepsilon) \geq 4$ and $n(\pi_2 \otimes \varepsilon) \geq 4$ by the above inequality, since in our case $\pi^\vee \cong \pi$ follows from $(\Pi'_\infty)^\vee \cong \Pi'_\infty \otimes \omega_\infty^{-1}$ (which was a consequence of the second temporary assumption).

This finally completes the proof of Theorem 1.

Remark on orthogonal representations. For a weak endoscopic lift Π the statement of Theorem 1 is known by [W2], Hypothesis A. So in this case there was no need not go to the trace formula arguments required otherwise. The trace formula arguments, when applied for a weak endoscopic lift, nevertheless yield something interesting. Let Π be cuspidal irreducible representation Π of $\mathrm{GSp}(4, \mathbb{A})$, which is a weak endoscopic lift attached to a pair of cuspidal representations (σ_1, σ_2) of $\mathrm{GL}(2, \mathbb{A})$ with central character ω, but which is not CAP. Then we can still construct the irreducible automorphic representation $\tilde{\Pi}'$ of $\mathrm{GL}(5, \mathbb{A})$ from Π as above. However the representation $\tilde{\Pi}'$ will not be cuspidal any more, since its L-series $L^S(\tilde{\Pi}', s) = L^S(\sigma_1 \times \sigma_2 \otimes \omega^{-1}, s)\zeta^S(s)$ has a pole at $s = 1$. Hence the automorphic representation $\tilde{\Pi}'$ is Eisenstein, in fact induced from an automorphic irreducible representation (π_4, π_1) of the Levi subgroup $\mathrm{GL}(4, \mathbb{A}) \times \mathrm{GL}(1, \mathbb{A})$. This gives rise to an irreducible automorphic representation π_4 of $\mathrm{GL}(4, \mathbb{A})$ attached to (σ_1, σ_2), which after a character twist by ω will be denoted

$$\sigma_1 \times \sigma_2 .$$

The automorphic representation $\sigma_1 \times \sigma_2$ is uniquely determined by its partial L-series $L^S(\sigma_1 \times \sigma_2, s)$, which coincides with the partial L-series attached to the orthogonal four dimensional $\overline{E}_\lambda$-valued λ-adic Galois representation $\rho_{\sigma_1,\lambda} \otimes_{\overline{E}_\lambda} \rho_{\sigma_2,\lambda}$, the tensor product of the two dimensional Galois representations $\rho_{\sigma_i,\lambda}$ attached to σ_1 and σ_2. This four dimensional tensor product is an orthogonal Galois representation. It should not be confused with the symplectic four dimensional Galois representation obtained in [W2], Theorem I, which for weak endoscopic lifts is the direct sum of the Galois representations $\rho_{\sigma_i,\lambda}$. D.Ramakrishnan obtained a completely different and more general construction of the automorphic epresentation $\sigma_1 \times \sigma_2$ using converse theorems.

Bibliography

[AG]   Arthur J., Gelbart S., Lectures on automorphic L-functions. In: L-functions and Arithmetic. Eds. J. Coates and M.J. Taylor. London Mathematical Society Lecture Notes Series 153. pp. 1–59.

[Bo]   Borel A., Automorphic L-functions. In: Proc. of Symp. pure Math., vol. XXXIII, part 2, AMS 1979, pp. 27–91.

[B]   Borel A., Admissible representations of a semi-simple group over a local field with vectors fixed under a Iwahori group. Invent. Math. 35 (1976), pp. 233–259.

[BW]   Borel.A, Wallach N., Continuous cohomology, discrete subgroups, and representations of reductive groups. 2nd ed., Math. Surv. AMS vol. 67 (2000).

[BWW]   Ballmann J., Weissauer R., Weselmann U., Remarks on the fundamental lemma for stable twisted endoscopy of classical groups. Manuskripte der Forschergruppe Arithmetik, n. 2002-7, Heidelberg (2002). Available at http://www.mathi.uni-heidelberg.de/~weissaue/papers/bww.ps

[D]   Deligne, P., Formes modulaires et représentations l-adiques. In: Sem. Bourbaki 198/69, Springer Lecture Notes 79, pp. 347–363 (1971).

[Fl]   Flicker Y.F., Matching of orbital integrals on GL(4) and GSp(2). Mem. Amer. Math. Soc. 137 (1999).

[Fr]   Franke J., Harmonic analysis in weighted L_2-spaces. Ann. Sci. Ecole Norm. Sup. 31 (1998), pp. 181–279.

[Fu]   Furusawa M., On central critical values of the degree four L-functions for GSp(4): the fundamental lemma. Preprint (1999).

[GRS]   Ginzburg D., Rallis S., Soudry D., Periods, poles of L-functions and symplectic-orthogonal theta lifts. J. reine angew. Math. 487, pp. 85–114 (1997).

[GMP]   Goresky M., MacPherson R., Lefschetz numbers of Hecke correspondences. In: The zeta functions of Picard modular surfaces. Eds. Langlands R.P, Ramakrishnan D., Les Publications CRM, pp. 465–478 (1992).

[GMP2] Goresky M., MacPherson R., The topological trace formula. Journ. reine angew. Math. 560 (2003), pp. 71–150.

[Ha]   Hales Th., On the fundamental lemma for standard endoscopy: reduction to the unit element. Can. J. Math., vol 47 (1995), pp. 974–994.

[Ha2]   Hales Th., Shalika germs on GSp(4). In: Astérisque 171-172, Orbites unipotentes et représentations, II. Groupes p-adiques et réels, (1989), pp. 195–256.

[Ha3]   Hales Th., The twisted endoscopy of GL(4) and GL(5): transfer of Shalika germs. Duke Mathem. J. 76 (1994), pp. 595–632.

[H]   Howe R., Transcending classical invariant theory, Journal of the AMS. vol 2, no. 3 (1989), pp. 535–552.

[JPS] Jaquet H., Piateskii-Shapiro I., Shalika J., The θ-correspondence from GSp(4) to GL(4). Unpublished manuscript.

[JS] Jaquet H., Shalika J., On Euler Products and the classification of automorphic representations, I and II. Am. J. Math. 103 (1981), pp. 499–558 and pp. 777–815.

[JS2] Jaquet H., Shalika J., Exterior square L-functions. In: Automorphic forms, Shimura varieties, and L-functions, vol II, Proc. Conf., Ann Arbor/MI(USA) 1988, Persp. Math. 11, pp. 143–226 (1990).

[KV] Kashiwara M., Vergne M., On the Segal-Shale-Weil representation and harmonic polynomials. Invent. Math. 44, pp. 1–47 (1978).

[KS] Kottwitz R., Shelstad D., Foundations of twisted endoscopy. Astérisque 255 (1999).

[K2] Kottwitz R., Shimura varieties and λ-adic representations. In: Ann Arbor Proceedings (Eds. L. Clozel and J.S. Milne), Perspectives of Mathematics, Academic Press 1990, pp. 161–209.

[La] Labesse J.-P., Fonctions élementaires et lemme fondamental pour le changement de base stable. Duke Math. J. vol. 61, no 2 (1990), pp. 519–530.

[L] Langlands R., On the notion of an automorphic representation. A supplement to the preceding paper. In: Automorphic forms, representations, and L-functions, Proc. Symp. in Pure Math. vol 23, part 1, AMS (1979).

[LS] Langlands R., Shelstad D., Descent for transfer factors. In: The Grothendieck Festschrift, volume II, Prog. in Math. 87 (1990), pp. 485–563.

[P] Piatetski-Shapiro I.I., On the Saito-Kurokawa lifting. Inv. Math. 71, pp. 309–338 (1983).

[PSS] Piatetski-Shapiro I.I., Soudry D., On a correspondence of automorphic forms on orthogonal groups of order five. J. Math. Pures et Appl. 66, (1987), pp. 407–436.

[R] Rallis S., On the Howe duality conjecture. Compositio Math. 51 (1984), pp. 333–399.

[Ro] Rogawski J.D., Automorphic representations of unitary groups in three variables. Annals of Mathematics Studies 123, Princeton University Press (1990).

[Sha] Shalika J., The multiplicity one theorem for GL(n). Ann. Math. 100, pp. 171–193 (1974).

[S] Soudry D., The CAP representations of GSp(4). Crelles Journal 383, pp. 97–108 (1988).

[S2] Soudry D., A uniqueness theorem for representations of GSO(6) and the strong multiplicity one theorem for generic representations of GSp(4). Israel J. of Math. 58, pp. 257–287 (1987).

[S3] Jiang D., Soudry D., The multiplicity-one theorem for generic automorphic forms of GSp(4). Pacific J. Math. 229, no. 2, 381–388 (2007).

[V] Vignéras M.F., Correspondance entre représentations automorphes de GL(2) sur une extension quadratique de GSp(4) sur $\mathbb{Q}$, conjecture locale de Langlands pour GSp(4). Contemporary Mathematics, vol. 53, (1986).

[V2] Vignéras M.F., Caractérisation des intégrales orbitales sur un groupe réductif p-adique, J. Fac. Sci. Univ. Tokyo Sect. IA Math. 28 (1981), no. 3, pp. 945–961 (1982).

[Wa]   Waldspurger J.L., Une exercise sur GSp(4, F) et les représentations de Weil.
          Bull. Soc. Math. France 115, (1987), pp. 35–69.

[W1]   Weissauer R., Endoscopy for GSp(4). To appear in Springer Lecture Notes.

[W2]   Weissauer R., Four dimensional Galois Representations. In: Formes Auto-
          morphes (II), Le cas du groupe GSp(4). Edited by Tilouine, Carayol,
          Harris and Vignéras, Astérisque 302 (2006), pp. 67–150.

[W3]   Weissauer R., A remark on the existence of Whittaker models for L-
          packets of automorphic representations of GSp(4). Manuskripte der
          Forschergruppe Arithmetik n. 2000-26 (2000), http://www.mathi.uni-
          heidelberg.de/~weissaue/papers/whittaker.ps

[W4]   Weissauer R., Spectral approximation of twisted local κ-orbital integrals. In
          preparation.

[W5]   Weissauer R., Twisted SO(V)-invariant distibutions. Preprint.

[Wes]   Weselmann U., A twisted topological trace formula for Hecke operators and
          liftings from GSp(4) to general linear groups. Submitted to Comp. Math.

Multiplying Modular Forms

Martin H. Weissman

Abstract

The space of elliptic modular forms of fixed weight and level can be identified with a space of intertwining operators, from a holomorphic discrete series representation of $SL_2(\mathbb{R})$ to a space of automorphic forms. Moreover, multiplying elliptic modular forms corresponds to a branching problem involving tensor products of holomorphic discrete series representations.

In this paper, we explicitly connect the ring structure on spaces of modular forms with branching problems in the representation theory of real semisimple Lie groups. Furthermore, we construct a family of intertwining operators from discrete series representations into tensor products of other discrete series representations. This collection of intertwining operators provides the well-known ring structure on spaces of holomorphic modular forms.

An analytic subtlety prevents this collection of intertwining operators from directly yielding a ring structure on some spaces of modular forms. We discuss discrete decomposability, and its role in constructing ring structures.

Mathematics Subject Classification: 11F70, 22E46

Introduction

One of the easiest ways to construct a cuspidal elliptic modular form is by multiplying and adding Eisenstein series. For example, a cusp form of weight 12 is given by $\Delta = g_2^3 - 27g_3^2$, where g_2 and g_3 are Eisenstein series of weights 4 and 6, respectively. The fact that elliptic modular forms can be multiplied is practically obvious – but it relies on the fact that the product of holomorphic functions is again holomorphic. Multiplication of modular forms extends, without difficulty, to multiplying Siegel modular forms, and more generally, to multiplying scalar-valued holomorphic modular forms on tube domains.

311

In Remarque 2.1.4 of [Del73], Deligne notices a connection between multiplication of elliptic modular forms and the tensor product of two holomorphic discrete series representations. Motivated by this observation, we approach the problem of multiplying modular forms as a branching problem for tensor products of discrete series representations. In expanding on Deligne's remark, we attempt to answer two questions:

(i) What is the precise connection between branching problems in representation theory and multiplication of modular forms? More specifically, what statements from representation theory allow the construction of a ring of modular forms?

(ii) How, and when, can one find a canonical embedding of one discrete series representation into the tensor product of two other discrete series representations of a real semisimple Lie group?

It is far from clear whether one can multiply non-holomorphic modular forms. For example, one might try to multiply those modular forms on the split group of type G_2 considered by Gross, Gan, and Savin in [GGS02], or those on the split group of type D_4 in [Wei06]. Most generally, we are interested in multiplying modular forms, whose associated representation of a real semisimple group is in the discrete series. The "algebraic modular forms" of Gross [Gro99], where the associated real semisimple group is compact, can be multiplied by results of Khuri-Makdisi [KM01]. This paper can be seen as a generalization of his results, from finite-dimensional representations of compact Lie groups, to holomorphic discrete series representations of semisimple Lie groups, and an attempt to generalize to other discrete series.

One of our main results is a canonical embedding of a discrete series representation in the tensor product of two others, in the setting of unitary representations on Hilbert spaces. The existence of an embedding follows directly from work of Vargas [Var01]; using the analytic results of Ørsted-Vargas [ØV04], we are able to make such an embedding canonical, which is probably a result of independent interest.

The failure of discrete decomposability for tensor products of quaternionic discrete series seems to eliminate any hope of multiplying modular forms on G_2 or D_4. In fact, we suspect (after conversations with T. Kobayashi and others) that tensor products of discrete series representations never decompose discretely, outside of the finite-dimensional and holomorphic regimes (e.g., holomorphic and anti-holomorphic discrete series). The existence of embeddings in the unitary setting demonstrates that the failure of discrete decomposability is the only obstacle to multiplying modular forms. This obstacle demonstrates that one cannot hope for a naïve analogue of Shimura varieties, for (isotropic) groups without holomorphic discrete series.

We finish this paper by describing how our representation-theoretic multiplication corresponds to more classical multiplication of modular forms. This expands on Deligne's remark.

Acknowledgements. I would like to thank many people for their advice during the preparation of this paper. I exchanged a series of e-mails with J. Vargas while working on these results – I am very grateful for his advice. It also appears that the current work of Vargas has some overlap with mine, though I believe that this paper is of significant independent interest. I would also like to thank W. Schmid, H.-Y. Loke, W.-T. Gan, B. Gross, and D. Toledo, G. van der Geer, and L. Ji for some additional advice along the way.

The original draft of this paper contained a serious flaw; I am very thankful to Gordan Savin for pointing out this problem. Although this flaw curtails the inital ambitions of this paper, it is useful to present some negative results. In particular, results of Kobayashi [Kob98] on discrete decomposability explain why some modular forms cannot be multiplied.

Notation

G will always denote a connected semisimple algebraic group over $\mathbb{Q}$. Then we let $\mathfrak{g}$ denote its complexified Lie algebra: $\mathfrak{g} = \text{Lie}(\mathbf{G}) \otimes_{\mathbb{Q}} \mathbb{C}$. Let $\mathcal{U}$ denote the universal enveloping algebra of $\mathfrak{g}$, and $\mathcal{Z}$ the center of $\mathcal{U}$.

We write G for $\mathbf{G}(\mathbb{R})$, viewed as a semisimple real Lie group. Note that G is connected with finite center. Let K denote a maximal compact subgroup of G; K is then a connected compact Lie group. Γ will denote an arithmetic subgroup of $\mathbf{G}(\mathbb{Q})$.

All vector spaces will be complex vector spaces; all representations will be on complex vector spaces, and all tensor products will be over $\mathbb{C}$. If H is a group, and V, V' are representations of H, then we write $V \boxtimes V'$ for the "external tensor product" representation of $H \times H$. On the other hand, we write $V \otimes V'$ for the usual tensor product of representations, i.e., the restriction of $V \boxtimes V'$ to the diagonally embedded H in $H \times H$.

If U, V, W are (complex) vector spaces, then there are canonical isomorphisms which we name:

$$\text{can} \quad : \quad U \to \mathbb{C} \otimes U,$$
$$\text{comm} \quad : \quad U \otimes V \to V \otimes U,$$
$$\text{assoc} \quad : \quad (U \otimes V) \otimes W \to U \otimes (V \otimes W).$$

1 Automorphic and Modular Forms

The definition of an automorphic form varies slightly from one source to another. We follow the definition contained in the article of Borel and Jacquet [BJ79]. Our definition of modular forms is based on that of Gross, Gan, and Savin [GGS02].

1.1 $(\mathfrak{g}, K)$-modules

The definition of a $(\mathfrak{g}, K)$-module is taken from Borel-Wallach [BW00].

Definition 1.1. A $(\mathfrak{g}, K)$-module is a complex vector space W, endowed with simultaneous actions of $\mathfrak{g}$ and K (meaning a Lie algebra homomorphism $d\pi \colon \mathfrak{g} \to \mathrm{End}_{\mathbb{C}}(W)$ and a group homomorphism $\pi_K \colon K \to \mathrm{GL}(W)$), satisfying the following axioms

- W is locally finite and semisimple as a K-module.
- For all $X \in \mathfrak{g}$, all $k \in K$, and all $w \in W$,

$$\pi_K(k)(d\pi(X)w) = d\pi[Ad(k)(X)](\pi_K(k)w).$$

- If $W' \subset W$ is a K-stable finite-dimensional subspace of W, then the representation of K on W' is differentiable, and the resulting action $d(\pi_K)$ of $\mathfrak{k}$ agrees with the restriction $(d\pi)|_{\mathfrak{k}}$ of $d\pi$.

If W is a $(\mathfrak{g}, K)$-module, then the Lie algebra action of $\mathfrak{g}$ extends canonically to an action of the universal enveloping algebra $\mathfrak{U}$. We say that W has an *infinitesimal character* if there exists a homomorphism $\omega \colon \mathfrak{Z} \to \mathbb{C}$ such that $Zw = \omega(Z) \cdot w$, for all $Z \in \mathfrak{Z}$, and all $w \in W$.

Definition 1.2. A Harish-Chandra module is a $(\mathfrak{g}, K)$-module W, such that W is finitely generated as a $\mathfrak{U}$-module, and such that every irreducible representation of K occurs with finite multiplicity in W.

If W is an irreducible Harish-Chandra module, then W has an infinitesimal character.

Definition 1.3. A unitary structure on a $(\mathfrak{g}, K)$-module W is a positive non-degenerate scalar product $\langle \cdot, \cdot \rangle$ on W, which satisfies:

- For all $k \in K$, and all $v, w \in W$, $\langle \pi_K(k)v, \pi_K(k)w \rangle = \langle v, w \rangle$.
- For all $X \in \mathfrak{g}$, and all $v, w \in W$, $\langle d\pi(X)v, w \rangle + \langle v, d\pi(X)w \rangle = 0$.

Definition 1.4. A *weight* is an irreducible Harish-Chandra module W, endowed with a unitary structure.

As tensor products of $(\mathfrak{g}, K)$-modules arise often, we fix a few definitions here. First, there exists a unique associative algebra homomorphism $\Delta: \mathcal{U} \to \mathcal{U} \otimes \mathcal{U}$, which satisfies $\Delta(X) = 1 \otimes X + X \otimes 1$, for all $X \in \mathfrak{g}$. Thus, if W_1 and W_2 are $(\mathfrak{g}, K)$-modules, the tensor product $W_1 \otimes W_2$ is endowed with a natural $\mathcal{U}$ action as well as a K action via this diagonal embedding. This makes the tensor product $W_1 \otimes W_2$ a $(\mathfrak{g}, K)$ module as well.

1.2 Automorphic and Modular Forms

Here, we recall the definition of automorphic forms given by Borel-Jacquet in [BJ79]. Fix a finite-dimensional complex representation $\sigma: G \to \mathrm{GL}(E)$, with finite kernel and closed image. Fix a $\sigma(K)$-invariant Hilbert space structure on E. Define, for all $g \in G$,

$$\|g\| = \sqrt{|\mathrm{tr}(\sigma(g)^*\sigma(g))|}.$$

Here $\sigma(g)^*$ denotes the adjoint to $\sigma(g)$, with respect to the chosen Hilbert space structure.

Definition 1.5. An *automorphic form* on G (with respect to Γ) is a complex-valued function $f \in C^\infty(\Gamma\backslash G)$ such that:

(i) f has moderate growth. In other words, there exists a positive integer n, and a constant C such that:

$$|f(g)| \leq C \cdot \|g\|^n, \text{ for all } g \in G.$$

(ii) f is K-finite on the right.
(iii) f is $\mathcal{Z}$-finite on the right.

We write $\mathcal{A}(\Gamma\backslash G)$ for the complex vector space of automorphic forms on $\Gamma\backslash G$. This space is a $(\mathfrak{g}, K)$-module, by right-translation. Since it will be convenient later, we also define:

Definition 1.6. A *relaxed automorphic form* on G (wish respect to Γ) is a complex-valued function $f \in C^\infty(\Gamma\backslash G)$ such that conditions (i) and (ii), described above, are satisfied.

We write $\mathcal{A}_{\mathrm{rel}}(\Gamma\backslash G)$ for the complex vector space of relaxed automorphic forms on $\Gamma\backslash G$. This space is also a $(\mathfrak{g}, K)$-module, by right-translation.

Modular forms arise from irreducible $(\mathfrak{g}, K)$-submodules of $\mathcal{A}(\Gamma\backslash G)$. Let W be a weight, i.e., an irreducible unitary Harish-Chandra module.

Definition 1.7. A modular form on $\Gamma\backslash G$, of weight W, is a homomorphism F of $(\mathfrak{g}, K)$-modules from W to $\mathcal{A}(\Gamma\backslash G)$.

The space of modular forms of weight W is given by:

$$\mathcal{M}(\Gamma\backslash G, W) = \mathrm{Hom}_{(\mathfrak{g}, K)}(W, \mathcal{A}(\Gamma\backslash G)).$$

We may also define relaxed modular forms:

Definition 1.8. A relaxed modular form on $\Gamma\backslash G$, of weight W, is a homomorphism F of $(\mathfrak{g}, K)$-modules from W to $\mathcal{A}_{\mathrm{rel}}(\Gamma\backslash G)$.

The space of relaxed modular forms of weight W is given by:

$$\mathcal{M}_{\mathrm{rel}}(\Gamma\backslash G, W) = \mathrm{Hom}_{(\mathfrak{g}, K)}(W, \mathcal{A}_{\mathrm{rel}}(\Gamma\backslash G)).$$

However, it turns out that relaxed modular forms are no different than ordinary modular forms:

Proposition 1.9. *Every relaxed modular form is a modular form. In other words,* $\mathcal{M}_{\mathrm{rel}}(\Gamma\backslash G, W) = \mathcal{M}(\Gamma\backslash G, W).$

Proof. Suppose that $F \in \mathcal{M}_{\mathrm{rel}}(\Gamma\backslash G, W)$. We must show that, for all $w \in W$, the relaxed automorphic form $F(w)$ is $\mathcal{Z}$-finite. Since W is a weight, it has an infinitesimal character ω. Since F is a $(\mathfrak{g}, K)$-module homomorphism, it is $\mathcal{Z}$-intertwining. Therefore, for all $Z \in \mathcal{Z}$, we have:

$$ZF(w) = F(Zw) = \omega(Z) \cdot F(w).$$

Hence $F(w)$ is trivially $\mathcal{Z}$-finite.   □

2 Multiplication

Hereafter, we omit mention of Γ and G, when it does not reduce clarity.

Suppose that W_1 and W_2 are weights (irreducible unitary Harish-Chandra modules). We are interested in the following question: under what circumstances does there exist a weight W, and a natural $\mathbb{C}$-bilinear operation:

$$\mathcal{M}(W_1) \otimes \mathcal{M}(W_2) \to \mathcal{M}(W) \ ?$$

2.1 Multiplying automorphic forms

If $f_1, f_2 \in \mathcal{A} = \mathcal{A}(\Gamma\backslash G)$, let $m(f_1, f_2) = f_1 \cdot f_2$ denote the pointwise product. This extends to a bilinear map:

$$m: \mathcal{A} \otimes \mathcal{A} \to C^\infty(\Gamma\backslash G).$$

Proposition 2.1. *The map m is a homomorphism of $(\mathfrak{g}, K)$-modules. The image of m is contained in the space $\mathcal{A}_{\mathrm{rel}}$ of relaxed automorphic forms.*

Proof. The bilinear form m is a morphism of $(\mathfrak{g}, K)$-modules, since multiplication is compatible with right-translation. In other words, translation on the right by K is preserved by multiplication. Furthermore, if $X \in \mathfrak{g}$, then we have:

$$X(f_1 \cdot f_2) = (Xf_1) \cdot f_2 + f_1 \cdot (Xf_2) = m\left((X \otimes 1 + 1 \otimes X)(f_1 \otimes f_2)\right).$$

Now, we check that if $f_1, f_2 \in \mathcal{A}$, then $f = f_1 \cdot f_2 \in \mathcal{A}_{\mathrm{rel}}$. Since f_1, f_2 have moderate growth, we have:

$$|f(g)| = |f_1(g)||f_2(g)| \leq C_1 C_2 \cdot \|g\|^{n_1 + n_2},$$

where C_1, C_2, n_1, n_2 are the appropriate constants of moderate growth for f_1, f_2. It follows that f has moderate growth.

Finally, let Φ_1, Φ_2, Φ denote the complex vector spaces spanned by the K-translates of f_1, f_2, and f, respectively. Then $\Phi \subset m(\Phi_1 \otimes \Phi_2)$, and Φ_1 and Φ_2 are finite-dimensional. Hence Φ is finite-dimensional. $\qquad \square$

The above proposition states that the product of two automorphic forms is *almost* an automorphic form; the only condition missing is $\mathcal{Z}$-finiteness.

2.2 Multiplying modular forms

Suppose that we are given three weights W, W_1, W_2, and a homomorphism of $(\mathfrak{g}, K)$-modules:

$$\mu_{1,2} \colon W \to W_1 \otimes W_2.$$

Then we may define a bilinear map:

$$m_{1,2} = \mathrm{Co}(\mu_{1,2}) \colon \mathcal{M}(W_1) \otimes \mathcal{M}(W_2) \to \mathcal{M}_{\mathrm{rel}}(W),$$

by $m_{1,2}(F_1, F_2) = F$, where

$$F = m \circ [F_1 \otimes F_2] \circ \mu_{1,2},$$

following the sequence of maps:

$$W \xrightarrow{\mu_{1,2}} W_1 \otimes W_2 \xrightarrow{F_1 \otimes F_2} \mathcal{A} \otimes \mathcal{A} \xrightarrow{m} \mathcal{A}_{\mathrm{rel}}$$
$$\underset{F}{\longrightarrow}$$

Indeed, F is a $(\mathfrak{g}, K)$-module homomorphism, as a composite of three $(\mathfrak{g}, K)$-module homomorphisms. Since $\mathcal{M}_{\mathrm{rel}}(W) = \mathcal{M}(W)$, we arrive at:

Proposition 2.2. *If $F_1 \in \mathcal{M}(W_1)$, and $F_2 \in \mathcal{M}(W_2)$, then*

$$F = m_{1,2}(F_1, F_2) \in \mathcal{M}(W).$$

In this way, we get a linear map:

$$\text{Co}\colon \text{Hom}_{(\mathfrak{g},K)}(W, W_1 \otimes W_2) \to \text{Hom}_{\mathbb{C}}(\mathcal{M}(W_1) \otimes \mathcal{M}(W_2), \mathcal{M}(W)).$$

2.3 A graded ring

Suppose that C is a commutative monoid, with unit element 0. If $\lambda_1, \lambda_2 \in C$, we write $\lambda_1 + \lambda_2$ for the composition. A C-graded ring (over $\mathbb{C}$), is a commutative unital $\mathbb{C}$-algebra R, together with an isomorphism of vector spaces:

$$R = \bigoplus_{\lambda \in C} R_\lambda,$$

for subspaces R_λ of R, satisfying:

$$R_{\lambda_1} \cdot R_{\lambda_2} \subset R_{\lambda_1 + \lambda_2}$$

for all $\lambda_1, \lambda_2 \in C$. We also assume that $R_0 = \mathbb{C}$. To construct such an algebra, it suffices to define the collection of vector spaces R_λ, and the bilinear multiplication maps among them.

In order to construct a C-graded ring of modular forms, we require the following: suppose that for every $\lambda \in C$, one has an weight W_λ. Suppose that $W_0 = \mathbb{C}$ is the trivial Harish-Chandra module. Moreover, suppose we are given, for every pair λ_1, λ_2, a homomorphism of $(\mathfrak{g}, K)$-modules:

$$\mu_{\lambda_1, \lambda_2}\colon W_{\lambda_1 + \lambda_2} \to W_{\lambda_1} \otimes W_{\lambda_2}.$$

We refer to the set $\{\mu_{\lambda_1, \lambda_2}\}_{\lambda_1, \lambda_2 \in C}$ as the *family of comultiplications*. Define the graded vector space $\mathcal{M}_C = \bigoplus_{\lambda \in C} \mathcal{M}(W_\lambda)$. Then we have a set of multiplications on $\mathcal{M}_C$ via the bilinear maps $m_{\lambda_1, \lambda_2} = \text{Co}(\mu_{\lambda_1, \lambda_2})$ on each graded piece of $\mathcal{M}$. This makes $\mathcal{M}_C$ a C-graded ring if and only if the family of comultiplications satisfy the following axioms:

Co-identity: For all λ, there is an equality of maps:

$$W_\lambda \underset{\text{can}}{\overset{\mu_{0,\lambda}}{\rightleftarrows}} \mathbb{C} \otimes W_\lambda$$

Co-commutativity: For all $\lambda_1, \lambda_2 \in C$, and $\lambda = \lambda_1 + \lambda_2$, the following diagram commutes:

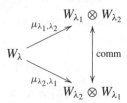

Co-associativity: For all $\lambda_1, \lambda_2, \lambda_3 \in C$, the following diagram commutes:

$$
\begin{array}{ccccc}
W_{\lambda_1+\lambda_2+\lambda_3} & \xrightarrow{\mu_{\lambda_1,\lambda_2+\lambda_3}} & W_{\lambda_1} \otimes W_{\lambda_2+\lambda_3} & \xrightarrow{\mathrm{Id}\otimes\mu_{\lambda_2,\lambda_3}} & W_{\lambda_1} \otimes (W_{\lambda_2} \otimes W_{\lambda_3}) \\
\parallel & & & & \Big\uparrow {\scriptstyle \text{assoc}} \\
W_{\lambda_1+\lambda_2+\lambda_3} & \xrightarrow{\mu_{\lambda_1+\lambda_2,\lambda_3}} & W_{\lambda_1+\lambda_2} \otimes W_{\lambda_3} & \xrightarrow{\mu_{\lambda_1,\lambda_2}\otimes\mathrm{Id}} & (W_{\lambda_1} \otimes W_{\lambda_2}) \otimes W_{\lambda_3}
\end{array}
$$

3 Compact Groups

Our first task will be to consider tensor products of representations of K, a maximal compact subgroup of G. In this section, we use the Borel-Weil realization of representations of K, in order to construct canonical embeddings of some representations of K in tensor products of others. While essentially contained in [KM01], we review these ideas thoroughly here as it will be important in what follows.

3.1 Representations of K

Recall that G is a connected semisimple Lie group, with finite center, and K is a maximal compact subgroup of G. In particular, K is also connected. Let T be a maximal torus in K. Assume hereafter that G has the same rank as K, so T is also a maximal torus in G.

We write Λ for the abelian group $\mathrm{Hom}_{\mathrm{alg}}(T_{\mathbb{C}}, \mathbb{C}^{\times})$. Fix a Borel subgroup $Q_{\mathbb{C}}$ of $K_{\mathbb{C}}$ containing $T_{\mathbb{C}}$, with unipotent racical $U_{\mathbb{C}}$. Let $U_{\mathbb{C}}^{-}$ denote the unipotent radical of the opposite Borel subgroup.

Let Δ_k^{+} denote the associated set of positive roots, for the adjoint action of $T_{\mathbb{C}}$ on $\mathfrak{q}_{\mathbb{C}}$. Let Λ_k^{+} denote the associated cone of dominant weights in Λ:

$$\Lambda_k^{+} = \{\lambda \in \Lambda \text{ such that } \forall \delta \in \Delta_k^{+}, \langle \lambda, \delta \rangle \geq 0\}.$$

We recall the famous Borel-Weil theorem that realizes the irreducible representations of K in the cohomology of a flag variety. This realization is particularly important, since we need some compatibilities among different representations.

The compact quotient $X = K/T$ may be identified with $K_{\mathbb{C}}/Q_{\mathbb{C}}$, and may thus be considered as a complex projective variety. Suppose that $\lambda \in \Lambda$. Extend λ to a homomorphism from $Q_{\mathbb{C}}$ to $\mathbb{C}^{\times}$, by having $U_{\mathbb{C}}$ sent to 1.

Let $\mathcal{O}(-\lambda)$ denote the line bundle on X associated to the character $-\lambda$. The holomorphic sections of $\mathcal{O}(-\lambda)$ consist of those holomorphic functions $h \colon K_{\mathbb{C}} \to \mathbb{C}$ satisfying:

$$h(kq) = \lambda(q) \cdot h(k) \text{ for all } k \in K_{\mathbb{C}}, q \in Q_{\mathbb{C}}.$$

We write $H^0(X, \mathcal{O}(-\lambda))$ for this space of holomorphic sections.

The complex vector space $H^0(X, \mathcal{O}(-\lambda))$ is a finite-dimensional representation of K, where K acts by left-translation $\tau = \tau_\lambda$:

$$[\tau(k)f](x) = f(k^{-1}x).$$

We recall the following theorem, which we believe originally appeared in [Ser95]; we refer to Knapp's book [Kna01] for a treatment as well:

Theorem 3.1 (Borel-Weil). *Suppose that $\lambda \in \Lambda_k^+$. Then $H_\lambda = H^0(X, \mathcal{O}(-\lambda))$ is an irreducible representation of K with highest weight λ.*

The Borel-Weil realization of the irreducible representations of K also fixes a number of auxiliary structures. First, the representations H_λ are endowed with natural Hilbert space structures. Namely, if $h_1, h_2 \in H_\lambda$, then we may define:

$$\langle h_1, h_2 \rangle_\lambda = \int_K h_1(k)\overline{h_2(k)}dk.$$

Here we normalize Haar measure so that K has measure 1. Note that this inner product is K-invariant.

The λ weight space of H_λ is one-dimensional. Moreover, the Borel-Weil realization yields two natural choices of highest-weight vectors.

Proposition 3.2. *There exists a unique highest-weight vector $h_\lambda \in H_\lambda$, which satisfies $h_\lambda(1) = 1$.*

Proof. Any highest-weight vector h_λ not only satisfies $h_\lambda(kq) = \lambda(q)h(k)$ for all $q \in Q_\mathbb{C}$, but also $h_\lambda(u^-k) = h_\lambda(k)$, for all u^- in $U_\mathbb{C}^-$. Thus the values of a highest-weight vector on the open Bruhat cell $U_\mathbb{C}^- Q_\mathbb{C}$ are nonzero complex multiples of the value at 1. It follows that any nonzero highest-weight vector h_λ satisfies $h_\lambda(1) \neq 0$. Since the highest-weight space is one-dimensional, there is a unique highest-weight vector $h_\lambda \in H_\lambda$, which additionally satisfies $h_\lambda(1) = 1$. $\qquad\square$

Proposition 3.3. *There exists a unique highest-weight vector $h_\lambda^\vee \in H_\lambda$, which satisfies $\langle h_\lambda^\vee, h \rangle = h(1)$, for all $h \in H_\lambda$.*

Proof. The map $h \mapsto h(1)$ is a linear functional on the finite-dimensional Hilbert space H_λ. Hence there exists a unique $h_\lambda^\vee \in H_\lambda$ which satisfies:

$$\langle h_\lambda^\vee, h \rangle = h(1), \quad \text{for all } h \in H_\lambda.$$

Moreover, the K-invariance of the inner product yields:

$$\langle \tau(t)h_\lambda^\vee, h \rangle = \langle h_\lambda^\vee, \tau(t)^{-1}h \rangle = h(t) = \lambda(t)h(1),$$

for all $t \in T$. Hence $h_\lambda^\vee$ is a highest-weight vector in H_λ. □

Note that $\langle h_\lambda^\vee, h_\lambda \rangle = 1$. We can also compute the following:

Proposition 3.4. *Let* $d_\lambda = \dim(H_\lambda)$. *Then we have:*

$$\langle h_\lambda, h_\lambda \rangle = d_\lambda^{-1}, \text{ and } \langle h_\lambda^\vee, h_\lambda^\vee \rangle = d_\lambda.$$

Proof. We may use the Schur orthogonality relations to compute this inner product:

$$
\begin{aligned}
\langle h_\lambda, h_\lambda \rangle &= \int_K h_\lambda(k) \overline{h_\lambda(k)} dk, \\
&= \int_K \langle h_\lambda^\vee, \tau(k)^{-1} h_\lambda \rangle \overline{\langle h_\lambda^\vee, \tau(k)^{-1} h_\lambda \rangle} dk, \\
&= \int_K \langle \tau(k) h_\lambda^\vee, h_\lambda \rangle \overline{\langle \tau(k) h_\lambda^\vee, h_\lambda \rangle} dk, \\
&= \frac{1}{d_\lambda} \langle h_\lambda^\vee, h_\lambda^\vee \rangle \overline{\langle h_\lambda, h_\lambda \rangle}.
\end{aligned}
$$

Since $\langle h_\lambda, h_\lambda \rangle$ is real and nonzero, we arrive at:

$$\langle h_\lambda^\vee, h_\lambda^\vee \rangle = d_\lambda.$$

Since h_λ is a scalar multiple of $h_\lambda^\vee$, and $\langle h_\lambda, h_\lambda^\vee \rangle = 1$, we deduce:

$$\langle h_\lambda, h_\lambda \rangle = d_\lambda^{-1}.$$

□

3.2 Tensor products of Borel-Weil representations

Consider the tensor product $H_{\lambda_1} \otimes H_{\lambda_2}$, for two dominant weights λ_1, λ_2. Let $\lambda = \lambda_1 + \lambda_2$. Then we find a natural map:

$$\mu_{\lambda_1, \lambda_2}^* : H_{\lambda_1} \otimes H_{\lambda_2} \to H_\lambda.$$

This map is defined by extending the following linearly:

$$\mu_{\lambda_1, \lambda_2}^*(h_1 \otimes h_2) = h_1 \cdot h_2.$$

This map is K-intertwining and nonzero. Moreover, evaluating at the identity yields:

$$\mu_{\lambda_1, \lambda_2}^*(h_{\lambda_1} \otimes h_{\lambda_2}) = h_\lambda.$$

Since we have natural K-invariant Hilbert structures on every space H_λ, we may dualize $\mu^*_{\lambda_1, \lambda_2}$ to obtain nonzero K-intertwining maps:

$$\mu_{\lambda_1, \lambda_2} \colon H_\lambda \to H_{\lambda_1} \otimes H_{\lambda_2}.$$

Let $\langle \cdot, \cdot \rangle_{\lambda_1 \otimes \lambda_2}$ denote the inner product on $H_{\lambda_1} \otimes H_{\lambda_2}$ given on simple tensors by:

$$\langle h_1 \otimes h_2, h'_1 \otimes h'_2 \rangle_{\lambda_1 \otimes \lambda_2} = \langle h_1, h'_1 \rangle_{\lambda_1} \cdot \langle h_2, h'_2 \rangle_{\lambda_2}.$$

Then, for all $h \in H_\lambda$, $\mu_{\lambda_1, \lambda_2}(h)$ satisfies, for all $h_1 \in H_{\lambda_1}$, and $h_2 \in H_{\lambda_2}$:

$$\langle \mu_{\lambda_1, \lambda_2}(h), h_1 \otimes h_2 \rangle_{\lambda_1 \otimes \lambda_2} = \langle h, h_1 \cdot h_2 \rangle_{\lambda_1 + \lambda_2}.$$

In particular, $\mu_{\lambda_1, \lambda_2}(h_\lambda^\vee)$ is a vector in the λ weight space of $H_{\lambda_1} \otimes H_{\lambda_2}$ (which has dimension 1). Moreover, we compute:

$$\langle \mu_{\lambda_1, \lambda_2}(h_\lambda^\vee), h_{\lambda_1} \otimes h_{\lambda_2} \rangle = \langle h_\lambda^\vee, h_\lambda \rangle = 1.$$

It follows that:

$$\mu_{\lambda_1, \lambda_2}(h_\lambda^\vee) = h_{\lambda_1}^\vee \otimes h_{\lambda_2}^\vee.$$

Theorem 3.5. *The family of K-intertwining maps $\{\mu_{\lambda_1, \lambda_2}\}$ satisfies the axioms of co-identity, co-commutativity, and co-associativity.*

Proof. **Co-Identity:** Note that $H_0 \cong \mathbb{C}$ consists only of the constant functions, since X is a projective variety and has no global holomorphic functions besides constants. Hence we have, for all $f \in H_\lambda$, all $C \in H_0$, and $f' \in H_\lambda$:

$$
\begin{aligned}
\langle \mu_{0, \lambda}(f), C \otimes f' \rangle_{0 \otimes \lambda} &= \langle f, Cf' \rangle_\lambda, \\
&= \langle 1 \otimes f, C \otimes f' \rangle_{0 \otimes \lambda}.
\end{aligned}
$$

Thus $\mu_{0, \lambda}(f) = 1 \otimes f$.

Co-Commutatitivity: Let $\lambda = \lambda_1 + \lambda_2$. For all $f \in H_\lambda$, all $f_1 \in H_{\lambda_1}$, and all $f_2 \in H_{\lambda_2}$, we have:

$$
\begin{aligned}
\langle \mu_{\lambda_1, \lambda_2}(f), f_1 \otimes f_2 \rangle_{\lambda_1 \otimes \lambda_2} &= \langle f, f_1 f_2 \rangle_\lambda, \\
&= \langle f, f_2 f_1 \rangle_\lambda, \\
&= \langle \mu_{\lambda_2, \lambda_1}(f), f_2 \otimes f_1 \rangle_{\lambda_2 \otimes \lambda_1}.
\end{aligned}
$$

It follows that $\mathrm{comm} \circ \mu_{\lambda_1, \lambda_2} = \mu_{\lambda_2, \lambda_1}$.

Co-Associativity: Suppose that $f \in H_{\lambda_1 + \lambda_2 + \lambda_3}$, and $f_i \in H_{\lambda_i}$ for $i = 1, 2, 3$. Define maps:

$$
\begin{aligned}
\mu_{\lambda_1, (\lambda_2, \lambda_3)} &= (\mathrm{Id} \otimes \mu_{\lambda_2, \lambda_3}) \circ \mu_{\lambda_1, \lambda_2 + \lambda_3}, \\
\mu_{(\lambda_1, \lambda_2), \lambda_3} &= (\mu_{\lambda_1, \lambda_2} \otimes \mathrm{Id}) \circ \mu_{\lambda_1 + \lambda_2, \lambda_3}.
\end{aligned}
$$

Then we compute:

$$\langle \mu_{\lambda_1,(\lambda_2,\lambda_3)}(f), f_1 \otimes (f_2 \otimes f_3) \rangle_{\lambda_1 \otimes (\lambda_2 \otimes \lambda_3)}$$
$$= \langle \mu_{\lambda_1,\lambda_2+\lambda_3}(f), f_1 \otimes f_2 f_3 \rangle_{\lambda_1 \otimes (\lambda_2+\lambda_3)},$$
$$= \langle f, f_1 f_2 f_3 \rangle_{\lambda_1+\lambda_2+\lambda_3}.$$

A similar computation yields:

$$\langle \mu_{(\lambda_1,\lambda_2),\lambda_3}(f), (f_1 \otimes f_2) \otimes f_3 \rangle_{(\lambda_1 \otimes \lambda_2) \otimes \lambda_3} = \langle f, f_1 f_2 f_3 \rangle_{\lambda_1+\lambda_2+\lambda_3}.$$

It follows that assoc $\circ \, \mu_{\lambda_1,(\lambda_2,\lambda_3)} = \mu_{(\lambda_1,\lambda_2),\lambda_3}$. $\qquad\square$

The above proposition immediately shows that the "algebraic modular forms", considered by Gross in [Gro99], form a graded ring. This was also proven, albeit in somewhat different language, but by similar techniques, by Khuri-Makdisi in [KM01].

Corollary 3.6. *Suppose that* **G** *is a connected semisimple algebraic group over* $\mathbb{Q}$, *and* $G = \mathbf{G}(\mathbb{R})$ *is compact. Then for any arithmetic (and hence finite) subgroup* $\Gamma \subset G$, *the space of modular forms* $\mathcal{M}(\Gamma \backslash G)$ *of all weights, forms a ring graded by the monoid of dominant weights of* **G**.

4 Noncompact Groups, Discrete series

In this section we find a family of comultiplications for discrete series representations of noncompact groups. However, these intertwining operators arise in the category of unitary representations on Hilbert spaces. A failure of discrete decomposability prevents these operators from yielding similar operators in the category of $(\mathfrak{g}, K)$-modules, except in some very special cases.

Just as the construction for compact groups relied on a geometric realization of representations, we rely on a geometric realization of discrete series representations here.

4.1 Realization of Discrete Series

Suppose that $\lambda \in \Lambda_k^+$. Let H_λ denote the irreducible representation of K with highest weight λ, as defined in the previous section. Recall that $\tau = \tau_\lambda$ denotes the action of K on H_λ by left-translation. This yields a vector bundle over the symmetric space $Y = G/K$ in the usual way; the smooth sections are defined by:

$$C^\infty(Y, H_\lambda) :=$$
$$\{ f \in C^\infty(G, H_\lambda) \mid \forall k \in K, g \in G, \ f(gk) = [\tau_\lambda(k^{-1})](f(g)) \}.$$

We may define the space of L^2 sections similarly, using the previously-defined Hilbert space structure on every H_λ. The Killing form yields a measure on $Y = G/K$, normalizing the inner product on L^2 sections. If Ω denotes the Casimir element of $\mathcal{Z}$, then define the space:

$$\mathcal{J}(Y, H_\lambda) = \left\{ f \in [C^\infty \cap L^2](Y, H_\lambda) \text{ such that } \Omega f = (\|\lambda\|^2 - \|\rho\|^2)f \right\}.$$

As explained in [ØV04], based on the original work of [HP74] and extended by [Wal76], we have:

Theorem 4.1. *Suppose that H_λ is isomorphic to the lowest K-type of a discrete series representation of G. Then $\mathcal{J}(Y, H_\lambda)$ is a closed subspace of $L^2(Y, H_\lambda)$. When endowed with the resulting Hilbert space structure and the left-translation action of G, $\mathcal{J}(Y, H_\lambda)$ is a irreducible unitary representation of G, in the discrete series, with lowest K-type isomorphic to H_λ.*

Hereafter, we define $\mathcal{J}_\lambda$ to be the subspace of K-finite vectors in $\mathcal{J}(Y, H_\lambda)$, viewed as an irreducible unitary Harish-Chandra module.

4.2 Tensor products of discrete series

Now, suppose that $\lambda_1, \lambda_2 \in \Lambda_k^+$, and $\lambda = \lambda_1 + \lambda_2$. Theorem 1 of [Var01] implies the following:

Proposition 4.2. *Suppose that H_{λ_1}, H_{λ_2}, and H_λ are all lowest K-types of discrete series representations of G. Then there exists a nonzero G-intertwining continuous embedding:*

$$\nu_{1,2} \colon \mathcal{J}(Y, H_\lambda) \hookrightarrow \mathcal{J}(Y, H_{\lambda_1}) \hat{\otimes} \mathcal{J}(Y, H_{\lambda_2}).$$

Proof. The completed tensor product $\mathcal{J}(Y, H_{\lambda_1}) \hat{\boxtimes} \mathcal{J}(Y, H_{\lambda_2})$ is a discrete series representation of $G \times G$, with lowest $(K \times K)$-type $H_{\lambda_1} \boxtimes H_{\lambda_2}$. Moreover, H_λ occurs in the restriction of $H_{\lambda_1} \boxtimes H_{\lambda_2}$ from $K \times K$ to K, as is well-known, and realized via the map $\mu_{\lambda_1, \lambda_2}$ from before. Theorem 1 of [Var01] applies, and yields the proposition immediately. $\qquad\square$

Unfortunately, the results and methods in [Var01] do not provide an explicit embedding of these representations. However, results of [ØV04] suggest that such an explicit embedding is possible.

We fix some specialized notation for restriction maps and tensor products. Suppose that H_1, H_2, H_3 are finite-dimensional representations of K. Suppose that j_1, j_2, j_3 are sections of the associated bundles: $j_i \in C^\infty(Y, H_i)$ for $i = 1, 2, 3$. Then we may take the fibrewise tensor product; we write $j_1 \odot j_2$ for the element of $C^\infty(Y, H_1 \otimes H_2)$, given by $[j_1 \odot j_2](y) = j_1(y) \otimes j_2(y)$. We use this notation to distinguish this operation from the "external" tensor product $j_1 \otimes j_2 \in C^\infty(Y, H_1) \otimes C^\infty(Y, H_2)$.

There are canonical isomorphsims:

$$\mathrm{comm}^Y : C^\infty(Y, H_1 \otimes H_2) \to C^\infty(Y, H_2 \otimes H_1),$$

$$\mathrm{assoc}^Y : C^\infty(Y, (H_1 \otimes H_2) \otimes H_3) \to C^\infty(Y, H_1 \otimes (H_2 \otimes H_3)).$$

We have the following properties of commutativity and associativity:

$$\mathrm{comm}^Y(j_1 \odot j_2) = j_2 \odot j_1,$$

$$\mathrm{assoc}^Y((j_1 \odot j_2) \odot j_3)) = j_1 \odot (j_2 \odot j_3).$$

Then, Theorem 1 of [ØV04], together with the density of the algebraic tensor product in the completed tensor product, directly implies the following:

Proposition 4.3. *The map $j_1 \otimes j_2 \mapsto j_1 \odot j_2$ extends uniquely to a continuous G-intertwining linear map of Hilbert spaces:*

$$r : \mathcal{J}(Y, H_{\lambda_1}) \hat{\otimes} \mathcal{J}(Y, H_{\lambda_2}) \to L^2(Y, H_{\lambda_1} \otimes H_{\lambda_2}).$$

Proof. The map $j_1 \otimes j_2 \mapsto j_1 \odot j_2$ can be seen as the restriction, to the diagonally embedded $G \subset G \times G$, of an appropriate element of the Hotta-Parthasarathy realization of the discrete series of $G \times G$. Thus, Theorem 1 of [ØV04] applies. $\qquad\square$

One has a G-intertwining continuous linear map, adjoint to restriction:

$$r^\star : L^2(Y, H_{\lambda_1} \otimes H_{\lambda_2}) \to \mathcal{J}(Y, H_{\lambda_1}) \hat{\otimes} \mathcal{J}(Y, H_{\lambda_2}).$$

Recall that there are canonical K-intertwining maps:

$$
\begin{aligned}
\mu_{\lambda_1, \lambda_2} &: & H_\lambda &\to H_{\lambda_1} \otimes H_{\lambda_2}, \\
\mu^*_{\lambda_1, \lambda_2} &: & H_{\lambda_1} \otimes H_{\lambda_2} &\to H_\lambda.
\end{aligned}
$$

Applying $\mu_{\lambda_1, \lambda_2}$ or $\mu^*_{\lambda_1, \lambda_2}$ fibrewise, It follows that there are continuous G-intertwining maps:

$$
\begin{aligned}
\mu^Y_{\lambda_1, \lambda_2} &: & L^2(Y, H_\lambda) &\to L^2(Y, H_{\lambda_1} \otimes H_{\lambda_2}), \\
\mu^{*Y}_{\lambda_1, \lambda_2} &: & L^2(Y, H_{\lambda_1} \otimes H_{\lambda_2}) &\to L^2(Y, H_\lambda).
\end{aligned}
$$

The maps $\mu^Y_{\lambda_1, \lambda_2}$ are co-commutative and co-associative in the natural way, following the same properties of $\mu_{\lambda_1, \lambda_2}$.

It follows that we may define:

Definition 4.4. There is a continuous G-intertwining map:

$$\nu_{\lambda_1,\lambda_2} = r^* \circ \mu^Y_{\lambda_1,\lambda_2} : \mathfrak{J}(Y, H_\lambda) \to \mathfrak{J}(Y, H_{\lambda_1}) \hat{\otimes} \mathfrak{J}(Y, H_{\lambda_2}).$$

If $j \in \mathfrak{J}(Y, H_\lambda)$ then $\nu_{\lambda_1,\lambda_2}(j)$ is the unique element of the space $\mathfrak{J}(Y, H_{\lambda_1}) \hat{\otimes} \mathfrak{J}(Y, H_{\lambda_2})$ that satisfies:

$$\langle \nu_{\lambda_1,\lambda_2}(j), j_1 \otimes j_2 \rangle = \langle \mu^Y_{\lambda_1,\lambda_2} j, j_1 \odot j_2 \rangle,$$

for all $j_1 \in \mathfrak{J}(Y, H_{\lambda_1})$ and all $j_2 \in \mathfrak{J}(Y, H_{\lambda_2})$.

4.3 Discrete decomposability

In the category of unitary representations of G on Hilbert spaces, we have constructed a continuous G-intertwining map:

$$\nu_{\lambda_1,\lambda_2} : \mathfrak{J}(Y, H_\lambda) \to \mathfrak{J}(Y, H_{\lambda_1}) \hat{\otimes} \mathfrak{J}(Y, H_{\lambda_2}).$$

However, one may not construct an intertwining map of $(\mathfrak{g}, K)$-modules in the naïve fashion. It is true that one may pass the the subspaces of K-finite vectors, to get a map:

$$\nu_{\lambda_1,\lambda_2} : \mathfrak{J}_\lambda \to \left(\mathfrak{J}(Y, H_{\lambda_1}) \hat{\otimes} \mathfrak{J}(Y, H_{\lambda_2}) \right)^K.$$

However, the target of this map rarely coincides with $\mathfrak{J}_{\lambda_1} \otimes \mathfrak{J}_{\lambda_2}$; this reflects the difference between K-finite vectors and $(K \times K)$-finite vectors. We are very thankful to Gordan Savin for pointing out this difficulty.

The seriousness of this difficulty is illustrated by the following, which easily follows from deeper results of Kobayashi [Kob98]:

Proposition 4.5. *The following conditions are equivalent:*

(i) *The unitary representation $\mathfrak{J}(Y, H_{\lambda_1}) \hat{\otimes} \mathfrak{J}(Y, H_{\lambda_2})$ of G decomposes discretely, i.e., as a Hilbert space direct sum of irreducible unitary representations of G.*

(ii) *The space of K-finite vectors coincides with the $K \times K$-finite vectors, in the sense that:*

$$\left(\mathfrak{J}(Y, H_{\lambda_1}) \hat{\otimes} \mathfrak{J}(Y, H_{\lambda_2}) \right)^K = \mathfrak{J}_{\lambda_1} \otimes \mathfrak{J}_{\lambda_2}.$$

(iii) *The intertwining map $\nu_{\lambda_1,\lambda_2}$ restricts to a homomorphism*

$$\mathfrak{J}_\lambda \to \mathfrak{J}_{\lambda_1} \otimes \mathfrak{J}_{\lambda_2}$$

of $(\mathfrak{g}, K)$-modules.

Proof. To check that (i) $\Rightarrow$ (ii), we assume that $\mathcal{J}(Y, H_{\lambda_1}) \hat{\otimes} \mathcal{J}(Y, H_{\lambda_2})$ decomposes discretely. Then, using Theorem 4.2 of [Kob98], and the fact that we work with discrete series representations, it follows that the restriction of $\mathcal{J}(Y, H_{\lambda_1}) \hat{\otimes} \mathcal{J}(Y, H_{\lambda_2})$ to K is admissible. From this, it follows from Proposition 1.6 of [Kob98] that the K-finite vectors coincide with the $K \times K$-finite vectors as needed.

The implication (ii) $\Rightarrow$ (iii) is obvious, following our previous comments.

For the implication (iii) $\Rightarrow$ (i), the irreducibility of $\mathcal{J}_\lambda$, the nonvanishing of $\nu_{\lambda_1,\lambda_2}$ (to be proven in Theorem 4.12), and Lemma 1.5 of Kobayashi [Kob98] imply that $\left(\mathcal{J}(Y, H_{\lambda_1}) \hat{\otimes} \mathcal{J}(Y, H_{\lambda_2})\right)^K$ decomposes discretely as a g-module (see Definition 1.1 of [Kob98]). By Lemma 1.3 of [Kob98], this implies a direct sum decomposition:

$$\left(\mathcal{J}(Y, H_{\lambda_1}) \hat{\otimes} \mathcal{J}(Y, H_{\lambda_2})\right)^K = \bigoplus_{i=1}^{\infty} X_i,$$

into orthogonal irreducible g-modules. By Theorem 4.2 of [Kob98], $\left(\mathcal{J}(Y, H_{\lambda_1}) \hat{\otimes} \mathcal{J}(Y, H_{\lambda_2})\right)$ is K-admissible, and therefore the Hilbert space $\mathcal{J}(Y, H_{\lambda_1}) \hat{\otimes} \mathcal{J}(Y, H_{\lambda_2})$ can be identified with the completion of its subspace of K-finite vectors. Thus, we may identify this Hilbert space with the Hilbert space direct sum of completions:

$$\mathcal{J}(Y, H_{\lambda_1}) \hat{\otimes} \mathcal{J}(Y, H_{\lambda_2}) = \widehat{\bigoplus}_{i=1}^{\infty} \widehat{X_i}.$$

The subspaces $\widehat{X_i}$ are irreducible unitary representations of G, since their underlying $(\mathfrak{g}, K)$-modules are irreducible. Thus (iii) $\Rightarrow$ (i). $\square$

The conditions of the previous proposition are quite restrictive. They are satisfied for holomorphic discrete series representations, for example. Kobayashi gives many sufficient and practical conditions for discrete decomposability, which we do not recall here. Instead, we simply give:

Definition 4.6. The pair λ_1, λ_2 is discretely decomposable, if the conditions of the previous proposition are satisfied.

4.4 The comultiplication data

We now prove that the family of comultiplications, given by $\nu_{\lambda_1,\lambda_2}$ is well-behaved:

Proposition 4.7. *Suppose that C is a submonoid of Λ_k^+, containing 0, such that for every $0 \neq \lambda \in C$, H_λ is isomorphic to the lowest K-type of a discrete* *series representation of G. Suppose that every pair $\lambda_1, \lambda_2 \in C$ is discretely decomposable. Then the family of comultiplications $\{\nu_{\lambda_1,\lambda_2}\}$, among the family of Harish-Chandra modules $\mathfrak{J}_\lambda$ ($\lambda \in C$) has the properties of co-identity, co-commutativity, and co-associativity.*

Proof. **Co-Identity:** We always fix $\mathfrak{J}_0 = \mathbb{C}$, and we define the comultiplications $\nu_{0,\lambda}$ to be the canonical maps. Thus the property of co-identity is vacuous.

Co-Commutativity: Let $\lambda = \lambda_1 + \lambda_2$, with $\lambda_1, \lambda_2 \in C$. Fix $j \in \mathfrak{J}_\lambda$, $j_1 \in \mathfrak{J}_{\lambda_1}$ and $j_2 \in \mathfrak{J}_{\lambda_2}$. Then we have:

$$
\begin{aligned}
\langle \nu_{\lambda_1,\lambda_2}(j), j_1 \otimes j_2 \rangle, &= \langle \mu^Y_{\lambda_1,\lambda_2} j, j_1 \odot j_2 \rangle, \\
&= \langle \mathrm{comm}^Y(\mu^Y_{\lambda_1,\lambda_2} j), \mathrm{comm}^Y(j_1 \odot j_2) \rangle, \\
&= \langle \mu^Y_{\lambda_2,\lambda_1} j, j_2 \odot j_1 \rangle, \\
&= \langle \nu_{\lambda_2,\lambda_1}(j), j_2 \otimes j_1 \rangle.
\end{aligned}
$$

Note that the middle step essentially relies on the co-commutativity of $\mu_{\lambda_1,\lambda_2}$. It follows that the family $\{\nu_{\lambda_1,\lambda_2}\}$ is co-commutative.

Co-Associativity: Suppose that $\lambda_1, \lambda_2, \lambda_3 \in C$. Suppose that $j \in \mathfrak{J}_{\lambda_1+\lambda_2+\lambda_3}$, and $j_i \in \mathfrak{J}_{\lambda_i}$ for $i = 1, 2, 3$. Define maps:

$$
\begin{aligned}
\nu_{\lambda_1,(\lambda_2,\lambda_3)} &= (\mathrm{Id} \otimes \nu_{\lambda_2,\lambda_3}) \circ \nu_{\lambda_1,\lambda_2+\lambda_3}, \\
\nu_{(\lambda_1,\lambda_2),\lambda_3} &= (\nu_{\lambda_1,\lambda_2} \otimes \mathrm{Id}) \circ \nu_{\lambda_1+\lambda_2,\lambda_3}.
\end{aligned}
$$

Then we compute:

$$
\begin{aligned}
\langle \nu_{\lambda_1,(\lambda_2,\lambda_3)}(j), j_1 \otimes (j_2 \otimes j_3) \rangle \\
= \Big\langle \big[\mathrm{Id} \otimes \mu^Y_{\lambda_2,\lambda_3}\big] (\nu_{\lambda_1,\lambda_2+\lambda_3} j), j_1 \otimes (j_2 \odot j_3) \Big\rangle, \\
= \Big\langle \nu_{\lambda_1,\lambda_2+\lambda_3} j, j_1 \otimes \mu^{*Y}_{\lambda_2,\lambda_3}(j_2 \odot j_3) \Big\rangle, \\
= \Big\langle \mu^Y_{\lambda_1,\lambda_2+\lambda_3} j, j_1 \odot \mu^{*Y}_{\lambda_2,\lambda_3}(j_2 \odot j_3) \Big\rangle, \\
= \Big\langle j, \mu^{*Y}_{\lambda_1,\lambda_2+\lambda_3} \big(j_1 \odot \mu^{*Y}_{\lambda_2,\lambda_3}(j_2 \odot j_3)\big) \Big\rangle, \\
= \Big\langle j, \mu^{*Y}_{\lambda_1+\lambda_2,\lambda_3} \big(\mu^{*Y}_{\lambda_1,\lambda_2}(j_1 \odot j_2) \odot j_3\big) \Big\rangle, \\
= \langle \nu_{(\lambda_1,\lambda_2),\lambda_3}(j), (j_1 \otimes j_2) \otimes j_3 \rangle.
\end{aligned}
$$

We are using the associativity of the bilinear operations μ^{*Y}, which follows from ordinary associativity of multiplication, and the Borel-Weil realization. It follows that the family $\{\nu_{\lambda_1,\lambda_2}\}$ is co-associative. $\square$

The above proposition yields the following:

Proposition 4.8. *Let C be any submonoid of Λ_k^+ containing 0, such that for all $0 \neq \lambda \in C$, the representation H_λ is isomorphic to the lowest K-type of some discrete series representation of G. Suppose that every pair in C is discretely decomposable. Let $\mathcal{M}_C = \bigoplus_{\lambda \in C} \mathcal{M}(\mathcal{J}_\lambda)$. Then the family of comultiplications $\{\nu_{\lambda_1, \lambda_2}\}$ endow $\mathcal{M}_C$ with the structure of a commutative ring.*

Of course, for the ring $\mathcal{M}_C$ to be interesting, we should expect the multiplication maps to be non-trivial. In other words, we must prove that the comultiplications $\nu_{\lambda_1, \lambda_2}$ are nonzero.

4.5 Nonvanishing of intertwining operators

In order to prove that the maps $\nu_{\lambda_1, \lambda_2}$ are nonzero, we use spherical trace functions. Essentially all of the facts we use about spherical trace functions can be found in the excellent treatment of [WW83].

First, we fix some notation. Let $\mathbf{i} \colon H_\lambda \hookrightarrow \mathcal{J}(Y, H_\lambda)$ denote the unique K-equivariant inclusion, which is adjoint to the evaluation map (at the identity) $\mathbf{ev} \colon \mathcal{J}(Y, H_\lambda) \to H_\lambda$. Let $\mathbf{p} \colon \mathcal{J}(Y, H_\lambda) \to \mathbf{i}(H_\lambda)$ denote the orthogonal projection onto the lowest K-type.

Similarly, let

$$\mathbf{I} \colon H_{\lambda_1} \boxtimes H_{\lambda_2} \to \mathcal{J}(Y \times Y, H_{\lambda_1} \boxtimes H_{\lambda_2})$$

denote the $K \times K$-equivariant inclusion adjoint to the evaluation map. Let

$$\mathbf{P} \colon \mathcal{J}(Y \times Y, H_{\lambda_1} \boxtimes H_{\lambda_2}) \to \mathbf{I}\left(H_{\lambda_1} \boxtimes H_{\lambda_2}\right)$$

denote the projection onto the lowest $(K \times K)$-type.

There is a close relationship between projecting onto the lowest K-type (or $(K \times K)$-type), and evaluating at the identity. Namely, for all $j \in \mathcal{J}(Y, H_\lambda)$, we have:

$$[\mathbf{i}^{-1} \circ \mathbf{p}](j) = \frac{1}{\mathbf{d}} \cdot j(1),$$

where $\mathbf{d}$ is the formal degree of $\mathcal{J}(Y, H_\lambda)$. A corresponding formula holds for $J \in \mathcal{J}(Y \times Y, H_{\lambda_1} \boxtimes H_{\lambda_2})$:

$$[\mathbf{I}^{-1} \circ \mathbf{P}](J) = \frac{1}{\mathbf{D}} \cdot J(1).$$

Let $\mathbf{l}$ denote the left-translation action of G on $\mathcal{J}(Y, H_\lambda)$. Let $\mathbf{L}$ denote the left-translation action of $G \times G$ on $\mathcal{J}(Y \times Y, H_{\lambda_1} \boxtimes H_{\lambda_2})$.

Let $h^1, \ldots, h^a$ denote an orthonormal basis of H_λ. Let $e^s = \mu_{\lambda_1, \lambda_2}(h^s)$ for all $1 \leq s \leq a$. Let F denote the orthogonal complement of $\mu_{\lambda_1, \lambda_2}(H_\lambda)$ in $H_{\lambda_1} \otimes H_{\lambda_2}$. Let $f^1, \ldots, f^b$ denote an orthonormal basis of F. Thus the set

$\{e^1, \ldots, e^a, f^1, \ldots, f^b\}$ is an orthogonal basis of $H_{\lambda_1} \otimes H_{\lambda_2}$. For all $1 \le s \le a$, we have $\langle e^s, e^s \rangle = c$, for some fixed positive number c.

With these choices, we may write the spherical trace functions quite explicitly. On G, the spherical trace function for $\mathcal{J}(Y, H_\lambda)$ is given by:

$$
\begin{aligned}
\phi(g) &= \mathrm{Tr}(\mathbf{p} \circ \mathbf{l}(g) \circ \mathbf{p}) = \mathrm{Tr}(\mathbf{i}^{-1} \circ \mathbf{p} \circ \mathbf{l}(g) \circ \mathbf{p} \circ \mathbf{i}) \\
&= \sum_{s=1}^{a} \langle \mathbf{i}^{-1} \circ \mathbf{p} \circ \mathbf{l}(g) \circ \mathbf{p} \circ \mathbf{i} h^s, h^s \rangle \\
&= \frac{1}{\mathbf{d}} \cdot \sum_{s=1}^{a} \langle [\mathbf{l}(g) \circ \mathbf{p} \circ \mathbf{i} h^s](1), h^s \rangle \\
&= \frac{1}{\mathbf{d}} \cdot \sum_{s=1}^{a} \langle [\mathbf{p} \circ \mathbf{i} h^s](g^{-1}), h^s \rangle = \frac{1}{\mathbf{d}} \cdot \sum_{s=1}^{a} \langle [\mathbf{i} h^s](g^{-1}), h^s \rangle.
\end{aligned}
$$

Similarly, on $G \times G$, we have an explicit expresssion for the spherical trace function. Restricting to the diagonal $G \subset G \times G$ yields:

$$
\begin{aligned}
\Phi(g) &= \mathrm{Tr}\,(\mathbf{P} \circ \mathbf{L}(g) \circ \mathbf{P}) \\
&= \frac{1}{c\mathbf{D}} \cdot \left(\sum_{s=1}^{a} \langle \mathbf{I} e^s(g^{-1}), e^s \rangle + \sum_{t=1}^{b} \langle \mathbf{I} f^t(g^{-1}), f^t \rangle \right).
\end{aligned}
$$

We recall Lemma 2 of [Var01]

Lemma 4.9. *The product* $\phi(g) \cdot \Phi(g)$ *is an integrable function on* G *(with respect to a Haar measure). Moreover,*

$$
0 < \int_G \phi(g) \overline{\Phi(g)} dg \in \mathbb{R}.
$$

Proof. This is precisely the statement of Lemma 2 of [Var01], for the pair of groups $G \subset G \times G$. $\qquad\square$

Using Schur's orthogonality relations, we may prove:

Lemma 4.10. *For all* $1 \le s \le a$, *and for all* $1 \le t \le b$, *we have:*

$$
\int_G \langle [\mathbf{i} h^s](g^{-1}), h^s \rangle \overline{\langle \mathbf{I} f^t(g^{-1}), f^t \rangle} dg = 0.
$$

Proof. Using the decomposition $G = KA^+K$, we can express the above integral as:

$$
\int_K \int_{A^+} \Delta(a) \int_K \langle [\mathbf{i} h^s](k_1 a k_2)^{-1}, h^s \rangle \overline{\langle \mathbf{I} f^t (k_1 a k_2)^{-1}, f^t \rangle} dk_2 \, da \, dk_1.
$$

Consider the innermost integral above:

$$\int_K \langle [\mathbf{l}(k_2)[ih^s]](k_1a)^{-1}, h^s \rangle \overline{\langle [\mathbf{L}(k_2)[\mathbf{I}f^t]](k_1a)^{-1}, f^t \rangle} dk_2.$$

Since ih^s, and $\mathbf{I}f^t$ are lowest K-type and $(K \times K)$-type vectors, respectively, the integral above equals:

$$\int_K \langle \tau_\lambda(k_2)h, h^s \rangle \overline{\langle (\tau_{\lambda_1} \otimes \tau_{\lambda_2})(k_2)f, f^t \rangle} dk_2,$$

where $h = [ih^s](k_1a)^{-1}$ and f is the orthogonal projection of $[\mathbf{I}f^t](k_1a)^{-1}$ in F; note that $h \in H_\lambda$ and $f \in F$.

The first inner product is a matrix coefficient for the representation H_λ of K. The second inner product is a matrix coefficient for the representation F of K. Since H_λ occurs with multiplicity one in $H_{\lambda_1} \otimes H_{\lambda_2}$, and F is complementary to the embedding $\mu_{\lambda_1,\lambda_2}(H_\lambda)$, these matrix coefficients are orthogonal by Schur's orthogonality. The lemma follows immediately. $\qquad\square$

Again, using Schur's orthogonality relations, we prove:

Lemma 4.11. *For all $1 \le s, s' \le a$, we have:*

$$\int_G \langle [ih^s](g^{-1}), h^s \rangle \overline{\langle [\mathbf{I}e^{s'}](g^{-1}), e^{s'} \rangle} dg = 0, \ \text{if } s \ne s'.$$

If $1 \le s \le a$, we have:

$$\int_G \langle [ih^s](g^{-1}), h^s \rangle \overline{\langle [\mathbf{I}e^s](g^{-1}), e^s \rangle} dg$$
$$= \int_G \langle \mu_{\lambda_1,\lambda_2} \left([ih^s](g^{-1}) \right), [\mathbf{I}e^s](g^{-1}) \rangle.$$

Proof. We use the decomposition $G = KA^+K$ as before, and compute the above integral as:

$$\int_K \int_{A^+} \Delta(a) \int_K \langle [ih^s](k_1ak_2)^{-1}, h^s \rangle \overline{\langle [\mathbf{I}e^{s'}](k_1ak_2)^{-1}, e^{s'} \rangle} dk_2 \, da \, dk_1.$$

The innermost integral is then:

$$\int_K \langle [\mathbf{l}(k_2)[ih^s]](k_1a)^{-1}, h^s \rangle \overline{\langle [\mathbf{L}(k_2)[\mathbf{I}e^{s'}]](k_1a)^{-1}, e^{s'} \rangle} dk_2.$$

By Schur's orthogonality relations, using the embedding $\mu_{\lambda_1,\lambda_2}$, the above integral equals:

$$\left\langle \mu_{\lambda_1,\lambda_2} \left([ih^s](k_1a)^{-1} \right), [\mathbf{I}e^{s'}](k_1a)^{-1} \right\rangle \overline{\langle e^s, e^{s'} \rangle}.$$

Since the inner product is K-invariant, the above equals:

$$\overline{\langle e^s, e^{s'} \rangle} \cdot \int_K \langle \mu_{\lambda_1, \lambda_2} \left([\mathbf{i}h^s](k_1 a k_2)^{-1} \right), [\mathbf{I}e^{s'}](k_1 a k_2)^{-1} \rangle dk_2.$$

From this, we compute the whole integral:

$$\int_G \langle [\mathbf{i}h^s](g^{-1}), h^s \rangle \overline{\langle [\mathbf{I}e^{s'}](g^{-1}), e^{s'} \rangle} dg$$

$$= \overline{\langle e^s, e^{s'} \rangle} \cdot \int_G \langle \mu_{\lambda_1, \lambda_2} \left([\mathbf{i}h^s](g^{-1}) \right), [\mathbf{I}e^{s'}](g^{-1}) \rangle dg.$$

The lemma follows immediately. $\square$

The previous three lemmas allow us to prove:

Theorem 4.12. *The map $\nu_{\lambda_1, \lambda_2}$ is nonzero.*

Proof. By Lemma 4.9, we have:

$$\int_G \phi(g) \overline{\Phi(g)} dg \neq 0.$$

Expanding this, and disregarding the formal degree constants, yields:

$$0 \neq \left(\sum_{s,s'} \int_G \langle [\mathbf{i}h^s](g^{-1}), h^s \rangle \overline{\langle [\mathbf{I}e^{s'}](g^{-1}), e^{s'} \rangle} \right) dg$$

$$+ \left(\sum_{s,t} \int_G \langle [\mathbf{i}h^s](g^{-1}), h^s \rangle \overline{\langle [\mathbf{I}f^t](g^{-1}), f^t \rangle} \right) dg.$$

The previous two lemmas imply the vanishing of all but a few terms above. We are left with:

$$0 \neq \sum_s \int_G \langle [\mathbf{i}h^s](g^{-1}), h^s \rangle \overline{\langle [\mathbf{I}e^s](g^{-1}), e^s \rangle} dg.$$

Finally, the previous lemma implies that there exists an s such that:

$$0 \neq \int_G \langle \mu_{\lambda_1, \lambda_2} \left([\mathbf{i}h^s](g^{-1}) \right), [\mathbf{I}e^s](g^{-1}) \rangle dg.$$

Hence, if $j = \mathbf{i}h^s \in \mathcal{J}(Y, H_\lambda)$, and $J = \mathbf{I}e^s \in \mathcal{J}(Y \times Y, H_{\lambda_1} \boxtimes H_{\lambda_2})$, then:

$$\langle \nu_{\lambda_1, \lambda_2} j, J \rangle \neq 0.$$

In particular, $\nu_{\lambda_1, \lambda_2}(j) \neq 0$. $\square$

In fact, the final steps in this proof illustrate something a bit stronger:

Corollary 4.13. *Suppose that j is a nonzero lowest K-type vector in $\mathcal{J}(Y, H_\lambda)$. Then the projection of $\nu_{\lambda_1, \lambda_2}(j)$ onto the lowest $(K \times K)$-type of $\mathcal{J}(Y \times Y, H_{\lambda_1} \boxtimes H_{\lambda_2})$ is nonzero.*

Proof. The proof of the previous theorem illustrates that the result is true for some element $\mathbf{i}h^s$ of the lowest K-type of $\mathcal{J}(Y, H_\lambda)$. Since $\nu_{\lambda_1, \lambda_2}$ is K-intertwining, and H_λ is irreducible, the result is true for every nonzero lowest K-type vector. $\qquad\square$

5 Classical Modular Forms

With G, Γ, and a discrete series Harish-Chandra module $\mathcal{J}_\lambda$ as before, we have defined a space of modular forms $\mathcal{M}(\mathcal{J}_\lambda)$. In this section, we discuss these modular forms in a more classical language.

5.1 Transition to classical modular forms

Definition 5.1. The space of *classical modular forms* of weight H_λ, denoted $\mathcal{M}_{\mathrm{cl}}(H_\lambda)$ is the subspace of functions $f \in C^\infty(\Gamma \backslash G, H_\lambda)$, satisfying the additional properties:

- $f(gk) = \tau(k^{-1})(f(g))$ for all $g \in G, k \in K$.
- $\Omega f = (\|\lambda\|^2 - \|\rho\|^2)f$.
- f has moderate growth.

Then we can view our modular forms as classical modular forms:

Proposition 5.2. *For all $F \in \mathcal{M}(\mathcal{J}_\lambda)$, there exists a unique classical modular form $f \in \mathcal{M}_{\mathrm{cl}}(H_\lambda)$, which satisfies:*

$$\langle f(g), h \rangle = F(\mathbf{i}(h))(g), \text{ for all } h \in H_\lambda, g \in G.$$

This yields an injective linear map $\mathcal{M}(\mathcal{J}_\lambda) \hookrightarrow \mathcal{M}_{\mathrm{cl}}(H_\lambda)$.

Proof. Since H_λ is a Hilbert space, the values of $f(g)$ are completely determined by the formula $\langle f(g), h \rangle = F(\mathbf{i}(h))(g)$. Fixing $F \in \mathcal{M}(\mathcal{J}_\lambda)$, we may compute:

$$\begin{aligned}
\langle f(gk), h \rangle &= F(\mathbf{i}(h))(gk), \\
&= F(\mathbf{l}(k)\mathbf{i}(h))(g), \\
&= F(\mathbf{i}(\tau(k)h))(g), \\
&= \langle f(g), \tau(k)h \rangle, \\
&= \langle \tau(k^{-1})f(g), h \rangle.
\end{aligned}$$

Next, if $Z \in \mathcal{Z}$, and ω is the infinitesimal character of $\mathcal{J}_\lambda$, then we compute:

$$\langle Zf(g), h \rangle = Z[F(\mathbf{i}(h))](g) = \omega(Z) \cdot F(i(h))(g).$$

In particular, this holds for $\Omega \in \mathcal{Z}$, so that f is an eigenfunction of Ω with the same eigenvalue as $F(\mathbf{i}(h))$.

Finally, f has moderate growth, since fixing an orthonormal basis $h_1, \ldots, h_a$ of H_λ, we compute:

$$
\begin{aligned}
|f(g)|^2_\lambda &= \sum_{s=1}^{a} |\langle f(g), h_s \rangle|^2, \\
&= \sum_{s=1}^{a} |F(\mathbf{i}(h_s))(g)|^2, \\
&\leq aC\|g\|^{2n},
\end{aligned}
$$

where C and n are the suprema of the constants of moderate growth for $F(\mathbf{i}(h_s))$.

Thus we have a linear map from $\mathcal{M}(\mathcal{J}_\lambda)$ to $\mathcal{M}_{\mathrm{cl}}(H_\lambda)$ as desired. It is injective, since the classical modular form f determines the restriction of a modular form F to its lowest K-type, and $\mathcal{J}_\lambda$ is generated as a $\mathcal{U}$-module by its lowest K-type. $\qquad\qquad\Box$

5.2 Holomorphic Modular Forms

Suppose that G/K is a Hermitian symmetric domain. In this case, multiplication of scalar-valued holomorphic modular forms is well-understood. Representation theoretically, this multiplication can be associated to properties of the holomorphic discrete series of G.

Let $\mathfrak{g} = \mathfrak{k} \oplus \mathfrak{p}$ denote the Cartan decomposition. There exists a decomposition of the roots of $(\mathfrak{g}, \mathfrak{t})$ into positive and negative roots, in such a way that the sum of any two noncompact positive roots is not a root. This yields a natural decomposition $\mathfrak{p} = \mathfrak{p}^+ \oplus \mathfrak{p}^-$. The elements of $\mathfrak{p}^-$ are called the "lowering operators". Holomorphic functions on G/K are precisely those functions which are annihilated by all lowering operators.

The scalar-valued holomorphic discrete series is a family of discrete series representations $\mathcal{J}_w$ of G, parameterized by the nonzero elements $w \in C \subset \mathbb{N}$ in a submonoid of $\mathbb{N}$, which satisfy the following properties:

- The lowest K-type of $\mathcal{J}_w$ is one-dimensional, denoted $\mathbb{C}_w$. The action of K on $\mathbb{C}_w$ will be denoted χ_w.

- Every element $j \in \mathbb{C}_w$, the lowest K-type, is annihilated by the lowering operators.
- Every pair $w_1, w_2 \in C$ is discretely decomposable (cf. [Rep79]).

For consistency, we realize the holomorphic discrete series, as all discrete series representations, as sections of a bundle over G/K, which satisfy a growth condition and the second-order differential equation from the Casimir operator. In this realization, we have the "evaluation at the identity" operator $\mathbf{ev}_w : \mathfrak{J}_w \to \mathbb{C}_w$, and the adjoint inclusion $\mathbf{i}_w : \mathbb{C}_w \to \mathfrak{J}_w$. Define $j_w^\vee = \mathbf{i}_w(1)$; then

$$j_w^\vee(1) = \langle j_w^\vee, j_w^\vee \rangle = \delta_w,$$

where δ_w is the formal degree of $\mathfrak{J}_w$. Let $j_w = \frac{1}{\delta_w} j_w^\vee$, so that j_w is the unique lowest K-type vector satisfying

$$j_w(1) = \langle j_w, j_w^\vee \rangle = 1.$$

When $w_1 + w_2 = w$, and $w_1, w_2 \in C$, we have constructed intertwining operators

$$\nu_{w_1, w_2} : \mathfrak{J}_w \to \mathfrak{J}_{w_1} \otimes \mathfrak{J}_{w_2}.$$

We prove the following fact about these intertwining operators:

Lemma 5.3. *Let ν_{w_1, w_2} be the canonical intertwining operator from $\mathfrak{J}_w$ to $\mathfrak{J}_{w_1} \otimes \mathfrak{J}_{w_2}$. Then*

$$\nu_{1,2}(j_w^\vee) = j_{w_1}^\vee \otimes j_{w_2}^\vee.$$

Proof. First note that $j_{w_1} \odot j_{w_2}$ is a holomorphic element of $C^\infty(Y, \mathbb{C}_w)$, since holomorphy is preserved under products. As such, it is annihilated by the Dirac operator, and hence is an eigenfunction of the Casimir operator, of the appropriate eigenvalue for $\mathfrak{J}_w$ by Proposition 3.2 of Parthasarathy [Par72]. It is also in $L^2(Y, \mathbb{C}_w)$, by Proposition 4.3. Finally, $(j_{w_1} \odot j_{w_2})(1) = 1$, by simply multiplying. Hence $j_{w_1} \odot j_{w_2}$ is a normalized lowest K-type vector in $\mathfrak{J}_w$, and from uniqueness of such, it follows that:

$$j_{w_1} \odot j_{w_2} = j_w.$$

To prove the adjoint statement, we note that $\nu_{1,2}(j_w^\vee)$ is a lowest K-type vector in $\mathfrak{J}_{w_1} \otimes \mathfrak{J}_{w_2}$, since the appropriate K-type has multiplicity one (as noted by Repka [Rep79], for example). It follows that for some nonzero constant C, we have:

$$\nu_{1,2}(j_w^\vee) = C j_1^\vee \otimes j_2^\vee.$$

To prove that this constant equals one, we compute:

$$
\begin{aligned}
\langle \nu_{1,2}(j_w^\vee), j_{w_1} \otimes j_{w_2} \rangle &= \langle \mu^Y(j_w^\vee), j_{w_1} \odot j_{w_2} \rangle, \\
&= \langle j_w^\vee, j_w \rangle, \\
&= 1.
\end{aligned}
$$

On the other hand, we compute:

$$
\begin{aligned}
\langle \nu_{1,2}(j_w^\vee), j_{w_1} \otimes j_{w_2} \rangle &= \langle C j_{w_1}^\vee \otimes j_{w_2}^\vee, j_{w_1} \otimes j_{w_2} \rangle, \\
&= C \langle j_{w_1}^\vee, j_{w_1} \rangle \langle j_{w_2}^\vee, j_{w_2} \rangle, \\
&= C.
\end{aligned}
$$

It follows that $C = 1$. $\qquad\qquad\qquad\qquad\qquad\qquad\qquad\qquad\qquad\square$

When working with holomorphic discrete series, we may consider modular forms that satisfy a first-order differential equation, instead of a second-order equation.

Definition 5.4. The space of holomorphic modular forms of weight w, denoted $\mathcal{M}_{\mathrm{hol}}(w)$ is the space of smooth functions $f: \Gamma\backslash G \to \mathbb{C}$ such that:

- $f(gk) = \chi_w(k^{-1})(f(g))$ for all $g \in G, k \in K$.
- $Xf = 0$, for all $X \in \mathfrak{p}^-$.
- f has moderate growth.

Holomorphic modular forms form a ring. Indeed, it is a fortunate fact that if $f_1 \in \mathcal{M}_{\mathrm{hol}}(w_1)$ and $f_2 \in \mathcal{M}_{\mathrm{hol}}(w_2)$, then $f_1 \cdot f_2 \in \mathcal{M}_{\mathrm{hol}}(w_1 + w_2)$; not only is the pointwise product naturally a section of the bundle $\mathbb{C}_w$, but it also satisfies the appropriate differential equations for holomorphy. Thus, multiplication of holomorphic modular forms is "easy".

Again, by computations in [Par72], elements $f \in \mathcal{M}_{\mathrm{hol}}(w)$ also satisfy $\Omega f = (\|\lambda\|^2 - \|\rho\|^2)f$, for some λ corresponding to w. It follows that $\mathcal{M}_{\mathrm{hol}}(w) \subset \mathcal{M}_{\mathrm{cl}}(\mathbb{C}_w)$, for all $w \in C$. Moreover, we have:

Proposition 5.5. *Suppose that* $F \in \mathcal{M}(\mathfrak{J}_w)$ *is a modular form, where* $\mathfrak{J}_w$ *is the scalar-valued holomorphic discrete series with lowest K-type* $\mathbb{C}_w$. *Let* $f \in \mathcal{M}_{\mathrm{cl}}(\mathbb{C}_w)$ *be the associated classical modular form. Then* $f \in \mathcal{M}_{\mathrm{hol}}(w)$.

Proof. Every lowest K-type vector in $\mathfrak{J}_w$ is annihilated by the "lowering operators" $X \in \mathfrak{p}^-$. If $F \in \mathcal{M}(\mathfrak{J}_w)$ is a modular form, then the associated classical modular form is given by a simple formula:

$$
f(g) = F(\mathbf{i}(1))(g) = F(j_w^\vee)(g).
$$

It follows that:

$$Xf(g) = XF(j_w^\vee)(g) = F(Xj_w^\vee)(g) = 0,$$

for every $X \in \mathfrak{p}^-$. $\qquad\square$

It is not clear that "easy multiplication" of holomorphic modular forms is compatible with the multiplication induced from the intertwining operator v_{w_1,w_2}. For $w_1, w_2 \in C$, and $w = w_1 + w_2$, we have two bilinear maps:

$$m_{\mathrm{hol}} \ : \ \mathcal{M}_{\mathrm{hol}}(w_1) \otimes \mathcal{M}_{\mathrm{hol}}(w_2) \to \mathcal{M}_{\mathrm{hol}}(w),$$
$$m_{1,2} \ : \ \mathcal{M}(\mathcal{J}_{w_1}) \otimes \mathcal{M}(\mathcal{J}_{w_2}) \to \mathcal{M}(\mathcal{J}_w).$$

We also have linear maps, for all $w \in C$, given by "classicalization":

$$\mathrm{cl}_w : \mathcal{M}(\mathcal{J}_w) \hookrightarrow \mathcal{M}_{\mathrm{hol}}(w).$$

We now prove compatibility:

Proposition 5.6. *The following diagram commutes, for any pair w_1, $w_2 \in C$:*

$$
\begin{array}{ccc}
\mathcal{M}(\mathcal{J}_{w_1}) \otimes \mathcal{M}(\mathcal{J}_{w_2}) & \xrightarrow{\ m_{1,2}\ } & \mathcal{M}(\mathcal{J}_w) \\
{\scriptstyle \mathrm{cl}_{w_1} \otimes \mathrm{cl}_{w_2}}\Big\downarrow & & \Big\downarrow{\scriptstyle \mathrm{cl}_w} \\
\mathcal{M}_{\mathrm{hol}}(w_1) \otimes \mathcal{M}_{\mathrm{hol}}(w_2) & \xrightarrow{\ m_{\mathrm{hol}}\ } & \mathcal{M}_{\mathrm{hol}}(w)
\end{array}
$$

Proof. Suppose that $F_1 \in \mathcal{M}(\mathcal{J}_{w_1})$ and $F_2 \in \mathcal{M}(\mathcal{J}_{w_2})$. Let f_1, f_2 be the corresponding holomorphic classical modular forms. Then we have:

$$f_1(g) = F_1(j_{w_1}^\vee)(g), \ \text{ and } f_2(g) = F_2(j_{w_2}^\vee)(g).$$

Hence, we have:

$$[f_1 \cdot f_2](g) = F_1(j_{w_1}^\vee)(g) \cdot F_2(j_{w_2}^\vee)(g).$$

On the other hand, let F be the product of F_1 and F_2 via $m_{1,2}$. Let f be the corresponding holomorphic classical modular form. Then we have:

$$
\begin{aligned}
f(g) &= F(j_w^\vee)(g), \\
&= m([F_1 \otimes F_2]v_{w_1,w_2}(j_w^\vee))(g), \\
&= m([F_1 \otimes F_2](j_{w_1}^\vee \otimes j_{w_2}^\vee))(g), \\
&= F_1(j_{w_1}^\vee)(g) \cdot F_2(j_{w_2}^\vee)(g).
\end{aligned}
$$

$\qquad\square$

Therefore, our multiplication of modular forms, via intertwining operators v_{w_1, w_2}, agrees with ordinary multiplication of holomorphic modular forms.

When G/K is a rational tube domain, the quotient $Y(\Gamma) = \Gamma \backslash G/K$ is a locally symmetric space. Its Satake compactification may be defined, as a projective algebraic variety, by:

$$\overline{Y(\Gamma)} = \text{Proj}(\mathcal{M}) = \text{Proj} \bigoplus_{w=0}^{\infty} \mathcal{M}(\mathcal{J}_w),$$

where $\mathcal{M}$ is the ring of scalar-valued holomorphic modular forms. Thus the Satake compactification may be constructed completely within the framework of representation theory. We are very interested in whether the geometric properties of $\overline{Y(\Gamma)}$ may be recovered via representation theory as well. For example, it can be shown that the formal degree δ_w of $\mathcal{J}_w$ satisfies $\delta_w = O(w^n)$, where $n = \dim(G/K)$. Heuristically, $\dim(\mathcal{M}(\mathcal{J}_w)) = O(\delta_w)$ – this is precisely true when considering cusp forms, or L^2 forms, in the limit as Γ becomes smaller. Therefore, the graded ring $\mathcal{M}$ has dimension n, and so we may accurately predict that:

$$\dim(\overline{Y(\Gamma)}) = \dim(G/K).$$

5.3 Multiplication of Vector-Valued Modular Forms

Consider, as previously, the case when G/K is a rational tube domain. In addition to scalar-valued modular forms, we may consider vector-valued modular forms.

Definition 5.7. Suppose that H_λ is an irreducible unitary representation of K. The space of holomorphic modular forms of weight H_λ, denoted $\mathcal{M}_{\text{hol}}(H_\lambda)$ is the space of smooth functions $f \colon \Gamma \backslash G \to H_\lambda$ such that:

- $f(gk) = \tau_\lambda(k^{-1})(f(g))$ for all $g \in G, k \in K$.
- $Xf = 0$, for all $X \in \mathfrak{p}^-$.
- f has moderate growth.

As before, if H_λ is the lowest K-type of a holomorphic discrete series representation $\mathcal{J}_\lambda$, then there is an injective classicalization map:

$$\mathcal{M}(\mathcal{J}_\lambda) \to \mathcal{M}_{\text{hol}}(H_\lambda).$$

As tensor products of holomorphic discrete series representations are discretely decomposable, one arrives at a ring of modular forms using intertwining operators among the $\mathcal{J}_\lambda$. Classically, multiplication of vector-valued modular forms is not as well-understood as for scalar-valued modular forms. However, our ring structure suggests (is compatible with) the following multiplication:

Definition 5.8. Suppose that $f_1 \in \mathcal{M}_{\text{hol}}(H_{\lambda_1})$ and $f_2 \in \mathcal{M}_{\text{hol}}(H_{\lambda_2})$ are two holomorphic modular forms; thus H_{λ_1}, H_{λ_2}, and H_λ ($\lambda = \lambda_1 + \lambda_2$) are lowest K-types of discrete series representations. Define the product by:

$$[f_1 \cdot f_2](g) = \mu^*_{\lambda_1, \lambda_2}(f_1(g) \otimes f_2(g)) \in H_\lambda.$$

Here, we recall that $\mu_{\lambda_1, \lambda_2}$ is the canonical intertwining operator obtained via the Borel-Weil realization. The resulting function is in $\mathcal{M}_{\text{hol}}(H_\lambda)$, since holomorphicity is preserved.

The classical ring structure, defined above to correspond with our general theory, seems to be new. One arrives at a ring

$$\mathcal{M}^{\text{vec}} = \bigoplus_{\lambda \in \Lambda_k^+} \mathcal{M}_{\text{hol}}(H_\lambda),$$

graded by the abelian group $\Lambda_k = \Lambda$. Viewing elements $\lambda \in \Lambda$ as algebraic characters $T_{\mathbb{C}} \to \mathbb{C}^\times$, we have an action of $T_{\mathbb{C}}$ on the ring $\mathcal{M}$. Namely, if $f \in \mathcal{M}_{\text{hol}}(H_\lambda)$, and $t \in T_{\mathbb{C}}$, we define:

$$t[f] = \lambda(t)f.$$

This action yields a geometric action. The complex algebraic torus $T_{\mathbb{C}}$ acts on the $\mathbb{C}$-scheme $Y^{\text{vec}} = \text{Spec}(\mathcal{M}^{\text{vec}})$. It seems important to consider the following questions:

- Describe the geometry of the GIT quotient $Y^{\text{vec}}//T_{\mathbb{C}}$. Does it provide a compactification of the locally symmetric space $\Gamma \backslash G/K$?
- Does this quotient have a natural model over a finite extension of $\mathbb{Q}$? Does it coincide with a toroidal compactification?
- Describe the behavior of constant terms of vector-valued modular forms under multiplication. How does this reflect the cuspidal geometry of $Y^{\text{vec}}//T_{\mathbb{C}}$?

We hope to approach these questions further in a future paper.

Bibliography

[BJ79]   A. Borel and H. Jacquet, *Automorphic forms and automorphic representa- tions*, Automorphic forms, representations and *L*-functions (Proc. Sym- pos. Pure Math., Oregon State Univ., Corvallis, Ore., 1977), Part 1, Proc. Sympos. Pure Math., XXXIII, Amer. Math. Soc., Providence, R.I., 1979,

With a supplement "On the notion of an automorphic representation" by R. P. Langlands, pp. 189–207. MR MR546598 (81m:10055)

[BW00] A. Borel and N. Wallach, *Continuous cohomology, discrete subgroups, and representations of reductive groups*, second ed., Mathematical Surveys and Monographs, vol. 67, American Mathematical Society, Providence, RI, 2000. MR MR1721403 (2000j:22015)

[Del73] P. Deligne, *Formes modulaires et représentations de* GL(2), Modular functions of one variable, II (Proc. Internat. Summer School, Univ. Antwerp, Antwerp, 1972), Springer, Berlin, 1973, pp. 55–105. Lecture Notes in Math., Vol. 349. MR MR0347738 (50 #240)

[GGS02] Wee Teck Gan, Benedict Gross, and Gordan Savin, *Fourier coefficients of modular forms on* G_2, Duke Math. J. **115** (2002), no. 1, 105–169. MR MR1932327 (2004a:11036)

[Gro99] Benedict H. Gross, *Algebraic modular forms*, Israel J. Math. **113** (1999), 61–93. MR MR1729443 (2001b:11037)

[HP74] R. Hotta and R. Parthasarathy, *Multiplicity formulae for discrete series*, Invent. Math. **26** (1974), 133–178. MR MR0348041 (50 #539)

[KM01] Kamal Khuri-Makdisi, *On the curves associated to certain rings of automorphic forms*, Canad. J. Math. **53** (2001), no. 1, 98–121. MR MR1814967 (2002i:11045)

[Kna01] Anthony W. Knapp, *Representation theory of semisimple groups*, Princeton Landmarks in Mathematics, Princeton University Press, Princeton, NJ, 2001, An overview based on examples, Reprint of the 1986 original. MR MR1880691 (2002k:22011)

[Kob98] Toshiyuki Kobayashi, *Discrete decomposability of the restriction of* $A_q(\lambda)$ *with respect to reductive subgroups. III. Restriction of Harish-Chandra modules and associated varieties*, Invent. Math. **131** (1998), no. 2, 229–256. MR MR1608642 (99k:22021)

[ØV04] Bent Ørsted and Jorge Vargas, *Restriction of square integrable representations: discrete spectrum*, Duke Math. J. **123** (2004), no. 3, 609–633. MR MR2068970 (2005g:22016)

[Par72] R. Parthasarathy, *Dirac operator and the discrete series*, Ann. of Math. (2) **96** (1972), 1–30. MR MR0318398 (47 #6945)

[Rep79] Joe Repka, *Tensor products of holomorphic discrete series representations*, Canad. J. Math. **31** (1979), no. 4, 836–844. MR MR540911 (82c:22017)

[Ser95] Jean-Pierre Serre, *Représentations linéaires et espaces homogènes kählériens des groupes de Lie compacts (d'après Armand Borel et André Weil)*, Séminaire Bourbaki, Vol. 2, Soc. Math. France, Paris, 1995, pp. Exp. No. 100, 447–454. MR MR1609256

[Var01] J. A. Vargas, *Restriction of some discrete series representations*, Algebras Groups Geom. **18** (2001), no. 1, 85–99. MR MR1834195 (2002e:22019)

[Wal76] Nolan R. Wallach, *On the Enright-Varadarajan modules: a construction of the discrete series*, Ann. Sci. École Norm. Sup. (4) **9** (1976), no. 1, 81–101. MR MR0422518 (54 #10505)

[Wei06] Martin H. Weissman, D_4 *modular forms*, Amer. J. Math. **128** (2006), no. 4, 849–898. MR MR2251588

[WW83] Nolan R. Wallach and Joseph A. Wolf, *Completeness of Poincaré series for automorphic forms associated to the integrable discrete series*, Representation theory of reductive groups (Park City, Utah, 1982), Progr. Math., vol. 40, Birkhäuser Boston, Boston, MA, 1983, pp. 265–281. MR MR733818 (85k:22038)

On projective linear groups over finite fields as Galois groups over the rational numbers

Gabor Wiese

Abstract

Ideas from Khare's and Wintenberger's article on the proof of Serre's conjecture for odd conductors are used to establish that for a fixed prime l infinitely many of the groups $\mathrm{PSL}_2(\mathbb{F}_{l^r})$ (for r running) occur as Galois groups over the rationals such that the corresponding number fields are unramified outside a set consisting of l, the infinite place and only one other prime.

1 Introduction

The aim of this article is to prove the following theorem.

Theorem 1.1. *Let l be a prime and s a positive number. Then there exists a set T of rational primes of positive density such that for each $q \in T$ there exists a modular Galois representation*

$$\bar{\rho} : G_{\mathbb{Q}} \to \mathrm{GL}_2(\overline{\mathbb{F}}_l)$$

which is unramified outside $\{\infty, l, q\}$ and whose projective image is isomorphic to $\mathrm{PSL}_2(\mathbb{F}_{l^r})$ for some $r > s$.

Corollary 1.2. *Let l be a prime. Then for infinitely many positive integers r the groups $\mathrm{PSL}_2(\mathbb{F}_{l^r})$ occur as a Galois group over the rationals.*

Using $\mathrm{SL}_2(\mathbb{F}_{2^r}) \cong \mathrm{PSL}_2(\mathbb{F}_{2^r})$ one obtains the following reformulation for $l = 2$.

Corollary 1.3. *The group $\mathrm{SL}_2(\mathbb{F}_{2^r})$ occurs as a Galois group over the rationals for infinitely many positive integers r.*

This contrasts with work by Dieulefait, Reverter and Vila ([D1],[D2], [RV] and [DV]) who proved that the groups $\mathrm{PSL}_2(\mathbb{F}_{l^r})$ and $\mathrm{PGL}_2(\mathbb{F}_{l^r})$ occur as Galois groups over $\mathbb{Q}$ for fixed (small) r and infinitely many primes l.

Dieulefait ([D3]) has recently obtained a different, rather elementary proof of Corollary 1.2 under the assumption $l \geq 5$ with a different ramification behaviour, namely $\{\infty, 2, 3, l\}$. His proof has the virtue of working with a family of modular forms that does not depend on l.

In the author's PhD thesis [W] some computational evidence on the statement of Corollary 1.3 was exhibited. More precisely, it was shown that all groups $\mathrm{SL}_2(\mathbb{F}_{2^r})$ occur as Galois groups over $\mathbb{Q}$ for $1 \leq r \leq 77$, extending results by Mestre (see [S], p. 53), by computing Hecke eigenforms of weight 2 for prime level over finite fields of characteristic 2. However, at that time all attempts to prove the corollary failed, since the author could not rule out theoretically that all Galois representations attached to modular forms with image contained in $\mathrm{SL}_2(\mathbb{F}_{2^r})$ for r bigger than some fixed bound and not in any $\mathrm{SL}_2(\mathbb{F}_{2^a})$ for $a \mid r$, $a \neq r$, have a dihedral image.

The present paper has undergone some developments since the first writing. However, the core of the proof has remained the same. It uses a procedure borrowed from the ground breaking paper [KW] by Khare and Wintenberger. The representations that we will construct in the proof are almost - in the terminology of [KW] - good-dihedral. The way we make these representation by level raising is adopted from a part of the proof of [KW], Theorem 3.4. Although the paper [KW] is not cited in the proof, this paper owes its mere existence to it.

Meanwhile, the results of the present article have been generalised to the groups $\mathrm{PSp}_{2n}(\mathbb{F}_{l^r})$ for arbitrary n by Khare, Larsen and Savin (see [KLS]). Their paper [KLS] also inspired a slight strengthening of the main result of the present article compared to an early version.

Acknowledgements. The author would like to thank Bas Edixhoven for very useful discussions and Alexander Schmidt for an elegant argument used in the proof of Lemma 3.1.

Notation. We shall use the following notation. By $S_k(\Gamma_1(N))$ we mean the complex vector space of holomorphic cusp forms on $\Gamma_1(N)$ of weight k. The notation $S_k(N, \chi)$ is used for the vector space of holomorphic cusp forms of level N, weight k for the Dirichlet character χ. If χ is trivial, we write $S_k(N)$ for short. We fix an algebraic closure $\overline{\mathbb{Q}}$ of $\mathbb{Q}$ and for every prime p an algebraic closure $\overline{\mathbb{Q}}_p$ of $\mathbb{Q}_p$ and we choose once and for all an embedding $\overline{\mathbb{Q}} \hookrightarrow \overline{\mathbb{Q}}_p$ and a ring surjection $\overline{\mathbb{Z}}_p \to \overline{\mathbb{F}}_p$, subject to which the following constructions are made. By $G_{\mathbb{Q}}$ we denote the absolute Galois group of the rational numbers. For a rational prime q, we let D_q and I_q be the corresponding decomposition and inertia group of the prime q, respectively. Given an eigenform $f \in S_k(\Gamma_1(N))$ one can attach to it by work of Shimura and Deligne a Galois representation $\rho_{f,p} : G_{\mathbb{Q}} \to GL_2(\overline{\mathbb{Q}}_p)$ with well-known properties. Choosing a lattice, reducing and semi-simplifying, one also obtains a representation $\overline{\rho}_{f,p} : G_{\mathbb{Q}} \to GL_2(\overline{\mathbb{F}}_p)$. We denote the composition $G_{\mathbb{Q}} \to GL_2(\overline{\mathbb{F}}_p) \twoheadrightarrow PGL_2(\overline{\mathbb{F}}_p)$ by $\overline{\rho}_{f,p}^{\text{proj}}$. All Galois representations in this paper are continuous. By ζ_{p^r} we always mean a primitive p^r-th root of unity.

2 On Galois representations

The basic idea of the proof is to obtain the groups $PSL_2(\mathbb{F}_{l^r})$ as the image of some $\overline{\rho}_{g,l}^{\text{proj}}$. In order to determine the possible images of $\overline{\rho}_{g,l}^{\text{proj}}$, we quote the following well-known group theoretic result due to Dickson (see [Hu], II.8.27).

Proposition 2.1 (Dickson). *Let l be a prime and H a finite subgroup of $PGL_2(\overline{\mathbb{F}}_l)$. Then a conjugate of H is isomorphic to one of the following groups:*

- *finite subgroups of the upper triangular matrices,*
- *$PSL_2(\mathbb{F}_{l^r})$ or $PGL_2(\mathbb{F}_{l^r})$ for $r \in \mathbb{N}$,*
- *dihedral groups D_r for $r \in \mathbb{N}$ not divisible by l,*
- *A_4, A_5 or S_4.*

We next quote a result by Ribet showing that the images of $\overline{\rho}_{g,l}^{\text{proj}}$ for a non-CM eigenform g are "almost always" $PSL_2(\mathbb{F}_{l^r})$ or $PGL_2(\mathbb{F}_{l^r})$ for some $r \in \mathbb{N}$.

Proposition 2.2 (Ribet). *Let $f = \sum_{n \geq 1} a_n q^n \in S_2(N, \chi)$ be a normalised eigenform of level N and some character χ which is not a CM-form.*

Then for almost all primes p, i.e. all but finitely many, the image of the representation

$$\overline{\rho}_{f,p}^{\text{proj}} : G_{\mathbb{Q}} \to PGL_2(\overline{\mathbb{F}}_p)$$

attached to f is equal to either $PGL_2(\mathbb{F}_{p^r})$ or to $PSL_2(\mathbb{F}_{p^r})$ for some $r \in \mathbb{N}$.

Proof. Reducing modulo p, Theorem 3.1 of [R1] gives, for almost all p, that the image of $\overline{\rho}_{f,p}$ contains $SL_2(\mathbb{F}_p)$. This already proves that the projective image is of the form $PGL_2(\mathbb{F}_{p^r})$ or $PSL_2(\mathbb{F}_{p^r})$ by Proposition 2.1. $\square$

In view of Ribet's result, there are two tasks to be solved for proving Theorem 1.1. The first task is to avoid the "exceptional primes", i.e. those for which the image is not as in the proposition. The second one is to obtain at the same time that the field $\mathbb{F}_{p^r}$ is "big".

We will now use this result by Ribet in order to establish the simple fact that for a given prime l there exists a modular form of level a power of l having only finitely many "exceptional primes". The following lemma can be easily verified using e.g. William Stein's modular symbols package which is part of MAGMA ([Magma]).

Lemma 2.3. *In any of the following spaces there exists a newform without CM:* $S_2(2^7)$, $S_2(3^4)$, $S_2(5^3)$, $S_2(7^3)$, $S_2(13^2)$.

Proposition 2.4. *Let l be a prime. Put*

$$
N := \begin{cases}
2^7, & \text{if } l = 2, \\
3^4, & \text{if } l = 3, \\
5^3, & \text{if } l = 5, \\
7^3, & \text{if } l = 7, \\
13^2, & \text{if } l = 13, \\
l, & \text{otherwise.}
\end{cases}
$$

Then there exists an eigenform $f \in S_2(N)$ such that for almost all primes p, i.e. for all but finitely many, the image of the attached Galois representation $\overline{\rho}_{f,p}$ is of the form $\mathrm{PGL}_2(\mathbb{F}_{p^r})$ or $\mathrm{PSL}_2(\mathbb{F}_{p^r})$ for some $r \in \mathbb{N}$.

Proof. If $l \in \{2, 3, 5, 7, 13\}$, we appeal to Lemma 2.3 to get an eigenform f without CM. If l is not in that list, then there is an eigenform in $S_2(l)$ and it is well-known that it does not have CM, since the level is square-free. Hence, Proposition 2.2 gives the claim. $\qquad\square$

We will be able to find an eigenform with image as in the preceding proposition which is "big enough" by applying the following level raising result by Diamond and Taylor. It is a special case of Theorem A of [DT].

Theorem 2.5 (Diamond, Taylor). *Let $N \in \mathbb{N}$ and let $p > 3$ be a prime not dividing N. Let $f \in S_2(N, \chi)$ be a newform such that $\overline{\rho}_{f,p}$ is irreducible. Let, furthermore, $q \nmid N$ be a prime such that $q \equiv -1 \mod p$ and $\mathrm{tr}(\overline{\rho}_{f,p}(\mathrm{Frob}_q)) = 0$.*
Then there exists a newform $g \in S_2(Nq^2, \tilde{\chi})$ such that $\overline{\rho}_{g,p} \cong \overline{\rho}_{f,p}$.

Corollary 2.6. *Assume the setting of Theorem 2.5. Then the following statements hold.*

(a) *The mod p reductions of χ and $\tilde{\chi}$ are equal.*
(b) *The restriction of $\rho_{g,p}$ to I_q, the inertia group at q, is of the form $\begin{pmatrix} \psi & * \\ 0 & \psi^q \end{pmatrix}$ with a character $I_q \xrightarrow{\psi} \mathbb{Z}[\zeta_{p^r}]^\times \hookrightarrow \mathbb{Z}_p[\zeta_{p^r}]^\times$ of order p^r for some $r > 0$.*

(c) For all primes $l \neq p, q$, the restriction of $\overline{\rho}_{g,l}$ to D_q, the decomposition group at q, is irreducible and the restriction of $\overline{\rho}_{g,l}$ to I_q is of the form $\begin{pmatrix} \psi & * \\ 0 & \psi^q \end{pmatrix}$ with the character $I_q \xrightarrow{\psi} \mathbb{Z}[\zeta_{p^r}]^\times \twoheadrightarrow \mathbb{F}_l(\zeta_{p^r})^\times$ of the same order p^r as in (b).

Proof. (a) This follows from $\overline{\rho}_{g,p} \cong \overline{\rho}_{f,p}$.

(b) As q^2 precisely divides the conductor of $\rho_{g,p}$, the ramification at q is tame and the restriction to I_q is of the form $\begin{pmatrix} \psi_1 & * \\ 0 & \psi_2 \end{pmatrix}$ for non-trivial characters $\psi_i : I_q \to \overline{\mathbb{Z}}_p^\times$. Their order must be a power of p, since their reductions mod p vanish by assumption. Due to $q \equiv -1 \mod p$, local class field theory tells us that ψ_i cannot extend to $G_{\mathbb{Q}_q}$, as their orders would divide $q - 1$. Hence, the image of $\rho_{g,p}|_{D_q} : D_q \to \mathrm{GL}_2(\overline{\mathbb{Q}}_p) \to \mathrm{PGL}_2(\overline{\mathbb{Q}}_p)$ is a finite dihedral group. Consequently, $\rho_{g,p}|_{D_q}$ is an unramified twist of the induced representation $\mathrm{Ind}_{D_q}^{G_K}(\psi)$ with K the unramified extension of degree 2 of $\mathbb{Q}_q$ and $\psi : G_K \to \mathbb{Z}[\zeta_{p^r}]$ a totally ramified character of order p^r for some $r > 0$. Conjugation by $\rho_{g,p}(\mathrm{Frob}_q)$ exchanges ψ_1 and ψ_2. It is well-known that this conjugation also raises to the q-th power. Thus, we find $\psi_2 = \psi_1^q$ and without loss of generality $\psi = \psi_1$.

(c) With the notations and the normalisation used in [CDT], p. 536, the local Langlands correspondence reads (for $l \neq q$)

$$(\rho_{g,l}|W_q)^{ss} \cong \overline{\mathbb{Q}}_l \otimes_{\overline{\mathbb{Q}}} \mathrm{WD}(\pi_{g,q}),$$

if g corresponds to the automorphic representation $\pi_g = \otimes \pi_{g,p}$. In particular, knowing by (b) that $\rho_{g,p}|_{I_q}$ is $\begin{pmatrix} \psi & * \\ 0 & \psi^q \end{pmatrix}$, it follows that $\rho_{g,l}|_{I_q}$ is also of that form. As ψ is of p-power order, it cannot vanish under reduction mod l if $l \neq p$. Hence, $\overline{\rho}_{g,l}|_{I_q}$ is of the claimed form. The irreducibility follows as in (b). $\square$

3 Proof and remarks

In this section we prove Theorem 1.1 and comment on possible generalisations.

Lemma 3.1. *Let l be a prime and $s \in \mathbb{N}$. Then there is a set of primes p of positive density such that p is 1 mod 4 and such that if the group $\mathrm{GL}_2(\mathbb{F}_{l^t})$ possesses an element of order p for some $t \in \mathbb{N}$, then $2 \mid t$ and $t > s$.*

Proof. We take the set of primes p such that p is split in $\mathbb{Q}(i)$ and inert in $\mathbb{Q}(\sqrt{l})$. By Chebotarev's density theorem this set has a positive density. The first condition imposed on p means that 4 divides the order of $\mathbb{F}_p^\times$ and the second one that the order of the 2-Sylow subgroup of $\mathbb{F}_p^\times$ divides the order of l in $\mathbb{F}_p^\times$. If the order of $g \in \mathrm{GL}_2(\mathbb{F}_{l^t})$ is p, then $\mathbb{F}_{l^{2t}}^\times$ contains an element of order p (an eigenvalue of g). Thus, $l^{2t} - 1$ is divisible by p, whence the order of l in $\mathbb{F}_p^\times$ divides $2t$ and consequently 2 divides t. The condition $t > s$ can be met by excluding finitely many p. $\qquad\square$

Proof of Theorem 1.1. Let us fix the prime l and the number s from the statement of Theorem 1.1. We will now exhibit a modular form g such that $\overline{\rho}_{g,l}^{\mathrm{proj}}$ has image equal to $\mathrm{PSL}_2(\mathbb{F}_{l^t})$ with $t > s$. We start with the eigenform $f \in S_2(l^*)$ provided by Proposition 2.4. Next, we let p be any of the infinitely many primes from Lemma 3.1 such that the image of $\overline{\rho}_{f,p}^{\mathrm{proj}}$ is $\mathrm{PSL}_2(\mathbb{F}_{p^r})$ or $\mathrm{PGL}_2(\mathbb{F}_{p^r})$ with some r. We want to obtain g by level raising. The next lemma will yield a set of primes of positive density at which the level can be raised in a way suitable to us.

Lemma 3.2. *Under the above notations and assumptions, the set of primes q such that*

(i) $q \equiv -1 \mod p$,
(ii) q splits in $\mathbb{Q}(i, \sqrt{l})$ and
(iii) $\overline{\rho}_{f,p}^{\mathrm{proj}}(\mathrm{Frob}_q)$ lies in the same conjugacy class as $\overline{\rho}_{f,p}^{\mathrm{proj}}(c)$, where c is any complex conjugation,

has a positive density.

Proof. The proof is adapted from [KW], Lemma 8.2. Let $K/\mathbb{Q}$ be the number field such that $G_K = \ker(\overline{\rho}_{f,p}^{\mathrm{proj}})$. Define L as $\mathbb{Q}(\zeta_p, i, \sqrt{l})$. Conditions (i) and (ii) must be imposed on the field L and Condition (iii) on K. We know that $\mathrm{Gal}(K/\mathbb{Q})$ is either $\mathrm{PSL}_2(\mathbb{F}_{p^r})$ or $\mathrm{PGL}_2(\mathbb{F}_{p^r})$. In the former case the lemma follows directly from Chebotarev's density theorem, as the intersection $L \cap K$ is $\mathbb{Q}$, since $\mathrm{PSL}_2(\mathbb{F}_{p^r})$ is a simple group. In the latter case the intersection $L \cap K = M$ is an extension of $\mathbb{Q}$ of degree 2. As $p \equiv 1 \mod 4$ by assumption, $\overline{\rho}_{f,p}^{\mathrm{proj}}(c)$ is in $\mathrm{Gal}(L/M) \cong \mathrm{PSL}_2(\mathbb{F}_{p^r})$, since $\det(\overline{\rho}_{f,p}(c)) = -1$ is a square mod p. Consequently, any q satisfying Condition (iii) is split in $M/\mathbb{Q}$. Again as $p \equiv 1 \mod 4$, complex conjugation fixes the quadratic subfield of $\mathbb{Q}(\zeta_p)$, whence any prime q satisfying Conditions (i) and (ii) is also split in $M/\mathbb{Q}$. Hence, we may again appeal to Chebotarev's density theorem, proving the lemma. $\qquad\square$

To continue the proof of Theorem 1.1 we let T be the set of primes provided by Lemma 3.2. Let $q \in T$. Condition (iii) assures that $\overline{\rho}_{f,p}(\mathrm{Frob}_q)$ has trace 0, since it is a scalar multiple of the matrix representing complex conjugation, i.e. $\left(\begin{smallmatrix} 1 & 0 \\ 0 & -1 \end{smallmatrix}\right)$. Hence, we are in the situation to apply Theorem 2.5 and Corollary 2.6. We, thus, get an eigenform $g \in S_2(l^*q^2, \chi)$ with χ a Dirichlet character of order a power of p (its reduction mod p is trivial) such that $\overline{\rho}_{g,l}|_{D_q}$ is irreducible and $\overline{\rho}_{g,l}|_{I_q}$ is of the form $\begin{pmatrix} \psi & * \\ 0 & \psi^q \end{pmatrix}$ with ψ a non-trivial character of order p^r. Hence, in particular, the image $\overline{\rho}_{g,l}^{\mathrm{proj}}(G_{\mathbb{Q}})$ contains an element of order p, as $\begin{pmatrix} \psi & * \\ 0 & \psi^q \end{pmatrix}$ cannot be scalar due to $p \nmid q - 1$.

We next show that $\overline{\rho}_{g,l}^{\mathrm{proj}}(G_{\mathbb{Q}})$ is not a dihedral group. It cannot be cyclic either because of irreducibility. If it were dihedral, then $\overline{\rho}_{g,l}^{\mathrm{proj}}$ would be a representation induced from a character of a quadratic extension $R/\mathbb{Q}$. If R were ramified at q, then $\overline{\rho}_{g,p}^{\mathrm{proj}}(I_q)$ would have even order, but it has order a power of p. So, R would be either $\mathbb{Q}(\sqrt{l})$ or $\mathbb{Q}(\sqrt{-l})$. As by Condition (ii) q would split in R, this implies that $\overline{\rho}_{g,p}^{\mathrm{proj}}|_{D_q}$ would be reducible, but it is irreducible. Consequently, $\overline{\rho}_{g,l}^{\mathrm{proj}}(G_{\mathbb{Q}})$ is either $\mathrm{PSL}_2(\mathbb{F}_{l^t})$ or $\mathrm{PGL}_2(\mathbb{F}_{l^t})$ for some t. By what we have seen above, $\mathrm{PGL}_2(\mathbb{F}_{l^t})$ then contains an element of order p. Hence, so does $\mathrm{GL}_2(\mathbb{F}_{l^t})$ and the assumptions on p imply $t > s$ and $2 \mid t$.

We know, moreover, that the determinant of $\overline{\rho}_{g,l}(\mathrm{Frob}_w)$ for any prime $w \nmid lq$ is $w^{k-1}\chi(w)$ for some fixed k (the Serre weight of $\overline{\rho}_{g,l}$). As $\chi(w)$ is of p-power order, $\mathbb{F}_{l^t}$ contains a square root of it. The square roots of elements of $\mathbb{F}_l^\times$ are all contained in $\mathbb{F}_{l^2}$ and thus also in $\mathbb{F}_{l^t}$, as t is even. Hence, $\overline{\rho}_{g,l}^{\mathrm{proj}}(\mathrm{Frob}_w) \in \mathrm{PSL}_2(\mathbb{F}_{l^t})$. As every conjugacy class contains a Frobenius element, the proof of Theorem 1.1 is finished. $\qquad\square$

Remark 3.3. One can develop the basic idea used here further, in particular, in order to try to establish an analogue of Theorem 1.1 such that the representations ramify at a given finite set of primes S and are unramified outside $S \cup \{\infty, l, q\}$.

Remark 3.4. It is desirable to remove the ramification at l. For that, one would need that $\overline{\rho}_{g,l}^{\mathrm{proj}}$ is unramified at l. This, however, seems difficult to establish.

References

[Magma] W. Bosma, J.J. Cannon, C. Playoust. *The Magma Algebra System I: The User Language.* J. Symbolic Comput. **24** (1997), 235–265.

[CDT] B. Conrad, F. Diamond, R. Taylor. *Modularity of certain potentially Barsotti-Tate Galois representations.* J. Am. Math. Soc. **12(2)** (1999), 521–567.

[DT] F. Diamond, R. Taylor. *Non-optimal levels of mod l modular representations.* Invent. math. **115** (1994), 435–462.

[D1] L. V. Dieulefait. *Newforms, inner twists, and the inverse Galois problem for projective linear groups.* Journal de Théorie des Nombres de Bordeaux 13 (2001), 395–411.

[D2] L. V. Dieulefait. *Galois realizations of families of Projective Linear Groups via cusp forms.* This volume, pp. 85–92.

[D3] L. V. Dieulefait. *A control theorem for the images of Galois actions on certain infinite families of modular forms.* This volume, pp. 79–84.

[DV] L. V. Dieulefait, N. Vila. *Projective linear groups as Galois groups over* $\mathbb{Q}$ *via modular representations.* J. Symbolic Computation **30** (2000), 799–810.

[Hu] B. Huppert. *Endliche Gruppen I.* Grundlehren der mathematischen Wissenschaften **134**. Springer-Verlag, 1983.

[KLS] C. Khare, M. Larsen, G. Savin. *Functoriality and the inverse Galois problem.* Preprint, 2006. `arXiv:math.NT/0610860`

[KW] C. Khare, J.-P. Wintenberger. *Serre's modularity conjecture: the case of odd conductor (I).* Preprint, 2006.

[RV] A. Reverter, N. Vila. *Some projective linear groups over finite fields as Galois groups over* $\mathbb{Q}$. Contemporary Math. **186** (1995), 51–63.

[R1] K. A. Ribet. *On l-adic representations attached to modular forms II.* Glasgow Math. J. **27** (1985), 185–194.

[R2] K. A. Ribet. *Images of semistable Galois representations.* Pacific Journal of Mathematics **181** No. 3 (1997), 277–297.

[S] J.-P. Serre. *Topics in Galois theory.* Research Notes in Mathematics **1**. Jones and Bartlett Publishers, Boston, MA, 1992.

[W] G. Wiese. *Modular Forms of Weight One Over Finite Fields.* PhD thesis, Universiteit Leiden, 2005.